D0216191

Standard Normal Probabilities

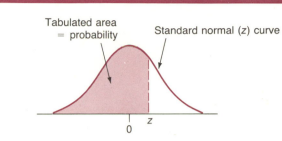

Tabulated area = probability

Standard normal (z) curve

z	.00	.01	.02	.03	.04	.05	.06	.07	.08	.09
0.0	.5000	.5040	.5080	.5120	.5160	.5199	.5239	.5279	.5319	.5359
0.1	.5398	.5438	.5478	.5517	.5557	.5596	.5636	.5675	.5714	.5753
0.2	.5793	.5832	.5871	.5910	.5948	.5987	.6026	.6064	.6103	.6141
0.3	.6179	.6217	.6255	.6293	.6331	.6368	.6406	.6443	.6480	.6517
0.4	.6554	.6591	.6628	.6664	.6700	.6736	.6772	.6808	.6844	.6879
0.5	.6915	.6950	.6985	.7019	.7054	.7088	.7123	.7157	.7190	.7224
0.6	.7257	.7291	.7324	.7357	.7389	.7422	.7454	.7486	.7517	.7549
0.7	.7580	.7611	.7642	.7673	.7704	.7734	.7764	.7794	.7823	.7852
0.8	.7881	.7910	.7939	.7967	.7995	.8023	.8051	.8078	.8106	.8133
0.9	.8159	.8186	.8212	.8238	.8264	.8289	.8315	.8340	.8365	.8389
1.0	.8413	.8438	.8461	.8485	.8508	.8531	.8554	.8577	.8599	.8621
1.1	.8643	.8665	.8686	.8708	.8729	.8749	.8770	.8790	.8810	.8830
1.2	.8849	.8869	.8888	.8907	.8925	.8944	.8962	.8980	.8997	.9015
1.3	.9032	.9049	.9066	.9082	.9099	.9115	.9131	.9147	.9162	.9177
1.4	.9192	.9207	.9222	.9236	.9251	.9265	.9279	.9292	.9306	.9319
1.5	.9332	.9345	.9357	.9370	.9382	.9394	.9406	.9418	.9429	.9441
1.6	.9452	.9463	.9474	.9484	.9495	.9505	.9515	.9525	.9535	.9545
1.7	.9554	.9564	.9573	.9582	.9591	.9599	.9608	.9616	.9625	.9633
1.8	.9641	.9649	.9656	.9664	.9671	.9678	.9686	.9693	.9699	.9706
1.9	.9713	.9719	.9726	.9732	.9738	.9744	.9750	.9756	.9761	.9767
2.0	.9772	.9778	.9783	.9788	.9793	.9798	.9803	.9808	.9812	.9817
2.1	.9821	.9826	.9830	.9834	.9838	.9842	.9846	.9850	.9854	.9857
2.2	.9861	.9864	.9868	.9871	.9875	.9878	.9881	.9884	.9887	.9890
2.3	.9893	.9896	.9898	.9901	.9904	.9906	.9909	.9911	.9913	.9916
2.4	.9918	.9920	.9922	.9925	.9927	.9929	.9931	.9932	.9934	.9936
2.5	.9938	.9940	.9941	.9943	.9945	.9946	.9948	.9949	.9951	.9952
2.6	.9953	.9955	.9956	.9957	.9959	.9960	.9961	.9962	.9963	.9964
2.7	.9965	.9966	.9967	.9968	.9969	.9970	.9971	.9972	.9973	.9974
2.8	.9974	.9975	.9976	.9977	.9977	.9978	.9979	.9979	.9980	.9981
2.9	.9981	.9982	.9982	.9983	.9984	.9984	.9985	.9985	.9986	.9986
3.0	.9987	.9987	.9987	.9988	.9988	.9989	.9989	.9989	.9990	.9990
3.1	.9990	.9991	.9991	.9991	.9992	.9992	.9992	.9992	.9993	.9993
3.2	.9993	.9993	.9994	.9994	.9994	.9994	.9994	.9995	.9995	.9995
3.3	.9995	.9995	.9995	.9996	.9996	.9996	.9996	.9996	.9996	.9997
3.4	.9997	.9997	.9997	.9997	.9997	.9997	.9997	.9997	.9997	.9998

INTRODUCTORY STATISTICS

INTRODUCTORY STATISTICS

California Polytechnic State University
San Luis Obispo

ROXY PECK

California Polytechnic State University
San Luis Obispo

WEST PUBLISHING COMPANY

St. Paul   New York   Los Angeles   San Francisco

Composition: Syntax International
Copyediting: Susan Gerstein
Cover and interior design: Janet Bollow
Illustrations: Scientific Illustrators

COPYRIGHT © 1990 By WEST PUBLISHING COMPANY
50 W. Kellogg Boulevard
P.O. Box 64526
St. Paul, MN 55164-1003

All rights reserved

Printed in the United States of America

97 96 95 94 93 92 91 90   8 7 6 5 4 3 2 1

Library of Congress Cataloging in Publication Data

Devore, Jay L.
    Introductory statistics/Jay Devore, Roxy Peck.
        p.   cm.
    ISBN 0-314-56884-0
    1. Statistics.   I. Peck, Roxy.   II. Title.
QA276. 12.D48   1990
519.5—dc20   89-36670
                                                    CIP

To the memory of my beloved brother Paul

J. D.

To Lygia and Kyle

R. P.

CONTENTS

PREFACE

Introductory Statistics provides traditional coverage of beginning probability and statistics with an emphasis on applications and data analysis. This book has been written with a one semester or two quarter course in mind. With judicious topic selection, it could also be used for a course as short as one quarter. Although the book does not presuppose the use of a statistical computer package, the role of the computer in data analysis is illustrated with some examples that show output from the more widely used computer packages, such as MINITAB, SPSS, BMDP, and SAS.

Throughout the text we have made an effort to avoid the use of contrived examples and exercises, and have spent a great deal of time seeking real applications taken from journals and other published sources in a wide variety of disciplines. We feel that this effort has been worthwhile in that it lends an air of credibility to the topics covered, and shows students that the techniques presented are, in fact, widely used.

There are a great many worked examples. An exercise set appears at the end of each section, and a supplementary exercise set appears at the end of each chapter. A summary of key concepts and formulas appears in each chapter just prior to the supplementary exercises. In addition, a student solutions manual containing worked solutions to the odd numbered problems, an instructor's manual (which includes worked solutions to all problems), transparency masters, and a test bank are all available from the publisher.

ACKNOWLEDGMENTS

Many people have made valuable contributions to the preparation of this book. We especially appreciated the help of our colleagues John Groves, who prepared the solutions manual and instructor's manual, and Susan Russell, who helped with typing and reviewed examples and exercise solutions for accuracy.

xi

We would also like to thank the following reviewers, who provided helpful comments and suggestions:

Steve Bajgier
  Drexel University
Warren Burch
  Brevard Community College
Dan Brunk
  Oregon State University
Veronica Czitrom
  University of Texas—San Antonio
James Hilton
  Grossmont College
K. G. Janardan
  Eastern Michigan University
Don Jeffries
  Orange Coast Community College
Robert Lacher
  South Dakota State University
Marvin Lentner
  Virginia Polytechnic Institute and State University
Maita Levine
  University of Cincinnati
John Martin
  Tarrant County Jr. College
Derek Mpinga
  Tarrant County Jr. College
Larry Ringer
  Texas A & M University
Gerald Rogers
  New Mexico State University
Helen Roberts
  Montclair State College
Mary Russo
  University of Iowa
Rich J. VanAmerongon
  Portland Community College
Timothy Wittig
  Mississippi State University

The staff at West Publishing Company were a pleasure to work with, and we especially thank our editor Peter Marshall, our developmental editor Maralene Bates, and our production editor Laura Nelson. Susan Gerstein did an excellent job of copyediting for us. Finally, the support of our families and friends throughout the ordeal of completing this book was appreciated more than they will ever know!

Jay Devore
Roxy Peck

A NOTE TO THE STUDENT

In all likelihood, you have started reading this book because it is the text for an introductory statistics course required of all students in your major. You may well be thinking to yourself that if it weren't for this requirement, you wouldn't be enrolled in a statistics course and could then spend your time in more interesting and productive ways. Perhaps you are even somewhat apprehensive about your ability to do well in the course, since you've probably heard through the grapevine that mastering statistics requires some facility for mathematical reasoning and manipulation. If you are indeed ambivalent about studying statistics and a bit fearful of what lies ahead, please realize that these feelings are shared by many other students. We hope to lay these fears to rest in short order by convincing you that statistics is important for gaining a better understanding of the world around you, relevant to your particular interests and field of study, and accessible even if you have a very modest mathematical background. To this end, the book emphasizes concepts and an intuitive presentation of the core methodology used in a wide variety of applications. Statistics does rest on a mathematical foundation, but we have tried to keep the notation and mathematical development simple. We hope the result is a friendly and informal survey that will help you in various ways long after the course is finished.

The key to success in your statistics course, as in so many endeavors, is to start with a positive attitude and resolve to invest a reasonable amount of time and effort. It won't always be easy and may occasionally be frustrating. (We ourselves sometimes get quite frustrated when attempting to learn new material.) But with the right attitude and commitment of your resources, we think that understanding, enjoyment, and a sense of accomplishment will quickly follow.

INTRODUCTION

Statistical methods for summary and analysis provide investigators with powerful tools for making sense out of data. Statistical techniques are being employed with increasing frequency in business, medicine, agriculture, social sciences, natural sciences, and the applied sciences (such as engineering). The pervasiveness of statistical analyses in such diverse fields has led to increased recognition that statistical literacy—a familiarity with the goals and methods of statistics—is a basic component of a well-rounded educational program. In this chapter we begin our march toward such literacy by first introducing two basic components of most statistical problems, *population* and *sample*, and then discussing various types of data that arise in statistical analyses.

1.1 POPULATIONS, SAMPLES, AND STATISTICS

For hundreds of years, individuals have been using statistical tools to organize and summarize data. Many of these tools—bar charts, tabular displays, various plots of economic data, averages and percentages—appear regularly in newspapers, magazines, and technical journals. Methods that organize and summarize data aid in effective presentation and increased understanding; such methods constitute a branch of the discipline called **descriptive statistics**.

Often the individuals or objects studied by an investigator come from a much larger collection, and the researcher's interest goes beyond just data summarization. It is frequently the larger collection about which the investigator wishes to draw conclusions. The entire collection of individuals or objects about which information is desired is called the **population** of interest. A **sample** is a subset of the population selected in some prescribed manner for study. The second major branch of statistics, **inferential statistics**, involves generalizing from a sample to the population from which it was selected. This type of generalization involves some risk, since a conclusion about the population will be reached on the basis of available, but incomplete, information. It may happen that the sample is, in some sense, unrepresentative of the population from which it came. An important aspect in the development of inference techniques involves quantifying the associated risks.

Considering some examples will help you to develop a preliminary appreciation for the scope and power of statistical methodology. We describe here three problems that can be handled using techniques to be presented in this text. First, suppose that a university has just implemented a new phone registration system. Students interact with the computer by entering information from a Touch-Tone® phone to select classes for the term. In order to assess student opinion regarding the effectiveness of the system, a survey of students is to be undertaken. Each student in a sample of 400 will be asked a variety of questions (such as the number of units received, the number of attempts required to get a phone connection, etc.) The result of such a survey will be a rather large and unwieldy data set. In order to make sense out of the raw data and to describe student responses, it is desirable to summarize the data. This would also make the results more accessible to others. Descriptive techniques to be presented in Chapters 2 and 3 could be used to accomplish this task. In addition, inferential methods from Chapters 8 and 9 could be employed in order to use the sample information to draw various conclusions about the experiences of all students at the university who used the registration system.

As a second example, consider a business application. Suppose that a publisher of college textbooks has two different machines that are used to bind the printed pages. One characteristic that affects the overall quality of the finished book is the strength of the binding. The publisher would like to determine if there is a significant difference between the two machines with respect to average binding strength. Strength could be measured for one sample of books bound by the first machine and for a second sample of books bound by the second machine. Hypothesis testing techniques (to be introduced in Chapters 9 and 10) could then be used to analyze the resulting data and provide the publisher with an answer to the question posed.

A final example comes from the discipline of forestry. When a fire occurs in a forested area, decisions must be made as to the best way to combat the fire. One possibility is to try to contain the fire by building a fire line. If building a fire line requires four hours, the decision as to where the line should be built involves making a prediction of how far the fire will spread during this period. Many factors must be taken into account, including wind speed, temperature, humidity, and time elapsed since the last rainfall. Regression techniques (Chapter 12) will enable us to develop a model for the prediction of fire spread using information available from past fires.

Some individuals regard conclusions based on statistical analyses with a great deal of suspicion. Extreme skeptics, usually speaking out of ignorance, characterize the discipline as a subcategory of lying—something used for deception rather than for positive ends. However, we believe that statistical methods, used intelligently, constitute a set of powerful tools for gaining insight into the world around us. We hope that this text will help you to understand the logic behind statistical reasoning, prepare you to apply statistical methods appropriately, and enable you to recognize when others are not doing so.

EXERCISES 1.1–1.7　　　　**SECTION 1.1**

1.1 Give a brief definition of the terms *descriptive statistics* and *inferential statistics*.

1.2 Give a brief definition of the terms *population* and *sample*.

1.3 The student senate at a university with 15,000 students is interested in the proportion of students who favor a change in the grading system to allow for + and − grades (i.e., B−, B, B+, rather that just B). Two hundred students are interviewed to determine their attitude toward this proposed change. What is the population of interest? What group of students constitutes the sample in this problem?

1.4 The supervisors of a rural county are interested in the proportion of property owners who support the construction of a sewer system. Because it is too costly to contact all 7000 property owners, a survey of 500 (selected at random) is undertaken. Describe the population and sample for this problem.

1.5 Representatives of the insurance industry wished to investigate the monetary loss due to damage to single-family dwellings in Pasadena, California, resulting from an earthquake that occurred on December 3, 1988. One hundred homes were selected for inspection from the set of all single-family homes in Pasadena. Describe the population and sample for this problem.

1.6 A consumer group conducts crash tests of new model cars. To determine the severity of damage to 1989 Mazda 626s resulting from a 10-mph crash into a concrete wall, six cars of this type are tested and the amount of damage is assessed. Describe the population and sample for this problem.

1.7 A building contractor has a chance to buy an odd lot of 5000 used bricks at an auction. She is interested in determining the proportion of bricks in the lot that are cracked and therefore unusable for her current project, but she does not have enough time to inspect all 5000 bricks. Instead, she checks 100 bricks to determine whether each is cracked. Describe the population and sample for this problem.

1.2 TYPES OF DATA

The individuals or objects in any particular group typically possess many attributes that might be studied. Consider as an example a group of students currently enrolled in a statistics course. One attribute is the brand of calculator owned by each student (Sharp, Hewlett–Packard, Casio, and so on). Another attribute of potential interest is the number of courses for which each student is registered (1, 2, 3, . . .), and yet another is the distance from the university to each student's permanent residence. In this example, *calculator brand* is a categorical attribute, since each student's response to the query "What brand of calculator do you own?" is a category. The collection of responses from all these students forms a **categorical data set**. The other two attributes, *number of units* and *distance*, are both numerical in nature. Determining the value of such a numerical attribute (by counting or measuring) for each student results in a **numerical data set**.

EXAMPLE 1 A sample of 15 people who belong to a certain tennis club is selected. Each one is then asked which brand of racket he or she uses. The resulting set of responses is

{Head, Prince, Prince, Wilson, Yonex, Head, Yamaha, Head, Head, Prince, Yamaha, Kennex, Prince, Wilson, Yonex}.*

This a categorical data set.

EXAMPLE 2 A sample of 20 automobiles of a certain type is selected and the fuel efficiency (miles per gallon or mpg) is determined for each one. The resulting numerical data set is

{29.8, 27.6, 28.3, 28.7, 27.9, 29.9, 30.1, 28.0, 28.7, 27.9, 28.5, 29.5, 27.2, 26.9, 28.4, 27.9, 28.0, 30.0, 29.6, 29.1}.

In both of the preceding examples, the data sets consisted of observations (categorical responses or numbers) on a single attribute. Such data sets are called *univariate*.

---

* We will often use braces to enclose the members of a set.

DEFINITION

A data set consisting of observations on a single attribute is a **uni-variate data set**. A univariate data set is **categorical** (or **qualitative**) if the individual observations are categorical responses; it is **numerical** (or **quantitative**) if the observations are numbers.

In some studies, attention focuses simultaneously on two different attributes. For example, both the height (in.) and weight (lb) might be recorded for each individual in a group. The resulting data set consists of pairs of numbers, such as (68, 146). This is called a **bivariate data set**. **Multivariate data** results from obtaining a category or value for each of three or more attributes: for example, height, weight, pulse rate, and systolic blood pressure for each individual. Much of this book will focus on methods for analyzing univariate data. In the last several chapters we consider briefly the analysis of some bivariate and multivariate data.

TWO TYPES OF NUMERICAL DATA

With numerical data, it is useful to make a further distinction. Visualize a number line (Figure 1) for locating values of the numerical attribute being studied. To every possible number (2, 3.125, −8.12976, etc.) there corresponds exactly one point on the number line. Now suppose that the attribute of interest is the number of cylinders of an automobile engine. The possible values of 4, 6, and 8 are identified in Figure 2(a) by the dots at the points marked 4, 6, and 8. These possible values are isolated from one another on the line; around any possible value we can place an interval that is small enough so that no other possible value is included in the interval. On the other hand, the line segment in Figure 2(b) identifies a plausible set of possible values for quarter-mile time. Here the possible values comprise an entire interval on the number line, and no possible value is isolated from the other possible values.

FIGURE 1

A number line.

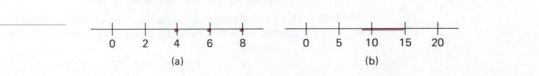

DEFINITION

Numerical data is **discrete** if the possible values are isolated points on the number line. Numerical data is **continuous** if the set of possible values forms an entire interval on the number line.

FIGURE 2

Possible values of a variable.
(a) Number of cylinders
(b) Quarter-mile time

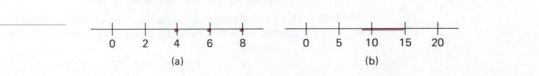

Discrete data usually arises when each observation is determined by counting (the number of classes for which a student is registered, the number of petals on a certain type of flower, etc.).

EXAMPLE 3

The number of telephone calls to a drug hotline is recorded for each different 24-hour period. The resulting data set is

$$\{3, \quad 0, \quad 4, \quad 3, \quad 1, \quad 0, \quad 6, \quad 2, \quad 0, \quad 0, \quad 1, \quad 2\}.$$

Possible values for the *number of calls* are 0, 1, 2, 3, . . . ; these are isolated points on the number line, so we have a sample consisting of discrete numerical data.

The sample of fuel efficiencies in Example 2 is an example of continuous data. A car's fuel efficiency could be 27.0, 27.13, 27.12796, or any other value in an entire interval. Other examples of continuous data arise when task completion times are observed, body temperatures are recorded, or packages are weighed. In general, data is continuous when observations involve making measurements, as opposed to counting.

In practice, measuring instruments do not have infinite accuracy. Thus possible measured values do not form a continuum on the number line. The distinction between discrete and continuous data will nevertheless be important in our discussion of probability models.

EXERCISES 1.8–1.11

SECTION 1.2

1.8 Classify each of the following attributes as either categorical or numerical. For those that are numerical, determine whether they are discrete or continuous.
  a. Number of students in a class of 35 who turn in a term paper before the due date *numeric, discrete*
  b. Sex of the next baby born at a particular hospital *categ.*
  c. Amount of fluid (oz) dispensed by a machine used to fill bottles with soda pop *numerical, continuous*
  d. Thickness of the gelatin coating of a vitamin E capsule *numerical, discrete*
  e. Birth classification (only child, first born, middle child, last born) of a math major *categorical*

1.9 Classify each of the following attributes as either categorical or numerical. For those that are numerical, determine whether they are discrete or continuous.
  a. Brand of personal computer purchased by a customer *categ.*
  b. State of birth for someone born in the United States *categ.*
  c. Price of a textbook *numerical (continuous)* *discr*
  d. Concentration of a contaminant (micrograms/cm³ or $\mu g/cm^3$) in a water sample *numerical, contin*
  e. Zip code (Think carefully about this one.) *categor*
  f. Actual weight of coffee in a 1-lb can *numerical, contin.*

1.10 For the following numerical attributes, state whether each is discrete or continuous.
  a. The number of checks received by a grocery store during a given month that bounce *numeric- discrete*

b. The amount by which a 1-lb package of ground beef decreases in weight (because of moisture loss) before purchase *continuous*

c. The number of New York Yankees during a given year who will not play for the Yankees the following year *discrete*

d. The number of students in a class of 35 who have purchased a used copy of the textbook *discrete*

e. The length of a 1-year-old rattlesnake *contin*

f. The altitude of a location in California selected randomly by throwing a dart at a map of the state *continuous*

g. The distance from the left edge at which a 12-in. plastic ruler snaps when bent sufficiently to cause a break *contin*

h. The price per gallon paid by the next customer to buy gas at a particular station *contin*

1.11 For each of the following situations, give some possible data values that might arise from making the observations described.

a. The country of manufacture for each of the next ten automobiles to pass through a given intersection is noted. *USA, Japan, Germany*

b. The grade point average for each of the 15 seniors in a statistics class is determined. *4.5, 8.7, 9.0, 8.1*

c. The number of gas pumps in use at each of 20 gas stations at a particular time is determined. *3, 5, 8, 1, 0*

d. The actual net weight of each of 12 bags of fertilizer having a labeled weight of 50 lb is determined. *50.1, 49.9, 50.2, 51.0*

e. Fifteen different radio stations are monitored during a 1-hour period and the amount of time devoted to commercials is determined for each one. *25 min, 15.4 min*

f. The brand of breakfast cereal purchased by each of 16 customers is noted. *Corn Fl. Spec. K, Kin mei*

g. The number of defective tires is determined for each of the next 20 automobiles stopped for speeding on a certain highway. *0, 1, 1, 0, 2*

CHAPTER 1   SUMMARY OF KEY CONCEPTS AND FORMULAS

TERM OR FORMULA	PAGE	COMMENT
Descriptive statistics	2	Numerical, graphical, and tabular methods for organizing and summarizing data
Population	2	The entire collection of individuals or measurements about which information is desired
Sample	2	That part of the population selected for study
Inferential statistics	2	Methods for generalizing from a sample to the population from which the sample was selected
Categorical data	5	Individual observations are categorical responses (nonnumerical).
Numerical data	5	Individual observations are numerical in nature.
Discrete numerical data	5	Possible values are isolated points along the number line.
Continuous numerical data	5	Possible values form an entire interval along the number line.

REFERENCES   Fairley, William, and Frederick Mosteller, eds. *Statistics and Public Policy*.
Reading, Mass.: Addison-Wesley, 1977. (Interesting articles focusing on how statistical methods are used in studying public policy questions and issues.)

Freedman, David, Robert Pisani, and Roger Purves. *Statistics*.
New York: W. W. Norton, 1978. (The first two chapters contain some interesting examples of both well-designed and poorly designed experimental studies.)

Moore, David. *Statistics: Concepts and Controversies*.
New York: W. H. Freeman, 1985. (Contains an excellent chapter on the advantages and pitfalls of experimentation and another one in a similar vein on sample surveys and polls.)

Tanur, Judith, ed. *Statistics: A Guide to the Unknown*.
Belmont, Calif.: Wadsworth, 1989. (Short articles by a number of well-known statisticians and users of statistics, all very nontechnical, on the application of statistics in various disciplines and subject areas.)

TABULAR AND PICTORIAL METHODS FOR DESCRIBING DATA

INTRODUCTION   An important first step in extracting information from a data set is to find insightful ways of organizing and summarizing the data. In this chapter we illustrate the use of stem-and-leaf displays for quantitative data, show how important aspects of qualitative or quantitative data can be summarized in a tabular format, and indicate how to construct a pictorial representation (called a histogram) of the tabulated information.

2.1 STEM-AND-LEAF DISPLAYS

Some preliminary organization of a data set will often reveal useful information and open up paths of inquiry. A **stem-and-leaf display** is an effective way to organize numerical data without expending much effort. Each observation is regarded as consisting of two pieces. One or more of the leading digits make up the **stem**, and the trailing digit or digits constitute the **leaf**. For example, the data set might consist of verbal SAT scores (whole numbers between 200 and 800). One choice of stems is the set of leading digits: 2, 3, 4, 5, 6, 7, 8. Then 641 has stem 6 (so the stem unit is hundreds) and leaf 41 (the leaf unit is ones, the place value of the digit on the far right). If all scores in the data set were between 500 and 599 (a very homogeneous group), a more sensible choice for stem is the first two digits. Thus 538 would have stem 53 (unit = tens, since the far right digit, 3, occupies the tens place) and leaf 8 (unit = ones).

STEPS FOR CONSTRUCTING A DISPLAY

1. Select one or more leading digits for the stem values. The trailing digit or digits become the leaves.
2. List possible stem values in a vertical column.
3. Record the leaf for every observation beside the corresponding stem value.
4. Indicate the units for stems and leaves someplace in the display.

EXAMPLE 1

The accompanying data (see Table 1) on elementary school student–teacher ratios for the 50 states and the District of Columbia appeared in the September 1987 issue of *NEA Today*.

TABLE 1

Student–Teacher Ratios

AL	19.84	IL	17.95	MT	15.70	RI	15.35
AK	20.83	IN	18.63	NB	15.18	SC	17.75
AZ	20.49	IA	15.83	NV	20.46	SD	15.57
AR	18.15	KS	15.46	NH	15.89	TN	19.85
CA	22.88	KY	18.66	NJ	14.86	TX	17.52
CO	18.19	LA	18.24	NM	18.01	UT	24.04
CT	13.96	ME	16.27	NY	14.66	VT	14.07
DE	16.05	MD	17.69	NC	18.87	VA	16.39
DC	15.34	MA	14.80	ND	15.02	WA	20.72
FL	17.17	MI	20.90	OH	18.65	WV	15.34
GA	18.57	MN	17.31	OK	16.47	WI	16.62
HI	19.68	MS	18.57	OR	17.41	WY	13.45
ID	20.69	MO	16.38	PA	16.40		

An obvious choice for stems is the two leading (tens and ones) digits. Thus the Alabama value 19.84 has stem 19 (19 ones) and leaf 84 (84 hundredths), Alaska's value has stem 20 and leaf 83, and so on. The leaves are entered in the display of Figure 1 in the order given in Table 1: first 84 beside the stem value 19, then 83 beside the stem value 20, and so on.

```
13 ‖ 96, 45
14 ‖ 80, 86, 66, 07
15 ‖ 34, 83, 46, 70, 18, 89, 02, 35, 57, 34
16 ‖ 05, 27, 38, 47, 40, 39, 62
17 ‖ 17, 95, 69, 31, 41, 75, 52
18 ‖ 15, 19, 57, 63, 66, 24, 57, 01, 87, 65
19 ‖ 84, 68, 85
20 ‖ 83, 49, 69, 90, 46, 72
21 ‖
22 ‖ 88
23 ‖           Stems:  Ones
24 ‖ 04        Leaves:  Hundredths
```

FIGURE 1

Stem-and-leaf display of student–teacher ratios.

The display shows that the smallest and largest values are 13.45 and 24.04, respectively. The observations 22.88 (California) and 24.04 (Utah) exceed the remaining values by a substantial amount. Most of the values are between 15 and 19, and a "typical" value (central value) is about 17.5.

A stem-and-leaf display is useful for locating a central (typical) value and for assessing the extent to which observations spread out about the center. **Outliers** (unusually small or large values) are easy to spot, and regions of high concentration become apparent.

REPEATED STEMS

Sometimes a natural choice of stems gives a display in which too many observations are concentrated in just a few stems. A more informative picture may be obtained by writing each stem value twice, once for low leaves and once for high leaves.

EXAMPLE 2

The paper "The Acid Rain Controversy: The Limits of Confidence" (*Amer. Statistician* (1983): 385–94) presented data on average sulfur dioxide emission rates (lb/million Btu) for industrial and utility boilers in 47 states (data from Alaska, Hawaii, and Idaho was not given). A stem-and-leaf display of the data appears as Figure 2. There are only five stems, and each leaf is a single digit (tenths).

```
0 ‖ 3  6  4  5  2  7  2  7  7  5  1  6  9  6
1 ‖ 5  5  3  2  2  0  4  0  7  5  5  4  9  0  7  8  7  8  4
2 ‖ 3  7  2  5  7  9  1  9
3 ‖ 8  6  4  7                    Stems:  Ones
4 ‖ 2  5                          Leaves:  Tenths
```

FIGURE 2

Stem-and-leaf display of emission rates.

Figure 3 displays the same data using stems 0L, 0H, 1L, 1H, . . . , 4L, and 4H. Any observation whose leaf was 0, 1, 2, 3, or 4 (low) in the original display goes in the corresponding L row, and observations with leaves 5, 6, 7, 8, or 9 are in H rows. This second display shows quite nicely how the 47 observations are distributed along the measurement scale.

```
0L   3  4  2  2  1
0H   6  5  7  7  7  5  6  9  6
1L   3  2  2  0  4  0  4  0  4
1H   5  5  7  5  5  9  7  8  7  8
2L   3  2  1
2H   7  5  7  9  9
3L   4
3H   8  6  7
4L   2                          Stems:  Ones
4H   5                          Leaves: Tenths
```

FIGURE 3

Display of emission rates using repeated stems.

COMPUTER-GENERATED STEM-AND-LEAF DISPLAYS

The computer is a very powerful tool for doing statistical analysis, because it can perform routine data organization and arithmetic calculations many times as fast as these tasks can be done by hand. Most such analysis is done using prepared packages of statistical computer programs. With a package of this sort, the user takes advantage of a program that has already been written. It is necessary only to enter the data properly and then give the computer a command that causes the desired operation to be performed. The most frequently used packages are MINITAB, BMDP (Biomedical Computer Programs), SAS (Statistical Analysis System), and SPSS (Statistical Package for the Social Sciences). Almost all the methods of analysis we discuss in this book can be carried out using any one of these four packages.

Figure 4 pictures a MINITAB stem-and-leaf display for the emissions data of Example 2. Repeated stems were used (the user has no choice in this), and within each stem the leaves were automatically ordered from smallest to largest.

```
Stem-and-leaf of EMISS   N=47
Leaf unit=0.10

       0   12234
       0   556667779
       1   0000223444
       1   555777889
       2   123
       2   57799
       3   4
       3   678
       4   2
       4   5
```

FIGURE 4

A computer-generated display using MINITAB.

COMPARATIVE DISPLAYS

Frequently an analyst has two groups of data and wishes to see if they differ in some fundamental way. A comparative stem-and-leaf display, in which the leaves from one group extend to the right of the stem values and those from the other group to the left, can give preliminary visual impressions and insights.

EXAMPLE 3

The Institute of Nutrition of Central America and Panama (INCAP) has carried out extensive dietary studies in various parts of Central America. One such study reported on in the paper "The Blood Viscosity of Various Socioeconomic Groups in Guatemala" (*Amer. J. of Clinical Nutrition* (Nov. 1964): 303–7) determined values of various physiological characteristics for several groups of Guatemalans. The stem-and-leaf display pictured in Figure 5 gives serum total cholesterol values (mg/l) both for a sample of high-income urban individuals and for a sample of low-income rural Indians. The first sample contains one value, 330, that is far above the rest of the data. Rather than extend the stems to capture this value at the bottom of the display, it is marked at the bottom with the symbol "HI." This is routinely done in computer-generated displays for unusually high or low values.

```
I                           9 ‖ 5                         II
                           10 ‖ 8,8
                           11 ‖ 5,4
                           12 ‖ 9,9,4
                    3,4    13 ‖ 5,1,6,6,1,9
                           14 ‖ 0,6,4,5,2,3,8,3,4,2
                      5    15 ‖ 2,8,7,2,5,8
                           16 ‖ 6,5,2
                  9,5,0    17 ‖ 5,4,3,2,1
                1,4,8,9    18 ‖ 0,9,1
                9,7,0,6    19 ‖ 2,4,7
    1,5,4,5,5,0,1,0,6      20 ‖ 4
                    4,7    21 ‖
              2,7,8,7,2    22 ‖ 3,6,0
              4,6,4,9      23 ‖ 1
              2,9,4,1      24 ‖
                      2    25 ‖
                           26 ‖
                    9,3    27 ‖         Stems: Tens
        HI: 330   4,4,4    28 ‖         Leaves: Ones
```

FIGURE 5

Comparative stem-and-leaf display of serum total cholesterol values.
(I) Urban Guatemalans
(II) Low-income rural Indian Guatemalans

A first impression is one of great variability of cholesterol values when each sample is considered separately. For low-income rural Indians (group II), values run from 95 to 231, a range of $231 - 95 = 136$. Disregarding for the moment the one outlying value on the high end of the display, the range for high-income urban individuals is $284 - 133 = 151$, not greatly different from the range for group II. A reasonable central value for group I is one with a stem of 20, and for group II a central value would be in the high 140s or low 150s. Notice that the shapes of the two sides of the display are rather similar, each rising to a peak near a central value and then declining. Roughly speaking, the main difference between the two groups is in location. If we were to push the display for group II down five or six stem values (50 or 60 cholesterol units), the two halves would be quite similar. The suggested explanation for this difference in location is the presence of more fats in group I diets. A formal statistical analysis would yield more precise information on the size of the shift in location.

SECTION 2.1

2.1 The Bureau of Justice's *Statistics Bulletin* on jail inmates for 1982 reported the following inmate population sizes for 40 of the smaller federal prisons:

644	512	448	730	401	450	419	647	792	885
501	458	755	569	417	405	509	440	402	624
603	599	791	407	433	559	777	856	492	400
484	554	634	553	723	565	424	417	524	468

Using stems 4, 5, 6, 7, and 8, construct a stem-and-leaf display for this data.

2.2 Consider the accompanying batch of exam scores. First construct a stem-and-leaf display in which each stem occurs just once. Then construct a display that repeats each stem. What feature of the data is highlighted by this second display?

74, 89, 80, 93, 64, 67, 72, 70, 66, 85, 89, 81, 81, 71,
74, 82, 85, 63, 72, 81, 81, 95, 84, 81, 80, 70, 69, 66,
60, 83, 85, 98, 84, 68, 90, 82, 69, 72, 87, 88

2.3 The accompanying observations are yardages for a sample of golf courses recently listed by *Golf Magazine* as being among the most challenging in the U.S. Construct a stem-and-leaf display, and explain why your choice of stems seems preferable to any of the other possible choices.

6526	6770	6936	6770	6583	6464	7005	6927
6790	7209	7040	6850	6700	6614	7022	6506
6527	6470	6900	6605	6873	6798	6745	7280
7131	6435	6694	6433	6870	7169	7011	7168
6713	7051	6904	7105	7165	7050	7113	6890

2.4 Soil pH, a measure of the extent to which soil is acidic or basic, is one characteristic that plays an important role in the suitability of soil to support vegetation at mine reclamation sites. The article "A Dual-Buffer Titration Method for Lime Requirement of Acid Minesoils" (*J. Environ. Qual.* (1988): 452–56) reported the following data on pH for 26 minesoil specimens.

3.59	4.36	3.86	4.25	4.46	4.53	2.62	6.79	6.49
4.27	3.84	4.78	4.65	2.91	3.90	6.00	2.83	3.58
4.43	4.58	4.75	3.49	4.11	3.58	5.21	4.41	

Construct a stem-and-leaf display. Based on your display, does the data set appear to contain any "outliers," i.e., observations far removed from the bulk of the data?

2.5 The accompanying values are rental rates per foot for boat storage at the 19 marinas in Marina del Rey (Calif.) and the 17 marinas at the Los Angeles–Long Beach Harbor (Source: *Los Angeles Times*, June 5, 1983).

MARINA DEL REY				LOS ANGELES–LONG BEACH			
$6.37	$6.60	$6.27	$6.49	$4.60	$4.75	$4.70	$8.75
$6.64	$6.82	$7.16	$6.45	$4.50	$5.40	$6.00	$6.00
$5.60	$5.95	$4.50	$6.60	$6.50	$6.00	$5.00	$5.00
$6.00	$6.82	$7.04	$5.30	$5.50	$4.35	$4.50	$5.20
$7.05	$7.05	$6.96		$4.95			

Construct a comparative stem-and-leaf display for rent per foot for the two areas. What conclusions can you draw from the stem-and-leaf display concerning differences between the two locations?

2.6 The Los Angeles Board of Education has enacted a policy that prohibits students who do not have a C average from participating in extracurricular activities. The *Los Angeles Times* (May 17, 1983) reported the percentages of ineligible students for 47 Los Angeles high schools. Figures were reported separately for athletes and nonathletes.

Percent Ineligible									
Athletes	27	12	15	15	21	15	17	14	21	27
Nonathletes	24	25	48	17	3	3	14	18	52	22
Athletes	15	10	36	19	29	35	16	13	17	18
Nonathletes	4	14	29	18	13	17	14	23	25	15
Athletes	13	28	24	40	35	10	16	37	26	18
Nonathletes	15	6	34	48	45	12	38	28	29	34
Athletes	8	21	17	21	20	15	39	37	10	23
Nonathletes	14	18	17	7	24	34	44	13	16	14
Athletes	14	12	16	8	29	17	9			
Nonathletes	15	20	18	9	30	25	8			

Construct a comparative stem-and-leaf display of the percent of ineligible students for athletes and nonathletes. Based on your stem-and-leaf display, do you think there is evidence that the percentage of disqualified students tends to be smaller for nonathletes than for athletes? Justify your answer.

2.7 The 1982 American Statistical Association Proceedings of the Social Statistics Section included a paper entitled "The Cost of Tenure/Promotion." In that paper the authors reported the age for each tenured faculty member at the University of Richmond in 1980, as presented here.

34	26	48	28	32	34	45	49	41	55	39	30	32
34	42	43	53	59	66	59	55	52	46	66	50	47
52	48	35	44	43	33	34	41	35	34	43	49	56
52	58	48	50	60	54	53	41	36	40	33	30	39
30	34	45	34	59	42	44	43	46	34	65	37	49
34	48	36	53	42	43	32	59	43	33	28	67	31
37	39	43	54	58	45	44	52	44	43	44	48	42
38	35	43	34	34	40	34	37	36	52	43	46	58
48	45	43	47	56	49	34	35	42	44	33	30	37
41	57	57	59	50	58	53	59	62	43	53	43	57
34	37	43	44	39	49	39	37	54	38	59	41	44
32	37	58	70	62	63							

a. Construct a stem-and-leaf display.

b. Use your display to determine the percentage of faculty who were eligible to retire during the subsequent five-year period. (Assume that a person is eligible to retire at age 65.)

2.2 FREQUENCY DISTRIBUTIONS

A stem-and-leaf display is not always an effective summary technique. It cannot be used for categorical data. In addition, a stem-and-leaf display is unwieldy when the data set contains a great many observations. A frequency distribution is useful for summarizing even a very large data set in a compact fashion.

FREQUENCY DISTRIBUTIONS FOR CATEGORICAL DATA

A **frequency distribution** is a table that displays the categories, frequencies, and relative frequencies. The **frequency** for a particular category is the number of observed responses that fall into that category. The corresponding **relative frequency** is the fraction or proportion of observed responses in the category. Suppose, for example, that 26 of the 80 tennis players in a sample use a Wilson racket. Then the frequency for the category *Wilson* is 26, and the relative frequency is 26/80 = .325. (Thus 32.5% of the observed responses are Wilson.)

EXAMPLE 4

A sample of 40 individuals who recently joined a certain travel club yielded the following responses on occupation (C = clerical, M = manager/executive, P = professional, R = retired, S = sales, T = skilled tradesman, O = other):

P R R M S R P R C P M R T O R S R S P M
R P C R R S T P P C M S R R P R S R R P

The frequency distribution appears as Table 2. A tally column has been included so that only one pass through the data is necessary. The two most frequently occurring occupational categories, retired and professional, account for 60% of the responses. The club managers might want to consider directing more advertising at these groups.

TABLE 2

Frequency Distribution for Occupational Category

Category	Tally	Frequency	Relative Frequency
Clerical	|||	3	.075
Manager	||||	4	.100
Professional	⊮||||	9	.225
Retired	⊮⊮⊮	15	.375
Sales	⊮|	6	.150
Tradesman	||	2	.050
Other			1	.025
	Totals:	40	1.000

The sum of relative frequencies should be 1, but there may be a slight discrepancy due to rounding. If only two-decimal-place accuracy had been used in Example 4, the sum would have been 1.02.

FREQUENCY DISTRIBUTIONS FOR DISCRETE NUMERICAL DATA

Discrete numerical data almost always results from counting. In such cases, each observation is a whole number. If, for example, possible values are 0, 1, 2, 3, . . . , then these are listed in a column. A running tally is kept as a single pass is made through the data. The number of tally marks beside each value gives the frequency of that value. Dividing each frequency by the total number of observations gives the corresponding relative frequency.

EXAMPLE 5 A sample of 708 bus drivers employed by public corporations was selected and the number of traffic accidents in which each was involved during a 4-year period was determined ("Application of Discrete Distribution Theory to the Study of Noncommunicable Events in Medical Epidemiology," *Random Counts in Biomedical and Social Sciences*, G. P. Patil, ed. University Park, PA: Penn. State Univ. Press, 1970). A listing of the 708 sample observations would look something like this:

$$3, 0, 6, 0, 0, 2, 1, 4, 1, \ldots, 6, 0, 2$$

The frequency distribution (Table 3) shows that 117 of the 708 drivers had no accidents, a relative frequency of $117/708 = .165$ (or 16.5%). Similarly, the proportion of sampled drivers with one accident is .222 (or 22.2%). The largest sample observation was 11 (presumably this driver was not at fault for most of these).

TABLE 3

Frequency Distribution for Number of Accidents by Bus Drivers

Number of Accidents	Frequency	Relative Frequency
0	117	.165
1	157	.222
2	158	.223
3	115	.162
4	78	.110
5	44	.062
6	21	.030
7	7	.010
8	6	.008
9	1	.001
10	3	.004
11	1	.001
Totals:	708	.998

We can easily calculate other quantities of interest from the relative frequencies.

(i) $\left(\begin{array}{c}\text{proportion with at}\\\text{most 1 accident}\end{array}\right) = \left(\begin{array}{c}\text{proportion}\\\text{with 0}\end{array}\right) + \left(\begin{array}{c}\text{proportion}\\\text{with 1}\end{array}\right)$

$$= .165 + .222$$
$$= .387 \quad (\text{or } 38.7\%)$$

(ii) $\left(\begin{array}{c}\text{proportion with at}\\\text{least 6 accidents}\end{array}\right) = \left(\begin{array}{c}\text{proportion}\\\text{with 6}\end{array}\right) + \left(\begin{array}{c}\text{proportion}\\\text{with 7}\end{array}\right) + \cdots + \left(\begin{array}{c}\text{proportion}\\\text{with 11}\end{array}\right)$

$$= .030 + .010 + \cdots + .001$$
$$= .054 \quad (\text{or } 5.4\%)$$

(iii) $\left(\begin{array}{c}\text{proportion with}\\\text{between 4 and 7}\\\text{(inclusive)}\end{array}\right) = \left(\begin{array}{c}\text{proportion}\\\text{with 4}\end{array}\right) + \cdots + \left(\begin{array}{c}\text{proportion}\\\text{with 7}\end{array}\right)$

$$= .110 + .062 + .030 + .010$$
$$= .212$$

Frequently a data set contains a few large values that are significantly separated from the bulk of the observations. For example, consider adding two more drivers, one with 16 accidents and one with 21 accidents, to the 708 drivers of Example 5. Rather than list individual count values all the way to 21, we might stop listing at 10 and add one further category, *at least 11* (often written ≥ 11). Then 3 of the 710 observations (the 11, the 16, and the 21) would belong in this category. Table 4 presents the resulting computer-generated frequency distribution.

TABLE 4

Computer-Generated Frequency Distribution for Number of Accidents (using SPSS)

Category Label	Code	Absolute Freq	Relative Freq (PCT)
0	0	117	16.5
1	1	157	22.1
2	2	158	22.3
3	3	115	16.2
4	4	78	11.0
5	5	44	6.2
6	6	21	3.0
7	7	7	1.0
8	8	6	.8
9	9	1	.1
10	10	3	.4
AT LEAST 11	11	3	.4
	TOTAL	710	100.0

FREQUENCY DISTRIBUTIONS FOR CONTINUOUS DATA

The difficulty with continuous data, such as observations on reaction time (sec) or fuel efficiency (mpg), is that there are no natural categories. Before we can compute and list frequencies, we need something analogous to categories. The way out of this dilemma is to define our own categories. For fuel efficiency data, suppose that we mark off some intervals on a horizontal miles-per-gallon measurement axis, as pictured in Figure 6. Each data value should fall in exactly one of these intervals. If the smallest observation were 25.3 and the largest were 29.8, we might use intervals of width .5, with the first one starting at 25.0 and the last one ending at 30.0. The resulting intervals are called **class intervals**, or just *classes*. The classes play the same role that the categories played earlier, with frequencies and relative frequencies tabulated as before.

FIGURE 6

Suitable class intervals for miles-per-gallon data.

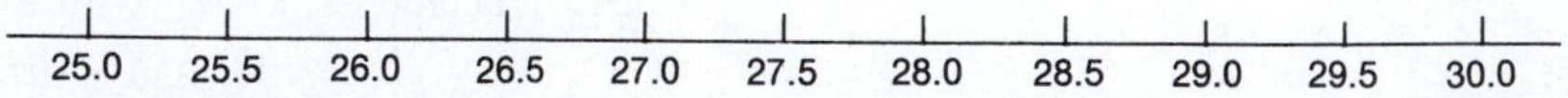

25.0	25.5	26.0	26.5	27.0	27.5	28.0	28.5	29.0	29.5	30.0

There is one further difficulty: where should we place an observation such as 27.0, which falls on a boundary between classes? Our convention will be to define intervals so that such an observation is placed in the upper rather than the lower class interval. Thus, in our frequency distribution, a typical class will be 26.5–<27.0, where the symbol < is a substitute for the phrase *less than*. The observation 27.0 would then fall in the class 27.0–<27.5.

EXAMPLE 6   An interesting problem in microbiology is to determine the lengths of macromolecules in order to obtain information about molecular weights. The paper "Estimation of the True Length of Broken Molecules" (*Biometrics* (1982): 201–13) reported data on segment lengths of a certain type of denatured RNA molecule observed with the aid of an electron micrograph. A frequency distribution appears in Table 5. The class intervals all have width .20. The first class begins at .05 rather than 0, since very small fragments cannot be counted using the electron micrograph. An interesting aspect of the frequency distribution is that the frequencies decline in a reasonably smooth fashion through the interval $1.65 - < 1.85$, but then there is a substantial increase in the next two intervals. The authors provided an explanation for this jump.

TABLE 5

Frequency Distribution for Segment Lengths of RNA Molecules (Microns)

Class	Class Interval	Frequency	Relative Frequency
1	.05 – < .25	298	.2326
2	.25 – < .45	211	.1647
3	.45 – < .65	149	.1163
4	.65 – < .85	116	.0906
5	.85 – <1.05	84	.0656
6	1.05 – <1.25	78	.0609
7	1.25 – <1.45	61	.0476
8	1.45 – <1.65	60	.0468
9	1.65 – <1.85	56	.0437
10	1.85 – <2.05	63	.0492
11	2.05 – <2.25	79	.0617
12	2.25 – <2.45	20	.0156
13	2.45 – <2.65	5	.0039
14	2.65 – <2.85	1	.0008
	Totals:	1281	1.0000

Again other quantities of interest can be calculated from this information:

(i) $\left(\begin{array}{c} \text{proportion of} \\ \text{segments with} \\ \text{length} < .85 \end{array} \right) = \left(\begin{array}{c} \text{proportion} \\ \text{in 1st class} \end{array} \right) + \left(\begin{array}{c} \text{proportion} \\ \text{in 2nd class} \end{array} \right)$

$$+ \left(\begin{array}{c} \text{proportion} \\ \text{in 3rd class} \end{array} \right) + \left(\begin{array}{c} \text{proportion} \\ \text{in 4th class} \end{array} \right)$$

$$= .2326 + .1647 + .1163 + .0906$$

$$= .6042 \quad (\text{roughly } 60\%)$$

(ii) $\left(\begin{array}{c} \text{proportion of} \\ \text{segments with} \\ \text{length between} \\ 1.25 \text{ and } 2.25 \end{array} \right) = \left(\begin{array}{c} \text{proportion} \\ \text{in 7th class} \end{array} \right) + \cdots + \left(\begin{array}{c} \text{proportion} \\ \text{in 11th class} \end{array} \right)$

$$= .0476 + \cdots + .0617$$

$$= .2490 \quad (\text{roughly } 25\%)$$

There are no strict guidelines for selecting either the number of class intervals or the interval lengths. Using a few relatively wide intervals will bunch the data, whereas using a great many relatively narrow intervals may spread the data out too much over the intervals, so that no interval contains more than a few observations. Neither type of distribution will give an informative picture of how values are distributed over the range of measurement. Generally speaking, with a small amount of data there should be relatively few intervals, perhaps between five and ten, whereas with a large amount of data, a distribution based on 15 to 20 (or even more) intervals is often recommended. Two people making reasonable and similar choices for the number of intervals, their width, and the starting point should obtain very similar pictures of the data.

CLASS INTERVALS OF UNEQUAL WIDTH

Figure 7 pictures a data set in which there are a great many observations concentrated near one another at the center of the set and just a few outlying, or stray, values both below and above the main body of data. If a frequency distribution is based on short intervals of equal width, a great many intervals will be required to capture all observations, and many of them will contain no observations (zero frequency). On the other hand, only a few wide intervals will capture all values, but then most of the observations will be grouped into a very few intervals. Neither choice will yield an informative distribution. In such a situation it is best to use a few relatively wide class intervals at the ends of the distribution and some shorter intervals in the middle.

FIGURE 7

Three different choices of class intervals for a data set with outliers.
(a) Many short intervals of equal width
(b) A few wide intervals of equal width
(c) Intervals of unequal width

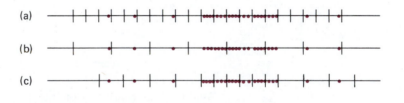

CUMULATIVE RELATIVE FREQUENCIES

Suppose that the first four class intervals in a frequency distribution are $0-<25$, $25-<50$, $50-<75$, $75-<100$, with corresponding relative frequencies .05, .10, .12, and .18. Consider the cumulative sums of these relative frequencies.

$$.05 = \text{proportion of observations less than 25}$$
$$.05 + .10 = .15 = \text{proportion of observations less than 50}$$
$$.05 + .10 + .12 = .27 = \text{proportion of observations less than 75}$$
$$.05 + .10 + .12 + .18 = .45 = \text{proportion of observations less than 100}$$

Each such cumulative sum is the proportion of observations less than the upper limit of the corresponding class interval. These cumulative sums are called **cumulative relative frequencies**; they are often displayed in a column next to the relative frequencies. Notice that each cumulative relative frequency is the sum of the previous one and the current relative frequency.

For example,

$$\left(\begin{array}{c}\text{fourth cumulative}\\\text{relative frequency}\end{array}\right) = \left(\begin{array}{c}\text{third cumulative}\\\text{relative frequency}\end{array}\right) + \left(\begin{array}{c}\text{fourth relative}\\\text{frequency}\end{array}\right)$$

$$= .27 + .18$$

$$= .45.$$

EXAMPLE 7 The strength of welds used in aircraft construction has been of great concern to aeronautical engineers in recent years. Table 6 gives a frequency distribution for shear strengths (lb) of ultrasonic spot welds made on a certain type of alclad sheet ("Comparison of Properties of Joints Prepared by Ultrasonic Welding and Other Means," *J. Aircraft* (1983): 552–56). The cumulative relative frequencies are .01, .01 + .02 = .03, .03 + .09 = .12, .12 + .14 = .26, and so on.

TABLE 6

Frequency Distribution with Cumulative Relative Frequencies

Class Interval	Frequency	Relative Frequency	Cumulative Relative Frequency
4000–<4200	1	.01	.01
4200–<4400	2	.02	.03
4400–<4600	9	.09	.12
4600–<4800	14	.14	.26
4800–<5000	17	.17	.43
5000–<5200	22	.22	.65
5200–<5400	20	.20	.85
5400–<5600	7	.07	.92
5600–<5800	7	.07	.99
5800–<6000	1	.01	1.00
Totals:	100	1.00	

Thus the proportion of welds with strength values less than 5400 is .85 (that is, 85% of the observations are below 5400). Notice also that any particular relative frequency is the difference between two consecutive cumulative relative frequencies. For example,

$$\left(\begin{array}{c}\text{relative frequency}\\\text{for 5000–<5200}\end{array}\right) = \left(\begin{array}{c}\text{cum. rel. freq.}\\\text{for 5000–<5200}\end{array}\right) - \left(\begin{array}{c}\text{cum. rel. freq.}\\\text{for 4800–<5000}\end{array}\right)$$

$$= .65 - .43$$

$$= .22.$$

Sometimes a published article will contain only cumulative relative frequencies, from which the relative frequencies themselves are easily calculated.

Cumulative relative frequencies can also be computed and displayed in a frequency distribution for discrete data.

2.8 Each of 25 students was asked to identify the dictionary he or she uses. The resulting responses were as follows (with A = American Heritage, F = Funk and Wagnalls, M = Macmillan, R = Random House, W = Webster's).

A R A W W M W R R F A W R
R R M W A W R R F W W A

a. Construct a relative frequency distribution for the type of dictionary used.
b. Suppose that the 25 students polled constitute a sample selected from all students at a certain state university. Use the frequency distribution to estimate the proportion of all students at the university who use a Webster's dictionary.

2.9 The *J. of Marketing Research* (Feb. 1975) published the results of a study in which 22 consumers reported the number of times that they had purchased a particular brand of a product during the previous 48-week period. The results were as follows.

0 2 5 0 3 1 8 0 3 1 1
9 2 4 0 2 9 3 0 1 9 8

a. Construct a relative frequency distribution for the number of purchases.
b. What proportion of the shoppers in this study never bought the brand under investigation?
c. Suppose that each of the 22 shoppers in this study had made exactly nine purchases of the product during the previous 48 weeks. What proportion of the shoppers purchased the brand under investigation more that half of the time? All of the time?

2.10 In the paper "Reproduction in Laboratory Colonies of Bank Vole" (*Oikos* (1983): 184), the authors presented the results of a study on litter size. (According to Webster's, a vole is a small rodent with a stout body, blunt nose, and short ears!) As each new litter was born, the number of babies was recorded, and the following results were obtained.

SIZE OF LITTER

3	6	5	6	5	7	5	7	6	6	6
4	6	5	6	4	3	5	6	4	5	9
6	5	6	1	9	7	8	3	7	4	5
5	6	7	3	6	6	9	4	5	7	5
6	8	6	4	7	5	7	4	5	8	6
7	2	7	7	3	3	5	4	6	4	6
3	7	8	5	7	7	7	7	9	8	7
6	7	6	4	7	10	5	2	3	6	6
4	7	6	7	5	5	5	7	5	8	8
4	9	7	5	4	6	5	8	4	5	6
6	3	6	8	6	8	6	5	8	6	11
4	7	6	8	9	7	3	8	3	4	6
4	5	7	5	6	5	7	6	9	3	5
9	7	5	6	7	5	8	6	8	8	6
5	7	4	8	7	7	7	5	3	8	6
10	4	5	5	5						

a. Construct a relative frequency distribution for this data.
b. What proportion of the litters had more than 6 babies? Between 3 and 8 (inclusive)?
c. Is it easier to answer questions like those posed in (b) using the relative frequency distribution than it would be using the raw data given in the table? Explain.

2.11   Compute the cumulative relative frequencies for the data of Exercise 2.10 and use them to answer the following questions.
   a. What proportion of observations are at most 8? At least 8?
   b. What proportion of litters contain between 5 and 10 (inclusive) offspring?

2.12   Is it really the case, as it might seem to an unsuccessful and frustrated angler, that 10% of those fishing reel in 90% of the fish caught? More generally, how is the number of fish caught distributed among those who are trying to catch them? The accompanying table presents data from a survey of 911 anglers done during a particular time period on the lower Current River in Canada ("Fisherman's Luck," *Biometrics* (1976): 265–71).

Number of Fish Caught	Number of Anglers (Frequency)
0	515
1	65
2	60
3	66
4	53
5	55
6	27
7	25
8	25
9	20
	911

   a. Calculate the relative frequencies. (Express each one using four digits of decimal accuracy.)
   b. What proportion of those in the sample caught no fish? One fish? At most one fish?
   c. What proportion of the 911 anglers caught at least five fish? More than five fish?
   d. Calculate the cumulative relative frequencies for the sample of 911 fisherman. Then use them to answer the questions posed in part (c).
   e. Suppose the sample had included an additional four anglers, with numbers of fish caught being 12, 14, 19, and 25 for these four. Construct a frequency distribution that has exactly one more row than the one just displayed.

2.13   The concentration of suspended solids in river water is an important environmental characteristic. The paper "Water Quality in Agricultural Watershed: Impact of Riparian Vegetation during Base Flow" (*Water Resources Bull.* (1981): 233–39) reported on concentration (in parts per million, or ppm) for several different rivers. Suppose that the following 50 observations had been obtained for a particular river:

55.8	60.9	37.0	91.3	65.8	42.3	33.8	60.6	76.0	69.0
45.9	39.1	35.5	56.0	44.6	71.7	61.2	61.5	47.2	74.5
83.2	40.0	31.7	36.7	62.3	47.3	94.6	56.3	30.0	68.2
75.3	71.4	65.2	52.6	58.2	48.0	61.8	78.8	39.8	65.0
60.7	77.1	59.1	49.5	69.3	69.8	64.9	27.1	87.1	66.3

   a. Why can't you base a frequency distribution on the class intervals 0–10, 10–20, 20–30, 30–40, . . . , 90–100?
   b. Construct a frequency distribution using class intervals 20–<30, 30–<40, . . . , 90–<100. (The resulting distribution agrees with that for one of the rivers discussed in the paper.)
   c. What proportion of the concentration observations were less than 50? At least 60?

    d. Just from the frequency distribution, can you determine the proportion of sample observations less than 65? Explain. Can you *estimate* this proportion from the distribution? How does your estimate compare with the actual value of the proportion?

2.14 Refer to the concentration data given in Problem 2.13.
    a. Obtain the cumulative relative frequencies for the class intervals specified in part (b) of that problem.
    b. Use the cumulative relative frequencies to calculate the proportions described in part (c) of Problem 2.13.
    c. Use the cumulative relative frequencies to calculate the proportion of observations in the interval 40–<70.
    d. Just from the frequency distribution, can you determine the proportion of observations that are at most 40? Explain.

2.15 In a study of warp breakage during the weaving of fabric (*Technometrics* (1982): 63), 100 pieces of yarn were tested. The number of cycles of strain to breakage was recorded for each yarn sample. The resulting data is given here.

86	146	251	653	98	249	400	292	131	169
175	176	76	264	15	364	195	262	88	264
157	220	42	321	180	198	38	20	61	121
282	224	149	180	325	250	196	90	229	166
38	337	65	151	341	40	40	135	597	246
211	180	93	315	353	571	124	279	81	186
497	182	423	185	229	400	338	290	398	71
246	185	188	568	55	55	61	244	20	284
393	396	203	829	239	236	286	194	277	143
198	264	105	203	124	137	135	350	193	188

    a. Using class intervals 0–<100, 100–<200, and so on, construct a relative frequency distribution for breaking strength.
    b. If weaving specifications require a breaking strength of at least 110 cycles, approximately what proportion of the yarn samples would be considered unsatisfactory? Answer using your relative frequency distribution.

2.16 The results of the 1980 census included a state-by-state listing of population density. The following values are the number of people per square mile for each of the 50 states.

New Jersey	986.2	Rhode Island	897.8	Massachusetts	733.3
Connecticut	637.8	Maryland	428.7	New York	370.6
Delaware	307.6	Pennsylvania	264.3	Ohio	263.3
Illinois	205.3	Florida	180.0	Michigan	162.6
Indiana	152.8	California	151.4	Hawaii	150.1
Virginia	134.7	North Carolina	120.4	Tennessee	111.6
South Carolina	103.4	New Hampshire	102.4	Louisiana	94.5
Georgia	94.1	Kentucky	92.3	Wisconsin	86.5
West Virginia	80.8	Alabama	76.7	Missouri	71.3
Washington	62.1	Vermont	55.2	Texas	54.3
Mississippi	53.4	Iowa	52.1	Minnesota	51.2
Oklahoma	44.1	Arkansas	43.9	Maine	36.3
Kansas	28.9	Colorado	27.9	Oregon	27.4
Arizona	23.9	Nebraska	20.5	Utah	17.8
Idaho	11.5	New Mexico	10.7	North Dakota	9.4
South Dakota	9.1	Nevada	7.3	Montana	5.4
Wyoming	4.8	Alaska	0.7		

a. Construct a relative frequency distribution for state population density.

b. In your relative frequency distribution of part (a), did you use class intervals of equal widths? Why or why not?

c. Use the relative frequency distribution to give an approximate value for the proportion of states that have a population density of more than 100 people per square mile. Is the approximate value close to the actual value?

2.17 The paper "Lessons from Pacemaker Implantations" (*J. Amer. Med. Assoc.* (1965): 231–32) gave the results of a study that followed 89 heart patients who had received electronic pacemakers. The time (in months) to the first electrical malfunction of the pacemaker was recorded:

24	20	16	32	14	22	2	12	24	6
10	20	8	16	12	24	14	20	18	14
16	18	20	22	24	26	28	18	14	10
12	24	6	12	18	16	34	18	20	22
24	26	18	2	18	12	12	8	24	10
14	16	22	24	22	20	24	28	20	22
26	20	6	14	16	18	24	18	16	6
16	10	14	18	24	22	28	24	30	34
26	24	22	28	30	22	24	22	32	

a. Summarize this data in the form of a frequency distribution using class intervals of 0–<6, 6–<12, and so on.

b. Compute the relative frequencies and cumulative relative frequencies for each class interval of the frequency distribution of part (a).

c. Show how the relative frequency for the class interval 12–<18 could be obtained from the cumulative relative frequencies.

Use the cumulative relative frequencies to give approximate answers to the following questions.

d. What proportion of those who participated in the study had pacemakers that did not malfunction within the first year?

e. If the pacemaker must be replaced as soon as the first electrical malfunction occurs, approximately what proportion required replacement between 1 and 2 years after implantation?

f. Estimate the time at which about 50% of the pacemakers had failed.

g. Estimate the time at which only about 10% of the pacemakers initially implanted were still functioning.

2.18 Birth weights for 302 eighth-born Chinese males born in Singapore are summarized in the accompanying frequency distribution (*Ann. Human Genetics* (1954): 58–73).

Weight (in ounces)	Frequency
72–<80	4
80–<88	5
88–<96	19
96–<104	52
104–<112	55
112–<120	61
120–<128	48
128–<136	39
136–<144	19

a. Construct the cumulative relative frequency distribution for this data.

   b. What proportion of observed birth weights are less than 96? At least 96? Can you use the given information to determine what proportion of birth weights are at most 96? Explain.

   c. Roughly what proportion of birth weights are less than 100? In answering this question, what assumption are you making about the 52 observations in the 96– <104 class interval?

   d. Approximately what birth weight is such that 50% of the observed weights are less than that weight value?

2.19 A student obtained data on fuel efficiency (mpg) and constructed a frequency distribution using the eight class intervals 27.0–<27.5, 27.5–<28.0, . . . , 30.5– <31.0. He then calculated and reported the following cumulative relative frequencies: .09, .23, .38, .35, .72, .80, .93, 1.00. Comment.

2.20 Referring to Exercise 2.8, would it make sense to calculate cumulative relative frequencies? Explain.

2.3 HISTOGRAMS

A **histogram** is a pictorial representation of the information in a frequency distribution. Pictures often have more impact and stay with us longer than tabulated numerical information. The general idea is to represent each relative frequency by a rectangle. The *area* of the rectangle is proportional to the corresponding relative frequency. To see what this means, consider the following partial listing of frequencies and relative frequencies.

class:	1	2	3	. . .
frequency:	30	60	75	. . .
relative frequency:	.06	.12	.15	. . .

The relative frequency for the second class is twice that for the first class, and the same relationship holds for the frequencies. Therefore the area of the rectangle for the second class must be twice the area of the rectangle for the first class. Similarly, .15/.06 = 2.5, so

$$\left(\begin{array}{c} \text{area of rectangle} \\ \text{for class 3} \end{array}\right) = 2.5 \left(\begin{array}{c} \text{area of rectangle} \\ \text{for class 1} \end{array}\right)$$

CONSTRUCTING A HISTOGRAM (BAR CHART) FOR CATEGORICAL DATA

1. Draw a horizontal line and write the category names at regularly spaced intervals.
2. Draw a vertical line and scale it using relative frequency values (frequencies themselves can also be used).
3. Above each category label, draw a rectangle whose height is the corresponding relative frequency (alternatively, frequency). All rectangles should have the same base width.

EXAMPLE 8

In many surveys, a group of individuals is selected, and then one or more attempts are made to contact each individual. The paper "I Hear You Knocking But You Can't Come In: The Effects of Reluctant Respondents and Refusers on Sample Survey Estimates" (*Sociological Methods and Research* (Aug. 1982): 3–32) reported on a study of how group composition changed as the number of attempts to contact increased. Table 7 gives relative frequencies for labor force categories.

TABLE 7

Relative Frequencies for Labor Force Categories

Category	After One Attempt to Contact (234 responses)	After Ten Attempts to Contact (1049 responses)
Full-time	.286	.524
Part-time	.103	.108
Looking	.051	.048
Retired	.226	.132
Not working	.333	.189
Totals:	.999	1.001

Figure 8 presents the corresponding histograms. They are obviously quite different. Making only a single attempt to contact might yield sample information that is quite misleading.

FIGURE 8

Histograms for the labor force frequency distribution.
(a) After one attempt
(b) After ten attempts

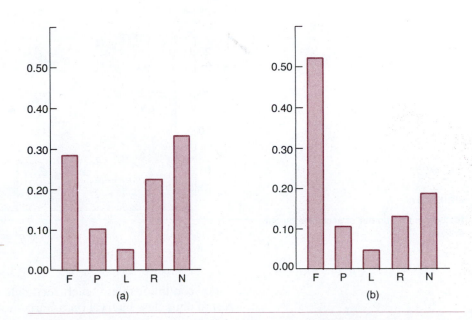

(a) (b)

NUMERICAL DATA

As we did for frequency distributions, we distinguish here between discrete and continuous data.

CONSTRUCTING A HISTOGRAM FOR DISCRETE DATA

1. Draw a vertical scale marked with either relative frequencies or frequencies. The height of each rectangle will then match the corresponding relative frequency or frequency.
2. Mark possible values on a horizontal scale. Each rectangle should be centered at the value to which it refers, and the widths should be identical. If possible values are whole numbers, each base width should be 1 (so the rectangle will extend .5 to either side of the value).

EXAMPLE 9

Figure 9 shows a histogram corresponding to the frequency distribution for the number of accidents given earlier, in Table 3. There is a peak at the values 1 and 2, and then a smooth decline in relative frequencies as we move to the right. A computer-generated histogram (using SPSS) appears in Figure 10. Such histograms are typically not as attractive as hand-drawn versions.

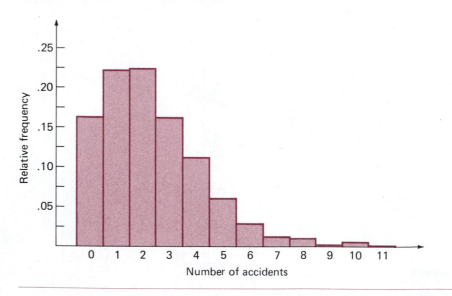

FIGURE 9

Histogram for number of accidents by bus drivers.

For continuous data, each rectangle in a histogram sits above the corresponding class interval on a horizontal measurement axis.

CONSTRUCTING A HISTOGRAM FOR CONTINUOUS DATA

1. Mark boundaries of the class intervals on a horizontal axis. The rectangle corresponding to a particular interval is drawn directly above that interval.
2. If the class intervals have identical widths, either relative frequencies or frequencies can be used on the vertical scale.
3. When class widths are different, the vertical scale *should not* be marked with frequencies or relative frequencies. Instead, a correct picture (one for which area is proportional to relative frequency) results if the height of each rectangle is calculated according to the following formula:

$$\text{height} = \frac{\text{relative frequency}}{\text{interval width}}$$

```
CODE
     I
  0  ****************************** (    117)
     I
  1  ******************************************* (    157)
     I
  2  ******************************************** (    158)
     I
  3  ***************************** (    115)
     I
  4  ********************* (     78)
     I
  5  ************ (     44)
     I
  6  ****** (     21)
     I
  7  *** (      7)
     I
  8  *** (      6)
     I
  9  * (      1)
     I
 10  ** (      3)
     I
 11  * (      1)
     I
     I.........I.........I.........I.........I.........I
     0        40        80       120       160       200
     FREQUENCY
```

FIGURE 10

Histogram from SPSS for number of accidents by bus drivers.

EXAMPLE 10 Mercury contamination is a serious environmental concern. Mercury levels are particularly high in certain types of fish. Citizens of the Republic of Seychelles, a group of islands in the Indian Ocean, are among those who consume the most fish in the world. The paper "Mercury Content of Commercially Important Fish of the Seychelles, and Hair Mercury Levels of a Selected Part of the Population" (*Environ. Research* (1983): 305–12) reported the following observations on mercury content (ppm) in the hair of 40 fishermen.

13.26	32.43	18.10	58.23	64.00	68.20	35.35
33.92	23.94	18.28	22.05	39.14	31.43	18.51
21.03	5.50	6.96	5.19	28.66	26.29	13.89
25.87	9.84	26.88	16.81	37.65	19.63	21.82
31.58	30.13	42.42	16.51	21.16	32.97	9.84
10.64	29.56	40.69	12.86	13.80		

A reasonable choice for class intervals is to start the first interval at zero and let each one have width 10. The resulting frequency distribution is displayed in Table 8, and the corresponding histogram appears in Figure 11.

TABLE 8

Frequency Distribution for Hair Mercury Content of Seychelles Fishermen (ppm)

Class Interval	Frequency	Relative Frequency
0–<10	5	.125
10–<20	11	.275
20–<30	10	.250
30–<40	9	.225
40–<50	2	.050
50–<60	1	.025
60–<70	2	.050
Totals:	40	1.000

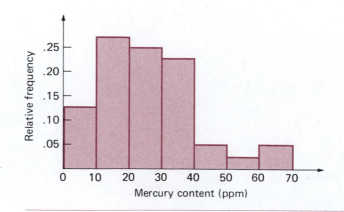

FIGURE 11

Histogram for hair mercury content of Seychelles fishermen.

EXAMPLE 11 Individuals asked for the values of their characteristics such as age or weight sometimes shade the truth in their responses. The paper "Self-Reports of Academic Performance" (*Soc. Methods and Research* (Nov. 1981): 165–85) focused on such characteristics as SAT scores and grade point average. For each student in a sample, the difference in GPA (reported − actual) was determined. Positive differences resulted from individuals reporting grade point averages larger than the correct values. Most differences were close to zero, but there were some rather gross errors. Because of this, a frequency distribution based on unequal class widths gives an informative yet concise summary. Table 9 displays such a distribution based on classes with boundaries at −2.0, −.4, −.2, −.1, 0, .1, .2, .4, and 2.0.

TABLE 9

Frequency Distribution for Errors in Reported GPA

Class Interval	Relative Frequency	Width	Height
−2.0−<−.4	.023	1.6	.014
−.4−<−.2	.055	.2	.275
−.2−<−.1	.097	.1	.970
−.1−<0	.210	.1	2.100
0−<.1	.189	.1	1.890
.1−<.2	.139	.1	1.390
.2−<.4	.116	.2	.580
.4−<2.0	.171	1.6	.107

Figure 12 displays two histograms based on this frequency distribution. The histogram of part (a) is correctly drawn, in that height = relative frequency/interval width. The histogram of part (b) has height = relative frequency and is therefore not correct. In particular, this second histogram considerably exaggerates the incidence of grossly overreported and underreported values—the areas of the two most extreme rectangles are much too large.

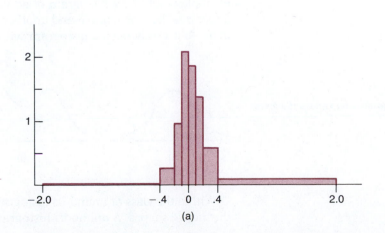

FIGURE 12

Histograms for errors in reporting GPA.
(a) A correct picture
    (height = relative frequency/width)

(a)

FIGURE 12 (continued)

Histograms for errors in reporting GPA.
(b) An incorrect picture
   (height = relative frequency)

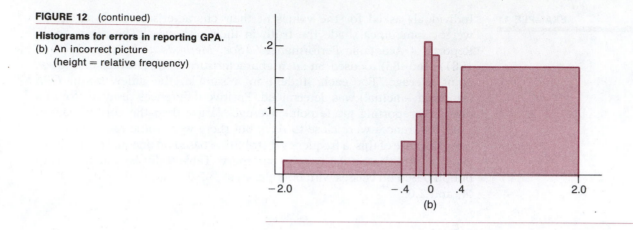

HISTOGRAM SHAPES

It is often desirable to describe the general shape of a histogram. For this purpose, it suffices to consider a *smoothed histogram* obtained by approximating the histogram itself with a smooth curve. This is illustrated in Figure 13.

FIGURE 13

Approximating a histogram by a smooth curve.

One characterization of general shape relates to the number of peaks, or **modes**. A histogram is said to be **unimodal** if it has a single peak, **bimodal** if it has two peaks, and **multimodal** if it has more than two peaks. These shapes are illustrated in Figure 14. Many numerical data sets give rise to a unimodal histogram. Occasionally we encounter a bimodal histogram—an example would be a histogram of adult heights, with one peak at roughly 5 feet 6 inches for women and another peak at roughly 5 feet 9 inches for men—but rarely does a histogram with more than two peaks occur.

FIGURE 14

Smoothed histograms with various numbers of modes.
(a) Unimodal
(b) Bimodal
(c) Multimodal

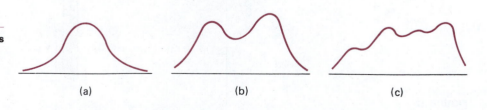

(a)   (b)   (c)

Within the class of unimodal histograms, there are still several important distinctive shapes. A unimodal histogram is **symmetric** if there is a vertical *line of symmetry* such that the part of the histogram to the left of the line is a mirror image of the part to the right (bimodal and multimodal histograms can also be symmetric in this way). Several different symmetric smoothed histograms are pictured in Figure 15.

FIGURE 15

Several symmetric unimodal smoothed histograms.

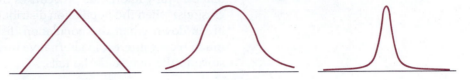

Proceeding to the right from the peak of a unimodal histogram, we move into what is called the **upper tail** of the histogram. Going in the opposite direction moves us into the **lower tail**. For symmetric histograms, the rate of decrease of the curve is the same as we move into either tail.

A unimodal histogram that is not symmetric is said to be **skewed**. If the upper tail of the histogram stretches out farther than the lower tail, then the distribution of values is **positively skewed**. If, on the other hand, the lower tail is longer than the upper tail, the histogram is **negatively skewed**. These two types of skewness are illustrated in Figure 16.

FIGURE 16

Two examples of skewed smoothed histograms.
(a) Positive skew
(b) Negative skew

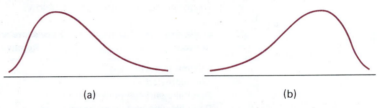

(a)　　　　　　　　　　　(b)

Positive skewness is much more frequently encountered than is negative skewness. An example giving rise to positive skewness is the distribution of single-family home prices in Los Angeles County, where most homes are moderately priced (at least for California), whereas the relatively few homes in Beverly Hills and Malibu have much higher price tags.

One rather specific shape, a *normal curve*, arises more frequently than any other in statistical applications. Many histograms can be well approximated by a normal curve, for example, characteristics such as blood pressure, brain weight, adult male heights, adult female heights, and I.Q. scores. Here we mention briefly several of the most important qualitative properties of such a curve, postponing a more detailed discussion until Chapter 6. A normal curve is not only symmetric but also bell-shaped; it looks like the curve in Figure 17(a). However, not all bell-shaped curves are normal. Starting from the top of the bell, the height of the curve decreases at a well-defined rate when moving out into either tail. (This rate of decrease is specified by a certain mathematical function.)

FIGURE 17

Three examples of bell-shaped histograms.
(a) Normal
(b) Heavy-tailed
(c) Light-tailed

(a)　　　　　　　　(b)　　　　　　　　(c)

A curve with tails that do not decline as rapidly as the tails of a normal curve is said to specify a **heavy-tailed** distribution (compared to the normal curve). Similarly, a curve with tails that decrease more rapidly than the normal tails is called **light-tailed**. Figure 17(b) and (c) illustrate these possibilities. Many inferential procedures that work well (result in accurate conclusions) when the population distribution is approximately normal tend to break down when the population distribution is heavy-tailed, prompting much recent interest in alternative methods of analysis that are not so sensitive to the nature of the tails.

EXERCISES 2.21–2.32   **SECTION 2.3**

2.21 A common problem facing researchers who rely on mail questionnaires is that of nonresponse. In the paper ''Reasons for Nonresponse on the Physicians' Practice Survey'' (*Proc. Social Statistics Section Amer. Stat. Assoc.* (1980): 202), 811 doctors who did not respond to the A.M.A. Survey of Physicians were contacted about the reason for their nonparticipation. The results are summarized in the relative frequency distribution given below.

Reason	Relative Frequency
No time to participate	.264
Not interested	.300
Don't like surveys in general	.145
Don't like this particular survey	.025
Hostility toward the government	.054
Desire to protect privacy	.056
Other reason for refusal	.053
No reason given	.103

Draw the histogram corresponding to the frequency distribution.

2.22 Many researchers have speculated that a relationship exists between birth order and vocational preferences. This theory is investigated in the paper ''Birth Order and Vocational Preference'' (*J. of Experimental Educ.* (1980): 15–18). In this study, 244 New York City high school students were given the Self-Directed Search Test. This test is designed to identify occupational preferences by classifying the student into one of six categories: realistic, investigative, artistic, social, enterprising, and conventional. The results are summarized in the two accompanying frequency distributions.

Vocational Class	Firstborn	Later Born
Conventional	38	9
Realistic	26	19
Enterprising	24	15
Social	12	15
Artistic	12	21
Investigative	10	43

a. Construct two histograms to represent occupational preference—one for first-borns and one for those born later.

b. What inferences would you make based on a comparison of the two histograms?

2.23 The article "Associations between Violent and Nonviolent Criminality" (*Multivariate Behavioral Research* (1981): 237–42) reported the number of previous convictions for 283 adult males arrested for felony offenses. The following frequency distribution is a summary of the data given in the paper.

Number of Previous Convictions	Frequency
0	0
1	16
2	27
3	37
4	46
5	36
6	40
7	31
8	27
9	13
10	8
11	2

Draw the histogram corresponding to this frequency distribution.

2.24 In a study of author productivity ("Lotka's Test," *Collection Mgmt.* (1982): 111–18) a large number of authors were classified according to the number of papers they had written and the results were presented in the following frequency distribution.

Number of Papers	Number of Authors (Frequency)
1	784
2	204
3	127
4	50
5	33
6	28
7	19
8	19
9	6
10	7
11	6
12	7
13	4
14	4
15	5
16	3
17	3

a. Construct a histogram for this frequency distribution.

b. Suppose the five 15s, three 16s, and three 17s had been lumped into a single row labeled "≥15". Would you be able to draw a histogram? Explain.

c. Suppose that instead of the last three rows, there had been a single row, labeled 15–17, with frequency 11. Would you be able to draw a histogram? Explain.

2.25 The accompanying histogram, based on data in the paper "Service Frequency, Schedule Reliability, and Passenger Wait Times at Transit Stops" (*J. of Trans. Research* (1981): 465–71) shows the time (in minutes) that people had to wait for the next schelduled bus when buses were running on a 20-minute schedule. Suppose that the histogram is based on a sample of 300 waiting times. Construct the corresponding frequency distribution.

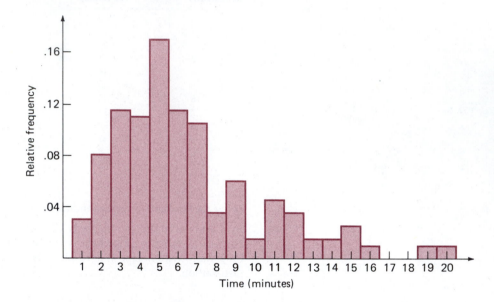

2.26 The mileage traveled before the first major motor failure for each of 191 buses was reported in an article that appeared in *Technometrics* ((Nov. 1980): 588). The frequency distribution appearing in that paper is given here.

Distance Traveled (Thousands of Miles)	Frequency
0–< 20	6
20–< 40	11
40–< 60	16
60–< 80	25
80–<100	34
100–<120	46
120–<140	33
140–<160	16
160–<180	2
180–<200	2

a. Draw the histogram corresponding to this frequency distribution.

b. Use the histogram to estimate the proportion of all buses of this type that operate for more than 100,000 miles before the first major motor failure.

c. Use the histogram to estimate the proportion of all buses that have the first major motor failure after operating for between 50,000 and 125,000 miles.

2.27 Suppose the information in Problem 2.26 had been summarized in the following table:

Distance	Frequency
0–< 40	17
40–< 60	16
60–< 80	25
80–<100	34
100–<120	46
120–<140	33
140–<180	18
180–<220	2

a. What are the widths of the eight class intervals? What are the corresponding relative frequencies?
b. Divide each relative frequency calculated in (a) by the corresponding width to obtain the heights of the rectangles for a histogram.
c. The tallest rectangle has a height of about .012. Mark 0, .002, .004, .006, .008, .010, .012, and .014 on the vertical axis, and then draw the histogram.
d. What is the area of the rectangle above the fifth class interval? What is the total area of all rectangles in the histogram?

2.28 The 1974 edition of *Accident Facts* reported the state-by-state accidental death rates (deaths per 100,000 population) for 1973 as shown.

Alaska	137.8	Washington	51.5	Oregon	59.1
California	57.2	Idaho	77.9	Nevada	88.5
Arizona	84.3	Utah	64.7	Wyoming	106.2
Montana	85.9	Colorado	57.9	New Mexico	93.0
Texas	55.8	Oklahoma	61.4	Kansas	56.2
Nebraska	58.7	South Dakota	78.1	North Dakota	69.8
Minnesota	53.1	Iowa	56.6	Missouri	59.3
Arkansas	102.8	Louisiana	67.6	Mississippi	81.9
Alabama	73.6	Tennessee	70.1	Kentucky	64.0
Illinois	39.3	Wisconsin	49.2	Michigan	50.3
Indiana	54.2	Ohio	44.2	Maine	57.8
New Hampshire	44.8	Vermont	64.2	Massachusetts	45.6
Rhode Island	40.8	Connecticut	34.4	New York	35.5
New Jersey	40.0	Pennsylvania	42.0	West Virginia	64.2
Delaware	49.3	Maryland	36.1	Virginia	57.7
North Carolina	74.2	South Carolina	77.8	Georgia	59.7
Florida	68.5	Hawaii	37.7		

Construct a relative frequency distribution for this data using equal interval widths. Draw the histogram corresponding to your frequency distribution.

2.29 The paper "Paraquat and Marijuana Risk Assessment" (*Amer. J. of Public Health* (1983): 784–88) reported the results of a 1978 telephone survey on marijuana usage. The accompanying frequency distribution gives the amount of marijuana (in grams) smoked per week for those respondents who indicated that they did use the drug.

Grams Smoked per Week	Frequency
0–<3	94
3–<11	269
11–<18	70
18–<25	48
25–<32	31
32–<39	10
39–<46	5
46–<53	0
53–<60	1
60–<67	0
67–<74	1

a. Display the information given in the frequency distribution in the form of a histogram.

b. What proportion of respondents smoked 25 or more grams per week?

c. Use the histogram to estimate the proportion of respondents who smoked more than 15 grams per week.

2.30 An exam is given to students in an introductory statistics course. What is likely to be true of the shape of the histogram of scores if

a. The exam is quite easy?

b. The exam is quite difficult?

c. Half the students in the class have had calculus, the other half have had no prior college math courses, and the exam emphasizes mathematical manipulation?

Explain your reasoning in each case.

2.31 Construct a histogram corresponding to each of the five frequency distributions given in the following table, and state whether each histogram is symmetric, bimodal, positively skewed, or negatively skewed.

Class Interval	Frequency				
	I	II	III	IV	V
0–<10	5	40	30	5	6
10–<20	10	25	10	25	5
20–<30	20	10	8	8	6
30–<40	30	8	7	7	9
40–<50	20	7	7	20	9
50–<60	10	5	8	25	23
60–<70	5	5	30	10	42

2.32 Using the following class intervals, devise a frequency distribution based on 70 observations whose histogram could be described as

a. Symmetric;   **c.** Positively skewed;

b. Negatively skewed;   **d.** Bimodal.

Class Interval
100–<120
120–<140
140–<160
160–<180
180–<200

CHAPTER 2 SUMMARY OF KEY CONCEPTS AND FORMULAS

TERM OR FORMULA	PAGE	COMMENT
Stem-and-leaf display	10	A method of organizing quantitative data in which the stem values (leading digit(s) of the observations) are listed in a column, and the leaf (trailing digit(s)) for each observation is then listed beside the corresponding stem. Sometimes stems are repeated to stretch the display.
Frequency distribution	16	A table that displays frequencies and relative frequencies (proportions) for the different categories (when data is categorical), values of a counting variable, or class intervals (when observations are on a continuous variable).
Histogram	26	A picture of the information in a frequency distribution. A rectangle is drawn above each category label, value of a counting variable, or class interval. The rectangle's area is proportional to the corresponding relative frequency (or, equivalently, frequency).
Histogram shapes	32	A (smoothed) histogram may be unimodal (a single peak), bimodal (two peaks), or multimodal. A unimodal histogram may be symmetric, positively skewed (a long right or upper tail), or negatively skewed. A frequently occurring shape is that of a normal curve.

SUPPLEMENTARY EXERCISES   2.33–2.45

2.33   The accompanying frequency distribution of the number of years of continuous service at the time of resignation from a job with an oil company appeared in the paper "The Role of Performance in the Turnover Process" (*Academy of Management J.* (1982): 137–47). Construct the histogram corresponding to this frequency distribution. Which terms introduced in this chapter (symmetric, skewed, etc.) would you use to describe the histogram?

Years of Service	Frequency
0–<1	4
1–<2	41
2–<3	67
3–<4	82
4–<5	28
5–<6	43
6–<7	14
7–<8	17
8–<9	11
9–<10	7
10–<11	14
11–<12	6
12–<13	14
13–<14	5
14–<15	2

2.34   A random sample of 60 preschool-age children who were participating in a day-care program was used to obtain information about parental work status. The two given frequency distributions appeared in the paper "Nutritional Understanding of Preschool Children Taught in the Home of a Child Development Laboratory" (*Home Econ. Research J.* (1984): 52–60).

Work Status of Mother	Frequency
Unemployed	41
Employed part time	13
Employed full time	6

Father's Occupation	Frequency
Professional	21
Craftsman	19
Manager	6
Proprietor	7
Other	7

a. Construct a relative frequency distribution for the mothers' work statuses. What proportion of the mothers work outside the home?

b. Draw the histogram corresponding to the frequency distribution for the fathers' occupations.

2.35 Data on engine emissions for 46 vehicles is given here (*Technometrics* (Nov. 1980): 487).

Vehicle	HC	CO	Vehicle	HC	CO
1	.50	5.01	2	.65	14.67
3	.46	8.60	4	.41	4.42
5	.41	4.95	6	.39	7.24
7	.44	7.51	8	.55	12.30
9	.72	14.59	10	.64	7.98
11	.83	11.53	12	.38	4.10
13	.38	5.21	14	.50	12.10
15	.60	9.62	16	.73	14.97
17	.83	15.13	18	.57	5.04
19	.34	3.95	20	.41	3.38
21	.37	4.12	22	1.02	23.53
23	.87	19.00	24	1.10	22.92
25	.65	11.20	26	.43	3.81
27	.48	3.45	28	.41	1.85
29	.51	4.10	30	.41	2.26
31	.47	4.74	32	.52	4.29
33	.56	5.36	34	.70	14.83
35	.51	5.69	36	.52	6.35
37	.57	6.02	38	.51	5.79
39	.36	2.03	40	.49	4.62
41	.52	6.78	42	.61	8.43
43	.58	6.02	44	.46	3.99
45	.47	5.22	46	.55	7.47

a. Construct a frequency distribution and histogram for the hydrocarbon (HC) emissions data.

b. Construct a frequency distribution and histogram for the carbon monoxide (CO) data.

c. Are the HC and CO histograms symmetric or skewed? If they are both skewed, is the direction of the skew the same?

2.36 The following data refers to active repair times (hours) for an airborne communication receiver.

1.1	4.0	0.5	5.4	2.0	0.5	0.8	9.0	5.0	3.3
0.3	0.7	2.2	22.0	4.0	2.7	1.0	3.0	1.0	1.5
24.5	1.5	3.3	2.5	1.0	0.8	1.5	0.6	10.3	1.3
0.7	2.0	4.7	3.0	0.8	1.0	8.8	0.6	4.5	7.0
0.7	5.4	1.5	1.5	0.2	7.5	0.5			

a. Construct a stem-and-leaf display in which the two largest values are displayed separately in a row labeled HI.

b. Construct a histogram based on six class intervals with zero as the lower limit of the first interval and interval lengths of 2, 2, 2, 4, 10, and 10, respectively.

2.37 Suppose that the accompanying observations are heating costs for a sample of two-bedroom apartments in Southern California for the month of November.

HEATED BY GAS

25.42	26.12	25.22	23.60	27.77	28.52
21.60	29.49	26.22	25.52	20.19	23.99
26.32	23.38	26.77	31.56	25.54	22.72
27.58	29.96	26.20	23.97	28.17	18.01
22.98					

HEATED WITH ELECTRICITY

33.52	51.01	41.99	33.80	25.93	30.32
32.06	39.86	24.62	31.80	48.58	44.65
31.30	35.04	19.24	40.78	43.39	34.78
25.43	33.82	26.47	34.62	32.02	27.98
30.92					

Construct a comparative stem-and-leaf display contrasting heating costs for gas and electricity. How do the two sides of the display compare?

2.38 The two accompanying frequency distributions appeared in the paper "Aqueous Humour Glucose Concentration in Cataract Patients and its Effect on the Lens" (*Exp. Eye Research* (1984): 605–9). The first is a frequency distribution of lens sodium concentration (in mM) for nondiabetic cataract patients, while the second is for diabetic cataract patients. Draw the histogram corresponding to each frequency distribution. Do you think that the distributions for the population of all diabetic patients and for the population of nondiabetic cataract patients are similar? Explain.

Sodium Concentration (mM)	Nondiabetic Frequency	Diabetic Frequency
0–<20	7	0
20–<40	12	0
40–<60	5	1
60–<80	1	2
80–<100	0	3
100–<120	1	2
120–<140	1	1
140–<160	4	0
160–<180	8	1
180–<200	3	0
200–<220	2	0
220–<240	1	0

2.39 Referring to Exercise 2.38, construct the cumulative relative frequencies corresponding to the frequency distribution of lens sodium concentration (in mM) of nondiabetic cataract patients. Use them to answer the following questions.

a. What proportion of the nondiabetic cataract patients had a lens sodium concentration below 100 mM?

b. What proportion of the nondiabetic cataract patients had a lens sodium concentration between 100 and 200 mM?

c. What proportion of the nondiabetic cataract patients had a lens sodium concentration of at least 140 mM?

d. Find a sodium concentration value for which approximately half of the observed sodium concentrations are smaller than this value.

2.40 The two given frequency distributions of storm duration (in minutes) are based on data appearing in the article "Lightning Phenomenology in the Tampa Bay Area" (*J. of Geophysical Research* (1984): 11, 789–805). Construct a histogram for each of the frequency distributions and discuss the similarities and differences between the two with respect to shape.

Storm Duration	Single-Peak Storms: Frequency	Multiple-Peak Storms: Frequency
0–<25	1	0
25–<50	17	1
50–<75	14	1
75–<100	11	3
100–<125	8	2
125–<150	8	2
150–<175	5	1
175–<200	4	3
200–<225	3	1
225–<250	2	6
250–<275	0	4
275–<300	1	2

2.41 The paper "The Acid Rain Controversy: The Limits of Confidence" (*Amer. Statistician* (1983): 385–94) gave the accompanying data on average SO_2 (sulfur dioxide) emission rates from utility and industrial boilers (lb/million Btu) for 47 states. (Data from Idaho, Alaska, and Hawaii was not given.)

```
2.3   2.7   1.5   1.5   0.3   0.6   4.2   1.3   1.2   0.4
0.5   2.2   4.5   3.8   1.2   0.2   1.0   0.7   0.2   1.4
0.7   3.6   1.0   0.7   1.7   0.5   0.1   0.6   2.5   2.7
1.5   1.4   2.9   1.0   3.4   2.1   0.9   1.9   1.0   1.7
1.8   0.6   1.7   2.9   1.8   1.4   3.7
```

a. Summarize this set of data by constructing a relative frequency distribution.

b. Draw the histogram corresponding to the frequency distribution in part (a). Would you describe the histogram as symmetric or skewed?

c. Use the relative frequency distribution of part (a) to compute the cumulative relative frequencies.

d. Use the cumulative relative frequencies to give the approximate proportion of states with SO_2 emission rates that
   i. were below 1.0 lb/million Btu;
   ii. were between 1.0 and 2.0 lb/million Btu;
   iii. were at least 2.0 lb/million Btu.

2.42 The Los Angeles Unified School District includes 49 public high schools. The *Los Angeles Times* (January 20, 1985) published the average math SAT exam score for each of the 49 schools. Use several of the methods described in this chapter to summarize this data.

AVERAGE SAT MATH SCORE FOR 49 PUBLIC HIGH SCHOOLS

341	477	461	456	349	481	499
328	471	436	440	414	448	503
399	335	332	422	356	375	488
406	375	458	341	468	404	482
464	475	398	317	466	470	463
409	478	469	404	487	439	459
464	502	472	480	481	402	339

2.43 The soil stability index (SSI) of eroded topsoil was recorded for 41 randomly selected sites under dry conditions and for 39 randomly selected sites under green conditions ("Use of Landsat Radiance Parameters to Distinguish Soil Erosion, Stability, and Deposition in Arid Central Australia," *Remote Sensing of Environment* (1984): 195–209). Construct and interpret a comparative stem-and-leaf display.

Soil Stability Index

Dry Conditions						Green Conditions					
31	44	44	44	36	36	20	20	20	20	21	21
36	45	37	45	45	38	21	24	24	24	24	25
39	39	39	39	39	39	25	25	25	25	27	27
39	39	39	39	40	40	27	27	27	28	28	28
40	40	40	40	40	40	30	30	30	41	41	41
41	41	41	41	42	42	42	42	50	50	50	50
42	42	43	43	43		50	50	59			

2.44 Americium 241 (^{241}Am) is a radioactive material used in the manufacture of smoke detectors. The article "Retention and Dosimetry of Injected ^{241}Am in Beagles" (*Radiation Research* (1984): 564–75) described a study in which 55 beagles were injected with a dose of ^{241}Am (proportional to the animals' weights). Skeletal retention of ^{241}Am (μCi/kg) was recorded for each beagle, resulting in the given data.

.196	.451	.498	.411	.324	.190	.489	.300	.346	.448
.188	.399	.305	.304	.287	.243	.334	.299	.292	.419
.236	.315	.447	.585	.291	.186	.393	.419	.335	.332
.292	.375	.349	.324	.301	.333	.408	.399	.303	.318
.468	.441	.306	.367	.345	.428	.345	.412	.337	.353
.357	.320	.354	.361	.329					

a. Construct a frequency distribution for this data and draw the corresponding histogram.

b. Write a short description of the important features of the shape of the histogram.

2.45 A transformation of data values by means of some mathematical function, such as $\sqrt{x}$ or $1/x$, can often yield a set of numbers which has "nicer" statistical properties than the original data. In particular, it may be possible to find a function for which the histogram of transformed values is more symmetric (or, even better, more like a normal curve) than the original data. As an example, the paper "Time Lapse Cinematographic Analysis of Beryllium–Lung Fibroblast Interactions" (*Environ. Research* (1983): 34–43) reported the results of experiments designed to study the behavior of certain individual cells that had been exposed to beryllium. An important characteristic of such an individual cell is its interdivision time (IDT). IDTs were determined for a large number of cells both in exposed (treatment) and unexposed (control) conditions. The authors of the paper used a logarithmic transformation, i.e., transformed value = log(original value). Consider the following representative IDT data.

IDT	$\log_{10}$(IDT)	IDT	$\log_{10}$(IDT)	IDT	$\log_{10}$(IDT)
28.1	1.45	60.1	1.78	21.0	1.32
31.2	1.49	23.7	1.37	22.3	1.35
13.7	1.14	18.6	1.27	15.5	1.19
46.0	1.66	21.4	1.33	36.3	1.56
25.8	1.41	26.6	1.42	19.1	1.28
16.8	1.23	26.2	1.42	38.4	1.58
34.8	1.54	32.0	1.51	72.8	1.86
62.3	1.79	43.5	1.64	48.9	1.69
28.0	1.45	17.4	1.24	21.4	1.33
17.9	1.25	38.8	1.59	20.7	1.32
19.5	1.29	30.6	1.49	57.3	1.76
21.1	1.32	55.6	1.75	40.9	1.61
31.9	1.50	25.5	1.41		
28.9	1.46	52.1	1.72		

Use class intervals 10–<20, 20–<30, . . . to construct a histogram of the original data. Use intervals 1.1–<1.2, 1.2–<1.3, . . . to do the same for the transformed data. What is the effect of the transformation?

REFERENCES

Chambers, John, William Cleveland, Beat Kleiner, and Paul Tukey. *Graphical Methods for Data Analysis.*
Belmont, Calif.: Wadsworth, 1983. (This is an excellent survey of methods, illustrated with numerous interesting examples.)

Cleveland, William. *The Elements of Graphing Data.*
Belmont, Calif.: Wadsworth, 1985 (An informal and informative introduction to various aspects of graphical analysis.)

Koopmans, Lambert H. *An Introduction to Contemporary Statistics.*
Boston: Duxbury, 1986. (The first part of this book contains an interesting presentation of both traditional descriptive methods and the more recently developed exploratory techniques.)

McNeil, Donald R. *Interactive Data Analysis.*
New York: John Wiley, 1977. (A very informal brief introduction to some of the most useful recently developed methods for doing exploratory data analysis.)

Velleman, Paul, and David Hoaglin. *Applications, Basics, and Computing of Exploratory Data Analysis.*
Boston: Duxbury, 1981. (Subtitled "ABC's of EDA," this book contains a good treatment of exploratory methods; it is a bit more comprehensive than the McNeil book.)

CHAPTER 3

NUMERICAL SUMMARY MEASURES

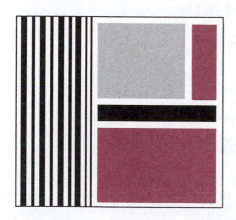

INTRODUCTION Stem-and-leaf displays, frequency distributions, and histograms are effective in conveying general impressions about the distribution of values in a data set. To gain deeper insights and develop methods for further analysis, however, we need compact and precise ways of describing and characterizing data. Here we will see how we can summarize and convey some important features of a data set by using just a few numerical summary quantities computed from the data. In Section 1 we introduce the mean and median as measures of the center of a numerical data set. The variance and standard deviation are presented in Section 2 as measures of the extent to which data spreads out about the center. The last section illustrates how these and other summarizing quantities and techniques convey information about prominent features of a data set.

3.1 DESCRIBING THE CENTER OF A DATA SET

An informative way to describe the location of a numerical data set is to report a central value, one that is representative of the observations in the set. The two most popular measures of center are the *mean* and the *median*.

THE MEAN

The **mean** of a set of numerical observations is just the familiar arithmetic average, the sum of observations divided by the number of observations. Values in such a data set are observations on a numerical variable. The variable might be the number of traffic accidents, the number of pages in a book, reaction time, yield from a chemical reaction, etc. For the case of sample data, it is helpful to have concise notation for the variable, sample size, and individual observations. Let

x = the variable for which we have sample data;

n = the number of sample observations (sample size);

x_1 = the first sample observation;

x_2 = the second observation;

$\vdots$

x_n = the nth (last) sample observation.

As an example, we might have a sample consisting of $n = 4$ observations on x = battery lifetime (hr): $x_1 = 5.9$, $x_2 = 7.3$, $x_3 = 6.6$, $x_4 = 5.7$. Notice that the value of the subscript on x has no relationship to the magnitude of the observation. In this example, x_1 is not the smallest observation, only the first one obtained by an investigator, and x_4 is not the largest observation.

The sum of $x_1, x_2, \ldots, x_n$ can be denoted by $x_1 + x_2 + \cdots + x_n$, but this is cumbersome. The Greek letter $\sum$ is traditionally used in mathematics to denote summation. In particular, $\sum x$ will denote the sum of the x values in the data set under consideration.

▌▌ DEFINITION

The **sample mean** of a numerical sample $x_1, x_2, \ldots, x_n$, denoted by $\bar{x}$, is

$$\bar{x} = \frac{x_1 + x_2 + \cdots + x_n}{n} = \frac{\sum x}{n}$$

EXAMPLE 1

Physical anthropologists frequently use fossil measurements to provide evidence for various anthropological theories. The paper ''A Reconsideration of the Fontechevade Fossils'' (*J. of Phys. Anthropology* (1973): 25–36) reported the following data on height of the left frontal sinus (in mm) for a sample of $n = 14$ fossils known to be Neanderthals.

$$x_1 = 42 \quad x_2 = 27 \quad x_3 = 25 \quad x_4 = 40 \quad x_5 = 33 \quad x_6 = 31 \quad x_7 = 42$$
$$x_8 = 34 \quad x_9 = 35 \quad x_{10} = 25 \quad x_{11} = 29 \quad x_{12} = 30 \quad x_{13} = 29 \quad x_{14} = 35$$

The sum of these sample values is $42 + 27 + \cdots + 35 = 457$, so the sample mean height is

$$\bar{x} = \frac{\sum x}{n} = \frac{457}{14} = 32.6$$

Thus we could report 32.6 mm as a representative sinus height value for this sample (even though there is no skull in the sample with this value).

The data values in Example 1 were all integers, yet the mean was given as 32.6. It is common practice to report one extra digit of decimal accuracy for the mean.* This allows the value of the mean to fall between possible observable values (e.g., the average number of children per family could be 1.8, whereas no single family will have 1.8 children). It could also be argued that the mean carries with it more precision than does any single observation.

The sample mean $\bar{x}$ is computed from sample observations, so it is a characteristic of the particular sample in hand. It is customary to use a Roman letter to denote such a sample characteristic, as we have done with $\bar{x}$. The **population mean** (average of all population values) is a characteristic of the entire population. Statisticians commonly use Greek letters to denote population characteristics. In particular, we will let μ denote the population mean. For example, the true average fuel efficiency for all cars of a certain type under specified conditions might be $\mu = 27.5$ mpg. A sample of $n = 5$ cars might yield efficiencies

27.3, 26.2, 28.4, 27.9, 26.5,

from which we obtain $\bar{x} = 27.26$ (somewhat smaller than μ). However, a second sample might give $\bar{x} = 28.52$, a third $\bar{x} = 26.85$, and so on. The value of $\bar{x}$ varies from sample to sample, whereas there is just one value for μ. Later on we shall see how the value of $\bar{x}$ from a particular sample can be used to draw various conclusions about the value of μ.

A potential drawback to the mean is that its value can be greatly affected by the presence of even a single outlier (an unusually large or small observation) in the data set. Suppose the page lengths of five randomly selected library books are 217, 312, 196, 258, and 185, from which we calculate $\bar{x} = 233.6$. If, however, $x_5 = 985$ rather than 185, then $\bar{x} = 393.6$. Many would argue that 393.6 is not a very representative value for this sample since it is larger than all but one of the observations. We now turn our attention to a measure of center that is quite insensitive to outliers.

---

* Often $\bar{x}$ is used in the calculation of other quantities. More digits may then be necessary to ensure accurate results. A first instance of this will appear in the next section.

THE MEDIAN The median strip of a highway divides the highway in half, and the median of a numerical data set performs an analogous function. Once the numbers in the set have been listed in order from smallest to largest, the **median** is the middle value in the list and divides the list into two equal parts. Let us first consider the case of a sample containing n observations. When n is an odd number (say, 5), the sample median is the single middle value. But when n is even (say, 6), there are two middle values in the ordered list, so we average them to obtain the sample median.

▌▌▌ **DEFINITION**

The **sample median** is obtained by first ordering the n observations from smallest to largest (with any repeated values included, so that every sample observation appears in the ordered list). Then

$$\text{sample median} = \begin{cases} \text{the single middle value if } n \\ \quad \text{is odd} \\ \text{the average of the two middle} \\ \quad \text{values if } n \text{ is even} \end{cases}$$

EXAMPLE 2 Durability of materials is a major concern of engineers. The accompanying observations are on $x =$ lifetime of power apparatus insulation (hr) under specified experimental conditions ("On the Estimation of Life of Power Apparatus Insulation under Combined Electrical and Thermal Stress", *I.E.E.E. Trans. on Elec. Insulation* (1985): 70–78):

$x_1 = 501$	$x_2 = 1072$	$x_3 = 1905$	$x_4 = 282$	$x_5 = 1122$
$x_6 = 2138$	$x_7 = 1202$	$x_8 = 851$	$x_9 = 741$	$x_{10} = 1585$

Since $n = 10$ is even, the sample median is the average of the two middle values in the ordered list (the fifth and sixth values). Ordering gives us

282, 501, 741, 851, 1072, 1122, 1202, 1585, 1905, 2138

from which

$$\text{sample median} = \frac{1072 + 1122}{2} = 1097.0$$

The sample mean is $\bar{x} = 11399/10 = 1139.9$, somewhat larger than the sample median.

The **population median** plays the same role for the population that the sample median plays for the sample. It is the middle value in the ordered list consisting of all population observations.

We previously noted that the mean—population or sample—is very sensitive to even a single value that lies far above or below the rest of the data. The value of the mean is pulled out toward such an outlying value or values. The median, on the other hand, is quite insensitive to outliers. For example, the largest sample observation (2138) in Example 2 can be increased by an arbitrarily large amount without changing the value of the median. Similarly, an increase in the second or third largest observations does not affect the median, nor would a decrease in several of the smallest observations.

This stability of the median is what sometimes justifies its use as a measure of center. Income distributions are commonly summarized by reporting the median rather than the mean, since otherwise a few $100,000 salaries would distort the resulting typical salary.

COMPARING THE MEAN AND THE MEDIAN

Figure 1 presents several smoothed histograms that might represent either a distribution of sample values or a population distribution. Pictorially, the median is the value on the measurement axis that separates the histogram into two parts with .5 (50%) of the area under each part of the curve. The mean is a bit harder to visualize. If the histogram were placed on a fulcrum with a sharp point, it would tilt unless the fulcrum were positioned exactly at the mean. The mean is the balance point for the distribution.

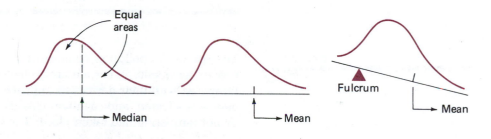

FIGURE 1

Picturing the mean and the median.

When the histogram is symmetric, the point of symmetry is both the dividing point for equal areas and the balance point, so the mean and median are identical. However, when the histogram is unimodal (single-peaked) with a longer upper tail, the relatively few outlying values in the upper tail pull the mean up, so that it lies above the median. For example, an unusually high exam score raises the mean but does not affect the median. Similarly, when a unimodal histogram is negatively skewed, the mean is smaller than the median (see Figure 2).

FIGURE 2

Relationship between the mean and the median.

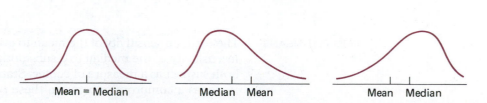

CATEGORICAL DATA

The natural numerical summary quantities for a categorical data set are the relative frequencies for the various categories. Each relative frequency is the proportion (fraction) of responses that are in the corresponding category. Often there are only two possible responses (a *dichotomy*)—male or female, does or does not have a driver's license, did or did not vote in the last election, and so on. It is convenient in such situations to label one of the two possible responses "S" (for success) and the other "F" (for failure). As long as further analysis is consistent with the labeling, it is immaterial which category is assigned the S label. When the data set comprises a sample, the fraction of S's in the sample is called the *sample proportion of successes.*

▌▌▌ DEFINITION

The **sample proportion of successes**, denoted by p, is

$$p = \left(\begin{array}{c}\text{sample proportion}\\\text{of successes}\end{array}\right) = \frac{\text{number of S's in the sample}}{n}$$

EXAMPLE 3

The use of antipollution equipment on automobiles has substantially improved air quality in certain areas. Unfortunately, many car owners have tampered with their devices to improve performance. A sample of $n = 15$ cars was selected, and each was classified as S or F, according to whether or not tampering had taken place. The resulting data was

S, F, S, S, S, F, F, S, S, F, S, S, S, F, F

This sample contains nine S's, so

$$p = \frac{9}{15} = .60$$

That is, 60% of the sample responses are S's, so in 60% of the cars sampled, there has been tampering with the air pollution control devices.

The Greek letter π will be used to denote the **population proportion of S's**. We shall see later how the value of p from a particular sample can be used to make inferences about π.

TRIMMED MEANS

The extreme sensitivity of the mean to even a single outlier and the extreme insensitivity of the median to a substantial proportion of outliers can make both suspect as a measure of center. Statisticians have proposed a *trimmed mean* as a compromise between these two extremes.

DEFINITION

A **trimmed mean** is computed by first ordering the data values from smallest to largest, deleting a selected number of values from each end of the ordered list, and finally averaging the remaining values. The **trimming percentage** is the percentage of values deleted from each end of the ordered list.

EXAMPLE 4

The paper "Snow Cover and Temperature Relationships in North America and Eurasia" (*J. of Climate and Appl. Meteorology* (1983): 460–69) used statistical techniques to relate amount of snow cover on each continent to average continental temperature. Data presented there included the following ten observations on October snow cover for Eurasia during the years 1970–1979 (in million km²):

6.5   12.0   14.9   10.0   10.7   7.9   21.9   12.5   14.5   9.2

The ordered values are

6.5   7.9   9.2   10.0   10.7   12.0   12.5   14.5   14.9   21.9

Since 20% of 10 is 2, the 20% trimmed mean results from deleting the two observations at each end of the ordered list and averaging the remaining six values:

$$20\% \text{ trimmed mean} = \frac{9.2 + 10.0 + 10.7 + 12.0 + 12.5 + 14.5}{6} = 11.48$$

The mean, 12.01, is a 0% trimmed mean—no deletion before averaging. In this example, the median, 11.35, is a 40% trimmed mean, the largest possible trimming percentage when $n = 10$. The 20% trimmed mean is a good choice here for a representative October snow cover value.

A trimmed mean with a small-to-moderate trimming percentage—between 5% and 25%—is less affected by outliers than the mean, yet it is not as insensitive as the median. Trimmed means are therefore being used with increasing frequency.

EXERCISES 3.1–3.14

SECTION 3.1

3.1 The accompanying data on concentration of lead (ppm) in core samples taken at 17 Texaco drilling sites appeared in the paper "Statistical Comparison of Heavy Metal Concentrations in Various Louisiana Sediments" (*Environ. Monitoring and Assessment* (1984): 163–70).

55   53   55   59   58   50   63   50   50
48   56   63   54   53   56   50   55

a. Compute the sample mean.

b. Compute the sample median. How do the values of the sample mean and median compare?

c. The paper also gave the accompanying data on zinc concentration (ppm). Compute and interpret the value of the sample mean.

```
86   77    91    86    81   87   94   90   70
92   90   108   112   101   88   99   98
```

3.2 The paper "Penicillin in the Treatment of Meningitis" (*J. Amer. Med. Assoc.* (1984): 1870–73) reported the body temperatures (°F) of patients hospitalized with meningitis. Ten of the observations were as follows.

```
104.0   104.8   101.6   108.0   103.8   100.8   104.2   100.2   102.4   101.4
```

a. Compute the sample mean.

b. Do you think the 10% trimmed mean would differ much from the sample mean computed in part (a)? Why? Answer without actually computing the trimmed mean.

3.3 In anticipation of the 1984 Olympics, the *Los Angeles Times* (August 15, 1983) reported the ozone levels at several sites that were to be used for Olympic events the following summer. Listed below are the ozone readings (in ppm) taken at noon from July 28 to August 12 at East Los Angeles College.

```
10   14   13   18   12   22   14   19
22   13   14   16    3    6    7   19
```

a. Compute the sample mean.

b. Compute the sample median.

c. Compute the 6.25% trimmed mean (by deleting the smallest and largest values before averaging). Is this a more representative measure than the sample mean or median? Explain.

d. Consider now the ozone levels for a second Olympic site, the Coliseum.

```
 8   13   10    9   16   12   13   14
17   13    9   16    5    9    8   12
```

Describe the center of this data set using the mean and the median. How do the values of the mean and median for the Coliseum compare to those for East Los Angeles College?

3.4 One of the problems with which health service administrators must deal is patient dissatisfaction. One common complaint focuses on the amount of time that a patient must wait in order to see a doctor. In a survey to investigate waiting times, medical-clinic secretaries were asked to record the waiting times (measured from arrival at the clinic until doctors are seen) for a sample of patients. The data (from *Statistics and Public Policy*, cited in the Chapter 1 References) for one day is given in the accompanying table.

WAITING TIME (min)

```
40   30   40   55   30   60   35   55   40
35    5   10   65   35   35   30   30   60
35   25   65   30   30   45   85   25   25
10   10   15
```

a. Describe a typical waiting time by using both the mean and the median. Which do you think is the most representative measure of the center of the data set? Why?

b. Compute a 10% trimmed mean. How does the trimmed mean compare in value to the mean and median calculated in part (a)?

3.5 The article "Drug Screening: The Never-Ending Search for New and Better Drugs" (from *Statistics: A Guide to the Unknown,* cited in the Chapter 1 References) reports actual tumor weights (in grams) observed in three animals treated with a test drug and in six untreated control animals:

Treated: 0.96   1.59   1.14
Untreated: 1.29   1.60   2.27   1.31   1.88   2.21

Calculate the mean and the median for each of the two samples. How do these values compare?

3.6 Consider the following statement: Over 65% of the residents of Los Angeles earn less than the average wage for that city. Could this statement be correct? If so, how? If not, why not?

3.7 Five experimental animals were put on a certain diet for several weeks, and another five animals were put on a second diet. The resulting weight gains (in pounds) at the end of the period were as follows:

Diet #1:   12   13   7   5   15
Diet #2:   11   14   3   13   4

Without doing any calculation, say which group of animals achieved the larger average weight gain. Hint: Base your reasoning on the ordered values from each sample.

3.8 Reconsider the situation described in the previous problem, but suppose now that the weight gains for the second diet are 5, 13, 11, 10, and 3 pounds. Without doing any calculation, say how the average gain for the first diet compares to that for the second diet. Hint: Again order the observations from smallest to largest within each sample.

3.9 A sample consisting of four pieces of luggage was selected from among those checked at an airline counter, yielding the following data on x = weight (lb):

$x_1 = 33.5, x_2 = 27.3, x_3 = 36.7, x_4 = 30.5.$

Suppose one more piece is selected; denote its weight by x_5. Find a value of x_5 so that $\bar{x}$ = sample median.

3.10 Refer to Exercise 3.9, and suppose that the weights for the first four pieces of luggage are

$x_1 = 30, x_2 = 49, x_3 = 29, x_4 = 31.$

If x_5 is between 31 and 49, can it be the case that $\bar{x}$ = sample median? Explain your reasoning. Hint: Let $x_5 = 31 + k$, where $k \geq 0$; now what can be said about $\bar{x}$?

3.11 Suppose that the ten patients whose temperatures were given in Exercise 3.2 received treatment with large doses of penicillin. Three days later, temperatures were again recorded, and the treatment was considered successful if there had been a reduction in a patient's temperature. Denoting success by S and failure by F, the ten observations are

S   S   F   S   S   S   F   F   S   S

a. What is the value of the sample proportion of successes?
b. Replace each S with a 1 and each F with a 0. Then calculate $\bar{x}$ for this numerically coded sample. How does $\bar{x}$ compare to p?

c. Suppose that it is decided to include 15 more patients in the experiment. How many of these would have to be S's to give $p = .80$ for the entire sample of 25 patients?

3.12 An experiment to study the lifetime (in hours) for a certain type of component involved putting ten components into operation and observing them for 100 hours. Eight of the components failed during that period, and those lifetimes were recorded. Denote the lifetimes of the two components still functioning after 100 hours by "100+". The resulting sample observations were

48   79   100+   35   92   86   57   100+   17   29

Which of the measures of center discussed in this section can be calculated, and what are the values of those measures? Note: The data from this experiment is said to be "censored on the right."

3.13 An instructor has graded 19 exam papers submitted by students in a certain class of 20 students, and the average so far is 70 (the maximum possible score is 100). How high would the score on the last paper have to be to raise the class average by one point? By two points?

3.14 A certain college has two sections of introductory statistics during a particular semester. One section has 20 students and the other has 100 students.
a. What is the mean number of students per section (the average over sections)?
b. Let x_1 denote the number of students in the first student's class, let x_2 be the number of students in the second student's class, and so on (up through x_{120}). What is the average of these x values? (This is the mean class size when averaged over students rather than classes.)
c. From a student perspective, which of the two averages, the one calculated in part (a) or the one in part (b), is more pertinent?

3.2 DESCRIBING VARIABILITY IN A DATA SET

Reporting a measure of center gives only partial information about a data set. It is also important to describe the spread of values about the center. The three different samples displayed in Figure 3 all have mean = median = 45. There is much variability in the first sample compared to the extent of spread in the third sample. The second sample shows less variability than

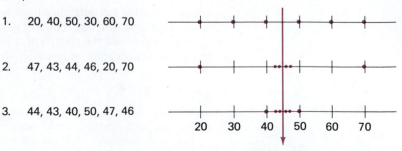

Sample

1.   20, 40, 50, 30, 60, 70

2.   47, 43, 44, 46, 20, 70

3.   44, 43, 40, 50, 47, 46

FIGURE 3

Three samples with the same center and different amounts of variability.

Mean = median

the first and more than the third; most of the variability in the second sample is due to the two extreme values being so far from the center.

The simplest measure of variability is the **range**, which is the difference between the largest and smallest values. Generally speaking, more variability will be reflected in a larger range. However, variability depends on more than just the distance between the two most extreme values. The first two samples of Figure 3 both have a range of 50, but there is substantially less dispersion in the second sample.

DEVIATIONS FROM THE MEAN   Let us focus on sample data. Our primary measures of variability will depend on the extent to which each sample observation deviates from the sample mean $\bar{x}$. Subtracting $\bar{x}$ from each observation gives us the set of deviations from the mean.

 DEFINITION

The n **deviations from the sample mean** are the differences

$$x_1 - \bar{x}, \ x_2 - \bar{x}, \ \ldots, \ x_n - \bar{x}.$$

A particular deviation will be positive if the x value exceeds $\bar{x}$ and negative if the x value is less than $\bar{x}$.

EXAMPLE 5   The sample mean of the $n = 10$ insulation lifetimes given in Example 2 was 1139.9. The observations and corresponding deviations are displayed in Table 1.

TABLE 1

Deviations of Insulation Lifetimes from the Mean

Observation Number	Observation (x)	Deviation (x − x̄)
1	501	−638.9
2	1072	−67.9
3	1905	765.1
4	282	−857.9
5	1122	−17.9
6	2138	998.1
7	1202	62.1
8	851	−288.9
9	741	−398.9
10	1585	445.1

The sixth deviation, $x_6 - \bar{x} = 2138 - 1139.9 = 998.1$, is positive and quite large because $x_6 = 2138$ greatly exceeds the mean. The fifth observation, $x_5 = 1122$, is just a bit below $\bar{x}$ on the measurement scale, yielding a negative deviation, -17.9, that is small in magnitude.

Generally speaking, the larger the magnitudes of the deviations, the greater the amount of variability in the sample. Thus the magnitudes of the deviations for the first sample in Figure 3 substantially exceed those for the third sample, implying more dispersion in sample #1.

We now consider how to combine the deviations into a single numerical measure. A first thought is to calculate the average deviation. That is, add the deviations together—this sum can be denoted compactly by $\sum (x - \bar{x})$—and divide by n. This does not work, though, because negative and positive deviations counteract one another in the summation.

> Except for the effects of rounding in computing the deviations, it is always true that $\sum (x - \bar{x}) = 0$.* Thus the average deviation is always zero and so cannot be used as a measure of variability.

EXAMPLE 6

It is easily verified that the sum of the $n = 10$ deviations in Example 5 is exactly zero. The sample mean for the $n = 14$ Neanderthal sinus heights given in Example 1 is (to one decimal place) 32.6. The corresponding deviations are

9.4   −5.6   −7.6   7.4   0.4   −1.6   9.4
1.4   2.4   −7.6   −3.6   −2.6   −3.6   2.4

The sum of these deviations is .60. The discrepancy between this and zero is due entirely to rounding. Using $\bar{x} = 32.64$ gives $\sum (x - \bar{x}) = .04$. Carrying more digits of decimal accuracy in $\bar{x}$ gives a sum even closer to zero.

THE VARIANCE AND STANDARD DEVIATION

The standard way to prevent negative and positive deviations from counteracting one another is to square them before combining. Then deviations with opposite signs but the same magnitude, such as $+20$ and -20, will make identical contributions to variability. The squared deviations are $(x_1 - \bar{x})^2, (x_2 - \bar{x})^2, \ldots, (x_n - \bar{x})^2$, and their sum is

$$(x_1 - \bar{x})^2 + \cdots + (x_n - \bar{x})^2 = \sum (x - \bar{x})^2.$$

Dividing this sum by the sample size n gives the average squared deviation. However, for a reason to be explained shortly, a divisor slightly smaller than n will be used.

---

*$\sum (x - \bar{x}) = (x_1 - \bar{x}) + (x_2 - \bar{x}) + \cdots + (x_n - \bar{x})$
$= \sum x - n\bar{x} = \sum x - n(\sum x/n) = \sum x - \sum x = 0$

DEFINITION

The **sample variance**, denoted by s^2, is the sum of squared deviations from the mean divided by $n - 1$. That is,

$$s^2 = \frac{\sum (x - \bar{x})^2}{n - 1}$$

The **sample standard deviation** is the square root of the variance and is denoted by s.

A large amount of variability in the sample is indicated by a relatively large value of s^2 or of s, while a small value of s^2 or s goes along with a small amount of variability. For most statistical purposes, s is the desired quantity, but s^2 must be computed first. Notice that whatever unit is used for x (such as lb or sec), the squared deviations and therefore s^2 are in squared units. Taking the square root gives a measure expressed in the same units as x.

EXAMPLE 7   (*Example 1 continued*) Table 2 presents squared deviations for the Neanderthal sinus height data using $\bar{x} = 32.64$.

TABLE 2

Deviations and Squared Deviations for the Neanderthal Sinus Height Data

Observation	$(x - \bar{x})$	$(x - \bar{x})^2$
42	9.36	87.6096
27	−5.64	31.8096
25	−7.64	58.3696
40	7.36	54.1696
33	.36	.1296
31	−1.64	2.6896
42	9.36	87.6096
34	1.36	1.8496
35	2.36	5.5696
25	−7.64	58.3696
29	−3.64	13.2496
30	−2.64	6.9696
29	−3.64	13.2496
35	2.36	5.5696
	Sum:	427.2144 $= \Sigma(x - \bar{x})^2$

Thus,

$$s^2 = \frac{\sum (x - \bar{x})^2}{n - 1} = \frac{427.2144}{13} = 32.8626 \text{ cm}^2,$$

$$s = \sqrt{32.8626} = 5.73 \text{ cm}$$

A Note Concerning Computation

The computation of s^2 using the defining formula can be a bit tedious. An alternative expression for the numerator of s^2 simplifies the arithmetic by eliminating the need to calculate the deviations.

> A computational formula for the sum of squared deviations is
>
> $$\sum (x - \bar{x})^2 = \sum x^2 - n\bar{x}^2$$
>
> Thus a **computational formula for the sample variance** is
>
> $$s^2 = \frac{\sum x^2 - n\bar{x}^2}{n - 1}$$

According to this formula, after squaring each x value and adding these to obtain $\sum x^2$, the single quantity $n\bar{x}^2$ (that is, $n \cdot \bar{x} \cdot \bar{x}$) is subtracted from the result. Instead of the n subtractions required to obtain the deviations, just one subtraction now suffices.

The computed value of s^2, whether obtained from the defining formula or the computational formula, can sometimes be greatly affected by the way in which $\bar{x}$ is rounded. Protection against adverse rounding effects can virtually always be achieved by using four or five digits of decimal accuracy beyond the decimal accuracy of the data values themselves.

EXAMPLE 8

(*Example 7 continued*) For efficient computation, it is convenient to place the x values in a single column (or row) and the x^2 values just beside (or below) them. Adding the numbers in these two columns (or rows) then gives $\sum x$ and $\sum x^2$, respectively.

Observation Number	x	x^2
1	42	1,764
2	27	729
3	25	625
4	40	1,600
5	33	1,089
6	31	961
7	42	1,764
8	34	1,156
9	35	1,225
10	25	625
11	29	841
12	30	900
13	29	841
14	35	1,225
Sums:	$\sum x = 457$	$\sum x^2 = 15,345$

Thus $\bar{x} = 457/14 = 32.64286$ and

$$n\bar{x}^2 = (14)(32.64286)(32.64286) = 14{,}917.788$$

so

$$s^2 = \frac{15,345 - 14,917.788}{14 - 1} = \frac{427.212}{13} = 32.862$$

$$s = \sqrt{32.862} = 5.73 \text{ mm}$$

Interpretation and Properties

A standard deviation may be informally interpreted as the size of a "typical" deviation from the mean. Thus, in Example 8, a typical deviation from $\bar{x} = 32.6$ is about 5.73.

We computed $s = 5.73$ in Example 8 without saying whether this value indicated a large or small amount of variability. At this point it is better to use s for comparative purposes than for an absolute assessment of variability. If we obtained a sample of skulls of a second type and computed $s = 2.1$ for those skulls, then we would conclude that there is more variability in our original sample than in this second sample. A particular value of s can be judged large or small only in comparison to something else.

There are measures of variability for the entire population that are analogous to s^2 and s for a sample. These measures are called the **population variance** and **population standard deviation** and are denoted by σ^2 and σ, respectively. (We again use a lowercase Greek letter for a population characteristic.) The population standard deviation σ is expressed in the same units of measurement as are the values in the population. As with s, the value of σ can be used for comparative purposes.

In many statistical procedures we would like to use the value of σ, but unfortunately it is not usually available. We therefore have to use in its place a value computed from the sample that we hope is close to σ (that is, a good *estimate* of σ). This is why the divisor $n - 1$ is used in s^2: the value of s^2 as we defined it tends to be a bit closer to σ^2 than if s^2 were defined using a divisor of n. We will say more about this in Chapter 7.

THE INTERQUARTILE RANGE

As with $\bar{x}$, the value of s can be greatly affected by the presence of even a single unusually small or large observation. The *interquartile range* is a measure of variability that is resistant to the effects of outliers. It is based on quantities called *quartiles*. The *lower quartile* separates the bottom 25% of the data set from the upper 75%, and the *upper quartile* separates the top 25% from the bottom 75%. The *middle quartile* is the median. Figure 4 illustrates the locations of these quartiles for a smoothed histogram.

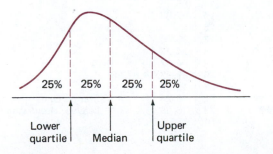

FIGURE 4

The quartiles for a smoothed histogram.

The quartiles for sample data are obtained by dividing the n ordered observations into a lower half and an upper half; if n is odd, the median is included in both halves. The two extreme quartiles are then the medians of the two halves.

||| **DEFINITION**

Lower quartile = median of the lower half of the sample

Upper quartile = median of the upper half of the sample

(If n is odd, the median of the entire sample is included in both halves.)

The **interquartile range (iqr)**, a resistant measure of variability, is given by

iqr = upper quartile − lower quartile

EXAMPLE 9

Cardiac output and maximal oxygen uptake typically decrease with age in sedentary individuals, but these decreases are at least partially arrested in middle-aged individuals who engage in a substantial amount of physical exercise. To understand better the effects of exercise and aging on various circulatory functions, the paper "Cardiac Output in Male Middle-Aged Runners" (*J. of Sports Medicine* (1982): 17–22) presented data from a study of 21 male middle-aged runners. Figure 5 is a stem-and-leaf display of oxygen uptake values (mL/kg·min) while pedaling at 100 watts on a bicycle ergometer. The sample size $n = 21$ is an odd number, so the median, 20.88, is included in both halves of the sample:

lower half:	12.81	14.95	15.83	15.97	17.90	18.27
	18.34	19.82	19.94	20.62	20.88	
upper half:	20.88	20.93	20.98	20.99	21.15	22.16
	22.24	23.16	23.56	35.78	36.73	

```
12 || 81
13 ||
14 || 95
15 || 97, 83
16 ||
17 || 90
18 || 34, 27
19 || 94, 82
20 || 99, 93, 98, 62, 88
21 || 15
22 || 16, 24          Stems: Ones
23 || 16, 56          Leaves: Hundreths
HI: 35.78, 36.73
```

FIGURE 5

Stem-and-leaf display of oxygen uptake (mL/kg·min).

Each half of the sample contains 11 observations, so each quartile is the sixth value in from either end of the corresponding half. This gives us:

lower quartile = 18.27

upper quartile = 22.16

and

$$iqr = 22.16 - 18.27 = 3.89$$

The sample mean and standard deviation are 21.10 and 5.75, respectively. If we were to change the two largest values from 35.78 and 36.73 to 25.78 and 26.73 (so that they are still the two largest values), the median and the interquartile range would not be affected, whereas the mean and standard deviation would change to 20.14 and 3.44, respectively.

The resistant nature of the iqr follows from the fact that up to 25% of the smallest sample observations and up to 25% of the largest can be made more extreme without affecting its value.

The **population interquartile range** is the difference between the upper and lower population quartiles. If a histogram of the data set under consideration (whether a population or a sample) can be reasonably well approximated by a normal curve, then the relationship between the standard deviation (sd) and interquartile range is roughly sd = iqr/1.35. A value of the standard deviation much larger than iqr/1.35 suggests a histogram with heavier (or longer) tails than a normal curve. For the data of Example 9 we had $s = 5.75$, while iqr/1.35 = 3.89/1.35 = 2.88. This suggests that the distribution of sample values is indeed heavy-tailed compared to a normal curve.

EXERCISES   3.15–3.26   **SECTION 3.2**

3.15   The paper "Improving Fermentation Productivity with Reverse Osmosis" (*Food Technology* (1984): 92–96) summarized the results of an investigation into glucose concentration (g/l) for a particular blend of malt liquor. Eight batches were analyzed, resulting in the given glucose concentrations.

74   54   52   51   52   53   58   71

a. Using $\bar{x} = 58.125$, calculate the deviations from the mean and verifty that their sum is zero.
b. Use the deviations in part (a) to calculate s^2 and s.

3.16   Iron status in athletes is important because of the central role of this mineral in the synthesis of hemoglobin and enzymes fundamental to energy production. An investigation of such matters was reported in the paper "Iron Status in Winter Olympic Sports" (*J. of Sports Sciences* (1987): 261–71). Consider the following sample observations on hemoglobin level (g/dl) for female alpine skiers:

14.6   14.3   15.1   12.7   11.8   13.4   13.8

Use the computational formula to calculate s^2 and then obtain the sample standard deviation.

3.17 A study of the relationship between age and various visual functions (such as acuity and depth perception) reported the following observations on area of scleral lamina (mm^2) from human optic nerve heads ("Morphometry of Nerve Fiber Bundle Pores in the Optic Nerve Head of the Human," *Experimental Eye Research* (1988): 559–68):

2.75	2.62	2.74	3.85	2.34	2.74	3.93	4.21	3.88
4.33	3.46	4.52	2.43	3.65	2.78	3.56	3.01	

a. Calculate Σx and Σx^2.
b. Use the values calculated in part (a) to compute the sample variance s^2 and then the sample standard deviation s. How can the value of s be interpreted?

3.18 **a.** Give two sets of five numbers that have the same mean but different standard deviations.
b. Give two sets of five numbers that have the same standard deviation but different means.

3.19 Although bats are not known for their eyesight, they are able to locate prey (mainly insects) by emitting high-pitched sounds and listening for echoes. A paper appearing in *Animal Behavior* ("The Echolation of Flying Insects by Bats" (1960): 141–54) gave the following distances (in cm) at which a bat first detected a nearby insect.

62	23	27	56	52	34	42	40	68	45	83

a. Compute the sample mean distance at which a bat first detects an insect.
b. Compute the sample variance and standard deviation for this data set. How would you interpret these values?

3.20 The paper "Evaluating Variability of Filling Operations" (*Food Technology* (1984): 51–55) gave data on the actual amount of fluid dispersed by a machine designed to disperse 10 ounces. Ten observed values were used to compute $\Sigma x = 100.2$ and $\Sigma x^2 = 1004.4$. Compute the sample variance and standard deviation.

3.21 For the data in Exercise 3.19, add -10 to each sample observation. (This is the same as subtracting 10.) For the new set of values, compute the mean and the deviations from the mean. How do these deviations compare to the deviations from the mean for the original sample? How does s^2 for the new values compare to s^2 for the old values? In general, what effect does adding the same number to each observation have on s^2 and s? Explain.

3.22 For the data of Exercise 3.19, multiply each data value by 10. How does s for the new values compare to s for the original values?

3.23 The first four deviations from the mean in a sample of $n = 5$ reaction times were .3, .9, 1.0, and 1.3. What is the fifth deviation from the mean? Give a sample for which these are the five deviations from the mean.

3.24 Refer to the data in Exercise 3.17 on scleral lamina areas.
a. Determine the lower and upper quartiles.
b. Calculate the value of the interquartile range.
c. If the two largest sample values, 4.33 and 4.52, had instead been 5.33 and 5.52, how would this affect the iqr? Explain.
d. By how much could the observation 2.34 be increased without affecting the iqr? Explain.
e. If an eighteenth observation, $x_{18} = 4.60$, is added to the sample, what is the iqr?

3.25 The following numbers are salinity values for water specimens taken from North Carolina's Pamlico Sound.

7.6	7.7	4.3	5.9	5.0	10.5	7.7	9.5	12.0	12.6
6.5	8.3	8.2	13.2	12.6	13.6	14.1	13.5	11.5	12.0
10.4	10.8	13.1	12.3	10.4	13.0	14.1	15.1		

a. Calculate the value of s.
b. Calculate iqr/1.35. How does this compare to s, and what does it suggest about the shape of the histogram of this data?

3.26 The standard deviation alone does not measure relative variation. For example, a standard deviation of $1 would be considered large if it is describing the variability from store to store in the price of an ice cube tray. On the other hand, a standard deviation of $1 would be considered small if it is describing store-to-store variability in the price of a particular brand of freezer. A quantity designed to give a relative measure of variability is the *coefficient of variation*. Denoted by CV, the coefficient of variation expresses the standard deviation as a percent of the mean. It is defined by the formula

$$CV = (s/\bar{x})(100)$$

Consider the two given samples. Sample 1 gives the actual weight (in ounces) of the contents of cans of pet food labeled as having a net weight of 8 oz. Sample 2 gives the actual weight (in pounds) of the contents of bags of dry pet food labeled as having a net weight of 50 lb.

Sample 1:	8.3	7.1	7.6	8.1	7.6	8.3	8.2	7.7	7.7	7.5
Sample 2:	52.3	50.6	52.1	48.4	48.8	47.0	50.4	50.3	48.7	48.2

a. For each of the given samples, calculate the mean and the standard deviation.
b. Compute the coefficient of variation for each sample. Do the results surprise you? Why or why not?

3.3 SUMMARIZING A DATA SET

A measure of center can be combined with a measure of variability to obtain informative statements about how values in a data set are distributed along the measurement scale. Often statements of interest refer to observations that are a specified number of standard deviations from the mean.

EXAMPLE 10 Consider a data set of IQ scores with mean and standard deviation 100 and 15, respectively. We can make the following statements:

(a) Because $100 - 85 = 15$, we say that a score of 85 is "one standard deviation *below* the mean." Similarly, 115 is "one standard deviation *above* the mean."
(b) Since $(2)(15) = 30$ and $100 + 30 = 130$, $100 - 30 = 70$, the scores $70, 71, 72, \ldots, 129, 130$ are those *within* two standard deviations of the mean.
(c) Because $100 + (3)(15) = 145$, the scores $146, 147, 148, \ldots$ exceed the mean by more than three standard deviations.

CHEBYSHEV'S RULE Without knowing anything more about the data set than just the mean and standard deviation, **Chebyshev's Rule** gives information about the proportion of observations that fall within a specified number of standard deviations of the mean. The rule applies to both a sample and a population.

CHEBYSHEV'S RULE

Consider any number k which is at least 1. Then *the proportion of observations that are within k standard deviations of the mean is at least* $1 - (1/k^2)$.

Substituting selected values of k gives the following table.

Number of Standard Deviations, k	Proportion within k Standard Deviations
2	At least $1 - \frac{1}{4} = .75$
3	At least $1 - \frac{1}{9} = .89$
4	At least $1 - \frac{1}{16} = .94$
4.472	At least $1 - \frac{1}{20} = .95$
5	At least $1 - \frac{1}{25} = .96$
10	At least $1 - \frac{1}{100} = .99$

EXAMPLE 11 Tensile strength is one of the most important properties of various wire products used for industrial purposes. The article "Fluidized Bed Patenting of Wire Rods" (*Wire J.* (June 1977): 56–61) reported on a sample of tensile strength observations (in kg/mm^2), for specimens of a certain type of wire. Summary quantities were

$$n = 129 \qquad \bar{x} = 123.6 \text{ kg/mm}^2 \qquad s = 2.0 \text{ kg/mm}^2$$

Figure 6 displays values that are 1, 2, and 3 standard deviations from the mean. Chebyshev's rule allows us to assert the following:

(a) The proportion of observations between 119.6 and 127.6 is at least .75. Since 75% of the sample size, 129, is 96, at least 96 of the values are in the interval from 119.6 to 127.6.

(b) The interval from 117.6 to 129.6 includes all values within three standard deviations of the mean. Because at least 89% of the observations are in this interval, at most 11% of the observations are either less than 117.6 or exceed 129.6.

FIGURE 6

Measurement scale for the tensile strength data.

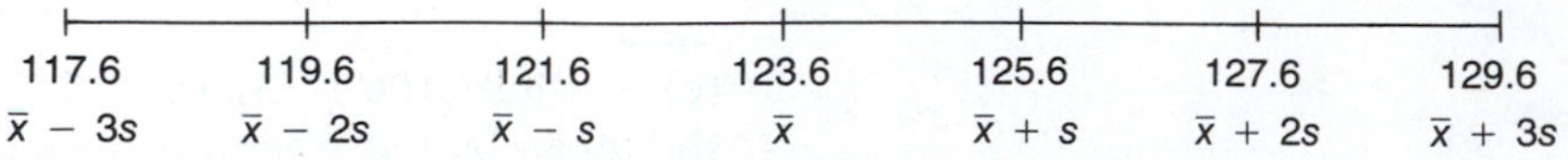

117.6	119.6	121.6	123.6	125.6	127.6	129.6
$\bar{x} - 3s$	$\bar{x} - 2s$	$\bar{x} - s$	$\bar{x}$	$\bar{x} + s$	$\bar{x} + 2s$	$\bar{x} + 3s$

(c) The values 116.6 and 130.6 are 7 kg/mm² from the mean, so they are $7/2 = 3.5$ standard deviations from $\bar{x}$. Substituting $k = 3.5$ into the expression in Chebyshev's Rule gives

$$1 - 1/k^2 = 1 - 1/(3.5)^2$$
$$= 1 - .082$$
$$= .918$$

Therefore at least 91.8% (or 118) of the observations are in this interval. (d) From part (b), at most 11% of the observations are outside the interval from 117.6 to 129.6, but they are *not* necessarily equally divided between the two ends of the sample. Thus we *cannot* say that at most 5.5% exceed 129.6. The distribution of values may be quite skewed. The best we can say is that at most 11% exceed 129.6.

Because Chebyshev's Rule is applicable to any data set (distribution), whether symmetric or skewed, we must be careful when making statements about the proportion above a particular value, below a particular value, or inside or outside an interval that is not centered at the mean. The rule must be used in a conservative fashion. There is another side to this conservatism. Whereas the rule states that at least 75% of the observations are within two standard deviations of the mean, in many data sets, substantially more than 75% of the values satisfy this condition. The same sort of understatement is frequently encountered for other values of k (numbers of standard deviations).

EXAMPLE 12

Figure 7 gives a stem-and-leaf display of IQ scores of 112 children in one of the early studies that used the Stanford revision of the Binet–Simon intelligence scale. (See the well-known book *The Intelligence of School Children* by L. M. Terman. Boston: Houghton Mifflin Company, 1919.)

```
 6 ║ 1
 7 ║ 2,5,6,7,9
 8 ║ 0,0,0,0,1,2,4,5,5,5,6,6,8
 9 ║ 0,0,0,0,1,1,2,3,3,3,4,4,6,6,6,6,7,7,8,8,8,9
10 ║ 0,0,0,1,1,2,2,2,2,2,3,3,3,5,6,6,6,7,7,7,7,8,8,9,9,9,9,9
11 ║ 0,0,0,0,1,1,2,2,3,3,3,3,4,4,4,4,4,4,7,7,8,9,9
12 ║ 0,1,1,1,1,1,2,3,4,4,5,6,6,9
13 ║ 0,0,6
14 ║ 2,6            Stems: Tens
15 ║ 2              Leaves: Ones
```

FIGURE 7

Stem-and-leaf display of IQ scores.

Summary quantities include

$$\bar{x} = 104.5 \qquad s = 16.3 \qquad 2s = 32.6 \qquad 3s = 48.9$$

Table 3 shows how Chebyshev's Rule considerably understates actual percentages.

TABLE 3

Summarizing the Distribution of IQ Scores

$k = $ # of sd's	$\bar{x} \pm ks$	Chebyshev	Actual
2	71.9 to 137.1	at least 75%	96% (108)
2.5	63.7 to 145.3	at least 84%	97% (109)
3	55.6 to 153.4	at least 89%	100% (112)

THE EMPIRICAL RULE

The fact that statements deriving from Chebyshev's Rule are frequently very conservative suggests that we should look for rules that are less conservative and more precise. The most useful such rule is the **Empirical Rule**, which can be applied whenever the distribution of data values can be reasonably well described by a normal curve. The word *empirical* means deriving from practical experience, and practical experience has shown that a normal curve gives a reasonable fit to many data sets.

EMPIRICAL RULE

If the histogram of values in a data set can be reasonably well approximated be a normal curve, then:

roughly 68% of the observations are within one standard deviation of the mean;

roughly 95% of the observations are within two standard deviations of the mean;

roughly 99.7% of the observations are within three standard deviations of the mean.

The Empirical Rule makes "approximately" instead of "at least" statements, and the percentages for $k = 1$, 2, and 3 standard deviations are much higher than those allowed by Chebyshev's Rule.

EXAMPLE 13

One of the earliest papers to argue for the wide applicability of the normal distribution was "On the Laws of Inheritance in Man. I. Inheritance of Physical Characters" (*Biometrika* (1903): 375–462). Among the data sets discussed in the paper was one consisting of 1052 measurements of mothers' statures. The mean and standard deviation were

$$\bar{x} = 62.484 \text{ in.}, \qquad s = 2.390 \text{ in.}$$

A normal curve did provide a good fit to the data. Table 4 contrasts actual percentages with those from Chebyshev's Rule and the Empirical Rule. Clearly, the Empirical Rule here is much more successful and informative than Chebyshev's Rule would have been.

TABLE 4

Summarizing the Distribution of Mothers' Statures

# of sd's	Interval	Actual	Empirical	Chebyshev
1	60.094 to 64.874	72.1%	68%	$\geq 0\%$
2	57.704 to 67.264	96.2%	95%	$\geq 75\%$
3	55.314 to 69.554	99.2%	99.7%	$\geq 89\%$

Our detailed study of the normal distribution and areas under normal curves in Chapter 6 will enable us to make statements analogous to those of the Empirical Rule for values other than $k = 1$, 2, or 3 standard deviations. For now, note that it is rather rare to see an observation from a normally distributed population that is further than two standard deviations from the mean (only 5%), and it is shocking to see one more than three standard deviations away. If you encountered a mother whose stature was 72 inches, you would probably conclude that she was not part of the population described by the data set in Example 13.

MEASURES OF RELATIVE STANDING

When you obtain your score after taking an achievement test, you probably want to know how it compares to scores of others who have taken the test. Is your score above or below the mean and by how much? Does your score place you in the top 5% of those who took the test, or only among the top 25%? Questions of this sort are answered by finding ways to measure the position of a particular value in a data set relative to all values in the set. One such measure involves calculating a *z score* (sometimes called a *standard score*).

||| **DEFINITION**

The **z score** corresponding to a particular observation in a data set is

$$z \text{ score} = \frac{\text{observation} - \text{mean}}{\text{standard deviation}}$$

The z score gives the distance in standard deviations between the observation and the mean. It is positive or negative according to whether the observation lies above or below the mean.

EXAMPLE 14

Suppose that two graduating seniors, one a marketing major and the other an accounting major, are comparing job offers. The marketing student has an offer for $18,000 per year, and the accounting major has one for $20,000 per year. The mean and standard deviation of accounting offers that year are $21,000 and $1500, respectively. The corresponding quantities for marketing offers are \$17,500 and \$1000. Thus

$$\text{accounting } z \text{ score} = \frac{20,000 - 21,000}{1500} = -.67$$

(so \$20,000 is .67 standard deviations below the mean), whereas

$$\text{marketing } z \text{ score} = \frac{18,000 - 17,500}{1000} = .5$$

Relative to the appropriate data sets, the marketing offer is actually more attractive than the accounting offer (though this may not offer much solace to the marketing major).

The z score is particularly useful when the distribution of observations is approximately normal. In this case, by the Empirical Rule, a z score outside the interval from -2 to $+2$ will occur in about 5% of all cases, while a z score outside the interval from -3 to $+3$ will occur only about .3% of the time.

A particular observation can be located even more precisely by giving the percent of observations that fall at or below that observation. If, for example, 95% of all test scores are at or below 650, whereas only 5% are above 650, then 650 is called the *95th percentile* of the data set (or of the distribution of scores). Similarly, if 10% of all scores are at or below 400 and 90% are above 400, then the value 400 is the 10th percentile.

▌▌▌ **DEFINITION**

For any particular number r between 0 and 100, the **rth percentile** is the value such that r percent of the observations in the data set fall at or below that value.

Figure 8 illustrates the 90th percentile of a data set. We have already met several percentiles in disguise. The median is the 50th percentile,

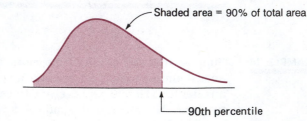

FIGURE 8

90th percentile from a smoothed histogram.

and the lower and upper quartiles are the 25th and 75th percentiles, respectively.

EXAMPLE 15

The frequency distribution in Table 5 was constructed from information in the paper "An Alternative Procedure for Estimating Speed Distribution Parameters on Motorways and Similar Roads" (*Transportation Research* (1976): 25–29). The data values are speeds (mph) of randomly selected vehicles traveling on a major British highway.

TABLE 5

Relative and Cumulative Relative Frequencies for Vehicle Speed Data

Class	30–35	35–40	40–45	45–50	50–55	55–60	60–65
Relative frequency	.003	.017	.055	.105	.124	.206	.185
Cumulative relative frequency	.003	.020	.075	.180	.304	.510	.695

Class	65–70	70–75	75–80	80–85	85–90	90–95	95–100
Relative frequency	.129	.096	.047	.015	.013	.004	.001
Cumulative relative frequency	.824	.920	.967	.982	.995	.999	1.000

Because exactly 51% of the observations are at most 60, the 51st percentile for this data set is 60. Similarly, since the cumulative relative frequency for the 70–75 class interval is .920, the 92nd percentile is 75.

There is no class for which the cumulative relative frequency is .90. We shall therefore approximate the 90th percentile. Table 5 shows that the desired percentile is between 70 and 75. The cumulative relative frequency up to the value 70 is .824, so the additional relative frequency needed is .900 − .824 = .076. The relative frequency of the entire 70–75 class is .096. This suggests starting at 70 and moving the fraction .076/.096 of the way to the next class boundary:

$$\text{90th percentile} \approx 70 + \frac{.076}{.096}(5) = 73.958 \approx 74$$

Thus roughly 90% of the area under the corresponding histogram lies to the left of 74. That is, about 90% of the drivers on this particular highway drive at speeds below 74 mph.

BOX PLOTS (Optional)

It would be nice to have a method of summarizing data that gives more detail than just a measure of center and spread and yet less detail than a stem-and-leaf display or histogram. A *box plot* is one such technique. It is compact, yet it provides information about center, spread, symmetry versus skewness of the data, and the presence of outlying values. To make the plot resistant to the presence of outliers, it is based on the median and interquartile range rather than the mean and standard deviation.

CONSTRUCTION OF A BOX PLOT

1. Draw a rectangular box whose left edge is at the lower quartile and whose right edge is at the upper quartile. (So the box width is iqr.) Draw a vertical line segment inside the box at the median.
2. Place marks at distances 1.5 iqr from either end of the box—these are the **inner fences**. Similarly, place marks for the **outer fences** at distances 3 iqr from either end.
3. Extend horizontal line segments (''whiskers'') from each end of the box out to the most extreme observations that are still within the inner fences.
4. **Mild outliers** are observations between the inner and outer fences. Show them as shaded circles. **Extreme outliers**, those observations beyond the outer fences, are shown as open circles.

The regions associated with mild and extreme outliers are illustrated in Figure 9. The presence of such outliers, especially the extreme ones, can cause trouble for many standard inferential procedures. Such observations should be investigated to see if they resulted from errors or exceptional behavior of some sort. In a large data set whose histogram is well approximated by a normal curve, only about .7% of the observations (7 out of every 1000) will be outliers, and only about .0002% (2 out of every 1,000,000) will be extreme outliers. So, if a small sample from some population contains an extreme outlier, it is likely that a normal curve will not give a good approximation to the population histogram.

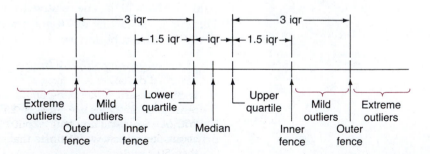

FIGURE 9

Regions for mild and extreme outliers in a box plot.

EXAMPLE 16 The accompanying data came from an anthropological study of rectangular shapes (*Lowie's Selected Papers in Anthropology*, Cora Dubois, ed. Berkeley, Calif.: Univ. of Calif. Press, 1960, pp. 137–142). Observations were made on the variable x = width/length for a sample of $n = 20$ beaded rectangles used in Shoshoni Indian leather handicrafts.

.553 .570 .576 .601 .606 .606 .609 .611 .615 .628
.654 .662 .668 .670 .672 .690 .693 .749 .844 .933

The quantities needed for constructing a box plot are:

median = .641   lower quartile = .606   upper quartile = .681
iqr = .681 − .606 = .075   1.5 iqr = .1125   3 iqr = .225
inner fences: .606 − .1125 = .4935,   .681 + .1125 = .7935
outer fences: .606 − .225 = .381,   .681 + .225 = .906

The most extreme observations on either side of the median that are within the inner fences are .553 and .749. Thus the whiskers extend out to these two values. There are no outliers of either type on the lower end of the sample. The second largest sample observation, .844, is a mild outlier (it falls between the inner and outer fences), and the largest, .933, is an extreme outlier. The box plot appears as Figure 10. The median line is not at the center of the box, so there is a slight asymmetry in the middle half of the data. However, the most striking feature is the presence of the two outliers. These two x values considerably exceed the "golden ratio" .618, used since antiquity as an aesthetic standard for rectangles.

FIGURE 10

Box plot for rectangle data.

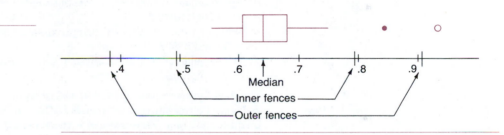

EXERCISES   3.27–3.45

SECTION 3.3

3.27 The average playing time of records in a large collection is 35 minutes, and the standard deviation is 5 minutes.
  a. Without assuming anything about the distribution of times, at least what percentage of the times are between 25 and 45 minutes?
  b. Without assuming anything about the distribution of times, what can be said about the percentage of times that are either less than 20 minutes or greater than 50 minutes?
  c. Assuming that the distribution of times is normal, approximately what percentage of times are between 25 and 45 minutes? Less than 20 minutes or greater than 50 minutes? Less than 20 minutes?

3.28 In a study to investigate the effect of car speed on accident severity, 5000 accident reports of fatal automobile accidents were examined and the vehicle speed at impact was recorded for each one. It was determined that the average speed was 42 mph and that the standard deviation was 15 mph. In addition, a histogram revealed that vehicle speed at impact could be described by a normal curve.
  a. Roughly what proportion of vehicle speeds were between 27 mph and 57 mph?
  b. Roughly what proportion of vehicle speeds exceeded 57 mph?

3.29 Mobile homes are very tightly constructed for energy conservation. This may lead to a buildup of pollutants generated indoors. The paper "A Survey of Nitrogen Dioxide Levels Inside Mobile Homes" (*J. Air Pollut. Control Assoc.* (1988): 647–51) discussed various aspects of NO_2 concentration in these structures.

a. In one sample of homes in the Los Angeles area, the mean NO_2 concentration in kitchens during the summer was 36.92 ppb and the standard deviation was 11.34. Making no assumptions about the shape of the histogram, what can be said about the percentage of observations between 14.24 and 59.60?

b. Inside what interval is it guaranteed that at least 89% of the concentration observations will lie?

c. In a sample of non–Los Angeles homes, the average kitchen concentration during the winter was 24.76 ppb and the standard deviation was 17.20. Do these values suggest that the histogram of sample observations did not closely resemble normal curve? Hint: What is $\bar{x} - 2s$?

3.30 A sample of concrete specimens of a certain type is selected and the compressive strength of each one is determined. The mean and standard deviation are calculated as $\bar{x} = 3000$ and $s = 500$, and the sample histogram is found to be very well approximated by a normal curve.

a. Approximately what percentage of the sample observations are between 2500 and 3500, and what result justifies your assertion?

b. Approximately what percentage of sample observations are outside the interval from 2000 to 4000?

c. What can be said about the approximate percentage of observations between 2000 and 2500?

d. Why would you not use Chebyshev's Rule to answer the questions posed in parts (a) through (c)?

3.31 The *Los Angeles Times* (Oct. 30, 1983) reported that a typical customer of the 7–Eleven convenience stores spends $3.24. Suppose that the average amount spent by customers of 7–Eleven stores is the reported value of $3.24 and that the standard deviation for amount of sale is $8.88.

a. Based on the given mean and standard deviation, do you think that the distribution of the variable *amount of sale* could have been normal in shape? Why or why not?

b. What can be said about the proportion of all customers that spend over $20.00 on a purchase at a 7–Eleven store?

3.32 Exercise 2.4 gave the accompanying 26 observations on pH for specimens of minesoil:

3.59	4.36	3.86	4.25	4.46	4.53	2.62	6.79	6.49
4.27	3.84	4.78	4.65	2.91	3.90	6.00	2.83	3.58
4.43	4.58	4.75	3.49	4.11	3.58	5.21	4.41	

a. Compute the upper and lower quartiles and then the interquartile range.

b. How large or small does an observation have to be in order to be considered an outlier? An extreme outlier? Are there any mild outliers or extreme outliers in the sample?

c. Construct a box plot for this data and comment on any interesting features.

3.33 The accompanying 16 ozone readings (ppm) were taken at noon at East Los Angeles College (*Los Angeles Times*, Aug. 18, 1983).

10	14	13	18	12	22	14	19
22	13	14	16	3	6	7	19

a. Compute the upper quartile, lower quartile, and the interquartile range.

b. Are there any observations that are mild outliers? Extreme outliers?

c. Construct a box plot for the ozone values.

3.34 Suppose the four additional ozone readings of 8, 13, 17, and 28 (ppm) are added to those of Exercise 3.33.

a. Does inclusion of these four values change the values of the quartiles or the interquartile range?

b. Are any of the observations mild or extreme outliers?

c. Construct a box plot for this new set of ozone values (use all 20 observations). In what way does this box plot differ from the plot of Exercise 3.33(c)?

3.35 A student took two national aptitude tests in the course of applying for admission to colleges. The national average and standard deviation for the first test were 475 and 100, respectively, whereas for the second test the average and standard deviation were 30 and 8, respectively. The student scored 625 on the first test and 45 on the second. Use z scores to determine on which exam the student performed better.

3.36 Suppose that your younger sister is applying for entrance to college and she has taken the SAT exams. She scored at the 83rd percentile on the verbal section of the test and at the 94th percentile on the math section of the test. Since you have been studying statistics, she asks you for an interpretation of these values. What would you tell her?

3.37 The paper "Modeling and Measurements of Bus Service Reliability" (*Trans. Research* (1978): 253–6) studied various aspects of bus service and presented data on travel times from several different routes. We give here a frequency distribution for bus travel times from origin to destination on one particular route in Chicago during peak morning traffic periods.

Class Interval	Frequency	Relative Frequency
15–<16	4	.02
16–<17	0	.00
17–<18	26	.13
18–<19	99	.49
19–<20	36	.18
20–<21	8	.04
21–<22	12	.06
22–<23	0	.00
23–<24	0	.00
24–<25	0	.00
25–<26	16	.08

a. Construct the corresponding histogram.

b. Compute the following (approximate) percentiles.

i. 86th **ii.** 15th **iii.** 90th **iv.** 95th **v.** 10th

3.38 Exercise 3.4 has data on waiting times for patients at a medical clinic. Construct a box plot of the data. Does it indicate that there are any extreme or unusual observations in the sample?

3.39 An advertisement for the "30-inch wonder" that appeared in the September 1983 issue of the journal *Packaging* claimed that the 30-inch wonder weighs cases and bags up to 110 lb and provides accuracy down to 1/4 oz. Suppose that a 50-oz weight was repeatedly weighed on this scale and the weight readings recorded. The mean value was 49.5 oz and the standard deviation was .1. What can be said about the proportion of the time that the scale actually showed a weight that was within 1/4 oz of the true value of 50 oz? (Hint: Try to make use of Chebyshev's Rule.)

3.40 Suppose your statistics professor returned your first midterm exam with only a z score written on it. She also tells you that a histogram of the scores was closely described by a normal curve. How would you interpret each of the following z scores?

a. 2.2   **b.** −.4   **c.** −1.8   **d.** 1.0   **e.** 0

3.41 The paper "Answer Changing on Multiple-Choice Tests" (*J. of Experimental Education* (1980): 18–21) reported that for a group of 162 college students, the average number of responses changed from the correct answer to an incorrect answer on a test containing 80 multiple choice items was 1.4. The corresponding standard deviation was reported to be 1.5. Based on this mean and standard deviation, what can you tell about the shape of the distribution of the variable, *number of answers changed from right to wrong*? What can you say about the number of students that changed at least six answers from correct to incorrect?

3.42 The article "Does Air Pollution Shorten Lives?" (from the book *Statistics and Public Policy* cited in the Chapter 1 References) states that when the sulfate level for 117 standard metropolitan statistical areas was recorded, the resulting mean and standard deviation were 47.2 mg/m^3 and 31.3 mg/m^3, respectively. Use this information to make a statement about the proportion of metropolitan areas that have sulfate levels below 109.8 mg/m^3.

3.43 The average reading speed of students completing a speed reading course is 450 words per minute (wpm). If the standard deviation is 70 wpm, find the z score associated with each reading speed.

a. 320 wpm   **b.** 475 wpm   **c.** 420 wpm   **d.** 610 wpm

3.44 The 1974 edition of *Accident Facts* reported the motor vehicle death rate for each of the fifty states. These rates, expressed in deaths per 100,000 population, are given here.

23.0	38.7	31.2	17.4	41.8	25.4	30.2	54.4	18.4	30.8
45.3	18.7	32.5	26.6	28.0	48.7	18.3	27.3	35.5	34.6
57.7	20.2	26.3	21.8	24.5	44.8	16.8	33.0	16.3	16.9
31.3	25.3	34.6	21.1	28.5	24.0	28.1	35.8	33.4	27.6
22.4	22.6	34.9	46.6	13.5	30.4	23.8	33.2	29.9	39.8

a. Summarize this data set with a frequency distribution. Construct the corresponding histogram.

b. Use the histogram in part (a) to find approximate values of the following percentiles
   i. 50th   **ii.** 70th   **iii.** 10th   **iv.** 90th   **v.** 40th

3.45 The accompanying table gives the mean and standard deviation of reaction times (sec) for each of two different stimuli.

	Stimulus 1	Stimulus 2
mean	6.0	3.6
standard deviation	1.2	.8

If your reaction time for the first stimulus is 4.2 sec and for the second stimulus is 1.8 sec, to which stimulus are you reacting (when compared to all others) relatively more quickly?

CHAPTER 3   SUMMARY OF KEY CONCEPTS AND FORMULAS

TERM OR FORMULA	PAGE	COMMENT
$x_1, x_2, \ldots, x_n$	46	Notation for sample data consisting of observations on a variable x, where n is the sample size.
Sample mean, $$\bar{x} = \frac{\sum x}{n}$$	46	The most frequently used measure of center of a sample. It can be very sensitive to the presence of a single outlier (unusually large or small observation).
Population mean, μ	47	The average x value in the entire population.
Sample median	48	The middle value in the ordered list of sample observations. (For n even, the median is the average of the two middle values.) It is very insensitive to outliers.
Trimmed mean	51	A measure of center in which the observations are first ordered from smallest to largest, one or more observations are deleted from each end, and the remaining ones are averaged. In terms of sensitivity to outliers, it is a compromise between the mean and median.
Deviations from the mean: $x_1 - \bar{x}, x_2 - \bar{x}, \ldots, x_n - \bar{x}$	55	Quantities used to assess variability in a sample. Except for roundoff effects, $\sum (x - \bar{x}) = 0$.
The sample variance, $$s^2 = \frac{\sum (x - \bar{x})^2}{n - 1},$$ and standard deviation, $$s = \sqrt{s^2}.$$	57	The most frequently used measures of variability for sample data.
$\sum x^2 - n\bar{x}^2$	58	The computing formula for the numerator of s^2.
The population variance σ^2 and standard deviation σ	59	Measures of variability for the entire population.
Quartiles and the interquartile range	60	The lower quartile separates the smallest 25% of the data from the remaining 75%, and the upper quartile separates the largest 25% from the smallest 75%. The interquartile range, a measure of variability less sensitive to outliers than s, is the difference between the upper and lower quartiles.
Chebyshev's Rule	64	This rule states that for any number $k \geq 1$, *at least* $100[1 - (1/k^2)]\%$ of the observations in *any* data set are within k standard deviations of the mean. It is typically conservative in that the actual percentages often considerably exceed $100[1 - (1/k^2)]$.
Empirical Rule	66	This rule gives the approximate percentage of observations within one standard deviation (68%), two standard deviations (95%), or three standard deviations (99.7%) of the mean when the histogram is well-approximated by a normal curve.
z score	67	This quantity gives the distance between an observation and the mean expressed as a certain number of standard deviations. It is positive (negative) if the observation lies above (below) the mean.
Box plot	70	A picture that conveys information about the most important features of a data set: center, spread, extent of skewness, and presence of outliers.

SUPPLEMENTARY EXERCISES   **3.46–3.61**

3.46 Five randomly selected normal rats were treated with an injection of HRP (horse-radish peroxidase). The total number of injured neurons in the fourth nerve nucleus was recorded. The resulting data was: 209, 187, 123, 184, and 194. (Source: "The Injury Response of Nerve Fibers in the Anterior Medullary Velum of the Adult Rat," *Brain Research* (1984): 257–68). Compute and interpret the values of the sample mean and standard deviation.

3.47 Strength is an important characteristic of materials used in prefabricated housing. Each of 11 prefabricated plate elements was subjected to a severe stress test and the maximum width (in millimeters) of the resulting cracks was recorded. The given data appeared in the paper "Prefabricated Ferrocement Ribbed Elements for Low-Cost Housing" (*J. of Ferrocement* (1984): 347–64). Compute the sample mean, median, and standard deviation for this data set.

.684 .598 .924 .483 3.520 3.130 2.650 2.540 1.497 1.285 1.038

3.48 Eleven sediment samples from Gannoway Lake in Texas were analyzed for concentration of iron (μg/g) and zinc (μg/g). The given data appeared in the paper "The Analysis of Aqueous Sediments for Heavy Metals" (*J. Environ. Science and Health* (1984): 911–24).

iron	2.5	4.5	1.5	3.2	3.3	1.8	3.4	3.4	4.0	3.9	2.9
zinc	62	66	39	67	50	220	89	110	68	66	69

a. Calculate the sample mean and median for the iron concentration data. Are the numerical values of the mean and median roughly equal?

b. Calculate the sample mean and median for the zinc concentration data. Which of the two would you recommend as a measure of location? Explain.

c. Which of the samples (iron or zinc) has a larger variance? Answer without actually computing the two sample variances. Explain the reason for your selection.

3.49 Age at death (in days) for each of 12 infants who died of sudden infant death syndrome was given in the paper "Post-Mortem Analysis of Neuropeptides in Brains from Sudden Infant Death Victims" (*Brain Research* (1984): 277–85). The resulting observations were 54, 55, 56, 60, 60, 60, 105, 120, 135, 140, 154, and 247.

a. Compute and interpret the values of the sample mean, median, variance, and standard deviation.

b. Calculate the upper and lower quartiles and the interquartile range. Is the value 247 a mild or extreme outlier?

c. Construct a box plot for this data set.

3.50 The air quality in major cities is monitored on a regular basis. The *Los Angeles Times* (October 25, 1984) reported that a first-stage smog alert occurs when the index of pollutants in the air is between 200 and 275. A second-stage smog alert occurs when the index exceeds 275. Suppose that the index of pollutants has a distribution with mean 125 and standard deviation 75. Without assuming anything about the shape of the distribution, what can be said about the proportion of days on which a first-stage smog alert is declared? What can be said about the proportion of days on which a second-stage alert is declared?

3.51 The paper "Sodium–Calcium Exchange Equilibria in Soils as Affected by Calcium Carbonate and Organic Matter" (*Soil Science* (1984): 109) gave ten observations on soil pH. The data resulted from analysis of ten samples of soil from the Central Soil Salinity Research Institute experimental farm.

Soil pH   8.53   8.52   8.01   7.99   7.93   7.89   7.85   7.82   7.80   7.72

    a. Calculate and interpret the values of the sample mean, variance, and standard deviation.

    b. Compute the 10% trimmed mean and the sample median. Do either of these values differ much from the value of the sample mean?

    c. Find the upper quartile, the lower quartile, and the interquartile range.

    d. Illustrate the location and spread of this sample using a box plot.

3.52 The *New York Times* News Service reported that the average price of a home in the United States in 1984 was $101,000, whereas the median price was $80,900. What do the relative sizes of the mean and median imply about the shape of the distribution of home prices?

3.53 Age at diagnosis for each of 20 patients under treatment for meningitis was given in the paper "Penicillin in the Treatment of Meningitis" (*J. Amer. Med. Assoc.* (1984): 1870–4). The ages (in years) were as follows.

18	18	25	19	23	20	69	18	21	18
18	20	18	18	20	18	19	28	17	18

    a. Calculate the values of the sample mean and the standard deviation.

    b. Calculate the 10% trimmed mean. How does the value of the trimmed mean compare to that of the sample mean? Which would you recommend as a measure of location? Explain.

    c. Compute the upper quartile, the lower quartile, and the interquartile range.

    d. Are there any mild or extreme outliers present in this data set?

    e. Construct the box plot for this data set.

3.54 Although blood pressure is a continuous variable, its value is often reported to the nearest 5 mm Hg (e.g., 100, 105, 110, and so on). Suppose that the actual blood pressure values for nine randomly selected individuals are

118.6	127.4	138.4	130.0	113.7	122.0	108.3	131.5	133.2

    a. If values are reported as suggested (i.e., rounded to the nearest 5 mm Hg), what is the sample of the *reported* values and what is the median of this sample?

    b. Suppose that the second individual's blood pressure is 127.6 rather than 127.4 (a small change in a single value). How does this change the median of reported values? What does this say about the median's sensitivity to rounding or grouping of the data?

3.55 The accompanying observations are carbon monoxide levels (ppm) in air samples obtained from a certain region.

9.3	10.7	8.5	9.6	12.2	16.6	9.2	10.5
7.9	13.2	11.0	8.8	13.7	12.1	9.8	

    a. If a trimmed mean is calculated by first deleting the smallest and largest observations, what is the corresponding trimming percentage? Answer this question if the two smallest and two largest observations are deleted.

    b. Calculate the two trimmed means referred to in part (a).

    c. Using the results of part (b), how might you calculate a measure of center for this sample that could be regarded as a 10% trimmed mean?

3.56 In recent years, many teachers have been subject to increased levels of stress that have contributed to disenchantment with teaching as a profession. The paper "Professional Burnout Among Public School Teachers" (*Public Personnel Mgmt.* (1988): 167–89) looked at various aspects of this problem. Consider the following information on total psychological effects scores for a sample of 937 teachers.

Burnout level	Range of scores	Frequency
Low	0–21	554
Moderate	22–43	342
High	44–65	41
	$\bar{x} = 19.93, \; s = 12.89$	

a. Draw a histogram.

b. Mark the value of $\bar{x}$ on the measurement scale, and use the histogram to locate the approximate value of the median. Why are the two measures of center not identical?

c. In what interval of values are we guaranteed to find at least 75% of the scores, irrespective of the histogram shape? From the histogram, roughly what percentage of the scores actually fall in this interval?

3.57 a. Suppose that n, the number of sample observations, is an odd number. Under what conditions on the sample will $\bar{x}$ and the sample median be identical?

b. If $n = 10$, when will $\bar{x}$ and the 10% trimmed mean be identical?

3.58 a. A statistics instructor informed her class that the median exam score was 78 and the mean was 70. Sketch a picture of what the smoothed histogram of scores might look like.

b. Suppose that the majority of students in the class studied for the exam, but a few students had not. How might this be reflected in a smoothed histogram of exam scores?

3.59 The *Los Angeles Times* (December 22, 1983) reported that the average cost per day of a semiprivate hospital room for California, Oregon, and Washington was $268, $210, and $220, respectively. Suppose that you were interested in the average cost per day of a semiprivate hospital room in these three West Coast states combined. If each state contained the same number of hospital rooms, we could average the three values given. However, this is not the case, so the three averages must be weighted in proportion to the respective number of hospital rooms in each state. In general, if $\bar{x}_1, \bar{x}_2, \bar{x}_3, \ldots, \bar{x}_n$ are n group means and $w_1, w_2, \ldots, w_n$ are the respective weights, the **weighted mean** $\bar{x}_w$ is defined to be

$$\bar{x}_w = \frac{w_1\bar{x}_1 + w_2\bar{x}_2 + \cdots + w_n\bar{x}_n}{w_1 + w_2 + \cdots + w_n}$$

The number of hospital rooms in California, Oregon, and Washington are 120,000, 30,000, and 45,000, respectively. Use the given formula with $w_1 = 120,000$, $w_2 = 30,000$, and $w_3 = 45,000$ to find the average cost of a hospital room on the West Coast.

3.60 Suppose that an auto dealership employs clerical workers, salespeople, and mechanics. The average monthly salaries of the clerical employees, salespeople, and mechanics are $1100, $1800, and $1900, respectively. If the dealership has three clerical employees, ten salespeople, and eight mechanics, find the average monthly wage for all employees of the dealership. (Hint: Refer to Exercise 3.59.)

3.61 Suppose that the distribution of scores on an exam is closely described by a normal curve with mean 100. The 16th percentile of this distribution is 80.

a. What is the 84th percentile?

b. What is the approximate value of the standard deviation of exam scores?

c. What z score is associated with an exam score of 90?

d. What percentile corresponds to an exam score of 140?

e. Do you think there were many scores below 40? Explain.

REFERENCES See the references at the end of Chapter 2.

C H A P T E R 4

PROBABILITY

INTRODUCTION Almost all situations that we confront in our everyday activities involve some aspects of uncertainty. There is uncertainty concerning the number of cars in line at a bank's drive-up window, concerning whether or not an appliance will need repair while still under warranty, concerning the amount of weight one might lose on a particular diet, and so on. **Probability** is the scientific discipline with the objective of studying uncertainty in a systematic fashion.

The first growth spurt of probability occurred during the seventeenth century in attempts to answer questions concerning games of chance. Even today games of chance suggest many interesting questions that can be answered using methods from probability. For example, it used to be thought that the odds in blackjack virtually always favored the house (the dealer's employer), but in the 1960s probability methods were used to discover many situations (involving cards not yet dealt) in which the advantage lay with the individual bettor. In the twentieth century the scope of probability has enlarged considerably as investigators have attempted to come to grips with the pervasiveness of uncertainty in both scientific contexts and in everyday life. Probability methods have recently been used to increase understanding of such diverse phenomena as the spread of an epidemic through a population, the mechanism of memory recall, the operating characteristics of various computer time-sharing systems, the diffusion of particles through a membrane, changes in consumers' brand preferences over time, social class mobility through succeeding generations, and (of course) what tomorrow's weather might be like. Our goal in this chapter is to introduce you to just a few of the most important concepts and methods of probability. In later chapters, probability will be used to specify the degree of reliability for various inferential procedures.

4.1 CHANCE EXPERIMENTS AND EVENTS

The basic ideas and terminology of probability are most easily introduced in situations that are both familiar and reasonably simple. Thus, some of our initial examples will involve such mundane activities as tossing a coin once or several times, selecting one or more cards from a deck, and rolling a single die or several dice. Once the basics are in place, we shall move to more interesting and realistic situations.

CHANCE EXPERIMENTS

When a single coin is tossed, it can land with its head side up or its tail side up. The selection of a single card from a well-mixed deck could result in the ace of spades, the 5 of diamonds, the jack of hearts, or any one of the other 49 possibilities. Consider rolling both a red die and a green die. One possible outcome is (4, 1)—that is, the red die lands with four dots facing up and the green die shows one dot on its upturned face. Another outcome is (3, 3), and yet another is (1, 4). There are 36 possibilities (six in which the red die shows 1, another six in which it shows 2, and so on). Prior to carrying out any one of these activities, there is uncertainty as to which of the possible outcomes will result.

DEFINITION

A **chance experiment** is any activity or situation in which there is uncertainty concerning which of two or more possible outcomes will result.

Rolling a die might not seem like much of an experiment in the usual sense of the word, but our usage is broader than what is typically implied.

80

A chance experiment could also refer to determining whether or not each person in a sample supports the death penalty (an opinion poll or survey) or to an investigation carried out in a laboratory to study how varying the amount of a certain chemical input affects the yield of a product.

EXAMPLE 1

Consider the experiment in which two people are randomly selected from a list of all new homeowners in a certain area. The type of mortgage—fixed rate (F) or adjustable rate (A)—is then noted for each one. The first person can be in either the F or A category, and for each of these there are two further possibilities for the second person. Thus there are four possible experimental outcomes:

FF, FA, AF, AA

The four outcomes of Example 1 can be displayed in a picture called a **tree diagram**, as illustrated in Figure 1. The two line segments (branches) leading out from the initial point on the left side of the diagram correspond to the two possibilities for the first individual selected; these are sometimes referred to as *first-generation branches*. There are two further *second-generation branches* emanating from the tip of each first-generation branch, corresponding to the possibilities for the second individual selected. Starting from the initial point, if we traverse a first-generation branch followed by a second-generation branch, a particular experimental outcome results. The four possible outcomes are shown on the right side of the diagram.

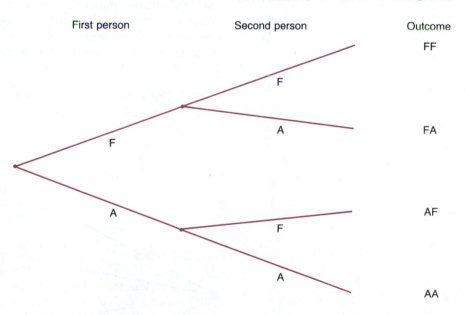

FIGURE 1

A tree diagram for Example 1.

If three people rather than two are selected, there are eight possible outcomes (four from adding an additional F to each outcome just considered and an additional four from adding an A). Figure 2 shows the corresponding tree diagram, which contains three generations of branches.

FIGURE 2

A tree diagram for Example 1 when three
people are selected.

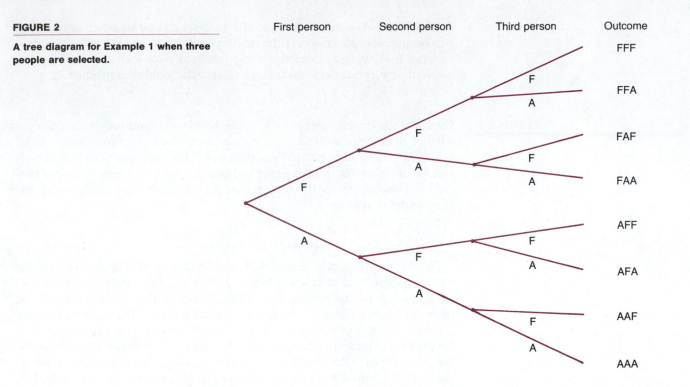

| First person | Second person | Third person | Outcome |

EXAMPLE 2 All but one of the jurors for a trial have been selected. There are four potential jurors—individuals *a, b, c,* and *d*—left to question. All four are acceptable to the prosecution, but two of the four, *a* and *b*, are unacceptable to the defense and so will be excused if questioned. Suppose that the potential jurors are selected for questioning in random order (by drawing

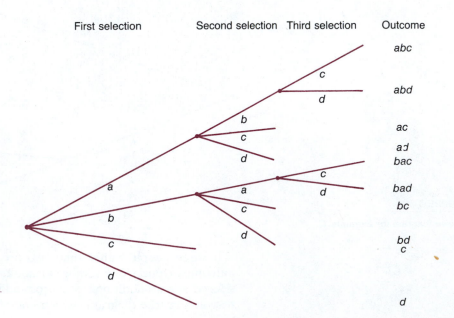

| First selection | Second selection Third selection | Outcome |

FIGURE 3

Tree diagram for Example 2.

slips of paper from a box), with the process terminating when an acceptable juror (*c* or *d*) is questioned. There are three different types of possible outcomes.

Type of Outcome	Outcomes
One potential juror is questioned	*c, d*
Two potential jurors are questioned	*ac, ad, bc, bd*
Three potential jurors are questioned	*abc, abd, bac, bad*

Figure 3 shows the corresponding tree diagram. Notice that this diagram lacks the symmetry exhibited by the diagrams of Example 1. This is because some outcomes result in earlier termination of the experiment than do others.

In statistical literature, the collection (or set) of all possible experimental outcomes is frequently referred to as the **sample space** of the experiment. When listing elements of a set, mathematicians customarily enclose the elements in braces. Thus the sample space for the two-person experiment of Example 1 is

sample space = {FF, FA, AF, AA}

and for Example 2 it is

sample space = {*c, d, ac, ad, bc, bd, abc, bac, abd, bad*}

EVENTS   An individual who tosses a pair of dice may be interested in which particular one of the 36 possible outcomes will result. Alternatively, interest may focus on whether the outcome will be in the group consisting of (1, 4), (2, 3), (3, 2), (4, 1), those outcomes having a sum value of five. In the situation summarized in Figure 2, the number of homeowners having adjustable-rate mortgages may be of primary concern. Outcomes in the group AAF, AFA, FAA are those for which exactly two homeowners have adjustable-rate mortgages. More generally, an investigator will often wish to know whether the resulting outcome belongs to some specified collection.

▌▌▌ **DEFINITION**

An **event** is any collection of possible outcomes from a chance experiment.

A **simple event** is an event consisting of exactly one outcome.

It is customary to denote events by upper-case letters, such as *A*, *B*, or *C*. Sometimes several events are denoted by using the same letter with different numerical subscripts: E_1, E_2, E_3, and so on. As with the sample space, when listing the outcomes in an event, it is customary to enclose the outcomes in braces.

EXAMPLE 3

Reconsider the situation of Example 1, in which each of two homeowners was categorized as having either an adjustable-rate mortgage (A) or a fixed-rate mortgage (F). We had

sample space = {FF, FA, AF, AA}

Because there are four outcomes, there are four simple events:

$$O_1 = \{FF\}, \quad O_2 = \{FA\}, \quad O_3 = \{AF\}, \quad O_4 = \{FF\}$$

One event of interest is

$$B = \{FF, FA, AF\}$$

A word description of B is that it consists of all outcomes for which at least one person had a fixed-rate loan. Another event is

$$C = \{AF, AA\}$$

In words, C is the event that the first person selected has an adjustable-rate loan.

EXAMPLE 4

The sample space for the jury selection situation of Example 2 was

{c, d, ac, ad, bc, bd, abc, abd, bac, bad}

Some events of potential interest are

$A = \{c, d\}$
   = the event that exactly one potential juror is questioned
$B = \{c, ac, bc, abc, bac\}$
   = the event that individual c is selected
$O_1 = \{c\}$,   a simple event
$O_2 = \{ad\}$,   another simple event

Because there are ten possible outcomes, there are ten simple events; the number of nonsimple events is quite large.

EXAMPLE 5

An observer stands at the bottom of a freeway offramp and records the turning direction (L or R) of each of four successive vehicles. The sample space contains 16 outcomes:

{LLLL, RLLL, LRLL, LLRL, LLLR, RRLL, RLRL, RLLR,
LRRL, LRLR, LLRR, LRRR, RLRR, RRLR, RRRL, RRRR}

These outcomes could also be displayed in a tree diagram containing four generations of branches. Each of these outcomes determines a simple event. For example,

$$O_1 = \{LLLL\}, \qquad O_2 = \{RLLL\}, \qquad \text{and so on.}$$

Other events include

$A = \{$RLLL, LRLL, LLRL, LLLR$\}$

    = the event that exactly one of the four cars turns right

$B = \{$LLLL, RLLL, LRLL, LLRL, LLLR$\}$

    = the event that at most one of the four cars turns right

$C = \{$LLLL, RRRR$\}$

    = the event that all four cars turn in the same direction

Suppose that the resulting outcome is RLLL. This outcome is in the simple event O_2. It is also in the nonsimple events A and B, but not in C, O_1, or any of the other 14 simple events.

In general, only one outcome, and thus one simple event, will occur when an experiment is performed. However, as Example 5 demonstrates, the resulting outcome will often be contained in many nonsimple events. We shall say that a given event has occurred whenever the resulting outcome is contained in the event. If the outcome in Example 5 is RLLL, then O_2 has occurred, so has A (exactly one car did turn right), so has B (at most one car turned right), and so has *any* other event containing RLLL. Typically, many events will simultaneously occur when an experiment is performed, though most will not be of interest.

FORMING NEW EVENTS

Once some events have been specified, there are several useful ways of manipulating them to create new events.

DEFINITION

Let A and B denote two events.

1. The event ***not A*** consists of all experimental outcomes that are not in A.

2. The event ***A or B*** consists of all experimental outcomes that are in at least one of the two events, that is, in A or in B or in both of these.

3. The event ***A and B*** consists of all experimental outcomes that are in both A and B, that is, the outcomes common to the two events.

EXAMPLE 6

Consider two four-sided (tetrahedral) dice, one red and one green. The four faces of each die are marked with the numbers 1, 2, 3, and 4, respectively. When such a die is rolled, the outcome is the number on the side facing down when the die comes to rest (the other three sides are facing up). If each die is rolled once, there are 16 possible outcomes; these are displayed in the accompanying rectangular table.

	Outcomes	Green Die			
		1	2	3	4
	1	(1, 1)	(1, 2)	(1, 3)	(1, 4)
Red	2	(2, 1)	(2, 2)	(2, 3)	(2, 4)
Die	3	(3, 1)	(3, 2)	(3, 3)	(3, 4)
	4	(4, 1)	(4, 2)	(4, 3)	(4, 4)

The outcomes can also be displayed in a tree diagram with four first-generation branches (for the four faces of the red die) and four second-generation branches leading out from each first-generation branch.

Define the following events:

A = the event that both faces show the same number
B = the event that at least one face shows a 1
C = the event that the sum of the numbers showing is 4
D = the event that the smallest number showing is a 3

Then

$A = \{(1, 1), (2, 2), (3, 3), (4, 4)\}$
$B = \{(1, 1), (1, 2), (1, 3), (1, 4), (2, 1), (3, 1), (4, 1)\}$
$C = \{(3, 1), (2, 2), (1, 3)\}$
$D = \{(3, 3), (3, 4), (4, 3)\}$

Here are some events that can be formed from those just defined:

not $A = \{(1, 2), (1, 3), (1, 4), (2, 1), (2, 3), (2, 4),$
$\qquad\qquad\qquad (3, 1), (3, 2), (3, 4), (4, 1), (4, 2), (4, 3)\}$
$\qquad$ = the event that the two faces show different numbers
not $B = \{(2, 2), (2, 3), (2, 4), (3, 2), (3, 3), (3, 4), (4, 2), (4, 3), (4, 4)\}$
$\qquad$ = the event that neither face shows a 1
A or $B = \{(1, 1), (2, 2), (3, 3), (4, 4), (1, 2),$
$\qquad\qquad\qquad (1, 3), (1, 4), (2, 1), (3, 1), (4, 1)\}$
C or $D = \{(3, 1), (2, 2), (1, 3), (3, 3), (3, 4), (4, 3)\}$
A and $B = \{(1, 1)\}$, a simple event (only this outcome
$\qquad\qquad$ is common to A and B)
B and $C = \{(3, 1), (1, 3)\}$

Notice also that C and D have no outcomes in common.

It frequently happens, as was the case for events C and D in Example 6, that two events have no common outcomes. There is special terminology for such situations.

DEFINITION

Two events *A* and *B* that have no common outcomes are said to be **disjoint** or **mutually exclusive**.

In particular, any two different simple events are disjoint, since each contains a single outcome and they are different.

It is sometimes useful to draw an informal picture of events in order to visualize relationships. In a **Venn diagram**, the collection of all possible outcomes, i.e., the sample space (itself an event), is shown as the interior of a rectangle. Other events are then identified with specified regions inside this rectangle. Figure 4 illustrates several Venn diagrams.

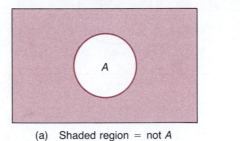

(a)   Shaded region = not *A*

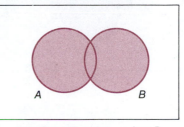

(b)   Shaded region = *A* or *B*

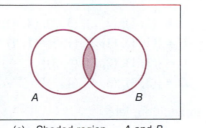

(c)   Shaded region = *A* and *B*

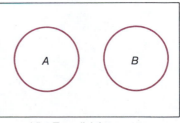

(d)   Two disjoint events

FIGURE 4

Venn diagrams.

The use of the *or* and *and* operations can be extended to form new events from more than two initially specified events.

 DEFINITION

Let $A_1, A_2, \ldots, A_k$ denote *k* events.

1.   The event **A_1 or A_2 or . . . or A_k** consists of all outcomes in at least one of the individual events $A_1, A_2, \ldots, A_k$.

2.   The event **A_1 and A_2 and . . . and A_k** consists of all outcomes in every one of the individual events $A_1, A_2, \ldots, A_k$.

These *k* events are **disjoint** if no two of them have any common outcomes.

Venn diagrams illustrating these concepts appear in Figure 5.

FIGURE 5

Venn diagrams illustrating:
(a) A_1 or A_2 or A_3
(b) A_1 and A_2 and A_3
(c) Three disjoint events.

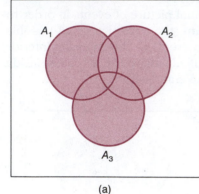

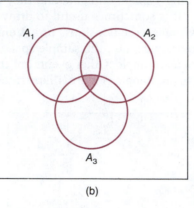

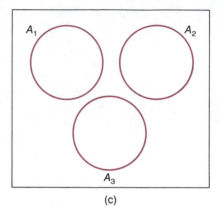

(a)   (b)   (c)

EXAMPLE 7

(*Example 6 continued*) In the experiment involving the rolling of two four-sided dice, we defined the events

$A = \{(1, 1), (2, 2), (3, 3), (4, 4)\}$
$B = \{(1, 1), (1, 2), (1, 3), (1, 4), (2, 1), (3, 1), (4, 1)\}$
$C = \{(3, 1), (2, 2), (1, 3)\}$
$D = \{(3, 3), (3, 4), (4, 3)\}$

In addition, let

$E = \{(1, 1), (1, 3), (2, 2), (3, 1), (4, 2), (3, 3), (2, 4), (4, 4)\}$
  = the event that the sum is an even number
$F = \{(1, 1), (1, 2), (2, 1)\}$
  = the event that the sum is at most 3

Then

A or C or $D = \{(1, 1), (2, 2), (3, 3), (4, 4), (3, 1), (1, 3), (3, 4), (4, 3)\}$

The outcome (3, 1) is contained in each of the events B, C, and E, as is the outcome (1, 3), and these are the only two common outcomes, so

B and C and $E = \{(3, 1), (1, 3)\}$

The events C, D, and F are disjoint because no outcome in any one of these events is contained in either of the other two events.

EXERCISES 4.1–4.9   **SECTION 4.1**

4.1   Consider the experiment in which the type of transmission—automatic (A) or manual (M)—is recorded for each of the next two cars purchased from a certain dealer.
a. What is the set of all possible outcomes (the sample space)? Hint: The experiment is similar to that of Example 1.
b. Display the possible outcomes in a tree diagram.
c. List the outcomes in each of the following events: B = the event that at least one car has an automatic transmission; C = the event that exactly one car has an automatic transmission; D = the event that neither car has an automatic transmission. Which of these events are simple events?
d. What outcomes are in the event B *and* C? In the event B *or* C?

4.2   Suppose that a (six-sided) die is rolled and then a coin is tossed.
a. Display the possible outcomes in a rectangular table similar to the one given in Example 6.
b. Display the possible outcomes on a tree diagram.
c. Let D_1 denote the event that the upturned face on the die shows a 1, let E denote the event that an even number results from the die toss, and let F be the event that the head side of the coin lands up. List the outcomes in these three events.
d. Referring to part (c), list the outcomes in the event *not E* and the outcomes in the event *not F*.
e. Referring to part (c), what outcomes are in the event *E or F*? In *E and F*?
f. Is any pair of the events defined in part (c) disjoint? Which one(s)?

4.3   A college library has four copies of a certain book; the copies are numbered 1, 2, 3, and 4. Two of these are randomly selected (using four slips of paper). The first book selected will be placed on two-hour reserve, and the second one may be checked out on an overnight basis.
a. Construct a tree diagram to display the 12 outcomes in the sample space.
b. Let A denote the event that at least one of the books selected is an even-numbered copy. What outcomes are in A?
c. Suppose copies 1 and 2 are first printings, whereas copies 3 and 4 are second printings. Let B denote the event that exactly one of the copies selected is a first printing. What outcomes are contained in B?

4.4   A library has five copies of a certain text on reserve, of which two copies (1 and 2) are first printings and the other three (3, 4, and 5) are second printings. A student examines these books in random order, stopping only when a second printing has been selected.
a. Display the possible outcomes in a tree diagram.
b. What outcomes are contained in the event A, that exactly one book is examined before the experiment terminates?
c. What outcomes are contained in the event C, that the experiment terminates with the examination of book 5?

4.5   Suppose that starting at a certain time, batteries coming off an assembly line are examined one-by-one to see whether they are defective (let D = defective and N = not defective). The experiment terminates as soon as a nondefective battery is obtained.
a. Give five possible experimental outcomes.
b. What can be said about the number of outcomes in the sample space?
c. What outcomes are in the event E, that the number of batteries examined is an even number?

4.6   Refer to Exercise 4.5, and now suppose that the experiment terminates only when *two* nondefective batteries are obtained.

a. Let A denote the event that at most three batteries must be examined in order to terminate the experiment. What outcomes are contained in A?

b. Let B be the event that exactly four batteries must be examined before the experiment terminates. What outcomes are in B?

c. What can be said about the number of possible outcomes?

4.7 A family consisting of three persons—P_1, P_2, and P_3—belongs to a medical clinic that always has a physician at each of stations 1, 2, and 3. During a certain week, each member of the family visits the clinic exactly once and is randomly assigned to a station. One experimental outcome is (1, 2, 1), which means that P_1 is assigned to station 1, P_2 to station 2, and P_3 to station 1.

a. There are 27 possible outcomes; list them. Hint: First list the nine outcomes in which P_1 goes to station 1, then the nine in which P_1 goes to station 2, and finally the nine in which P_1 goes to station 3; a tree diagram might help.

b. List all outcomes in the event A, that all three people go to the same station.

c. List all outcomes in the event B, that all people go to different stations.

d. List all outcomes in the event C, that no one goes to station 2.

e. Identify outcomes in the following events: *not B, not C, A or B, A and B, A and C*.

4.8 An engineering construction firm is currently working on power plants at three different sites. Define events E_1, E_2, and E_3 as follows:

E_1 = the plant at site 1 is completed by the contract date

E_2 = the plant at site 2 is completed by the contract date

E_3 = the plant at site 3 is completed by the contract date

The accompanying Venn diagram pictures the relationships among these events.

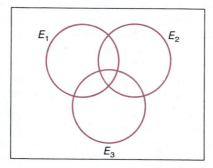

Shade the region in the Venn diagram corresponding to each of the following events (redraw the Venn diagram for each part of the problem).

a. At least one plant is completed by the contract date.

b. All plants are completed by the contract date.

c. None of the plants is completed by the contract date.

d. Only the plant at site 1 is completed by the contract date.

e. Exactly one of the three plants is completed by the contract date.

f. Either the plant at site 1 or both of the other two plants are completed by the contract date.

4.9 Consider a Venn diagram picturing two events A and B that are not disjoint.

a. Shade the event *not(A or B)*. On a separate Venn diagram, shade the event *(not A) and (not B)*. How are these two events related?

b. Shade the event *not(A and B)*. On a separate Venn diagram, shade *(not A) or (not B)*. How are these two events related?

Note: These two relationships together are called **DeMorgan's laws**.

4.2 THE DEFINITION OF PROBABILITY AND BASIC PROPERTIES

When any given chance experiment is performed, some events are relatively likely to occur, whereas others are not so likely to occur. For a specified event *E*, we want to assign a number to this event that gives a precise indication of how likely it is that *E* will occur. This number is called *the probability of the event E* and is denoted by *P(E)*. The value of *P(E)* will depend on how frequently *E* occurs when the experiment is performed repeatedly.

EXAMPLE 8
One of the simplest chance experiments involves tossing a coin just once. Let's define an event *H* for this chance experiment by

H = the event that the coin lands with its head side facing up

Frequently we hear a coin described as "fair," or we are told that there is a 50% chance of the coin landing head up—a 50% chance that the event *H* will occur. Such a description cannot refer to the result of a single toss, since a single toss cannot result in both a head and a tail. Might "fairness" and "50%" refer to ten successive tosses yielding exactly five heads and five tails? Not really, since it is easy to imagine a coin characterized as fair landing head up on only three or four of the ten tosses.

Suppose that we take such a coin and begin to toss it over and over. After each toss, we compute the relative frequency of heads observed so far, that is, the value of the ratio

$$\frac{\text{number of times the event } H \text{ occurs}}{\text{number of tosses}}$$

Suppose the results of the first ten tosses are as follows:

Toss Number	1	2	3	4	5	6	7	8	9	10
Outcome	T	H	H	H	T	T	H	H	T	T
Cumulative # of H's	0	1	2	3	3	3	4	5	5	5
Relative freq. of H	0	.5	.667	.75	.6	.5	.571	.625	.556	.5

Figure 6 illustrates how the relative frequency of heads fluctuates during a sample sequence of 50 tosses. Much empirical evidence suggests that *as the number of tosses increases, the relative frequency of heads does not continue to fluctuate wildly but instead stabilizes and approaches some fixed number (limiting value).* This stabilization is illustrated for a sequence of 1000 tosses in Figure 7.

Because each relative frequency is between 0 and 1, the limiting value is also. It is then natural to call the coin "fair" if the limiting value is .5. The 50% chance doesn't refer to exact results in some fixed number of tosses, such as 10 or 100, but to what happens to the relative frequency of heads as we repeat the chance experiment over and over and over again. In terms of probability, we say that the probability that the event H will occur is .5, and we write $P(H) = .5$.

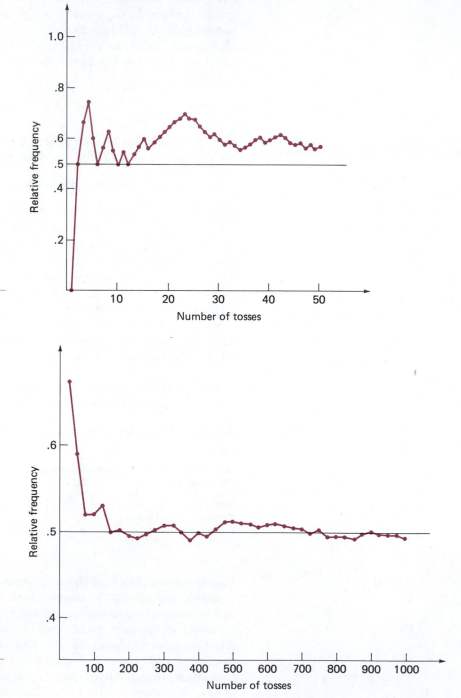

FIGURE 6

Relative frequency of heads in the first 50 of a long series of tosses.

FIGURE 7

Stabilization of the relative frequency of heads in coin tossing.

EXAMPLE 9   Consider selecting a single card from a well-mixed deck of 52 cards. Card players would surely say that there is a 25% chance of selecting a card whose suit is clubs (as opposed to spades, diamonds, or hearts). In other words, if the event C consists of all cards (outcomes) whose suit is clubs, there is a 25% chance that the event C will occur. To interpret this, think of performing this chance experiment over and over again: Select a first card, replace and shuffle, select a second card, replace and shuffle, and so on. If we examine the ratio

$$\text{relative frequency of } C = \frac{\text{number of times } C \text{ occurs}}{\text{number of replications}}$$

as the number of replications increases, the relative frequency stabilizes at the value .25; that is, $P(C) = .25$. This is why we say that there is a 25% chance of selecting a club. One implication of this is that the limiting relative frequency of the event *not C* is .75, that is, $P(\textit{not } C) = .75 = 1 - P(C)$.

Let's complicate matters by considering the experiment in which five cards are dealt from the deck (some people call this a poker hand). One interesting event is

$F =$ the event that all 5 cards are from the same suit

(called a flush in poker). Consider performing this experiment repeatedly—deal five cards, replace and shuffle, again deal five cards, replace and shuffle, etc.—and tracking the quantity

$$\text{relative frequency of } F = \frac{\text{number of times } F \text{ occurs}}{\text{number of replications}}$$

This relative frequency also stabilizes as the number of replications increases, but the limiting value is not obvious even to poker players. Probability methods can be used to show that it is .00198, so the chance of such a hand occurring is much less than 1%.

These examples motivate the following general definition of probability.

▌▌▌ DEFINITION

The **probability of an event E**, denoted by $P(E)$, is the value approached by the relative frequency of occurrence of E in a very long series of replications of a chance experiment.

Because $P(E)$ is the limiting value of E's relative frequency of occurrence, it is a number between 0 and 1. Thus E can be judged relatively unlikely or likely according to where the value of $P(E)$ lies, compared to the extremes of 0 and 1.

For a single toss of a fair coin, $P(\text{head side up}) = .5$. When a single card is selected from a well-mixed deck, $P(\text{selected card is a heart}) = .25$. When we informally speak of a 10% chance of occurrence, we mean a probability of .10.

Our definition of probability depends on being able to perform a chance experiment repeatedly under identical conditions. However, probability language and concepts are often used in situations in which replication is not feasible. For example, when a new product is introduced, the marketing manager might state that the probability of its being successful is .3. Or, a public utility executive may testify that the probability of a nuclear plant meltdown during the next decade is .00000001. The most common alternative to the definition of probability based on relative frequency is a subjective, or personal, interpretation. Here probability is a measure of how strongly a person believes that an event will occur. This interpretation permits two people with different opinions to assign different probabilities to the same event (reflecting differing strengths of belief). However, both subjective probabilities and those based on limiting relative frequencies do satisfy the same general rules of probability, so probabilities of complex events can be calculated once probabilities of simple events have been specified. We will not pursue subjective probabilities any further. The relative frequency definition is intuitive, very widely used, and most relevant for the inferential procedures that we present. Whenever you want to interpret a probability, we recommend that you think in terms of the limiting value of the relative frequency.

PROPERTIES OF PROBABILITY

There would seem to be a practical difficulty at this point: how can we find $P(E)$ without performing a long series of chance experiments? Consider, for example, the experiment in which a fair die is rolled once. Fairness implies a limiting relative frequency (probability) of 1/6 for each of the six outcomes (simple events). Also, if E is the event consisting of the even outcomes 2, 4, 6, then

$$\begin{array}{c}\text{relative freq.} \\ \text{of } E\end{array} = \begin{array}{c}\text{relative freq.} \\ \text{of } 2\end{array} + \begin{array}{c}\text{relative freq.} \\ \text{of } 4\end{array} + \begin{array}{c}\text{relative freq.} \\ \text{of } 6\end{array}$$

This implies that

$$P(E) = 1/6 + 1/6 + 1/6 = 3/6 = .5$$

More generally, prior experience with or careful thought about an experiment will often suggest appropriate probabilities for the simple events. Then some general rules can be used to calculate the probabilities of more complex events.

PROPERTIES OF PROBABILITY

1. For any event E, $0 \leq P(E) \leq 1$.
2. The sum of probabilities of all simple events must be 1 (analogous to the sum of all relative frequencies being 1).
3. For any event E, the probability $P(E)$ is the sum of the probabilities of all simple events corresponding to outcomes contained in E.
4. For any event E,

   $$P(E) + P(not\ E) = 1,$$

   so $P(not\ E) = 1 - P(E)$ and $P(E) = 1 - P(not\ E)$.

EXAMPLE 10

Customers at a certain department store pay for purchases either with cash or with one of four types of credit card. Store records indicate that 30% of all purchases involve cash, 25% are made with the store's own credit card, 18% with Mastercard, 15% with VISA, and the remaining 12% of purchases are made with an American Express card. The accompanying table displays the probabilities of the simple events for the experiment in which the mode of payment for the next transaction is observed.

Mode of payment	Cash	Store Card	MC	V	AE
Simple event	O_1	O_2	O_3	O_4	O_5
Probability	.30	.25	.18	.15	.12

Let E be the event that the next purchase is made with a nationally distributed credit card. This event consists of outcomes MC, V, and AE, so

$$P(E) = P(O_3) + P(O_4) + P(O_5) = .18 + .15 + .12 = .45$$

That is, in the long run, 45% of all purchases are made using one of the three national cards. Additionally,

$$P(not\ E) = 1 - P(E) = 1 - .45 = .55,$$

which could also have been obtained by noting that *not E* consists of outcomes corresponding to simple events O_1 and O_2.

The property $P(E) = 1 - P(not\ E)$ is surprisingly useful. This is because there are many situations in which calculation of $P(not\ E)$ is much easier than direct determination of $P(E)$.

EQUALLY LIKELY OUTCOMES

Experiments involving tossing fair coins, rolling fair dice, or selecting cards from a well-mixed deck have equally likely outcomes. For example, if a fair die is rolled once, each outcome (simple event) has probability 1/6. With E denoting the event that the outcome is an even number, we saw earlier that $P(E) = 3/6$. This is just the ratio of the number of outcomes in E to the total number of possible outcomes. The following box presents the generalization of this result.

CALCULATING PROBABILITIES WHEN OUTCOMES ARE EQUALLY LIKELY

Consider an experiment that can result in any one of N possible outcomes. Denote the corresponding simple events by $O_1, O_2, \ldots, O_N$. If these simple events are equally likely to occur, then

1. $P(O_1) = 1/N, \quad P(O_2) = 1/N, \quad \ldots, \quad P(O_N) = 1/N$
2. For *any* event E,

$$P(E) = \frac{\text{number of outcomes in } E}{N}$$

EXAMPLE 11

An academic department with eight faculty members has voted by secret ballot to elect from its ranks either candidate *A* or candidate *B* to serve on a grievance panel. Suppose that the votes are on identical slips of paper in a box and that *A* has actually received five votes. The slips are removed from the box one by one and a cumulative tally is kept. How likely is it that *A* remains ahead of *B* throughout the vote count?

One possible experimental outcome is the sequence *AABABAAB*. For this outcome, candidate *A* does remain ahead of *B* throughout the count. The outcome *AABBABAA* is one in which *B* catches up with *A* after the fourth vote is tallied.

Table 1 displays all 56 possible outcomes. If the slips have been well mixed before selection, no one of the outcomes is any more or less likely than any other. Thus the probability of each outcome is 1/56. If the ex-

TABLE 1

Possible Outcomes for the Ballot-Counting Problem

BBBAAAAA	BAABAAAB	ABABAABA	*AABABAAB
BBABAAAA	BAAABBAA	ABABAAAB	*AABAABBA
BBAABAAA	BAAABABA	ABAABBAA	*AABAABAB
BBAAABAA	BAAABAAB	ABAABABA	*AABAAABB
BBAAAABA	BAAAABBA	ABAABAAB	AAABBBAA
BBAAAAAB	BAAAABAB	ABAAABBA	*AAABBABA
BABBAAAA	BAAAAABB	ABAAABAB	*AAABBAAB
BABABAAA	ABBBAAAA	ABAAAABB	*AAABABBA
BABAABAA	ABBABAAA	AABBBAAA	*AAABABAB
BABAAABA	ABBAABAA	AABBABAA	*AAABAABB
BABAAAAB	ABBAAABA	AABBAABA	*AAAABBBA
BAABBAAA	ABBAAAAB	AABBAAAB	*AAAABBAB
BAABABAA	ABABBAAA	AABABBAA	*AAAABABB
BAABAABA	ABABABAA	*AABABABA	*AAAAABBB

* Denotes an outcome for which A remains ahead of B throughout the counting.

periment is repeated over and over, any particular outcome occurs roughly 1/56, or 1.79%, of the time. Let *E* denote the event that *A* remains ahead of *B* throughout the vote count. Table 1 shows that there are 14 outcomes in *E*, so

$$P(E) = 14/56 = .25$$

Similarly, define *F* to be the event that all three of *B*'s votes appear in succession. Since there are six outcomes in *F*,

$$P(F) = 6/56 \approx .107$$

As Example 11 shows, it can be quite tedious to determine both the total number of outcomes, *N*, and the number of outcomes in a specified event *E*. More advanced textbooks present a variety of counting techniques that facilitate determination of these numbers without listing outcomes.

ADDITION RULES   We have seen how the probability of an event can be calculated by adding together simple event probabilities. Simple events are by definition disjoint. This addition process is also legitimate when calculating certain probabilities involving events that are disjoint but not necessarily simple.

THE ADDITION RULE OF PROBABILITY FOR DISJOINT EVENTS

Let E and F be two disjoint events. Then

$$P(E \text{ or } F) = P(E) + P(F)$$

More generally, if events $E_1, E_2, \ldots, E_k$ are disjoint, then

$$P(E_1 \text{ or } E_2 \text{ or } \ldots \text{ or } E_k) = P(E_1) + P(E_2) + \cdots + P(E_k)$$

In words, the probability that any one of these k events occurs is the sum of the probabilities of the individual events.

EXAMPLE 12   Each movie available for rental at a large video store is classified into one of several different categories. Consider the movie rented by the next customer, and define events E_1, E_2, E_3 by

$$E_1 = \text{comedy}, \qquad E_2 = \text{horror film}, \qquad E_3 = \text{children's film}$$

Supposing that $P(E_1) = .25$, $P(E_2) = .18$, and $P(E_3) = .14$, the addition rule gives

$P(\text{comedy } or \text{ horror film } or \text{ children's film})$
$\quad = P(E_1) + P(E_2) + P(E_3)$
$\quad = .25 + .18 + .14$
$\quad = .57$

The probability that the next film rented is *not* one of these three types is

$$P(\text{not } (E_1 \text{ or } E_2 \text{ or } E_3)) = 1 - .57 = .43$$

The computation of $P(E \text{ or } F)$ when the two events are not disjoint is more complicated. Examining Figure 4(b) in Section 4.1, we see that in adding $P(E)$ and $P(F)$, the probability of each outcome in the event E *and* F (the overlap) is counted twice. This double counting is corrected by subtracting $P(E \text{ and } F)$.

A GENERAL ADDITION RULE FOR TWO EVENTS

For any two events E and F,

$$P(E \text{ or } F) = P(E) + P(F) - P(E \text{ and } F)$$

In words, to calculate the probability that at least one of the two events occurs, we add the two individual probabilities together and then subtract the probability that both events occur. When E and F are disjoint, the event E and F contains no outcomes. Then $P(E$ and $F) = 0$ and this addition rule reduces to the earlier one for disjoint events. This general rule can be used to determine any one of the four probabilities $P(E)$, $P(F)$, $P(E$ or $F)$, or $P(E$ and $F)$, provided that the other three are known.

EXAMPLE 13

Suppose that 60% of all customers of a large insurance agency have automobile policies with the agency, 40% have homeowner's policies, and 25% have both types of policies. If a customer is randomly selected, what is the probability that he/she has at least one of these two types of policies with the agency? Let

E = the event that the selected customer has auto insurance

F = the event that the selected customer has homeowner's insurance

The given information implies that

$$P(E) = .60, \qquad P(F) = .40,$$

and

$$P(E \text{ and } F) = .25,$$

from which we obtain

$P(\text{customer has at least one of the two types})$
$\qquad = P(E \text{ or } F) = P(E) + P(F) - P(E \text{ and } F)$
$\qquad = .60 + .40 - .25 = .75$

The event that the customer has neither type of policy is *not(E or F)*, so

$$P(\text{customer has neither type}) = 1 - P(E \text{ or } F) = .25$$

Now let us determine the probability that the selected customer has exactly one type of policy. The Venn diagram of Figure 8 shows that the event "at

FIGURE 8

Venn diagram for Example 13.

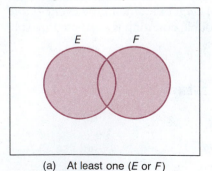

(a) At least one (E or F)

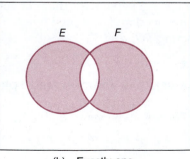

(b) Exactly one

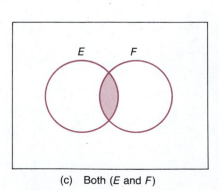

(c) Both (E and F)

least one," that is, *E or F*, consists of two disjoint parts: "exactly one" and "both." Thus

$$P(\text{at least one}) = P(E \text{ or } F)$$
$$= P(\text{exactly one}) + P(\text{both})$$
$$= P(\text{exactly one}) + P(E \text{ and } F)$$

so

$$P(\text{exactly one}) = P(E \text{ or } F) - P(E \text{ and } F)$$
$$= .75 - .25$$
$$= .50$$

The addition rule for more than two nondisjoint events is rather complicated. For example, in the case of three events,

$$P(E \text{ or } F \text{ or } G) = P(E) + P(F) + P(G) - P(E \text{ and } F)$$
$$- P(E \text{ and } G) - P(F \text{ and } G) + P(E \text{ and } F \text{ and } G)$$

A more advanced treatment of probability can be consulted for examples and extensions of these methods.

EXERCISES 4.10–4.25   **SECTION 4.2**

4.10   Insurance status—covered (C) or not covered (N)—is determined for each individual arriving for treatment at a hospital's emergency room. Consider the experiment in which this determination is made for two randomly selected patients. The simple events are $O_1 = \{CC\}$, $O_2 = \{CN\}$, $O_3 = \{NC\}$, and $O_4 = \{NN\}$. Suppose that probabilities are $P(O_1) = .81$, $P(O_2) = .09$, $P(O_3) = .09$, and $P(O_4) = .01$.
  a. What outcomes are contained in the event *A*, that at most one patient is covered, and what is $P(A)$?
  b. What outcomes are contained in the event *B*, that the two patients have the same status with respect to coverage, and what is $P(B)$?

4.11   Suppose the accompanying information on births in the United States over a given period of time is available to you.

Type of Birth	Number of Births
Single birth	41,500,000
Twins	500,000
Triplets	5,000
Quadruplets	100

Use this information to approximate the probability that a randomly selected pregnant woman who reaches full term:
  a. Delivers twins;
  b. Delivers quadruplets;
  c. Gives birth to more than a single child.

4.12   A mutual fund company offers its customers several different funds: a money-market fund, three different bond funds (short, intermediate, and long-term), two stock funds (moderate and high-risk), and a balanced fund. Among customers who own shares in just one fund, the percentages of customers in the different funds are as follows:

Money-market	20%	High-risk stock	18%
Short bond	15%	Moderate-risk stock	25%
Intermediate bond	10%	Balanced	7%
Long bond	5%		

A customer who owns shares in just one fund is randomly selected.
a. What is the probability that the selected individual owns shares in the balanced fund?
b. What is the probability that the individual owns shares in a bond fund?
c. What is the probability that the selected individual does not own shares in a stock fund?

4.13 A radio station that plays classical music has a "by request" program each Saturday evening. The percentages of requests for composers on a particular night are as follows:

Bach	5%	Mozart	21%
Beethoven	26%	Schubert	12%
Brahms	9%	Schumann	7%
Dvorak	2%	Tchaikovsky	14%
Mendelssohn	3%	Wagner	1%

Suppose that one of these requests is randomly selected.
a. What is the probability that the request is for one of the three B's?
b. What is the probability that the request is not for one of the two S's?
c. Neither Bach nor Wagner wrote any symphonies. What is the probability that the request is for a composer who wrote at least one symphony?
d. What is the probability that the request is for a piece by one of the main characters in the movie "Amadeus"? (If you don't know who this is, see the movie!)

4.14 The accompanying table gives the type of program and the percentage of households watching each program at 6 P.M. among subscribers to a certain television cable service.

Channel	Program	% Watching	Channel	Program	% Watching
2	national news	14	9	comedy	12
3	local news	7	10	national news	4
4	national news	10	11	drama	10
5	local news	8	12	game show	8
7	game show	15	13	movie	12

(The percentages are for those who are watching at this time, not for all subscribers.) A household is randomly selected from among those watching at this time.
a. What is the probability that the household is watching the local news? Is watching any news?
b. What is the probability that the household is watching either a game show or a comedy?
c. What is the probability that the household is not watching a game show?

4.15 A single card is randomly selected from a well-mixed deck. (A deck has 52 cards, with four suits—spades, hearts, diamonds, and clubs—and thirteen denominations—aces, twos, . . . , queens, and kings.)

a. How many simple events are there?

b. What is the probability of each simple event?

c. What is the probability that the selected card is a heart? A face card (jack, queen, or king)?

d. What is the probability that the selected card is both a heart and a face card?

e. Let A = the event that the selected card is a face card and B = the event that the selected card is a heart. What is $P(A \text{ or } B)$?

4.16 After mixing a deck of 52 cards very well, five cards are dealt out.

a. It can be shown that (disregarding the order in which the cards are dealt) there are 2,598,960 possible hands, of which only 1287 are hands consisting entirely of spades. What is the probability that a hand will consist entirely of spades? What is the probability that a hand will consist entirely of a single suit?

b. It can be shown that exactly 63,206 hands contain only spades and clubs, with both suits represented. What is the probability that a hand consists entirely of spades and clubs with both suits represented?

c. Using the result of part (b), what is the probability that a hand contains cards from exactly two suits?

4.17 After all students have left the classroom, a statistics professor notices that four copies of the text were left under desks. At the beginning of the next lecture, the professor distributes the four books in a completely random fashion to each of the four students (1, 2, 3, and 4) who claim to have left books. One possible outcome is that 1 receives 2's book, 2 receives 4's book, 3 receives his or her own book, and 4 receives 1's book. This outcome can be abbreviated (2, 4, 3, 1).

a. List the other 23 possible outcomes.

b. Which outcomes are contained in the event that exactly two of the books are returned to their correct owners? Assuming equally likely outcomes, what is the probability of this event?

c. What is the probability that exactly one of the four students receives his or her own book?

d. What is the probability that exactly three receive their own books?

e. What is the probability that at least two of the four students receive their own books?

4.18 An individual is presented with three different glasses of cola, labeled C, D, and P. He is asked to taste all three and then list them in order of preference. Suppose that the same cola has actually been put into all three glasses.

a. What are the simple events in this ranking experiment, and what probability would you assign to each one?

b. What is the probability that C is ranked first?

c. What is the probability that C is ranked first and D is ranked last?

4.19 The student council for a School of Science and Math has one representative from each of the five academic departments: Biology (B), Chemistry (C), Mathematics (M), Physics (P), and Statistics (S). Two of these students are to be randomly selected for inclusion on a university-wide student committee (by placing five slips of paper in a box, mixing, and drawing out two of them).

a. What are the ten possible outcomes (simple events)?

b. From the description of the selection process, all outcomes are equally likely; what is the probability of each simple event?

c. What is the probability that one of the committee members is the Statistics Department representative?

d. What is the probability that both committee members come from "laboratory science" departments?

4.20 A video store sells two different brands of VCR, each of which comes with either two heads or four heads. The accompanying table gives the percentages of recent purchasers buying each type of VCR.

	Number of Heads	
Brand	2	4
M	25%	16%
Q	32%	27%

Suppose that a recent purchaser is randomly selected and both the brand and the number of heads are determined.
a. What are the four simple events?
b. What is the probability that the selected purchaser bought brand Q, with two heads?
c. What is the probability that the selected purchaser bought brand M?

4.21 A library has five copies of a certain text, of which copies 1 and 2 are first printings, and copies 3, 4, and 5 are second printings. Two copies are to be randomly selected to be placed on two-hour reserve.
a. What is the probability that both selected copies are first printings?
b. What is the probability that both copies selected are second printings?
c. What is the probability that at least one copy selected is a first printing?
d. What is the probability that the selected copies are different printings?

4.22 An insurance company offers four different deductible levels—none, low, medium, and high—for its homeowner's policy holders, and three different levels—low, medium, and high—for its automobile policy holders. The accompanying table gives proportions for the various categories of policy holders who have both types of insurance. For example, the proportion of individuals with both low homeowner's deductible and low auto deductible is .06 (6% of all such individuals).

	Homeowner's			
Auto	N	L	M	H
L	.04	.06	.05	.03
M	.07	.10	.20	.10
H	.02	.03	.15	.15

Suppose that an individual having both types of policies is randomly selected.
a. What is the probability that the individual has a medium auto deductible and a high homeowner's deductible?
b. What is the probability that the individual has a low auto deductible? A low homeowner's deductible?
c. What is the probability that the individual is in the same category for both auto and homeowner's deductibles?
d. Based on your answer in part (c), what is the probability that the two categories are different?
e. What is the probability that the individual has at least one low deductible level?
f. Using the answer in part (e), what is the probability that neither deductible level is low?

4.23 There are two traffic lights on the route used by a certain individual to go from home to work. Let E denote the event that the individual must stop at the first light, and define the event F in a similar manner for the second light. Suppose that $P(E) = .4$, $P(F) = .3$, and $P(E \text{ and } F) = .15$.
  a. What is the probability that the individual must stop at at least one light; i.e., what is the probability of the event $E \text{ or } F$?
  b. What is the probability that the individual needn't stop at either light?
  c. What is the probability that the individual must stop at exactly one of the two lights?
  d. What is the probability that the individual must stop just at the first light? Hint: How is the probability of this event related to $P(E)$ and $P(E \text{ and } F)$? A Venn diagram might help.

4.24 A family with two automobiles is to be randomly selected. Let A be the event that the older car has an automatic transmission and let B be the event that the newer car has an automatic transmission. Suppose that $P(A) = .7$ and $P(B) = .8$.
  a. Can $P(A \text{ or } B)$ be calculated from the given information? Explain.
  b. Could it be the case that $P(A \text{ and } B) = .75$? Explain.
  c. If the probability is .9 that the selected family owns at least one car with an automatic transmission, what is the probability that both cars owned by the selected family have automatic transmissions? That at least one car has a manual transmission?

4.25 Suppose that a six-sided die is "loaded" so that any particular even-numbered face is twice as likely to be observed as any particular odd-numbered face.
  a. What are the probabilities of the six simple events? Hint: Denote these events by $O_1, \ldots, O_6$. Then $P(O_1) = p$, $P(O_2) = 2p$, $P(O_3) = p$, $\ldots$, $P(O_6) = 2p$. Now use a condition on the sum of these probabilities to determine p.
  b. What is the probability that the number showing is an odd number? Is at most 3?
  c. Now suppose the die is loaded so that the probability of any particular simple event is proportional to the number showing on the corresponding upturned face; i.e., $P(O_1) = c$, $P(O_2) = 2c$, $\ldots$, $P(O_6) = 6c$. What are the probabilities of the six simple events? Calculate the probabilities of part (b) for this die.

4.3 CONDITIONAL PROBABILITY AND INDEPENDENCE

Sometimes the knowledge that one event has occurred considerably changes the likelihood that another event will occur. One such situation involves a particular disease, with an incidence rate of .1% in a certain population. The presence of the disease cannot be discerned from outward appearances, but there is a diagnostic test available. Unfortunately, the test is not infallible: 80% of those with positive test results actually have the disease, whereas the other 20% showing positive results are false positives. To cast this in probability terms, consider the experiment in which an individual is randomly selected from the population. Define events by

E = the event that the individual has the disease

F = the event that the individual's diagnostic test shows positive

Let $P(E|F)$ denote the probability of the event E *given that* the event F has occurred. The preceding information then implies that

$$P(E) = .001, \qquad P(E|F) = .8$$

That is, prior to diagnostic test information, the occurrence of E is unlikely, whereas once it is known that the test result is positive, the likelihood of the disease increases dramatically. (If this were not so, the diagnostic test would not be very useful.)

The next example suggests how a *conditional* probability $P(E|F)$ can be calculated from other previously specified probabilities.

EXAMPLE 14

A GFI (ground fault interrupt) switch will turn off power to a system in the event of an electrical malfunction. A spa manufacturer currently has 25 spas in stock, each equipped with a single GFI switch. The switches are supplied by two different sources, and some of them are defective, as summarized in the accompanying table. For example, a total of 15 (= 10 + 5) of the 25 switches are from source 1, ten switches from source 1 are nondefective, two switches from source 2 are defective, and so on.

	Nondefective	Defective	
Source 1	10	5	15
Source 2	8	2	10
	18	7	25

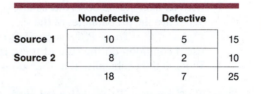

A spa is randomly selected for testing. Let

E = the event that the GFI switch in the chosen spa is from source 1

F = the event that the GFI switch in the chosen spa is defective

The tabulated information implies that

$P(E) = 15/25 = .60,$

$P(F) = 7/25 = .28,$

$P(E \text{ and } F) = 5/25 = .20$

Now suppose that testing reveals a defective switch. (Thus the chosen spa is one of the seven in the *Defective* column.) How likely is it that the switch came from the first source? Since five of the seven defectives are from source 1, intuition suggests that

$$P(E|F) = P(\text{source 1}|\text{defective}) = 5/7 \approx .714$$

Notice that this is larger than the "original" probability $P(E)$. This is because source 1 has a much higher defective rate than does source 2.

An alternative expression for the conditional probability is

$$P(E|F) = \frac{5}{7} = \frac{5/25}{7/25} = \frac{P(E \text{ and } F)}{P(F)}$$

That is, $P(E|F)$ is a ratio of two previously specified probabilities: the probability that both events occur divided by the probability of the "conditioning event" F. Additional insight comes from the Venn diagram of Figure 9. Once it is known that the outcome lies in F, the likelihood of E (also) occurring is the "size" of (E and F) relative to that of F.

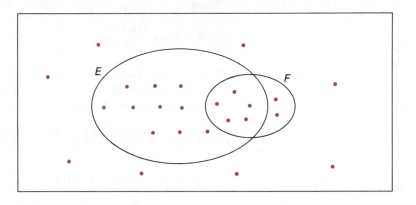

FIGURE 9

Venn diagram for Example 14 (each dot represents one switch).

The results of the previous example help motivate a general definition of conditional probability.

DEFINITION

Let E and F be two events with $P(F) > 0$. The **conditional probability of the event E given that the event F has occurred**, denoted by $P(E|F)$, is

$$P(E|F) = \frac{P(E \text{ and } F)}{P(F)}$$

EXAMPLE 15

A large lending institution gives both adjustable-rate mortgages and fixed-rate mortgages on residential property. It breaks residential property into three categories: single-family houses, condominiums, and multifamily dwellings. The accompanying table, sometimes called a *joint probability table*, displays probabilities appropriate to this situation.

	Single-Family	Condo	Multi-Family	
Adjustable	.40	.21	.09	.70
Fixed	.10	.09	.11	.30
	.50	.30	.20	

Thus 70% of all mortgages are adjustable rate, 50% of all mortgages are for single-family properties, 40% of all mortgages are adjustable rate for single-family properties (both adjustable-rate and single-family), and so on. Define events E and F by

E = the event that a mortgage is adjustable-rate

F = the event that a mortgage is for a single-family property

Then

$$P(E|F) = \frac{P(E \text{ and } F)}{P(F)} = \frac{.40}{.50} = .80$$

That is, if we focus only on loans made for single-family properties, 80% of them are adjustable-rate loans. Notice that $P(E|F)$ is larger than the original (unconditional) probability $P(E) = .70$. Also,

$$P(F|E) = \frac{P(E \text{ and } F)}{P(E)} = \frac{.40}{.70} = .571 > .5 = P(F)$$

Defining another event C by

C = the event that a mortgage is for a condominium

we have

$$P(E|C) = \frac{P(E \text{ and } C)}{P(C)} = \frac{.21}{.30} = .70$$

Notice that $P(E|C) = P(E)$, so if we are told that a mortgage is for a condominium, the likelihood that it is adjustable remains unchanged.

INDEPENDENCE   When two events E and F are such that $P(E|F) = P(E)$, the likelihood that event E will occur is the same after learning that F has occurred as it was prior to information about F's occurrence. That is, the chance of E occurring is unaffected by the knowledge that F has occurred. We then say that E and F are independent of one another.

DEFINITION

Two events E and F are said to be **independent** if

$P(E|F) = P(E)$.

If E and F are not independent, they are said to be **dependent** events.

The defining relationship for independence could just as well have been stated $P(F|E) = P(F)$, since either equality implies the other one. In words, two events are independent if the chance that one of them occurs is unchanged once it is known that the other has occurred. Moreover, independence of events E and F implies the following additional three relationships

4.3 CONDITIONAL PROBABILITY AND INDEPENDENCE

between conditional and unconditional (original probabilities):

$$P(not\ E|F) = P(not\ E)$$
$$P(E|not\ F) = P(E)$$
$$P(not\ E|not\ F) = P(not\ E)$$

The definition of conditional probability is

$$P(E|F) = \frac{P(E\ and\ F)}{P(F)}$$

and independence implies that

$$P(E|F) = P(E).$$

Combining these two relationships (when E and F are independent) gives

$$\frac{P(E\ and\ F)}{P(F)} = P(E).$$

Multiplying each side of this equation by $P(F)$ produces the **multiplication rule** for two independent events:

THE MULTIPLICATION RULE FOR TWO INDEPENDENT EVENTS

When E and F are independent,

$$P(E\ and\ F) = P(E) \cdot P(F)$$

Thus if E occurs 50% of the time when an experiment is repeatedly performed, F occurs 20% of the time, and E and F are independent, then E and F will occur together 10% of the time in the long run (since $(.5)(.2) = .1$).

EXAMPLE 16   Let E be the event that your statistics professor begins class on time and let F be the event that your philosophy professor does likewise. Suppose that E and F are independent (intuitively, a very reasonable assumption, since behavior of one professor should be unaffected by behavior of the other), with $P(E) = .9$ and $P(F) = .6$. Then

$$P(E\ and\ F) = P(\text{both professors begin class on time})$$
$$= P(E) \cdot P(F) = (.9)(.6) = .54$$

Also,

$$P(not\ E\ and\ not\ F) = P(\text{neither professor begins on time})$$
$$= P(not\ E) \cdot P(not\ F) = (.1)(.4) = .04$$

The probability that exactly one of the two begins on time is

$$1 - (.54 + .04) = .42.$$

The concept of independence extends to more than two events. Consider three events E_1, E_2, and E_3. Then independence means not only that

$$P(E_1 | E_2) = P(E_1)$$
$$P(E_3 | E_2) = P(E_3)$$

and so on, but also that

$$P(E_1 | E_2 \text{ and } E_3) = P(E_1)$$
$$P(E_1 \text{ and } E_3 | E_2) = P(E_1 \text{ and } E_3)$$

and so on. Furthermore, the multiplication rule remains valid.

Events E_1, E_2, . . . , E_k are **independent** if knowledge that some of the events have occurred does not change the probabilities that any particular one or more of the other events will occur. Independence implies that

$$P(E_1 \text{ and } E_2 \text{ and } \ldots \text{ and } E_k) = P(E_1) \cdot P(E_2) \cdot \ldots \cdot P(E_k).$$

Thus when events are independent, the probability that all occur together is the product of the individual probabilities. Furthermore, this relationship remains valid if one or more E_i's is replaced by the event *not E_i*.

EXAMPLE 17

A microcomputer system consists of a monitor, a disk drive, and the computer itself. Let

E_1 = the event that the newly purchased monitor operates properly

E_2 = the event that the newly purchased disk drive operates properly

E_3 = the event that the newly purchased computer operates properly

Suppose that these three events are independent, with

$$P(E_1) = .99, \qquad P(E_2) = .90, \qquad P(E_3) = .95$$

The probability that all three components operate properly (i.e., that the system functions) is

$$P(E_1 \text{ and } E_2 \text{ and } E_3) = P(E_1) \cdot P(E_2) \cdot P(E_3)$$
$$= (.99)(.90)(.95)$$
$$\approx .85$$

In the long run, roughly 85% of such systems will operate properly. The probability that only the monitor operates properly is

$$P(E_1 \text{ and not } E_2 \text{ and not } E_3) = P(E_1) \cdot P(\text{not } E_2) \cdot P(\text{not } E_3)$$
$$= (.99)(.10)(.05)$$
$$= .00495$$

A GENERAL MULTIPLICATION RULE   The definition of conditional probability states that

$$P(E|F) = \frac{P(E \text{ and } F)}{P(F)}$$

Multiplying both sides by $P(F)$ gives a useful expression for the probability that both events will occur.

GENERAL MULTIPLICATION RULE FOR TWO EVENTS

$$P(E \text{ and } F) = P(E|F) \cdot P(F)$$

The right-hand side of this equation is the product of two probabilities, the first conditional and the second unconditional. In the case of independence, $P(E|F) = P(E)$ and this multiplication rule reduces to the earlier one.

EXAMPLE 18   The accompanying table gives information on VCRs sold by a certain appliance store.

	Percentage of Customers Purchasing	Percentage Buying an Extended Warranty among Those Who Purchase
Brand 1	70%	20%
Brand 2	30%	40%

A purchaser is randomly selected from among all those having bought a VCR from the store. What is the probability that he/she purchased a brand #1 model and an extended warranty? What is the probability that the selected customer had purchased an extended warranty?
  To answer the first question, let

  B_1 = the event that brand #1 is purchased
  E = the event that an extended warranty is purchased

The tabulated information implies that

$$P(B_1) = P(\text{brand } \#1 \text{ purchased}) = .70$$
$$P(E|B_1) = P(\text{ext. warranty purchased}|\text{brand } \#1 \text{ purchased}) = .20$$

Notice that the 20% is identified with a *conditional* probability: among purchasers of brand #1, it is the percentage opting for an extended warranty. Substituting these numbers into the multiplication rule (with B_1 replacing F) yields

$$P(B_1 \text{ and } E) = P(E|B_1) \cdot P(B_1)$$
$$= (.20)(.70)$$
$$= .14$$

Thus 14% of all purchasers selected brand #1 and buy an extended warranty.

The tree diagram of Figure 10 gives a nice visual display of how the multiplication rule is used here. The two first-generation branches are labeled with events B_1 and B_2 along with their probabilities. There are two second-generation branches leading out from each first-generation branch. These correspond to the two events E and *not* E, and the *conditional* probabilities $P(E|B_1)$, $P(not\ E|B_1)$, $P(E|B_2)$, $P(not\ E|B_2)$ appear on these branches. The multiplication rule amounts to nothing more than multiplying probabilities across the tree diagram. For example,

$$P(B_2\ and\ E) = P(\text{brand } \#2\ and\ \text{warranty purchased})$$
$$= P(E|B_2) \cdot P(B_2)$$
$$= (.4)(.3)$$
$$= .12$$

and this probability is displayed to the right of the E branch that comes from the B_2 branch.

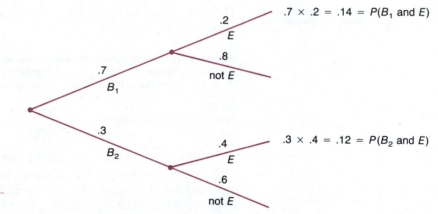

FIGURE 10

A tree diagram for probability calculations.

We can now easily answer the second question we posed earlier. There are two different ways a customer can purchase an extended warranty:

> buy brand #1 and warranty = B_1 *and* E
>
> buy brand #2 and warranty = B_2 *and* E

Furthermore, the events B_1 *and* E, B_2 *and* E are disjoint (each customer purchased a single VCR, so it can't be both brand #1 and brand #2). Thus

$$P(E) = P(B_1\ and\ E) + P(B_2\ and\ E)$$
$$= P(E|B_1) \cdot P(B_1) + P(E|B_2) \cdot P(B_2)$$
$$= (.2)(.7) + (.4)(.3)$$
$$= .14 + .12$$
$$= .26$$

This probability is the sum of the two probabilities shown on the right side of the tree diagram. Twenty-six percent of all VCR purchasers selected an extended warranty.

The multiplication rule can be extended to give an expression for the probability that several events occur together. In the case of three events E, F, and G, we have

$$P(E \text{ and } F \text{ and } G) = P(E|F \text{ and } G) \cdot P(F|G) \cdot P(G)$$

When the events are all independent, $P(E|F \text{ and } G) = P(E)$ and $P(F|G) = P(F)$, so the right-hand side is just the product of three unconditional probabilities.

EXAMPLE 19

Twenty percent of all passengers who fly from Los Angeles to New York do so on airline G. This airline misplaces at least one piece of luggage for 10% of its passengers, and 90% of this lost luggage is subsequently recovered. If a passenger who has flown from L.A. to N.Y. is randomly selected, what is the probability that the selected individual flew on airline G (event G), had at least one piece of luggage misplaced (event F) and subsequently recovered the misplaced piece(s) (event E)? The stated percentages imply that

$$P(G) = .20, \qquad P(F|G) = .10, \qquad P(E|F \text{ and } G) = .90$$

Thus

$$P(E \text{ and } F \text{ and } G) = (.90)(.10)(.20) = .018$$

SAMPLING WITH AND WITHOUT REPLACEMENT

Many applications of probability rules involve repeated sampling from a single population. Here is a simple example that introduces an important distinction.

EXAMPLE 20

Consider selecting three cards from a deck. This selection can be made either **without replacement** (dealing three cards off the top) or **with replacement** (replacing each card and shuffling before selecting the next card). You can probably already guess that one of these selection methods gives independence of successive selections, whereas the other does not. To see this more clearly, define these events:

E_1 = the event that the first card is a spade

E_2 = the event that the second card is a spade

E_3 = the event that the third card is a spade

For sampling with replacement, the probability of E_3 is .25, regardless of whether either E_1 or E_2 occurs, since replacing selected cards gives the same deck for the third selection as for the first two. Whether either of the first two cards is a spade has no bearing on the third card selected, so the three events E_1, E_2, and E_3 are independent.

When sampling is without replacement, the chance of a spade on the third draw very definitely depends on the results of the first two draws. If both E_1 and E_2 occur, only 11 of the 50 remaining cards are spades. Since any one of these 50 has the same chance of being selected, the probability of E_3 in this case is

$$P(E_3 | E_1 \text{ and } E_2) = 11/50 = .22$$

Alternatively, if neither of the first two cards is a spade, then all 13 spades remain in the deck for the third draw, so

$$P(E_3 | \text{not } E_1 \text{ and not } E_2) = 13/50 = .26$$

Information about the occurrence of E_1 and E_2 affects the chance that E_3 will occur, so for sampling without replacement, the events are not independent.

In opinion polls and other types of surveys, sampling is virtually always done without replacement. For this method of sampling, the results of successive selections are not independent of one another. This is unfortunate, because many results from probability and statistics are much easier to state and use when independence can be assumed. The next example suggests that under certain circumstances, the selections in sampling without replacement are approximately independent.

EXAMPLE 21

A lot of 10,000 industrial components consists of 2500 manufactured by one firm and 7500 manufactured by a second firm, all mixed together. Three components are to be randomly selected without replacement. Let E_1, E_2, and E_3 denote the events that the first, second, and third components selected were made by the first firm. Reasoning as in the card selection example,

$$P(E_3 | E_1 \text{ and } E_2) = 2498/9998 = .24985$$
$$P(E_3 | \text{not } E_1 \text{ and not } E_2) = 2500/9998 = .25005$$

While these two probabilities differ slightly, to three decimal places they are both .250. We conclude that the occurrence or nonoccurrence of E_1 or E_2 has virtually no effect on the chance that E_3 will occur. For practical purposes, the three events can be considered independent.

The essential difference between the situations of Example 19 and Example 20 is the size of the sample relative to the size of the population. In the former example, a relatively large proportion of the population was sampled (3 out of 52), whereas in the latter example, the proportion of the population sampled was quite small (only 3 out of 10,000).

If the individuals or objects in a sample are selected without replacement from a population and the sample size is small relative to the population size, the successive selections are approximately independent. As a reasonable rule of thumb, independence can be assumed if at most 5% of the population is sampled.

This result justifies the assumption of independence in many statistical problems.

EXERCISES 4.26–4.43   **SECTION 4.3**

4.26 Two different airlines have a flight from Los Angeles to New York that departs each weekday morning at a certain time. Let E denote the event that the first airline's flight is fully booked on a particular day, and let F denote the event that the second airline's flight is fully booked on that same day. Suppose that $P(E) = .7$, $P(F) = .6$, and $P(E \text{ and } F) = .54$.
   a. Calculate $P(E|F)$, the probability that the first airline's flight is fully booked given that the second airline's flight is fully booked.
   b. Calculate $P(F|E)$.
   c. Are E and F independent events? Explain.

4.27 A card is selected from a well-mixed deck. Let E be the event that the card is a heart, and let F be the event that the card is a face card.
   a. Determine $P(E)$ and $P(F)$.
   b. Determine $P(E|F)$ and $P(F|E)$.
   c. Are E and F independent events? Explain intuitively why this is so.

4.28 The probability that a randomly selected customer at a certain gas station checks the oil level is .10. The probability that a randomly selected customer checks tire pressure is .04. The probability that a randomly selected customer checks both oil level and tire pressure is .008.
   a. Given that a customer checks the oil level, what is the probability that the customer checks tire pressure?
   b. If the randomly selected customer checks tire pressure, what is the probability that the oil level is checked also?
   c. Are the events *checks oil level* and *checks tire pressure* independent? Explain.

4.29 Reconsider Example 6, in which red and green tetrahedral dice are tossed. Suppose that both dice are fair. Define events A, E, and F as follows:

   $A =$ the event that both dice show the same number

   $E =$ the event that the red die shows an even number

   $F =$ the event that the sum of the two numbers showing is at most 4

   Are A and E independent events? Are A and F independent? Are E and F independent?

4.30 A certain model of car comes in a two-door version, a four-door version, and a hatchback version. Each version can be equipped with either an automatic transmission or a manual transmission. The accompanying table gives the relevant proportions.

	Version		
Transmission Type	TD	FD	HB
A	.32	.27	.18
M	.08	.04	.11

A customer who has purchased one of these cars is randomly selected.

a. What is the probability that this customer purchased a car with an automatic transmission? A four-door car?

b. Given that the customer purchased a four-door car, what is the probability that it has an automatic transmission?

c. Given that the customer did not purchase a hatchback, what is the probability that the car has a manual transmission? How does this conditional probability compare to the (unconditional) probability that the car has a manual transmission?

4.31 Exercise 4.22 presented the accompanying table giving proportions of insured individuals with various automobile–homeowner deductible levels.

Auto	Homeowner's			
	N	L	M	H
L	.04	.06	.05	.03
M	.07	.10	.20	.10
H	.02	.03	.15	.15

Suppose an insured individual is randomly selected.

a. What is P(low auto deductible|low homeowner's deductible)?

b. Given that the selected individual does not have a high auto deductible, what is the probability that he/she does not have a high homeowner's deductible?

4.32 A certain university has ten cars available for use by faculty and staff upon request. Six of these are Chevrolets and four are Plymouths. On a particular day, only two requests for cars have been made. Suppose that the two cars to be assigned are chosen in a completely random fashion from among the ten.

a. Let E denote the event that the first car assigned is a Chevrolet. What is $P(E)$?

b. Let F denote the probability that the second car assigned is a Chevrolet. What is $P(F|E)$?

c. Use the results of (a) and (b) to calculate $P(E \text{ and } F)$.

4.33 A construction firm has bid on two different contracts. Let E_1 be the event that the bid on the first contract is successful, and define E_2 analogously for the second contract. Suppose that $P(E_1) = .4$, $P(E_2) = .3$, and that E_1 and E_2 are independent events.

a. Calculate the probability that both bids are successful (the probability of the event $E_1 \text{ and } E_2$).

b. Calculate the probability that neither bid is successful (the probability of the event (not E_1) and (not E_2)).

c. What is the probability that the firm is successful in at least one of the two bids?

4.34 Consider the system of four components connected as illustrated in the accompanying diagram. Let E_1 denote the event that component 1 functions properly, and define events E_2, E_3, and E_4 analogously for the other three components. Suppose that $P(E_1) = P(E_2) = P(E_3) = P(E_4) = .9$ and that E_1, E_2, E_3, and E_4 are independent events.

a. Because the two components in the 1–2 subsystem are connected in series, this subsystem will function if and only if the two components both function. What is the probability that this subsystem functions? What is the probability that the 3–4 subsystem functions?

b. What is the probability that both subsystems function?

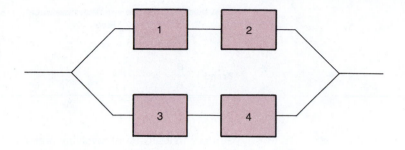

c. Because the subsystems are connected in parallel, the system will function if at least one of the two subsystems (1–2 or 3–4) functions. What is the probability that the system functions?

4.35 This case study is reported in the article "Parking Tickets and Missing Women," which appears in the book *Statistics: A Guide to the Unknown* (see the Chapter 1 References). In a Swedish trial on a charge of overtime parking, a policeman testified that he had noted the position of the two air valves on the tires of a parked car: to the closest hour, one was at the one o'clock position and the other was at the six o'clock position. After the allowable time for parking in that zone had passed, the policeman returned, noted that the valves were in the same position, and ticketed the car. The owner of the car claimed that he had left the parking place in time and had returned later. The valves just happened by chance to be in the same positions. An "expert" witness computed the probability of this occurring as $(\frac{1}{12})(\frac{1}{12}) = \frac{1}{144}$.

a. What reasoning did the expert use to arrive at the probability of 1/144?

b. Can you spot the error in the reasoning that leads to the stated probability of 1/144? What effect does this error have on the probability of occurrence? Do you think that 1/144 is larger or smaller than the correct probability of occurrence?

4.36 A particular airline has 10 A.M. flights from Chicago to New York, Atlanta, and Los Angeles. Let A denote the event that the New York flight is full, and define events B and C analogously for the Atlanta and Los Angeles flights. Suppose that $P(A) = .6$, $P(B) = .5$, $P(C) = .4$, and the three events are independent.

a. What is the probability that all three flights are full? That at least one flight is not full?

b. What is the probability that only the New York flight is full? That exactly one of the three flights is full?

4.37 A shipment of 5000 printed circuit boards contains 40 that are defective. Two boards are chosen at random, without replacement. Consider the two events

E_1 = the event that the first board selected is defective

E_2 = the event that the second board selected is defective

a. Are E_1 and E_2 dependent events? Explain in words.

b. Let *not* E_1 be the event that the first board selected is not defective (the event that E_1 does not occur). What is $P(\text{not } E_1)$?

c. How do the two probabilities $P(E_2 | E_1)$ and $P(E_2 | \text{not } E_1)$ compare?

d. Based on your answer to part (c), would it be reasonable to view E_1 and E_2 as approximately independent? Explain your answer.

4.38 A store sells two different brands of dishwasher soap, B_1 and B_2. Each brand comes in three different sizes: small (S), medium (M), and large (L). The proportions of the two brands and of the three sizes purchased are displayed as marginal totals in the accompanying table.

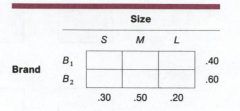

Suppose that any event involving brand is independent of any event involving size. What is the probability of the event that a randomly selected purchaser buys the small size of brand B_1 (the event B_1 *and* S)? What are the probabilities of the other brand–size combinations?

4.39 Let E_1 denote the event that the next fire to occur in a certain city is a single-alarm blaze. Similarly, let E_2 be the event that the next fire is in the two-alarm category. Suppose that $P(E_1) = .4$ and $P(E_2) = .3$. What is $P(E_1|E_2)$? What is $P(E_2|E_1)$? Are disjoint events independent? Explain.

4.40 Only one-tenth of one percent of the individuals in a certain population have a particular disease (an incidence rate of .001). Of those having the disease, 95% test positive when a certain diagnostic test is applied. Of those not having the disease, 90% test negative when the test is applied. Suppose that an individual from this population is randomly selected and given the test.
   a. Construct a tree diagram having two first-generation branches, for "has disease" and "doesn't have disease," and two second-generation branches leading out from each of these, for "positive test" and "negative test." Then enter appropriate probabilities on the four branches.
   b. Use the multiplication rule to calculate P(has disease *and* positive test).
   c. Calculate P(positive test).
   d. Use the definition of conditional probability along with the results of parts (b) and (c) to calculate P(has disease|positive test). Does the result surprise you? Give an intuitive explanation for the size of this probability.

4.41 A company uses three different assembly lines—A_1, A_2, and A_3—to manufacture a particular component. Of those manufactured by line A_1, there are 5% that need rework to remedy a defect, whereas 8% of A_2's components need rework and 10% of A_3's need rework. Suppose that 50% of all components are produced by line A_1, while 30% are produced by line A_2 and 20% come from line A_3.
   a. Construct a tree diagram with first-generation branches corresponding to the three lines. Leading from each one, draw one branch for rework (R) and another for no rework (N). Then enter appropriate probabilities on the branches.
   b. What is the probability that a randomly selected component came from line 1 and needed rework?
   c. What is the probability that a randomly selected component needed rework?

4.42 A certain company sends 40% of its overnight mail parcels via express mail service E_1. Of these parcels, 2% arrive after the guaranteed delivery time (denote by L the event "late delivery"). If a record of an overnight mailing is randomly selected from the company's file, what is the probability that the parcel went via E_1 and was late?

4.43 Return Exercise 4.42, and suppose that 50% of the overnight parcels are sent via express mail service E_2 and the remaining 10% are sent via E_3. Of those sent via E_2, only 1% arrive late, whereas 5% of the parcels handled by E_3 arrive late.
   a. What is the probability that a randomly selected parcel arrived late? Hint: A tree diagram should help.

b. Suppose that the selected record shows that the parcel arrived late, but the name of the service does not appear on the record. What is the probability that the parcel was handled by E_1? That is, what is the *posterior* probability of E_1 given L, denoted $P(E_1|L)$? What is $P(E_2|L)$? $P(E_3|L)$?

CHAPTER 4 SUMMARY OF KEY CONCEPTS AND FORMULAS

TERM OR FORMULA	PAGE	COMMENT
Chance experiment	80	Any experiment for which there is uncertainty concerning the resulting outcome
Sample space	83	The collection of all possible outcomes of a chance experiment
Event	83	Any collection of possible outcomes from a chance experiment
Simple event	83	An event that consists of a single outcome
Events 1. *not A* 2. *A or B* 3. *A and B*	85	1. The event consisting of all outcomes not in *A*; 2. the event consisting of all outcomes in at least one of the two events; 3. the event consisting of all outcomes in both events
Disjoint (mutually exclusive) events	87	Events that have no outcomes in common
Probability of an event *E*, *P(E)*	93	The long-run relative frequency of occurrence of *E* when the experiment is repeated indefinitely
Basic properties of probability	94	The probability of any event must be a number between zero and one. The sum of probabilities of all simple events in the sample space must be 1. $P(E)$ is the sum of the probabilities of all simple events in *E*. $P(E) + P(\text{not } E) = 1$
$P(E) = \dfrac{\text{\# of outcomes in } E}{N}$	95	$P(E)$ when the outcomes are equally likely, where *N* is the number of outcomes in the sample space
$P(E \text{ or } F) = P(E) + P(F)$ $P(E_1 \text{ or } \ldots \text{ or } E_k)$ $= P(E_1) + \cdots + P(E_k)$	97	Addition rules when events are disjoint
$P(E \text{ or } F)$ $= P(E) + P(F) - P(E \text{ and } F)$	97	The general addition rule for two events
$P(E|F) = \dfrac{P(E \text{ and } F)}{P(F)}$	105	The conditional probability of the event *E* given that the event *F* has occurred
Independence of events *E* and *F* $P(E|F) = P(E)$	106	Events *E* and *F* are independent if the probability that *E* will occur given *F* is the same as the probability that *E* will occur with no knowledge of *F*
$P(E \text{ and } F) = P(E) \cdot P(F)$ $P(E_1 \text{ and } \ldots \text{ and } E_k)$ $= P(E_1) \cdots P(E_k)$	107	Multiplication rules for independent events
$P(E \text{ and } F) = P(E|F) \cdot P(F)$	109	The general multiplication rule for two events

4.44 A student has a box containing 25 computer disks, of which 15 are blank and 10 are not. She randomly selects disks one by one and examines each one, terminating the process only when she finds a blank disk. What is the probability that she must examine at least two disks?

4.45 In a school machine shop, 60% of all machine breakdowns occur on lathes and 15% occur on drill presses. Let E denote the event that the next machine breakdown is on a lathe, and let F denote the event that a drill press is the next machine to break down. With $P(E) = .60$ and $P(F) = .15$, calculate:
a. $P(not\ E)$;
b. $P(E\ or\ F)$;
c. $P(not\ E\ and\ not\ F)$.

4.46 There are five faculty members in a certain academic department. These individuals have 3, 6, 7, 10, and 14 years of teaching experience, respectively. Two of these individuals are randomly selected to serve on a personnel review committee. What is the probability that the chosen representatives have a total of at least 15 years of teaching experience? Hint: Consider all possible committees.

4.47 Automobiles coming to a certain intersection can either turn left (L), turn right (R), or go straight (S). Suppose that for a randomly selected car, $P(L) = .1$, $P(S) = .7$, and $P(R) = .2$, and suppose that cars turn or go straight independently of one another.
a. Among four randomly selected cars, what is the probability that all go straight?
b. What is the probability that four randomly selected cars go in the same direction?
c. What is the probability that three randomly selected cars all go in a different direction?

4.48 The general addition rule for three events states that

$$P(A\ or\ B\ or\ C) = P(A) + P(B) + P(C) - P(A\ and\ B) - P(A\ and\ C) - P(B\ and\ C) + P(A\ and\ B\ and\ C)$$

A new magazine publishes columns entitled "Art" (A), "Books" (B), and "Cinema" (C). Suppose that 14% of all subscribers read column A, 23% read B, 37% read C, 8% read A and B, 9% read A and C, 13% read B and C, and 5% read all three columns. What is the probability that a randomly selected subscriber reads at least one of these three columns?

4.49 A theater complex is currently showing four R-rated movies, three PG-13 movies, two PG movies, and one G movie. The accompanying table gives the number of people at the first showing of each movie on a certain Saturday.

Theater	Rating	# Viewers
1	R	600
2	PG-13	420
3	PG-13	323
4	R	196
5	G	254
6	PG	179
7	PG-13	114
8	R	205
9	R	139
10	PG	87

Suppose that a single one of these viewers is randomly selected.

a. What is the probability that the selected individual saw a PG movie?

b. What is the probability that the selected individual saw a PG or a PG-13 movie?

c. What is the probability that the selected individual did not see an R movie?

4.50 Refer to Exercise 4.49, and suppose that two viewers are randomly selected. Let R_1 and R_2 denote the events that the first and second individuals, respectively, watched an R-rated movie. Are R_1 and R_2 independent events? Explain. From a practical point of view, can these events be regarded as independent? Explain.

4.51 A large department store sells sport shirts in three sizes (small, medium, and large), three patterns (plaid, print, and stripe), and two sleeve lengths (long and short). The accompanying tables give the proportions of shirts sold falling in the various category combinations.

SHORT-SLEEVED

Size	Pattern		
	Pl	Pr	St
S	.04	.02	.05
M	.08	.07	.12
L	.03	.07	.08

LONG-SLEEVED

Size	Pattern		
	Pl	Pr	St
S	.03	.02	.03
M	.10	.05	.07
L	.04	.02	.08

a. What is the probability that the next shirt sold is a medium, long-sleeved, print shirt?

b. What is the probability that the next shirt sold is a medium print shirt?

c. What is the probability that the next shirt sold is a short-sleeved shirt? A long-sleeved shirt?

d. What is the probability that the size of the next shirt sold is medium? That the pattern of the next shirt sold is a print?

4.52 Refer to Exercise 4.51.

a. Given that the shirt just sold was a short-sleeved plaid, what is the probability that its size was medium?

b. Given that the shirt just sold was a medium plaid, what is the probability that it was short-sleeved? Long-sleeved?

4.53 One box contains six red balls and four green balls, and a second box contains seven red balls and three green balls. A ball is randomly chosen from the first box and placed in the second box. Then a ball is randomly selected from the second box and placed in the first box.

a. What is the probability that a red ball is selected from the first box and a red ball is selected from the second box?

b. At the conclusion of the selection process, what is the probability that the numbers of red and green balls in the first box are identical to the numbers at the beginning?

4.54 A bowl contains four slips of paper. One says "win prize #1," one says "win prize #2," one says "win prize #3," and the last slip says "win prizes 1, 2, and 3." You randomly select a single slip. Let E_1 be the event that you win prize #1, and define E_2 and E_3 analogously for prizes 2 and 3, respectively.

a. Can the event E_1 *and* E_2 occur? What is $P(E_1 \text{ and } E_2)$? Are E_1 and E_2 independent events?

b. Are E_1 and E_3 independent events? Are E_2 and E_3 independent events?

c. Is it true that $P(E_1 \text{ and } E_2 \text{ and } E_3) = P(E_1) \cdot P(E_2) \cdot P(E_3)$? Note: This shows that events can be "pairwise independent" without being "mutually" independent, as required by our definition.

4.55 A quiz consists of five multiple choice questions. Suppose that in grading the quizzes, questions are marked independently of one another, and the probability that any particular question is marked correctly is .9.
  a. What is the probability that all five questions on a particular quiz are marked correctly?
  b. What is the probability that at least one marking error is made on a particular quiz?
  c. What is the probability that all five questions are incorrectly marked?

4.56 At a certain gas station, 40% of all customers fill their tanks. Of those who fill their tanks, 80% pay with a credit card.
  a. What is the probability that a randomly selected customer fills his/her tank and pays with a credit card?
  b. If three customers are randomly selected, what is the probability that all three fill their tanks and pay with a credit card?

4.57 Components of a certain type are shipped to a supplier in batches of ten. Suppose that 50% of all batches contain no defective components, 30% contain one defective component, and 20% contain two defective components. Two components from a batch are randomly selected and tested.
  a. If the batch from which the components were selected actually contains two defectives, what is the probability that neither of these is selected for testing?
  b. What is the probability that the batch contains two defectives and neither of these is selected for testing?
  c. What is the probability that neither component selected for testing is defective? Hint: This could happen with any one of the three types of batches. A tree diagram might help.

4.58 On Monday morning a loan officer at a credit union is given 15 auto loan applications to process. Three of these are for 36 months, five are for 48 months, and the other seven are for 60 months. Only two applications can be processed that morning. Suppose that the two to be processed are randomly selected from among the 15.
  a. Let A_1 be the event that the first application processed is for 36 months, and let A_2 be the event that the second application processed is for 36 months. What is $P(A_1)$? What is $P(A_2|A_1)$? Describe the event A_1 *and* A_2 in words, and calculate the probability of this event.
  b. What is the probability that both applications processed that morning are for loans of the same duration?

4.59 A company has just placed an order with a supplier for two different products. Let

  E = the event that the first product is out of stock

  F = the event that the second product is out of stock

  Suppose that $P(E) = .3$, $P(F) = .2$, and the probability that at least one product is out of stock is .4.

  a. What is the probability that both products are out of stock?
  b. Are E and F independent events?
  c. Given that the first product is in stock, what is the probability that the second is also?

4.60 An appliance dealer sells three different brands of refrigerator (B_1, B_2, and B_3). Each brand comes in either a side-by-side model (S) or a top-freezer model (T). The accompanying table provides information on probabilities of the various brand–model combinations for a randomly selected purchaser. In addition, $P(S) = .20$, $P(B_1|S) = .50$, and $P(T|B_2) = .76$.

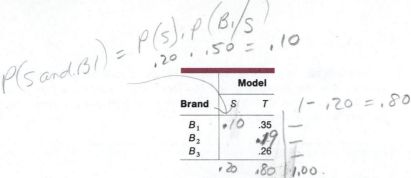

$$P(s \text{ and } B_1) = P(s) \cdot P(B_1/s)$$
$$\underset{.20}{} \cdot \underset{.50}{} = .10$$

Brand	Model S	Model T
B_1	•10	.35
B_2		.19
B_3		.26

$1 - .20 = .80$

.20 .80 1.00

a. Determine the probabilities of the remaining four brand–model combinations.

b. Given that a randomly selected purchaser chose a top-freezer model, what is the probability that it was not a brand B_1 refrigerator?

4.61 Suppose that three cards are drawn without replacement from a well-mixed deck. Let E denote the event that the first card is a face card, let F be the event that the second card is a face card, and let G be the event that the third card is a face card.
a. What is $P(E)$?
b. What is $P(F|E)$?
c. What is $P(G|E \text{ and } F)$?
d. What is the probability that all three cards selected are face cards?

4.62 Suppose three cards are selected from a well-mixed deck without replacement.
a. Use the general multiplication rule for three events to calculate the probability that all three cards are hearts.
b. Calculate the probability that all three cards come from the same suit.
c. If five cards are dealt from the deck (a poker hand), determine the probability that all are hearts. Hint: Extend the multiplication rule again.

4.63 A sales representative for a company that manufactures water treatment systems can make sales presentations to either one or two potential customers on any given day. The probability that there will be a single presentation is .6. If just one presentation is made, it is successful with probability .8. If two presentations are made, the first is successful with probability .8, the second is successful with probability .8, and these two events (*success on 1st* and *success on 2nd*) are independent.
a. What is the probability that just one presentation is made on a particular day and it is successful?
b. What is the probability that two presentations are made and both are successful?
c. What is the probability that two presentations are made and neither is successful?
d. What is the probability that the sales rep has no successful presentations on a given day? Hint: Construct a tree diagram with first-generation branches for one presentation (E_1) and two presentations (E_2). The branches leading out from the E_2 branch should specify all possibilities for each presentation.

4.64 A transmitter is sending a message by using a binary code, namely, a sequence of 0s and 1s. Each transmitted "bit" (0 or 1) must pass through three relays in order to reach the receiver. At each relay, the probability is .20 that the bit sent on will be different from the bit received (a reversal). Assume that the relays operate independently of one another.

transmitter ⟶ relay 1 ⟶ relay 2 ⟶ relay 3 ⟶ receiver

a. If a 1 is sent from the transmitter, what is the probability that a 1 is sent on by all three relays?
b. If a 1 is sent from the transmitter, what is the probability that a 1 is received by the receiver? Hint: The eight experimental outcomes can be displayed on a tree diagram with three generations of branches, one generation for each relay.

4.65 Referring to Exercise 4.64, suppose that 70% of all bits sent from the transmitter are 1s. If a 1 is received by the receiver, what is the probability that a 1 was sent?

4.66   Suppose that all outcomes in one event, A, are contained in another event, B (for example, A = jack is selected and B = face card is selected). The accompanying Venn diagram shows that B consists of two nonoverlapping pieces: one is A, and the other is *(not A) and B*.

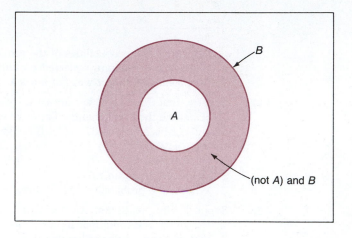

What is the relationship between $P(B)$, $P(A)$, and $P((not\ A)\ and\ B)$? What does this imply about the relationship between $P(A)$ and $P(B)$?

4.67   If $P(A) = .5$, $P(B) = .4$, and A and B are independent, could they possibly be disjoint? Explain.

REFERENCES   Devore, Jay L. *Probability and Statistics for Engineering and the Sciences*, 2nd Ed. Monterey, Calif. Brooks/Cole Publishing Co., 1987. (The treatment of probability in this source is more comprehensive and at a somewhat higher mathematical level than is ours.)

Mosteller, Frederick, Robert Rourke, and George Thomas. *Probability with Statistical Applications*. Reading, Mass.: Addison-Wesley Publishing Co., 1970. (Although a bit old, there is no more recently published book that provides a better in-depth coverage of probability at a very modest mathematical level.)

C H A P T E R 5

RANDOM VARIABLES AND DISCRETE PROBABILITY DISTRIBUTIONS

INTRODUCTION The focus of interest in a chance experiment is frequently on some numerical aspect of the outcome. An environmental scientist who obtains an air sample from a specified location might be especially concerned with the concentration of ozone (a major constituent of smog). A quality control inspector who must decide whether to accept a large shipment of components may base the decision on the number of defectives in a group of 20 components randomly selected from the shipment.

Prior to selection of the air sample, there is uncertainty as to what value of ozone concentration will result. Similarly, the number of defective components among the 20 selected might be any integer between zero and 20. Because the value of a variable quantity such as ozone concentration or number of defectives is subject to uncertainty, it is called a *random variable*.

In this chapter we first distinguish between two general types of random variables, *discrete* and *continuous*. We then consider probability models for discrete random variables. Any particular such model describes how probability is distributed among the possible values of the variable. Once a model or *probability distribution* has

123

been specified, it can be used to calculate probabilities of various events involving the variable, such as the probability that at most two of the selected components are defective. We next introduce the mean value and standard deviation as important summary quantities for a probability distribution. In the final section of the chapter we develop the *binomial distribution*, the most important discrete distribution in statistics. Continuous random variables and distributions are discussed in Chapter 6.

5.1 RANDOM VARIABLES

In most chance experiments there will be one or more variable quantities on which an investigator will focus attention. As an example, consider a management consultant who is studying the operation of a supermarket. The chance experiment might involve randomly selecting a customer leaving the store. One interesting numerical variable would be the number of items x purchased by the customer. Possible values of this variable are 0 (a frustrated customer), 1, 2, 3, and so on. Until a customer is selected and the number of items counted, there is uncertainty as to what value of x will result. Another variable of potential interest is the time y (min) spent in a checkout line. One possible y value is 3.0 and another is 4.0, but *any* other number between 3.0 and 4.0 is also a possibility. Whereas possible values of x are isolated points on the number line, possible y values form an entire interval (a continuum) on the number line.

DEFINITION

A numerical variable whose value depends on the outcome of a chance experiment is called a **random variable**. A random variable is **discrete** if its set of possible values is a collection of isolated points on the number line. The variable is **continuous** if its set of possible values is an entire interval on the number line.

In practice, a discrete random variable almost always arises in connection with counting (the number of items purchased, the number of checkstands open, the number of broken eggs in a carton, etc.). A continuous random variable is one whose value is typically obtained by measurement (the temperature in a freezer compartment, the weight of a pineapple, the amount of time spent in the store, etc.). Because there is a limit to the accuracy of any measuring instrument, such as a watch or a scale, it may seem that any variable should be regarded as discrete. However, when there are a very large number of very closely spaced values, the variable's behavior is most easily studied by conceptualizing it as continuous. (This is one reason that the branch of mathematics called *calculus* was developed.)

EXAMPLE 1

Consider an experiment in which the type of dryer—electric (E) or gas (G)—chosen by each of three successive purchasers at a large appliance store is noted. Define a random variable x by

x = the number of gas dryers purchased by the three customers

The experimental outcome in which the first and third customers purchase a gas dryer and the second one an electric dryer can be abbreviated GEG. The associated x value is 2, since two of the three customers selected a gas model. Similarly, the x value for the outcome GGG (all three purchase a gas model) is 3. We display each of the eight possible experimental outcomes and the corresponding value of x.

Outcome	EEE	GEE	EGE	EEG	GGE	GEG	EGG	GGG
x value	0	1	1	1	2	2	2	3

There are only four possible x values—0, 1, 2, and 3—and these are isolated points on the number line. Thus x is a discrete random variable.

In some situations the random variable of interest is discrete but the number of possible values is not finite. Here is an example.

EXAMPLE 2

Each trial run of a power generating unit can be either a success (S) or a failure (F). The unit's purchaser has specified that the unit will not be accepted until two consecutive successful trial runs have been observed. The random variable of interest is

x = the number of trial runs that must be made prior to acceptance of the unit

The simplest experimental outcome is SS, where the first two trial runs are both successful; in this case the associated x value is 2. There is also just one outcome for which $x = 3$, namely, FSS (for $x = 3$, the second and third runs must be S's; if the first were also S, the experiment would have terminated with $x = 2$). Some other possible x values and outcomes are as follows:

x value	Outcomes
4	FFSS, SFSS
5	FFFSS, SFFSS, FSFSS
6	FFFFSS, SFFFSS, FSFFSS, SFSFSS, FFSFSS
⋮	⋮
10	FFFFFFFFSS and many others
⋮	⋮

Any positive integer that is at least 2 (such as 2, 3, 4, 5, etc.) is a possible x value. For example, $x = 50$ if the first 48 trials result in F's and the 49th and 50th trials are S's (There are many other outcomes for which $x = 50$.) Similarly, one way to have $x = 127$ is to have 125 F's followed by two S's. Because the values 2, 3, 4, . . . are isolated points on the number line (x is determined by counting), x is a discrete random variable. But there is no upper limit to the number of possible x values.

Now consider an example of a continuous random variable.

EXAMPLE 3   A shot is fired at a circular target with a radius of one foot. Suppose that the shot is sure to land someplace on the target. As illustrated in Figure 1, the experimental outcome (landing point) can be described by a horizontal coordinate in combination with a vertical coordinate. Thus let x be the number on the horizontal axis that is directly below or above the landing point. Similarly, let y be the number on the vertical axis that is directly left or right of the landing point. Then the landing point is identified by the pair (x, y).

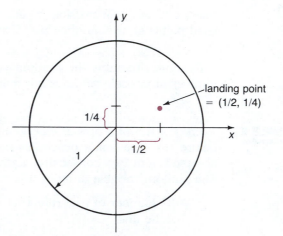

FIGURE 1

The circular target and experimental outcome $(\frac{1}{2}, \frac{1}{4})$.

Define a random variable z by

z = the distance from the target's center to the point at which the shot lands

The relation between the outcome (x, y) and the corresponding z value is

$$z = \sqrt{(\text{horiz. coord.})^2 + (\text{vert. coord.})^2}$$
$$= \sqrt{x^2 + y^2}$$

Thus the outcome $(\frac{1}{2}, \frac{1}{4})$ gives us

$$z = \sqrt{(\tfrac{1}{2})^2 + (\tfrac{1}{4})^2} = \sqrt{.3125} = .559$$

The smallest possible z value is 0, which occurs when the landing point is the center point. The maximum z value is 1, which occurs if the landing point is anywhere on the edge of the target. Relative to these two extremes, z can be *any* number between 0 and 1, comprising an entire interval (continuum) on the number line. Thus z is a continuous random variable.

___

In this chapter and the next we shall see how probability distributions for discrete random variables are quite different from those for continuous random variables.

EXERCISES 5.1–5.7

SECTION 5.1

5.1 Say whether each of the following random variables is discrete or continuous.
   a. The number of defective tires on a car *discr.*
   b. The body temperature of a hospital patient *Cont.*
   c. The number of pages in a book *discr*
   d. The number of draws with replacement from a deck of cards until a heart is selected *discr.*
   e. The lifetime of a light bulb *cont.*

5.2 Classify each of the following random variables as either discrete or continuous.
   a. The fuel efficiency (mpg) of an automobile *cont*
   b. The amount of rainfall at a particular location during the next year *cont.*
   c. The distance that a person throws a baseball *Cont*
   d. The number of questions asked during a one-hour lecture *discr.*
   e. The tension (lb/in.²) at which a tennis racket is strung *cont.*
   f. The amount of water used by a household during a given month *cont.*
   g. The number of traffic citations issued by the Highway Patrol in a particular county on a given day *discr.*

5.3 Starting at a particular time, each car entering an intersection is observed to see whether it turns left (L), right (R), or goes straight ahead (S). The experiment terminates as soon as a car is observed to go straight. Let y denote the number of cars observed. What are possible y values? List five different outcomes and their associated y values.

5.4 A point is randomly selected from the interior of a square, as pictured. Let x denote the distance from the lower left corner A of the square to the selected point. What are possible values of x? Is x a discrete or continuous variable?

B _____ C

| | |

A |_____| D

1 ft

5.5 A point is randomly selected on the surface of a lake that has maximum depth of 100 feet. Let $y =$ the depth of the lake at the randomly chosen point. What are possible values of y? Is y discrete or continuous?

5.6 A person stands at the corner marked A of the square pictured in Exercise 5.4 and tosses a coin. If it lands H, the person moves one corner clockwise, to B. If the coin lands T, the person moves one corner counterclockwise, to D. This process is then repeated until the person arrives back at A. Let y denote the number of coin tosses. What are possible values of y? Is y discrete or continuous?

5.7 A box contains four slips of paper marked 1, 2, 3, and 4. Two slips are selected without replacement. List the possible values for each of the following random variables.
   a. $x =$ the sum of the two numbers
   b. $y =$ the difference between the first and second numbers
   c. $z =$ the number of slips selected that show an even number
   d. $w =$ the number of slips selected that show a 4

5.2 PROBABILITY DISTRIBUTIONS FOR DISCRETE RANDOM VARIABLES

Let x be a discrete random variable associated with a particular chance experiment. The outcome that occurs when the experiment is performed determines which value of x is observed. The total proability for all outcomes is 1. The probability distribution of x describes how much of this probability is placed on each possible x value.

EXAMPLE 4 Six lots of components are ready to be shipped by a certain supplier. The number of defective components in each lot is as follows:

Lot:	1	2	3	4	5	6
Number of defectives:	0	2	0	1	2	0

One of these lots is to be randomly selected for shipment to a particular customer. Let x be the number of defectives in the selected lot. The three possible x values are 0, 1, and 2. Of the six equally likely simple events, three result in $x = 0$, one in $x = 1$, and the other two in $x = 2$. Thus

$P(x = 0) = P(\text{lot 1 or 3 or 6 is sent}) = 3/6 = .500$

$P(x = 1) = P(\text{lot 4 is sent}) = 1/6 = .167$

$P(x = 2) = P(\text{lot 2 or 5 is sent}) = 2/6 = .333$

That is, a probability of .500 is distributed to the x value 0, a probability of .167 is placed on the x value 1, and the remaining probability, .333, is associated with the x value 2.

▌▌▌ DEFINITION

The **probability distribution of a discrete random variable x** gives the probability associated with each possible x value. Each probability is the limiting relative frequency of occurrence of the corresponding x value when the experiment is repeatedly performed.

If 2 is one possible value of x, we will often write $p(2)$ in place of $P(x = 2)$. Similarly, $p(5)$ will denote the probability that $x = 5$, and so on.

EXAMPLE 5
Each of four randomly selected customers purchasing a dryer at a certain store chooses either an electric (E) or a gas (G) model. Assume that these customers make their choices independently of one another and that 40% of all customers select an electric model. This implies that for any particular one of the four customers, $P(E) = .4$ and $P(G) = .6$. One possible experimental outcome is EGGE, where the first and fourth customers select electric models and the other two choose gas models. Independence of the customers' choices implies that

$$P(\text{EGGE}) = P(\text{1st chooses E and 2nd chooses G and}$$
$$\text{3rd chooses G and 4th chooses E})$$
$$= P(E) \cdot P(G) \cdot P(G) \cdot P(E)$$
$$= (.4)(.6)(.6)(.4)$$
$$= .0576$$

Similarly,

$$P(\text{EGEG}) = P(E) \cdot P(G) \cdot P(E) \cdot P(G)$$
$$= (.4)(.6)(.4)(.6)$$
$$= .0576$$

and

$$P(\text{GGGE}) = (.6)(.6)(.6)(.4) = .0864$$

Now, let the random variable of interest be

$x =$ the number of electric dryers purchased by the four customers

Table 1 displays all 16 experimental outcomes, the probability of each one, and the x value associated with each outcome. The probability distribution of x is easily obtained from this information.

Consider the smallest possible x value, 0. The only outcome for which $x = 0$ is GGGG, so

$$p(0) = P(x = 0) = P(\text{GGGG}) = .1296$$

TABLE 1

Outcomes and Probabilities for Example 4

Outcome	Probability	x Value	Outcome	Probability	x Value
GGGG	.1296	0	GEEG	.0576	2
EGGG	.0864	1	GEGE	.0576	2
GEGG	.0864	1	GGEE	.0576	2
GGEG	.0864	1	GEEE	.0384	3
GGGE	.0864	1	EGEE	.0384	3
EEGG	.0576	2	EEGE	.0384	3
EGEG	.0576	2	EEEG	.0384	3
EGGE	.0576	2	EEEE	.0256	4

There are four different outcomes for which $x = 1$, so $p(1)$ results from summing the four corresponding probabilities:

$$p(1) = P(x = 1) = P(\text{EGGG or GEGG or GGEG or GGGE})$$
$$= P(\text{EGGG}) + P(\text{GEGG}) + P(\text{GGEG}) + P(\text{GGGE})$$
$$= .0864 + .0864 + .0864 + .0864$$
$$= 4(.0864)$$
$$= .3456$$

Similarly,

$$p(2) = P(\text{EEGG}) + \cdots + P(\text{GGEE}) = 6(.0576) = .3456$$
$$p(3) = 4(.0384) = .1536$$
$$p(4) = .0256$$

The probability distribution of x is summarized in the accompanying table.

x value	0	1	2	3	4
$p(x)$ = probability of value	.1296	.3456	.3456	.1536	.0256

To interpret $p(3) = .1536$, think of performing the experiment repeatedly, each time with a different group of four customers. In the long run, 15.36% of these groups will have exactly three customers purchasing an electric dryer.

The probability distribution can be used to determine probabilities of various events involving x. For example, the probability that at least two of the four customers choose electric models is

$$P(2 \leq x) = P(x = 2 \text{ or } 3 \text{ or } 4)$$
$$= p(2) + p(3) + p(4)$$
$$= .5248$$

Thus, in the long run, 52.48% of the time a group of four dryer purchasers will include at least two who select electric models.

In tabular form, the probability distribution for a discrete random variable looks exactly like a relative frequency distribution of the sort discussed in Chapter 2. There we introduced a histogram as a pictorial representation of a relative frequency distribution. An analogous picture for a discrete probability distribution is called a **probability histogram**. The picture has a rectangle centered above each possible value of x, and the area of each rectangle is the probability of the corresponding value. Figure 2 displays the probability histogram for the probability distribution of Example 5.

FIGURE 2

Probability histogram for the distribution of Example 5.

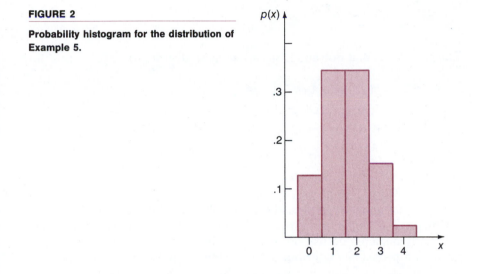

EXAMPLE 6

Concert tickets purchased by a group of three couples are for seats 1–6 in a single row. Suppose that the six tickets are distributed in a completely random fashion (by shuffling and dealing, as with cards). Define a discrete random variable y by

$y =$ the number of seats that separate Bob and Lygia

We wish to obtain the probability distribution of y.

Since the seat numbers occupied by the other four people are irrelevant, think of an experimental outcome as indicating just the seat numbers of Bob and Lygia. There are then 15 possible outcomes:

Outcome	y value	Outcome	y value
seats 1 and 2	0	seats 2 and 6	3
seats 1 and 3	1	seats 3 and 4	0
seats 1 and 4	2	seats 3 and 5	1
seats 1 and 5	3	seats 3 and 6	2
seats 1 and 6	4	seats 4 and 5	0
seats 2 and 3	0	seats 4 and 6	1
seats 2 and 4	1	seats 5 and 6	0
seats 2 and 5	2		

The manner in which tickets are distributed implies that each outcome has probability 1/15. The probability associated with the y value 2 is then

$$p(2) = P(y = 2)$$
$$= P((1 \text{ and } 4) \text{ or } (2 \text{ and } 5) \text{ or } (3 \text{ and } 6))$$
$$= 3/15$$

We can now tabulate the complete probability distribution.

y	0	1	2	3	4
$p(y)$	5/15	4/15	3/15	2/15	1/15

The probability that Bob and Lygia are separated by at most two seats is thus

$$P(y \leq 2) = P(y = 0 \text{ or } 1 \text{ or } 2)$$
$$= p(0) + p(1) + p(2)$$
$$= 12/15$$
$$= .80$$

In the foregoing examples, the probability distribution was derived by starting with a simple experimental situation and applying basic probability rules. Often such a derivation is not possible. Instead an investigator conjectures a probability distribution consistent with empirical evidence and prior knowledge. The only conditions that must be satisfied are

(i) $p(x) \geq 0$ for every x value;

(ii) $\sum\limits_{\substack{\text{all } x \\ \text{values}}} p(x) = 1$

EXAMPLE 7   A consumer organization that evaluates new automobiles customarily reports the number of major defects on each car examined. Let x denote the number of major defects on a randomly selected car of a certain type. One possible probability distribution is

x	0	1	2	3	4	5	6	7	8	9	10
$p(x)$	.041	.130	.209	.223	.178	.114	.061	.028	.011	.004	.001

The corresponding probability histogram appears in Figure 3. The probability that the number of major defects is between 2 and 5 inclusive is

$$P(2 \leq x \leq 5) = p(2) + p(3) + p(4) + p(5) = .724$$

If car after car of this type were examined, in the long run, 72.4% would have 2, 3, 4, or 5 major defects.

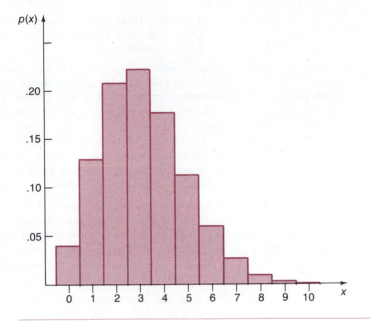

FIGURE 3

Probability histogram for the distribution of the number of major defects.

EXERCISES   5.8–5.19

SECTION 5.2

5.8   Let x be the number of courses for which a randomly selected student at a certain university is registered. The probability distribution of x appears in the accompanying table.

x	1	2	3	4	5	6	7
$p(x)$	.02	.03	.09	.25	.40	.16	.05

a. What is $P(x = 4)$? ⟨.25⟩
b. What is $P(x \leq 4)$? ⟨.39⟩
c. What is the probability that the selected student is taking at most five courses? ⟨.79⟩
d. What is the probability that the selected student is taking at least five courses? ⟨.61⟩ More than five courses? ⟨.21⟩
e. Calculate $P(3 \leq x \leq 6)$ and $P(3 < x < 6)$. Explain in words why these two probabilities are different. ⟨.9⟩ ⟨.65⟩

5.9   Airlines sometimes overbook flights. Suppose that for a plane with 100 seats, an airline takes 110 reservations. Define the variable x as the number of people who actually show up for a sold-out flight. From past experience, the probability distribution of x is given in the following table.

x	95	96	97	98	99	100	101	102	103	104	105	106	107	108	109	110
$p(x)$	.05	.10	.12	.14	.24	.17	.06	.04	.03	.02	.01	.005	.005	.005	.0037	.0013

a. What is the probability that the airline can accommodate everyone who shows up for the flight?
b. What is the probability that not all passengers can be accommodated?

c. If you are trying to get a seat on such a flight and you are number 1 on the standby list, what is the probability that you will be able to take the flight? What if you are number 3?

5.10 Many manufacturers have quality control programs that include inspection of incoming materials for defects. Suppose that a computer manufacturer receives computer boards in lots of five. Two boards are selected from each lot for inspection. We can represent possible outcomes of the selection process by pairs. For example, the pair (1, 2) represents the selection of boards 1 and 2 for inspection.
a. List the ten different possible outcomes.
b. Suppose that boards 1 and 2 are the only defective boards in a lot of five. Two boards are to be chosen at random. Define x to be the number of defective boards observed among those inspected. Find the probability distribution of x.

5.11 Simulate the experiment described in Exercise 5.10 using five slips of paper with two marked defective and three marked nondefective. Place the slips in a box, mix them well, and draw out two. Record the number of defectives. Replace the slips and repeat until you have 50 observations on the variable x. Construct a relative frequency distribution for the 50 observations and compare this with the probability distribution obtained in Exercise 5.10.

5.12 Suppose that on any given day, a particular stock has probability .6 of going up in price. If the stock is observed on three randomly chosen days and x is defined to be the number of days on which the stock went up, find the probability distribution of x. (Hint: One outcome is UDD, with $x = 1$; assuming independence, this outcome has probability $(.6)(.4)(.4) = .096$.)

5.13 Some parts of California are particularly earthquake-prone. Suppose that in one such area, 20% of all homeowners are insured against earthquake damage. Four homeowners are to be selected at random; let x denote the number among the four who have earthquake insurance.
a. Find the probability distribution of x. (Hint: Let S denote a homeowner who has insurance and F one who doesn't. Then one possible outcome is SFSS, with probability $(.2)(.8)(.2)(.2)$ and associated x value 3. There are 15 other outcomes.)
b. What is the most likely value for x?
c. What is the probability that at least two of the four selected have earthquake insurance?

5.14 Let $x =$ the number of underinflated tires on a randomly selected automobile.
a. Which of the following is a legitimate probability distribution for x, and why are the other two not allowed?

x	0	1	2	3	4
$p(x)$	.3	.2	.1	.05	.05
$p(x)$	.4	.1	.1	.1	.3
$p(x)$	.4	.1	.2	.1	.3

b. For the legitimate distribution of part (a), compute $P(2 \le x \le 4)$, $P(x \le 2)$, and $P(x \ne 0)$.
c. If $p(x) = c \cdot (5 - x)$ for $x = 0, 1, \ldots, 4$, what is the value of the constant c? Hint: $\sum p(x) = 1$.

5.15 A box contains five slips of paper. These slips are marked $1, $1, $1, $10, and $25. The winner of a contest will select two slips of paper at random and will then get the larger of the dollar amounts on the two slips. Define a random variable w by

$w =$ the amount awarded

Determine the probability distribution of w. Hint: Think of the slips as numbered 1, 2, 3, 4, and 5, so that an outcome of the experiment consists of two of these numbers.

5.16 Components coming off an assembly line are either free of defects (S, for success) or defective (F, for failure). Suppose that 70% of all such components are defect-free. Components are independently selected and tested one by one. Let y denote the number of components that must be tested until a defect-free component is obtained.

    a. What is the smallest possible y value, and what experimental outcome gives this y value? What is the second smallest y value, and what outcome gives rise to it?

    b. What is the set of all possible y values?

    c. Determine the probability of each of the five smallest y values. You should see a pattern that leads to a simple formula for $p(y)$, the probability distribution of y.

5.17 A fair die is rolled. If the outcome is 1, 2, or 3, the die is rolled a second time, and this terminates the experiment. If the outcome of the first roll is 4, 5, or 6, the experiment terminates immediately. Let x denote the number on the upturned face of the die when the experiment terminates. Obtain the probability distribution of x: Hint: A tree diagram might help; the initial branches should refer to the first roll, and some of these will have second-generation branches attached to them.

Bullshit.

5.18 A contractor is required by a county planning department to submit anywhere from one to five forms (depending on the nature of the project) in applying for a building permit. Let y = the number of forms required of the next applicant. The probability that y forms are required is known to be proportional to y; that is, $p(y) = ky$ for $y = 1, \ldots, 5$.

    a. What is the value of k? Hint: $\sum p(y) = 1$.

    b. What is the probability that at most three forms are required?

    c. What is the probability that between two and four forms (inclusive) are required?

    d. Could $p(y) = y^2/50$ for $y = 1, \ldots, 5$ be the probability distribution of y? Explain.

5.19 A library subscribes to two different weekly news magazines, each of which is supposed to arrive in Wednesday's mail. In actuality, each one could arrive on Wednesday, Thursday, Friday, or Saturday. Suppose that the two arrive independently of one another and that for each one,

$$P(W) = .4, \qquad P(T) = .3, \qquad P(F) = .2, \qquad P(S) = .1$$

Define a random variable y by y = the number of days beyond Wednesday that it takes for both magazines to arrive. For example, if the first arrives on Friday and the second on Wednesday, then $y = 2$, whereas $y = 1$ if both magazines arrive on Thursday. Obtain the probability distribution of y. Hint: Draw a tree diagram with two generations of branches, the first labeled with arrival days for magazine 1 and the second for magazine 2.

5.3 THE MEAN VALUE AND THE STANDARD DEVIATION

We study a random variable x, such as the number of insurance claims made by a homeowner (a discrete variable) or the tensile strength of wire (a continuous variable), to learn something about how its values are distributed along the measurement scale. The sample mean $\bar{x}$ and sample standard deviation s summarize center and spread in a sample of x values. Similarly, the mean value and standard deviation of a random variable describe where the variable's probability distribution is centered and the extent to which it spreads out about the center.

THE MEAN VALUE OF A DISCRETE RANDOM VARIABLE

Consider the experiment consisting of the random selection of an automobile licensed in a particular state. Let the discrete random variable x be the number of low-beam headlights on the selected car that need adjustment. Possible x values are 0, 1, and 2, and the probability distribution of x and the corresponding probability histogram might look as follows:

x value	0	1	2
Probability	.5	.3	.2

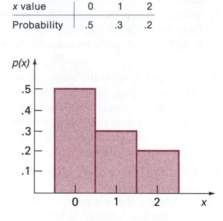

In a sample of 100 such cars, the sample relative frequencies might differ somewhat from these probabilities (which are the limiting relative frequencies). We might see 46 cars with $x = 0$ (relative frequency .46), 33 cars with $x = 1$, and 21 cars with $x = 2$. The sample average value of x for these 100 observations is then the sum of 46 zeros, 33 ones, and 21 twos, all divided by 100:

$$\bar{x} = \frac{(46)(0) + (33)(1) + (21)(2)}{100}$$

$$= \left(\frac{46}{100}\right)(0) + \left(\frac{33}{100}\right)(1) + \left(\frac{21}{100}\right)(2)$$

$$= \left(\begin{array}{c}\text{rel. freq.}\\\text{of 0}\end{array}\right)(0) + \left(\begin{array}{c}\text{rel. freq.}\\\text{of 1}\end{array}\right)(1) + \left(\begin{array}{c}\text{rel. freq.}\\\text{of 2}\end{array}\right)(2) = .75$$

As the sample size increases, each relative frequency will approach the corresponding probability. In a very long sequence of experiments, the value of $\bar{x}$ will approach

$$\left(\begin{array}{c}\text{probability}\\\text{that } x = 0\end{array}\right)(0) + \left(\begin{array}{c}\text{probability}\\\text{that } x = 1\end{array}\right)(1) + \left(\begin{array}{c}\text{probability}\\\text{that } x = 2\end{array}\right)(2)$$

$$= (.5)(0) + (.3)(1) + (.2)(2)$$

$$= .70 = \text{mean value of } x$$

Notice that the expression for $\bar{x}$ here is a weighted average of possible x values; the weight for each value is the observed relative frequency. Similarly, the mean value of x is a weighted average, but now the weights are the probabilities from the probability distribution.

▐▐▐ **DEFINITION**

The **mean value of a discrete random variable x,** denoted by μ_x (or just μ when the identity of x is obvious), is computed by first multiplying each possible x value by the probability of observing that value and then adding the resulting quantities. Symbolically,

$$\mu_x = \sum_{\substack{\text{all} \\ \text{possible} \\ x \text{ values}}} x \cdot (\text{probability of } x) = \sum x \cdot p(x)$$

The phrase "expected value" is sometimes used in place of mean value, and $E(x)$ is alternative notation for μ_x.

EXAMPLE 8

Individuals applying for a certain license are allowed up to four attempts to pass the licensing exam. Let x denote the number of attempts made by a randomly selected applicant. The probability distribution of x is as follows:

x	1	2	3	4
$p(x)$	.10	.20	.30	.40

Thus x has mean value

$$\begin{aligned}
\mu_x &= \sum_{x=1,\dots,4} x \cdot p(x) \\
&= (1) \cdot p(1) + (2) \cdot p(2) + (3) \cdot p(3) + (4) \cdot p(4) \\
&= (1)(.10) + (2)(.20) + (3)(.30) + (4)(.40) \\
&= .10 + .40 + .90 + 1.60 \\
&= 3.00
\end{aligned}$$

It is no accident that the symbol μ for the mean value is the same symbol used earlier for a population mean. When the probability distribution describes how x values are distributed among the members of a population (so probabilities are population relative frequencies), the mean value of x is exactly the average value of x in the population.

EXAMPLE 9

At one minute after birth and again at five minutes, each newborn child is given a numerical rating called an *Apgar score*. Possible values of this score are 0, 1, 2, . . . , 9, and 10. A child's score is determined by five factors: muscle tone, skin color, respiratory effort, strength of heartbeat, and reflex, with a high score indicating a healthy infant. Let the random variable x denote the Apgar score (at 1 min) of a randomly selected new-born infant at a particular hospital, and suppose that x has the following probability distribution.

x	0	1	2	3	4	5	6	7	8	9	10
$p(x)$	.002	.001	.002	.005	.02	.04	.17	.38	.25	.12	.01

The mean value of x is

$$\mu = (0) \cdot p(0) + (1) \cdot p(1) + \cdots + (9) \cdot p(9) + (10) \cdot p(10)$$
$$= (0)(.002) + (1)(.001) + \cdots + (9)(.12) + (10)(.01)$$
$$= 7.16$$

The average Apgar score for a sample of newborn children born at this hospital may be $\bar{x} = 7.05$, $\bar{x} = 8.30$, or any other number between 0 and 10. However, as child after child is born and rated, the average score will approach the value 7.16. This value can be interpreted as the mean Apgar score for the population of all babies born at this hospital.

THE STANDARD DEVIATION OF A DISCRETE RANDOM VARIABLE

The mean value μ provides only a partial summary of a probability distribution. Two different distributions may both have the same value of μ, yet a long sequence of sample values from one distribution may exhibit considerably more variability than a long sequence of values from the other distribution.

EXAMPLE 10

A television manufacturer receives certain components in lots of four from two different suppliers. Let x and y denote the number of defective components in randomly selected lots from the first and second suppliers, respectively. The probability distributions and associated probability histograms for x and y are given in Figure 4.

x	0	1	2	3	4
$p(x)$	.4	.3	.2	.1	0

y	0	1	2	3	4
$p(y)$	.2	.6	.2	0	0

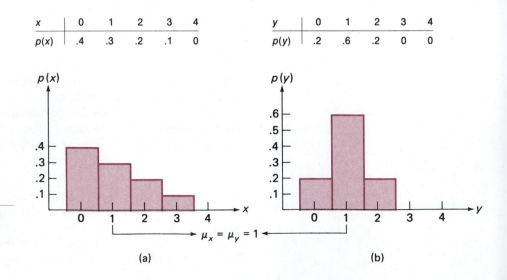

FIGURE 4

Probability distribution for the number of defective components:
(a) In a lot from supplier 1;
(b) In a lot from supplier 2.

The mean values of both x and y are 1, so for either supplier the long-run average number of defectives per lot is 1. However, the two probability histograms show that the probability distribution for the second supplier is concentrated nearer the mean value than is the first supplier's distribution. The greater spread of the first distribution implies that there will be more variability in a long sequence of observed x values than in an observed sequence of y values. For example, the y sequence will contain no 3s, whereas in the long run, 10% of the observed x values will be 3.

As with s^2 and s, the variance and standard deviation of x involve squared deviations from the mean. A value far from the mean results in a large squared deviation. However, such a value contributes substantially to variability in x only if the probability associated with that value is not too small. For example, if $\mu_x = 1$ and $x = 25$ is a possible value, the squared deviation is $(25 - 1)^2 = 576$. If, however, $P(x = 25) = .000001$, the value 25 will hardly ever be observed, so it won't contribute much to variability in a long sequence of observations. This is why each squared deviation is multiplied by the probability associated with the value (and thus weighted) to obtain a measure of variability.

▌▌▌ DEFINITION

The **variance of a discrete random variable x**, denoted by σ_x^2 or just σ^2, is computed by first subtracting the mean from each possible x value to obtain the deviations, then squaring each deviation and multiplying the result by the probability of the corresponding x value, and finally adding these quantities. Symbolically,

$$\sigma^2 = \sum_{\substack{\text{all} \\ \text{possible} \\ x \text{ values}}} (x - \mu)^2 \cdot p(x)$$

The **standard deviation of x**, denoted by σ_x or just σ, is the square root of the variance.

When the probability distribution describes how x values are distributed among members of a population (so probabilities are population relative frequencies), σ^2 and σ are the population variance and standard deviation (of x), respectively.

EXAMPLE 11 (*Example 10 continued*) For $x =$ the number of defectives in a lot from the first supplier,

$$\sigma_x^2 = (0 - 1)^2 \cdot p(0) + (1 - 1)^2 \cdot p(1) + (2 - 1)^2 \cdot p(2) + (3 - 1)^2 \cdot p(3)$$
$$= (1)(.4) + (0)(.3) + (1)(.2) + (4)(.1)$$
$$= 1.0$$

so $\sigma_x = 1.0$. For $y =$ the number of defectives in a lot from the second supplier,

$$\sigma_y^2 = (0 - 1)^2 \cdot (.2) + (1 - 1)^2(.6) + (2 - 1)^2(.2)$$
$$= .4$$

Then $\sigma_y = \sqrt{.4} = .632$. The fact that $\sigma_x > \sigma_y$ confirms the impression conveyed by Figure 4 concerning the variability of x and y.

EXERCISES 5.20–5.31

SECTION 5.3

5.20 A personal computer salesperson working on a commission receives a fixed amount for each system sold. Suppose that for a given month, the probability distribution of $x =$ the number of systems sold is given in the accompanying table.

x	1	2	3	4	5	6	7	8
$p(x)$	.05	.10	.12	.30	.30	.11	.01	.01

a. Find the mean value of x (the mean number of systems sold).
b. What is the probability that x is within 2 of its mean value?
c. Find the variance and standard deviation of x. How would you interpret these values?

5.21 A local television station sells 15-second, 30-second, and 60-second advertising spots. Let x denote the length of a randomly selected commercial appearing on this station, and suppose that the probability distribution of x is given in the table.

x	15	30	60
$p(x)$	.1	.3	.6

a. Find the average length for commercials appearing on this station.
b. If a 15-sec spot sells for $500, a 30-sec spot for $800, and a 60-sec spot for $1000, find the average amount paid for commercials appearing on this station. (Hint: Consider a new variable, $y =$ cost, and then find the probability distribution and mean value of y.)

5.22 An author has written a book and submitted it to a publisher. The publisher offers to print the book and gives the author the choice between a flat payment of $10,000 or a royalty plan. Under the royalty plan, the author would receive $1 for each copy of the book sold. The author thinks that the accompanying table gives the probability distribution of the variable $x =$ the number of books that will be sold. Which payment plan should the author choose? Why?

x	1,000	5,000	10,000	20,000
$p(x)$	.05	.3	.4	.25

5.23 A grocery store has an express line for customers purchasing at most five items. Let x be the number of items purchased by a randomly selected customer using this line. Give examples of two different assignments of probabilities such that the resulting distributions have the same mean but have standard deviations that are quite different.

5.24 Refer to Exercise 5.9 and compute the mean value of the number of people holding reservations who show up for the flight.

5.25   After shuffling a deck of 52 cards, five are dealt. Let x denote the number of suits represented in the five-card hand (the four suits are spades, hearts, diamonds, and clubs). Using some counting rules, it can be shown that the probability distribution of x is as given in the accompanying table. Compute the mean value and the standard deviation of x.

x	1	2	3	4
$p(x)$	.002	.146	.588	.264

5.26   Exercise 5.8 gave the following probability distribution for x, the number of courses for which a randomly selected student is registered.

x	1	2	3	4	5	6	7
$p(x)$	.02	.03	.09	.25	.40	.16	.05

a. Calculate the mean value of x.
b. Calculate the standard deviation of x.
c. What possible values of x are within two standard deviations of the mean value (that is, in the interval from $\mu - 2\sigma$ to $\mu + 2\sigma$)? What is the probability that the observed value of x is within two standard deviations of the mean value?

5.27   A gas station sells gasoline at the following prices (cents per gallon, depending on the type of gas and service) 95.9, 98.7, 109.9, 119.9, 124.5, and 139.7. Let y denote the price per gallon paid by a randomly selected customer.
a. Is y a discrete random variable? Explain.
b. Suppose that the probability distribution of y is as follows:

y	95.9	98.7	109.9	119.9	124.5	139.7
$p(y)$	.36	.24	.10	.16	.08	.06

What is the probability that a randomly selected customer has paid more than $1 per gallon? Less that $1.20 per gallon?
c. Refer back to part (b), and calculate the mean value and standard deviation of y.

5.28   A chemical supply company currently has in stock 100 lb of a certain chemical, which it sells to customers in 5-lb lots. Let x = the number of lots orderd by a randomly chosen customer. The probability distribution of x is as follows:

x	1	2	3	4
$p(x)$	.2	.4	.3	.1

a. Calculate the mean value of x.
b. Calculate the variance and standard devition of x.

5.29   Return to Exercise 5.28, and let y denote the amount of material left (lb) after the next customer's order is shipped.
a. What are possible values of y?
b. What is the probability associated with each possible y value (the probability distribution of y)?
c. Calculate μ_y.

5.30   Let x be a random variable with mean value μ_x and standard deviation σ_x. Define a new random variable $y = ax + b$, where a and b are numerical constants (for example, $y = 5x + 20$). Thus y is a linear function of x. It is then not difficult to

show that

$$\mu_y = a\mu_x + b, \qquad \sigma_y = |a| \cdot \sigma_x$$

(The absolute value of a is needed in case a is negative, since σ_y cannot be negative.)

a. Suppose x = the number of units sold, y = the resulting profit, where $y = 20x - 100$, and $\mu_x = 50$ and $\sigma_x = 10$. What is the mean value of the profit, and what is the standard deviation of the profit?

b. Return to Exercise 5.29, and use the fact that $\mu_x = 2.3$ to determine the mean value of the amount left after the next order has been shipped (x and y there *are* linearly related; what is the relationship?).

5.31 Frequently we are interested not so much in μ_x, the mean value of x, as in the mean value of a specified function of x. For example, x might be the number of units of a certain commodity sold; the function of interest would then be the resulting profit. Let $h(x)$ denote the function whose mean value is desired. Then

$$\mu_{h(x)} = \text{the mean value of the function } h(x) = \sum h(x) \cdot p(x)$$

That is, the mean value of x is a weighted average of x values, and the mean value of $h(x)$ is a weighted average of $h(x)$ values.

Example 8 gave the accompanying probability distribution for the variable x representing the number of attempts by a randomly selected applicant to pass a licensing exam. Suppose that someone who makes x attempts is charged x^2 dollars. Calculate the mean value of the amount charged, i.e., the mean value of $h(x) = x^2$.

x	1	2	3	4
$p(x)$	.1	.2	.3	.4

5.4 THE BINOMIAL DISTRIBUTION

Suppose we decide to record the sex of each of the next 25 newborn children at a particular hospital. What is the chance that at least 15 are female? What is the chance that between 10 and 15 are female? Or suppose that the same coin is to be tossed 20 times. What is the chance that at least half the tosses result in heads? How many heads can we expect to see? These and other similar questions can be answered by studying the *binomial probability distribution*. This distribution arises when the experiment of interest is a *binomial experiment*, one having the following characteristics.

PROPERTIES OF A BINOMIAL EXPERIMENT

1. It consists of a fixed number of smaller experiments called *trials*.
2. Each trial can result in one of only two outcomes, labeled success (S) and failure (F).
3. Outcomes of different trials are independent.
4. The probability that a trial results in S is the same for each trial.
5. The **binomial random variable** x is defined as

   x = the number of successes observed when the
   experiment is performed

The assignment of the S-F labels in any particular problem context is arbitrary. In coin tossing, for example, S can be identified with either a head or a tail, as long as subsequent calculations are consistent with the assignment.

Consider a binomial experiment based on five trials. An experimental outcome is a sequence consisting of five S's and F's. One such outcome is SFFSS—a success occurs on the first trial, the second and third trials result in failure, and successes are observed on the last two trials. The value of the binomial random variable for this outcome is $x = 3$. Another outcome for which $x = 3$ is SFSFS, whereas $x = 4$ for the outcomes FSSSS and SSFSS. With five trials, possible x values are 0, 1, 2, 3, 4, and 5. Since the number of possible x values is finite, x is a discrete random variable. (This is true for any binomial random variable.)

Once the probability of S is specified for a single trial, the probability of any particular outcome is easily computed. This is because the independence of trials allows us to multiply probabilities. When $P(S) = .7$, the probability of the outcome SFFSS is

$$P(\text{SFFSS}) = P(\text{S}) \cdot P(\text{F}) \cdot P(\text{F}) \cdot P(\text{S}) \cdot P(\text{S})$$
$$= (.7)(.3)(.3)(.7)(.7)$$
$$= (.7)^3(.3)^2$$
$$= .03087$$

This is also the probability of the outcome SFSFS:

$$P(\text{SFSFS}) = (.7)(.3)(.7)(.3)(.7)$$
$$= (.7)^3(.3)^2$$
$$= .03087$$

The probability of a particular outcome depends on the number of S's and F's but not on the order in which they appear.

THE BINOMIAL PROBABILITY DISTRIBUTION

The binomial probability distribution gives the probability associated with each possible x value. The easiest way to describe the distribution is by means of a compact formula. To better understand the basis for the formula, let's first consider a specific example.

EXAMPLE 12

Cars coming down a certain freeway offramp can turn either left (S) or right (F). Suppose that $P(S) = .6$ (in the long run, 60% of all cars turn left) and that the turning directions of four different cars are observed. Table 2 displays the 16 possible outcomes, the probability of each one, and the x value for each outcome.

The five possible values of x are 0, 1, 2, 3, and 4. Let's focus on $x = 3$. From Table 2, this value results from any of the four outcomes FSSS, SFSS, SSFS, or SSSF. Applying one of our basic rules of probability, we obtain

TABLE 2

Outcomes and Probabilities for a Binomial Experiment with Four Trials and $P(S) = .6$

Outcome	Probability	x Value	Outcome	Probability	x Value
FFFF	(.4)(.4)(.4)(.4)	0	FSSF	(.4)(.6)(.6)(.4)	2
SFFF	(.6)(.4)(.4)(.4)	1	FSFS	(.4)(.6)(.4)(.6)	2
FSFF	(.4)(.6)(.4)(.4)	1	FFSS	(.4)(.4)(.6)(.6)	2
FFSF	(.4)(.4)(.6)(.4)	1	FSSS	(.4)(.6)(.6)(.6)	3
FFFS	(.4)(.4)(.4)(.6)	1	SFSS	(.6)(.4)(.6)(.6)	3
SSFF	(.6)(.6)(.4)(.4)	2	SSFS	(.6)(.6)(.4)(.6)	3
SFSF	(.6)(.4)(.6)(.4)	2	SSSF	(.6)(.6)(.6)(.4)	3
SFFS	(.6)(.4)(.4)(.6)	2	SSSS	(.6)(.6)(.6)(.6)	4

$P(x = 3)$ by adding together the probabilities of these four outcomes:

$$p(3) = P(x = 3) = P(\text{FSSS or SFSS or SSFS or SSSF})$$
$$= P(\text{FSSS}) + P(\text{SFSS}) + P(\text{SSFS}) + P(\text{SSSF})$$
$$= (.6)^3(.4) + (.6)^3(.4) + (.6)^3(.4) + (.6)^3(.4)$$
$$= 4[(.6)^3(.4)]$$
$$= .3456$$

Similarly, $P(x = 2)$ is the sum of the probabilities of the six outcomes for which $x = 2$. Since each such outcome probability is $(.6)^2(.4)^2$,

$$p(2) = P(x = 2) = 6[(.6)^2(.4)^2] = .3456$$

Using the same reasoning for the other three x values, the probability distribution of x is as follows.

x	0	1	2	3	4
$p(x)$	.0256	.1536	.3456	.3456	.1296

Thus if we observe turning directions for successive groups of four cars at the chosen intersection, in the long run, 12.96% of the time all four will turn left, 34.56% of the time three out of four will turn left, and so on.

The five probabilities in the distribution of Example 12 can all be expressed in the same manner:

$$p(x) = P(x \text{ successes in 4 trials})$$
$$= \binom{\text{number of outcomes}}{\text{with } x \text{ successes}} \cdot (.6)^x(.4)^{4-x}$$

The exponent $4 - x$ is just the number of failures when there are x successes in the four trials. This expression remains valid if the number of trials is changed, provided that the 4 in the exponent $4 - x$ is changed accordingly. Suppose, for example, that there are five trials. Then there are 32 possible outcomes (16 from appending an S at the end of each outcome in Table 2, and 16 from appending an F), and ten of these have $x = 2$. Any one of these, such as SFFSF, has probability $(.6)(.4)(.4)(.6)(.4) = (.6)^2(.4)^3$. Summing the probabilities of these ten outcomes gives

$$p(2) = 10 \cdot (.6)^2(.4)^3 = 10 \cdot (.6)^2(.4)^{5-2} = .2304$$

It should now be clear that a general formula for $p(x)$ necessitates determining the number of outcomes that result in any particular x value. For this purpose it is convenient to introduce some widely used mathematical notation.

FACTORIAL NOTATION

Let m represent a positive whole number. Then the expression $m!$, read "m factorial," is defined to be

$$m! = m(m - 1)(m - 2)(m - 3) \cdots (3)(2)(1)$$

Thus

$$2! = (2)(1) = 2$$
$$6! = (6)(5)(4)(3)(2)(1) = 720$$
$$10! = (10)(9)(8) \cdots (2)(1) = 3,628,800$$

and so on.
In addition, zero factorial is defined by $0! = 1$.

The use of factorial notation makes it easy to write a concise formula for $p(x)$.

THE BINOMIAL PROBABILITY DISTRIBUTION

Denote the number of trials in a binomial experiment by n and the probability of S on any particular trial by π. Then

$$\begin{pmatrix} \text{number of outcomes} \\ \text{with } x \text{ successes} \end{pmatrix} = \frac{n!}{x!(n - x)!}$$

and the **binomial probability distribution** formula for $p(x) = P(x$ successes in n trials) is

$$p(x) = \frac{n!}{x!(n - x)!} \cdot \pi^x \cdot (1 - \pi)^{n-x} \qquad x = 0, 1, \ldots, n$$

The development of the expression for the number of outcomes with x successes requires a digression into counting techniques. We sketch the argument at the end of the section.

EXAMPLE 13

Sixty percent of all watches sold by a large discount store have a digital display and 40% have an analog display. The type of watch purchased by each of the next 12 customers will be noted. Define a random variable x by

x = the number of watches among these 12 that have a digital display

Letting S denote the sale of a digital watch, x is a binomial random variable with $n = 12$ and $\pi = P(S) = .60$. The probability distribution of x is given by

$$p(x) = \frac{12!}{x!(12-x)!} \cdot (.6)^x(.4)^{12-x} \qquad x = 0, 1, 2, \ldots, 12$$

The probability that exactly four watches are digital is

$$p(4) = P(x = 4) = \frac{12!}{4!8!} \cdot (.6)^4(.4)^8$$

$$= \frac{(12)(11)(10)(9)(8)(7)(6)\cdots(2)(1)}{(4)(3)(2)(1)\cdot 8!} \cdot (.6)^4(.4)^8$$

$$= \frac{(12)(11)(10)(9)}{(4)(3)(2)(1)} \cdot (.6)^4(.4)^8$$

$$= (495)(.6)^4(.4)^8$$

$$= .042$$

If group after group of 12 purchases is examined, the long-run percentage of those with exactly four digital watches will be 4.2%. According to this calculation, 495 of the possible outcomes (there are $2^{12} = 4096$) have $x = 4$. Notice also how the larger factorial in the denominator cancels with a portion of the numerator factorial.

The probability that between four and seven (inclusive) of these watches are digital is

$$P(4 \le x \le 7) = P(x = 4 \text{ or } 5 \text{ or } 6 \text{ or } 7)$$

$$= p(4) + p(5) + p(6) + p(7)$$

$$= \frac{12!}{4!8!} \cdot (.6)^4(.4)^8 + \cdots + \frac{12!}{7!5!} \cdot (.6)^7(.4)^5$$

$$= .042 + .101 + .177 + .227$$

$$= .547$$

Now suppose we want to know the probability that at least 75% of the watches purchased are digital. Since $(.75)(12) = 9$, that probability is

$$P(9 \le x) = P(x = 9 \text{ or } 10 \text{ or } 11 \text{ or } 12)$$

$$= p(9) + p(10) + p(11) + p(12)$$

$$= .225$$

Using the binomial distribution formula can be tedious unless n is very small. We have included in the appendixes a tabulation (Table I) of binomial probabilities for $n = 5, 10, 15, 20,$ and 25 in combination with various values of π. This should help you practice using the binomial distribution without getting bogged down in arithmetic. To obtain the probability of a particular x value, find the column headed by your value of π and move down to the row labeled with the desired value of x. Doing this for $n = 20$, $\pi = .8$ (a column), and $x = 15$ (a row) yields $p(15) = .175$. Although $p(x)$ is positive for every possible x value, many probabilities are zero to three decimal places, so they appear as .000 in the table. There are much more extensive binomial tables available. Alternatively, it is easy to program a computer or some calculators to calculate these probabilities.

Suppose that a population consists of N individuals or objects, each one classified as an S or an F. If sampling is carried out without replacement (as it almost always is), then successive draws are dependent. However, when the sample size n is much smaller than N, the extent of this dependence is minimal. In this case, x, the number of S's in the sample, has approximately a binomial distribution. The usual rule of thumb is that the binomial distribution gives accurate results if n is at most 5% of N.

EXAMPLE 14

An article in the *Los Angeles Times* (Dec. 11, 1983) discussed the environmental problems caused by underground gasoline storage tanks that leak. There are over 2 million tanks in the United States, and several studies have suggested that roughly 25% of them leak. Suppose that a random sample of $n = 20$ tanks is selected from the population of all tanks (so n is much less than 5% of N). Let S denote a tank that leaks (not what would ordinarily be thought a success!), and suppose that $\pi = .25$. Then the probability that *exactly* half of the tanks in the sample leak is

$$p(10) = P(x = 10)$$
$$= \text{the entry in the 10 row, .25 column of Table I for } n = 20$$
$$= .010$$

The probability that *at most* half of the sampled tanks leak is

$$P(x \leq 10) = p(0) + p(1) + \cdots + p(9) + p(10)$$
$$= .003 + .021 + \cdots + .027 + .010 = .996$$

The probability that *more than* half of the sampled tanks leak is

$$P(10 < x) = 1 - P(x \leq 10) = 1 - .996 = .004$$

Since this probability is so small, if more than half of the tanks in a sample did have leaks, this would cast considerable doubt on the validity of the value $\pi = .25$. In Chapter 9 we will show how hypothesis testing methods can be used to decide which claim, either $\pi = .25$ or $\pi > .25$, is more plausible.

THE MEAN VALUE AND THE STANDARD DEVIATION OF x

A binomial random variable x based on n trials has possible values 0, 1, 2, . . . , n, so the mean value is

$$\mu = \sum x \cdot p(x) = (0) \cdot p(0) + (1) \cdot p(1) + \cdots + n \cdot p(n)$$

Because the formula for $p(x)$ is somewhat complicated, evaluation of this sum looks especially unappealing. Fortunately, algebraic manipulation considerably simplifies this expression for any values of n and π. In fact, the result is simply $\mu = n\pi$. That is, the mean value of a binomial random variable is the product of the number of trials and the success probability.

The variance of x is

$$\sigma^2 = \sum (x - \mu)^2 \cdot p(x)$$
$$= (0 - n\pi)^2 \cdot p(0) + (1 - n\pi)^2 \cdot p(1) + \cdots + (n - n\pi)^2 \cdot p(n)$$

Again considerable algebraic simplification is possible, and the result is $\sigma^2 = n\pi(1 - \pi)$. The standard deviation of x is obtained by taking the square root of σ^2.

> The mean value and the standard deviation of a binomial random variable are
>
> $$\mu = n\pi, \qquad \sigma = \sqrt{n\pi(1 - \pi)}$$

EXAMPLE 15

(*Example 14 continued*) The number of leaking tanks in a sample of $n = 20$ gasoline storage tanks was assumed to have a binomial distribution with $\pi = .25$. The mean value of the number of leaking tanks in the sample is thus

$$\mu = n\pi = 20(.25) = 5$$

and the standard deviation of x is

$$\sigma = \sqrt{n\pi(1 - \pi)} = \sqrt{20(.25)(.75)} = \sqrt{3.75} \approx 1.94$$

If the sample had consisted of 2000 tanks (still small relative to the population size), then

$$\mu = 2000(.25) = 500,$$

$$\sigma = \sqrt{2000(.25)(.75)} = 19.4$$

Raising the sample size by a factor of 100 increases the standard deviation by a factor of 10.

The value of σ is zero when $\pi = 0$ or $\pi = 1$. In these two cases there is no uncertainty in x: we are sure to observe $x = 0$ when $\pi = 0$ and $x = n$ when $\pi = 1$. It is also easily verified that $\pi(1 - \pi)$ is largest when $\pi = .5$. Thus the binomial distribution spreads out the most when sampling from a 50-50 population. The farther π is from .5, the less spread out is the distribution.

COUNTING AND THE BINOMIAL DISTRIBUTION*

The formula for binomial probabilities requires an expression for the number of outcomes that have exactly x S's in n trials. Consider the case $n = 5$ and $x = 3$. Let's label the five trials 1, 2, 3, 4, 5. Then the number of outcomes with three S's is the number of ways of selecting three trials from among the five to be the S trials. Think of five labels resting in an urn. We reach in and select a first label, then a second, and then a third. There are five possibilities for the first selection and four for the second, so there is a total of $(5)(4) = 20$ for the first two selections (visualize a tree diagram with five first-generation branches and four second-generation branches emanating from the tip of each first-generation branch). For each of these 20 ways, there are three further ways to select the third label. Therefore, the total number of ways of selecting three labels out of the five in the urn is $(5)(4)(3) = 60$.

But there is a difficulty here. Among the 60 possibilities, we've counted

$$1, 2, 3 \qquad 1, 3, 2 \qquad 2, 1, 3 \qquad 2, 3, 1 \qquad 3, 1, 2 \qquad 3, 2, 1$$

That is, we've counted not only the number of different sets of three labels but every different ordering as well. These six possibilities all correspond to the same set of labels: $\{1, 2, 3\}$. Similarly, since there are $3 \cdot 2 \cdot 1 = 6$ ways to order the labels $\{2, 4, 5\}$, another six of our ordered sets correspond to the single set of labels $\{2, 4, 5\}$. In this case ($n = 5$ and $x = 3$), there are $3 \cdot 2 \cdot 1 = 6$ *ordered* sets corresponding to every set of interest to us. Therefore, the number of ways of selecting three out of the five labels (trials) without counting different orderings of the same set is

$$\frac{(5)(4)(3)}{(3)(2)(1)} = \frac{60}{6} = 10$$

It is easy to list these ten:

$$\{1, 2, 3\}, \quad \{1, 2, 4\}, \quad \{1, 2, 5\}, \quad \{1, 3, 4\}, \quad \{1, 3, 5\},$$
$$\{1, 4, 5\}, \quad \{2, 3, 4\}, \quad \{2, 3, 5\}, \quad \{2, 4, 5\}, \quad \{3, 4, 5\}$$

In the general case, the number of ways to select x labels from among the n trial labels, counting all orderings, is $n(n - 1)(n - 2) \cdots (n - x + 1)$. (We select a first from the n possibilities, a second from the remaining $n - 1$, and so on, until finally we select the xth from the remaining $n - x + 1$.) To eliminate different orderings of the same set of labels, we divide by $(x)(x - 1)(x - 2) \cdots (2)(1) = x!$, which is the number of ways of ordering a set of x labels. This gives us

$$\left(\begin{matrix} \text{the number of ways to} \\ \text{select } x \text{ labels} \end{matrix} \right) = \frac{n(n - 1)(n - 2) \cdots (n - x + 1)}{x!}$$

---

* The material in this subsection is optional.

Multiplying both the numerator and denominator by $(n - x)!$ (which doesn't change the value of the expression) yields

$$\frac{n!}{x!(n - x)!}$$

which appears in the binomial probability distribution formula.

EXERCISES 5.32–5.45　　　　　　　**SECTION 5.4**

5.32　　**a.** In a binomial experiment consisting of six trials, how many outcomes have exactly one S, and what are these outcomes?
　　　　b. In a binomial experiment consisting of 20 trials, how many outcomes have exactly ten S's? Exactly fifteen S's? Exactly five S's?

5.33　　Twenty-five percent of the customers entering a grocery store between 5 P.M. and 7 P.M. use an express checkout. Consider five randomly selected customers, and let x denote the number among the five who use the express checkout.
　　　　a. What is $p(2)$, that is, $P(x = 2)$?
　　　　b. What is $P(x \leq 1)$?
　　　　c. What is $P(2 \leq x)$? Hint: Make use of your computation in part (b).
　　　　d. What is $P(x \neq 2)$?

5.34　　The *Los Angeles Times* (Dec. 13, 1983) reported that only 58% of the tenth graders in Los Angeles high schools graduate from those schools three years later. (Of the 42% who did not graduate from Los Angeles schools, some moved to other school districts, but most are presumed to have dropped out.) Suppose that four tenth graders are randomly selected from Los Angeles schools. This is a binomial experiment with $\pi = .58$, $n = 4$, and $x = $ the number among the four who graduate.
　　　　a. What is the probability that all four students graduate three years later from a Los Angeles school (i.e., that $x = 4$)?
　　　　b. What is the probability that exactly three of the four graduate from Los Angeles schools?
　　　　c. What is the probability that at least three graduate from Los Angeles schools?
　　　　d. What is the probability that none of the four graduate from Los Angeles schools?

5.35　　A breeder of show dogs is interested in the number of female puppies in a litter. If a birth is equally likely to result in a male or female puppy, give the probability distribution of the variable

　　　　$x = $ the number of female puppies in a litter of size 5

5.36　　A manufacturer of camera flash bars notes that defective flash bulbs are sometimes produced. If the probability that any given bulb will be defective is .05, what is the probability that there are no defective flashes on a flash bar containing ten bulbs? What is the probability that at most one of the ten bulbs is defective?

5.37　　Industrial quality control programs often include inspection of incoming materials from suppliers. If parts are purchased in large lots, a typical plan might be to select 20 parts at random from a lot and inspect them. A lot might be judged acceptable if one or fewer defective parts are found among those inspected. Otherwise, the lot is rejected and returned to the supplier. Use Table I to find the probability of accepting lots that have each of the following.
　　　　a. 5% defective parts
　　　　b. 10% defective parts

c. 20% defective parts
(Hint: Identify success with a defective part.)

5.38 In an experiment to investigate whether a graphologist (handwriting analyst) could distinguish a normal person's handwriting from that of a psychotic, a well-known expert was given ten files, each containing handwriting samples from a normal person and from a person diagnosed as psychotic. The graphologist was then asked to identify the psychotic's handwriting. The graphologist made correct identifications in six of the ten trials (data taken from *Statistics in the Real World*, Larsen and Stroup. New York: MacMillan Publishing Co., 1976). Does this evidence indicate that the graphologist has an ability to distinguish the handwriting of psychotics? (Hint: What is the probability of correctly guessing six or more times out of ten? Your answer should depend on whether this probability is relatively small or large.)

5.39 If the temperature in Florida falls below 32°F during certain periods of the year, there is a chance that the citrus crop will be damaged. Suppose that the probability is .1 that any given tree will show measurable damage when the temperature falls to 30°F. If the temperature does drop to 30°F, what is the expected number of trees showing damage in an orchard of 2000 trees? What is the standard deviation of the number of trees that show damage?

5.40 Thirty percent of all automobiles undergoing a headlight inspection at a certain inspection station fail to pass.
  a. Among 15 randomly selected cars, what is the probability that at most five fail to pass inspection?
  b. Among 15 randomly selected cars, what is the probability that between five and ten (inclusive) fail to pass inspection?
  c. Among 50 randomly selected cars, what is the mean value of the number that pass inspection, and what is the standard deviation of the number that pass inspection?

5.41 You are to take a multiple choice exam consisting of 100 questions with five possible responses to each. Suppose that you have not studied and so must guess (select one of the five answers in a completely random fashion) on each question. Let x represent the number of correct responses on the test.
  a. What kind of probability distribution does x have?
  b. What is your expected score on the exam? (Hint: Your expected score is the mean value of the x distribution.)
  c. Compute the variance and standard deviation of x.
  d. Based on your answers to parts (b) and (c), is it likely that you would score over 50 on this exam? Explain the reasoning behind your answer.

5.42 Suppose that 20% of the 10,000 signatures on a certain recall petition are invalid. Would the number of invalid signatures in a sample of size 1000 have (approximately) a binomial distribution? Explain.

5.43 A coin is to be tossed 25 times. Let $x =$ the number of tosses that result in heads (H). Consider the following rule for deciding whether or not the coin is fair:

Judge the coin to be fair if $8 \leq x \leq 17$.

Judge it to be biased if either $x \leq 7$ or $x \geq 18$.

  a. What is the probability of judging the coin to be biased when it is actually fair?
  b. What the probability of judging the coin to be fair when $P(H) = .9$, so that there is a substantial bias? Repeat for $P(H) = .1$.
  c. What is the probability of judging the coin to be fair when $P(H) = .6$? When $P(H) = .4$? Why are these probabilities so large compared to the probabilities in part (b)?

    d. What happens to the "error probabilities" of parts (a) and (b) if the decision rule is changed so that the coin is judged fair if $7 \leq x \leq 19$ and unfair otherwise? Is this a better rule than the one first proposed?

5.44 A city ordinance requires that a smoke detector be installed in all residential housing. There is concern that too many residences are still without detectors, so a costly inspection program is being contemplated. Let π = the proportion of all residences that have a detector. A random sample of 25 residences will be selected. If the sample strongly suggests that $\pi < .80$ (fewer than 80% have detectors), as opposed to $\pi \geq .80$, the program will be implemented. Let x = the number of residences among the 25 that have a detector, and consider the following decision rule:

    Reject the claim that $\pi \geq .8$ and implement the program if $x \leq 15$.

    a. What is the probability that the program is implemented when $\pi = .80$?
    b. What is the probability that the program is not implemented if $\pi = .70$? If $\pi = .60$?
    c. How do the "error probabilities" of parts (a) and (b) change if the value 15 in the decision rule is changed to 14?

5.45 Exit polling has been a controversial practice in recent elections, since early release of the resulting information appears to affect whether or not those who have not yet voted will do so. Suppose that 90% of all registered California voters favor banning the release of information from exit polls in presidential elections until after the polls in California close. A random sample of 25 California voters is selected.
    a. What is the probability that more than 20 favor the ban?
    b. What is the probability that at least 20 favor the ban?
    c. What are the mean value and standard deviation of the number who favor the ban?
    d. If fewer than 20 in the sample favor the ban, is this at odds with the assertion that (at least) 90% of the populace favors the ban? Hint: Consider $P(x < 20)$ when $\pi = .9$.

CHAPTER 5 SUMMARY OF KEY CONCEPTS AND FORMULAS

TERM OR FORMULA	PAGE	COMMENT
Random variable: discrete or continuous	124	A numerical variable with a value determined by the outcome of a chance experiment: it is discrete if its possible values are isolated points on the number line and continuous if its possible values form an entire interval on the number line
Probability distribution $p(x)$ of a discrete random variable x	129	A formula, table, or graph that gives the probability associated of with each x value. Conditions on $p(x)$ are (i) $p(x) \geq 0$, and (ii) $\sum p(x) = 1$, where the sum is over all possible x values
$\mu_x = \sum x \cdot p(x)$	137	The mean value of a discrete random variable x: it locates the center of the variable's probability distribution
$\sigma_x^2 = \sum (x - \mu)^2 \cdot p(x)$ $\sigma_x = \sqrt{\sigma_x^2}$	139	The variance and standard deviation, respectively, of a discrete random variable: these are measures of the extent to which the variable's distribution spreads out about μ_x
Binomial experiment	142	An experiment consisting of n smaller experiments called *trials*. Each trial results in one of two outcomes, "success" or "failure"; the trials are independent; and the probability of success, π, is the same on each trial

(continued on next page)

CHAPTER 5 SUMMARY OF KEY CONCEPTS AND FORMULAS (continued)

TERM OR FORMULA	PAGE	COMMENT
Binomial probability distribution: $$p(x) = \frac{n!}{x!(n-x)!} \cdot \pi^x (1-\pi)^{n-x}$$	145	This formula gives the probability of observing x successes ($x = 0, 1, \ldots, n-1$, or n) among the n trials of a binomial experiment.
$\mu_x = n\pi, \quad \sigma_x = \sqrt{n\pi(1-\pi)}$	148	The mean and standard deviation of a binomial random variable

SUPPLEMENTARY EXERCISES 5.46–5.62

5.46 A mail-order computer software business has six telephone lines. Let x denote the number of lines in use at a specified time. The probability distribution of x is as follows:

x	0	1	2	3	4	5	6
$p(x)$	.10	.15	.20	.25	.20	.06	.04

Write each of the following events in terms of x, and then calculate the probability of each one.
a. At most three lines are in use.
b. Fewer than three lines are in use.
c. At least three lines are in use.
d. Between two and five lines (inclusive) are in use.
e. Between two and four lines (inclusive) are *not* in use.
f. At least four lines are *not* in use.

5.47 Return to Exercise 5.46.
a. Calculate the mean value and standard deviation of x.
b. What is the probability that the number of lines in use is farther than three standard deviations from the mean value?

5.48 A new battery's voltage may be acceptable (A) or unacceptable (U). A certain flashlight requires two batteries, so batteries will be independently selected and tested until two acceptable ones have been found. Suppose that 80% of all batteries have acceptable voltages, and let y denote the number of batteries that must be tested.
a. What is $p(2)$, i.e., $P(y = 2)$?
b. What is $p(3)$? Hint: There are two different outcomes that result in $y = 3$.
c. In order to have $y = 5$, what must be true of the fifth battery selected? List the four outcomes for which $y = 5$, and then determine $p(5)$.
d. Use the pattern in your answers for parts (a) through (c) to obtain a general formula for $p(y)$.

5.49 Each of the numbers 1, 2, 3, and 4 is to be assigned either a plus sign ($+$) or a minus sign ($-$). Once these have been assigned, the sum will be calculated. Let w denote the resulting sum. For example, if the "signed" numbers are -1, $+2$, -3, and -4, then $w = -1 + 2 - 3 - 4 = -6$. If the signs are assigned independently, and $P(+) = P(-) = .5$, determine the probability distribution of w. Hint: Construct a tree diagram with two first-generation branches for the two signs that 1 can receive. From each of these, draw two second-generation branches corresponding to

+2 and −2. Then draw third- and fourth-generation branches. Note: An important statistical procedure, the *signed-rank test*, is based on this type of distribution.

5.50 Two sisters, Allison and Teri, have agreed to meet between 1 and 6 P.M. on a particular day. In fact, Allison is equally likely to arrive at exactly 1 P.M., 2 P.M., 3 P.M., 4 P.M., 5 P.M., or 6 P.M. Teri is also equally likely to arrive at each of these six times, and Allison's and Teri's arrival times are independent of one another. There are thus 36 equally likely (Allison, Teri) arrival-time pairs, for example, (2, 3) or (6, 1). Suppose the first person to arrive waits until the second person does also; let w be the amount of time the first person has to wait.
a. What is the probability distribution of w?
b. How much time do you expect to elapse between the two arrivals?

5.51 Four people—a, b, c, and d—are waiting to give blood. Of these four, a and b have type AB blood, whereas c and d do not. An emergency call has just come in for some type *AB* blood. If blood samples are taken one-by-one from the four in random order for blood typing, and x is the number of samples taken to obtain an AB individual (so possible x values are 1, 2, and 3), what is the probability distribution of x? Hint: See the tree diagram of Example 2 in Chapter 4.

5.52 Bob and Lygia are going to play a series of Trivial Pursuit games. The first person to win four games will be declared the winner. (This is the World Series of Trivial Pursuit.) Suppose that outcomes of successive games are independent and that the probability of Lygia winning any particular game is .6. Define a random variable x as the number of games played in the series.
a. What is $p(4)$? Hint: Either Bob or Lygia could win four straight.
b. What is $p(5)$? Hint: For Lygia to win in exactly five games, what has to happen in the first four games and in game 5?
c. Determine the probability distribution of x.
d. How many games can you expect the series to last?

5.53 Refer to Exercise 5.52, and let y be the number of games won by the series loser. Determine the probability distribution of y.

5.54 A sporting goods store has a special sale on three brands of tennis balls, call them D, P, and W. Because the sale price is so low, only one can of balls will be sold to each customer. If 40% of all customers buy brand W, 35% buy brand P, 25% buy brand D, and x is the number among three randomly selected customers who buy brand W, what is the probability distribution of x?

5.55 A small drugstore orders copies of a news magazine for its magazine rack each week. Let x = the number of customers who come in to buy the magazine during a given week. The probability distribution of x is

x	1	2	3	4	5	6
$p(x)$	1/15	2/15	3/15	4/15	3/15	2/15

The store pays $.25 for each copy purchased, and the price of the magazine is $1. Any magazines unsold at the end of the week have no value.
a. If the store orders three copies of the magazine, what is the mean value of the profit? (profit = revenue − cost)
b. Answer the question posed in part (a) when four copies are ordered. Is it better to order three or four copies?

5.56 A student who must write a paper for a course has a choice of two topics. For the first topic, the student will have to request two books through interlibrary loan, or four books for the second topic. The student feels that there will be enough information only if at least half the books ordered arrive on time.

a. If books arrive independently of one another and the probability of a book arriving on time is .5, for which topic is there most likely to be sufficient information?

b. Answer the question posed in part (a) if the on-time probability is .9 rather than .5.

5.57 The n candidates for a job have been ranked 1, 2, 3, . . . , n. Let x be the rank of a randomly selected candidate, so that x has probability distribution

$$p(x) = \begin{cases} 1/n & \text{if } x = 1, 2, 3, \ldots, n \\ 0 & \text{otherwise} \end{cases}$$

a. Determine μ_x. Hint: The sum of the first n positive integers, $1 + 2 + \cdots + n$, is $n(n + 1)/2$.

b. A shortcut formula for the variance of a discrete random variable x is

$$\sigma_x^2 = \left[\sum x^2 \cdot p(x)\right] - \mu_x^2$$

(This is similar to the computational formula for the sample variance.) Use this to obtain σ_x^2 for x as defined at the beginning of the problem. Hint: The sum of the squares of the first n positive integers is

$$1^2 + 2^2 + \cdots + n^2 = \frac{n(n + 1)(2n + 1)}{6}$$

c. What are μ_x and σ_x when $n = 10$?

5.58 Consider a disease that can be diagnosed by carrying out a blood test. Let π denote the probability that a randomly selected individual has the disease. Given a sample of n independently selected individuals, one way to proceed is by carrying out a separate test on each of the n blood samples. A potentially more economical approach, *group testing*, was introduced during WWII to identify syphilitic men among army inductees. First a part of each blood sample is taken, these are combined, and then a single test is carried out. If no one has the disease, the result will be negative, and only this test is required. If at least one individual is diseased, the group test will give a positive result, in which case the n individual tests are then carried out. Let $y =$ the number of tests performed.

a. What are the two possible values of y?

b. What is $p(1)$, the probability that just one test will suffice? What is the probability of the other possible y value?

c. If $\pi = .1$ and $n = 3$, what is the mean value of the number of tests performed? How does it compare with the number of tests required without group testing?

5.59 A stereo store is offering a special price on a complete system of components. A purchaser is offered a choice of manufacturers for each component:

Receiver: JVC, Kenwood, Pioneer, Sansui, Sony

Turntable: BSR, Dual, Sony, Technics

Speakers: AR, KLH, JBL

Cassette deck: Advent, Sony, Teac

A switchboard in the store allows customers to hook together any selection of components (one of each type). Suppose that one component of each type is randomly selected in such a way that different selections are independent. Let x denote the number of Sony components selected.

a. What is $p(0)$, the probability that no Sony components are selected?

b. What is the probability that the only Sony component selected is the receiver? What is $p(1)$? Hint: The 1 could refer to either the receiver, the turntable, or the deck.

c. Obtain the probability distribution of x.

5.60 A plan for an executive traveler's club has been developed by an airline on the premise that 10% of its current customers would qualify for membership.

a. Assuming the validity of this premise, among 25 randomly selected current customers, what is the probability that between 2 and 6 (inclusive) qualify for membership?

b. Again assuming the validity of the premise, what are the expected number of customers who qualify and the standard deviation of the number who qualify in a random sample of 100 current customers?

c. Let x denote the number in a random sample of 25 current customers who qualify for membership. Consider rejecting the company's premise in favor of the claim that $\pi > .10$ if $x \geq 7$. What is the probability that the company's premise is rejected when it is actually valid?

d. Referring back to the decision rule introduced in part (c), what is the probability that the company's premise is not rejected even though $\pi = .20$ (that is, 20% qualify)?

5.61 A probability distribution that has been used with great success to model the number of events that occur during a specified time interval is the **Poisson distribution**. Here "events" could be accidents on a stretch of highway, telephone calls to a drug hotline, customers arriving at a service facility, occurrences of tornadoes, and so on. Let x denote the number of events that occur in a time interval of length t. The Poisson probability distribution is then

$$p(x) = P(x \text{ events occur in time } t)$$
$$= \frac{e^{-\alpha t}(\alpha t)^x}{x!}, \qquad x = 0, 1, 2, \ldots$$

where the letter e denotes a fixed number, called the *base of the natural logarithm system*, that has the approximate value 2.7182818. It can be shown that

$$\mu_x = \alpha t$$

so that when $t = 1$ (a one-unit time interval), $\mu_x = \alpha$. Thus α is the mean number of occurrences during a one-unit time interval, i.e., it is the *rate* at which events occur over time.

Suppose that airplanes arrive at a certain airport according to a Poisson distribution with rate $\alpha = 2$ per min.

a. What is the probability that exactly three planes arrive during a one-minute period? Hint: $e^{-2} \approx .1353353$.

b. What is the probability that at least two airplanes arrive during a one-minute period? Hint: Remember that $P(A) = 1 - P(\text{not } A)$.

c. What is the probability that exactly five planes arrive during a two-minute period?

d. What is the mean value of the number of airplanes that arrive during a half-hour period?

5.62 Shortly after being put into service, some buses of a certain type develop cracks on the underside of the mainframe. A particular city has 20 buses of this type, eight of which actually have cracks. Suppose that five buses are randomly selected for inspection in such a way that each sample of this size has the same chance of selection (by choosing five from among 20 numbered slips). Let x denote the number of buses in the sample that have cracks, and define the symbol $\binom{m}{k}$ by

$$\binom{m}{k} = \frac{m!}{k!(m-k)!}$$

Then counting rules can be used to show that

$$p(x) = P(x \text{ buses in the sample have cracks})$$

$$= \frac{\text{\# of samples for which } x \text{ buses have cracks}}{\text{number of samples of size } 5}$$

$$= \frac{\binom{8}{x} \cdot \binom{12}{5-x}}{\binom{20}{5}}$$

(This is called the **hypergeometric distribution**.)
a. What is $p(3)$?
b. What is $P(x \le 1)$?

REFERENCES   See the References at the end of Chapter 4.

C H A P T E R 6

CONTINUOUS PROBABILITY DISTRIBUTIONS

INTRODUCTION The previous chapter discussed probability distributions for discrete random variables. The other type of random variable is a continuous random variable. Here we will first present general properties of probability distributions for continuous random variables. We will then turn our attention to the most important distribution in all of statistics: the *normal distribution* and the corresponding *bell-shaped curve*.

6.1 PROBABILITY DISTRIBUTIONS FOR CONTINUOUS RANDOM VARIABLES

A continuous random variable is one that has as its set of possible values an entire interval on the number line. An example is the weight x (in lb) of a newborn child. Suppose for the moment that weight is recorded only to the nearest pound. Then possible x values are whole numbers, such as 4 or 9. The probability distribution can be pictured as a probability histogram in which the area of each rectangle is the probability of the corresponding weight value. The total area of all the rectangles is 1, and the probability that a weight (to the nearest pound) is between two values, such as 6 and 8, is the sum of the corresponding rectangular areas. Figure 1(a) illustrates this.

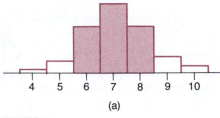 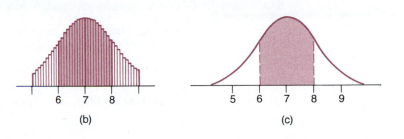

FIGURE 1

Probability distributions for birth weight: shaded area = $P(6 \leq \text{weight} \leq 8)$
(a) Weight measured to the nearest pound;
(b) Weight measured to the nearest tenth of a pound;
(c) Limiting curve as measurement accuracy increases.

Now suppose that weight is measured to the nearest *tenth* of a pound. There are many more possible weight values than before, such as 5.0, 5.1, 5.7, 7.3, 8.9, and so on. As shown in Figure 1(b), the rectangles in the probability histogram are much narrower and this histogram has a much smoother appearance than the first one. Again, this histogram can be drawn so that area equals probability and the total area of all the rectangles is 1.

Figure 1(c) shows what happens as weight is measured to a greater and greater degree of accuracy. The sequence of probability histograms approaches a smooth curve. The curve can't go below the horizontal measurement scale, and the total area under the curve is 1 (because this is true of each probability histogram). The probability that x falls in an interval such as $6 \leq x \leq 8$ is the area under the curve and above that interval.

▋▋ **DEFINITION**

A **probability distribution for a continuous random variable x** is specified by a smooth curve called the **density curve**, such that the total area under the curve is 1. Then the probability that x falls in any particular interval is the area under the density curve and above the interval.

For any two numbers a and b with $a < b$, define three events as follows:

(i) $a < x < b$ is the event that the value of x is between a and b.

(ii) $x < a$ is the event that the value of x is less than a.
(iii) $b < x$ is the event that the value of x exceeds b.

Figure 2 illustrates how the probabilities of these events are identified with areas under a density curve.

FIGURE 2

Probabilities as areas under a probability density curve.

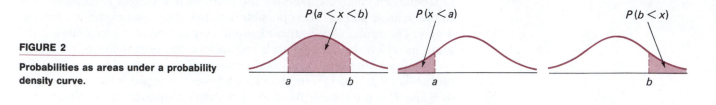

EXAMPLE 1

Define a continuous random variable x by

$x =$ the amount of time (min) taken by a clerk to process a certain type of application form

Suppose that the density curve is as pictured in Figure 3(a). This distribution, often referred to as *uniform*, is especially easy to use, since calculating probabilities requires only finding areas of rectangles using the formula

area = (base) · (height)

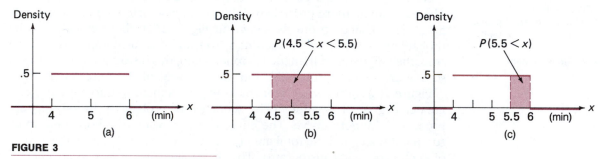

FIGURE 3

The uniform distribution for Example 1.

The curve has positive height only between $x = 4$ and $x = 6$, so according to this model, the smallest possible x value is 4 and the largest value is 6. The total area under the curve is just the area of the rectangle with base extending from 4 to 6 and with height .5. This gives

area = (6 − 4)(.5) = 1

as required.

As illustrated in Figure 3(b), the probability that x is between 4.5 and 5.5 is

$P(4.5 < x < 5.5)$ = area of shaded rectangle

= (base width) · (height)

= (5.5 − 4.5)(.5)

= .5

Similarly, because $x > 5.5$ is equivalent here to $5.5 < x < 6$, we have

$$P(5.5 < x) = (6 - 5.5)(.5) = .25$$

According to this uniform model, in the long run, 25% of all forms that are processed will have processing times that exceed 5.5 minutes.

The probability that a *discrete* random variable x lies in the interval between two limits a and b depends on whether or not either limit is included in the interval. Suppose, for example, that x is the number of major defects on a new automobile. Then

$$P(3 \leq x \leq 7) = p(3) + p(4) + p(5) + p(6) + p(7)$$

whereas

$$P(3 < x < 7) = p(4) + p(5) + p(6)$$

If, however, x is a *continuous* random variable, such as task completion time, then $P(3 \leq x \leq 7) = P(3 < x < 7)$. The reason is that the area under a density curve and above a single value such as 3 or 7 is zero. The area above an interval of values therefore does not depend on whether either endpoint is included.

For any two numbers a and b, with $a < b$,

$$P(a \leq x \leq b) = P(a < x \leq b) = P(a \leq x < b) = P(a < x < b)$$

when x is a continuous random variable.

A density curve is usually specified by giving a formula for the height of the curve above each possible value. Such a formula is called a *density function*. Once the density function is known, methods from mathematics (integral calculus) can be used to calculate various areas.

EXAMPLE 2 Let x denote the amount of time music is played during a one-hour period on a certain radio station. If x is measured in hours, it must be the case that $0 \leq x \leq 1$. (For example, 15 minutes corresponds to $x = .25$.) Suppose that for any such x, the density function is

$$f(x) = \left(\begin{array}{c} \text{height of density} \\ \text{curve above } x \end{array} \right) = 20x^3(1 - x)$$

Thus the height above .5 is $20(.5)^3(.5) = 1.25$, the height above .25 is $20(.25)^3(.75) = .234375 \approx .234$, and so on. The density curve is sketched in Figure 4. It is unimodal with the peak above .75 and it has a negative skew. The constant 20 in front of the "variable part" of the density function ensures that the total area under the curve is 1.

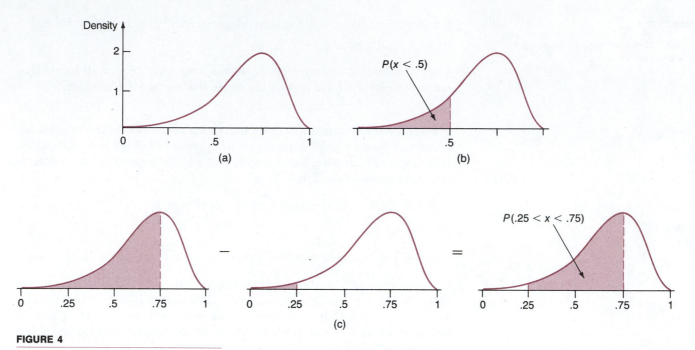

FIGURE 4

Density curve and probabilities for Example 2.

It can be shown (using calculus) that the area under the density curve to the left of a value x is

$$\text{area to the left of } x = 20\left(\frac{x^4}{4} - \frac{x^5}{5}\right)$$

Thus the probability that music is played for less than one-half hour is

$$P(x < .5) = \text{area under density curve to the left of .5}$$
$$= 20[\tfrac{1}{4}(.5)^4 - \tfrac{1}{5}(.5)^5] = .1875$$

This area is shown in Figure 4(b). Notice that if we start with the area to the left of .75 and subtract the area to the left of .25, the area above the interval from .25 to .75 remains. That is,

$$P(.25 < x < .75) = \left(\begin{matrix}\text{area to the}\\\text{left of .75}\end{matrix}\right) - \left(\begin{matrix}\text{area to the}\\\text{left of .25}\end{matrix}\right)$$

$$= 20[\tfrac{1}{4}(.75)^4 - \tfrac{1}{5}(.75)^5] - 20[\tfrac{1}{4}(.25)^4 - \tfrac{1}{5}(.25)^5]$$
$$= .617$$

An illustration appears in Figure 4(c). It is also the case that

$$P(.25 \le x \le .75) = .617.$$

THE MEAN AND THE STANDARD DEVIATION WHEN x IS CONTINUOUS

Figure 5 illustrates how the density curve for a continuous random variable can be approximated by a probability histogram of a discrete random variable. Computing the mean value and the standard deviation using this discrete distribution gives an approximation to μ and σ for the continuous random variable x. If an even more accurate approximating probability

FIGURE 5

Approximating a density curve by a probability histogram.

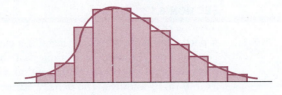

histogram is used (narrower rectangles), better approximations to μ and σ result.

In practice, such an approximation method is often unnecessary. Instead, μ and σ can be defined and computed using methods from calculus. The details need not concern us; what is important is that μ and σ play exactly the same role here as they did in the discrete case. The mean value μ locates the center of the continuous distribution and gives the approximate long-run average of many observed x values. The standard deviation σ measures to what extent the continuous distribution (density curve) spreads out about μ and gives information about the amount of variability that can be expected in a long sequence of observed x values.

EXAMPLE 3

A company receives concrete of a certain type from two different suppliers. Define random variables x and y by

x = compressive strength (psi) of a randomly selected batch from supplier #1

y = compressive strength of a randomly selected batch from supplier #2

Suppose that

$$\mu_x = 4650 \text{ psi}, \qquad \sigma_x = 200 \text{ psi},$$
$$\mu_y = 4500 \text{ psi}, \qquad \sigma_y = 275 \text{ psi}$$

The long-run average strength per batch for many, many batches from the first supplier will be roughly 4650 psi. This is 150 psi greater than the long-run average for batches from the second supplier. In addition, a long sequence of batches from supplier #1 will exhibit substantially less variability in compressive strength values than will a similar sequence from supplier #2. The first supplier is preferred to the second both in terms of average value and variability. Figure 6 displays density curves that are consistent with this information.

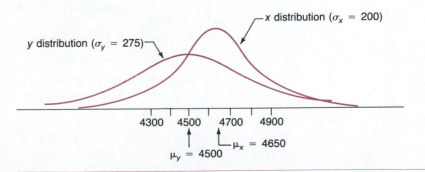

FIGURE 6

Density curves for Example 3.

EXERCISES 6.1–6.8   **SECTION 6.1**

6.1 A particular professor never dismisses class early (do you know anyone like this?). Let x denote the amount of time past the hour (minutes) that elapses before the professor dismisses class. Suppose that x has a uniform distribution on the interval from 0 to 10 minutes. The density curve is shown in the accompanying figure.

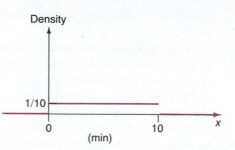

a. What is the probability that at most five minutes elapse before dismissal?
b. What is the probability that between three and five minutes elapse before dismissal?
c. What is the mean value of the time that elapses before dismissal? Explain the reasoning you used to obtain your answer.

6.2 If x has a uniform distribution on the interval from A to B, it can be shown that $\sigma = (B - A)/\sqrt{12}$.
a. Refer to Exercise 6.1. What is the standard deviation of elapsed time until dismissal?
b. Refer to Exercise 6.1. What is the probability that the elapsed time is within one standard deviation of its mean value?

6.3 The article "Modeling Sediment and Water Column Interactions for Hydrophobic Pollutants" (*Water Research* (1984): 1169–74) suggests the uniform distribution on the interval from 7.5 to 20 as a model for $x =$ depth (cm) of the bioturbation layer in sediment for a certain region. (Please don't ask us what the bioturbation layer is!)
a. Draw the density curve for x.
b. What must the height of the density curve be?
c. What is the probability that x is at most 12?
d. What is the probability that x is between 10 and 15? Between 12 and 17? Why are these two probabilities equal?
e. What is the mean value of x?
f. Use the formula for σ in Problem 6.2 to calculate the standard deviation of x.

6.4 Let x denote the amount of gravel sales (tons) during a randomly selected week at a particular sales facility. Suppose that the density curve has height $f(x)$ above the value x, where

$$f(x) = \begin{cases} 2(1 - x) & 0 \le x \le 1 \\ 0 & \text{otherwise} \end{cases}$$

The density curve (the graph of $f(x)$) is shown in the accompanying figure. Use the fact that the area of a triangle $= \frac{1}{2}(\text{base}) \cdot (\text{height})$ to calculate each of the following probabilities.
a. $P(x < \frac{1}{2})$
b. $P(x \le \frac{1}{2})$
c. $P(x < \frac{1}{4})$

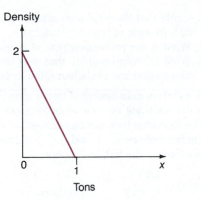

d. $P(\frac{1}{4} < x < \frac{1}{2})$ Hint: Use the results of parts (a) through (c).
e. The probability that sales exceed $\frac{1}{2}$ ton.
f. The probability that sales are at least $\frac{3}{4}$ ton.

6.5 Let x be the amount of time (min) that a particular San Francisco commuter must wait for a BART train. Suppose that the density curve is as pictured (a uniform distribution).

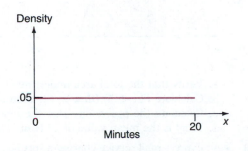

a. What is the probability that x is less than 10 min? More than 15 min?
b. What is the probability that x is between 7 and 12 min?
c. Find the value c for which $P(x < c) = .9$.

6.6 Referring to Exercise 6.5, let x and y be waiting times on two independently selected days. Define a new random variable w by $w = x + y$, the sum of the two waiting times. The set of possible values for w is the interval from 0 to 40 (since both x and y can range from 0 to 20). It can be shown that the density curve of w is as pictured. (It is called a *triangular distribution* for obvious reasons!)

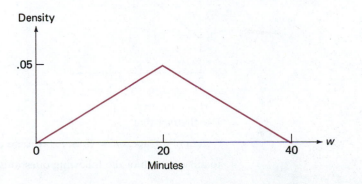

a. Verify that the total area under the density curve is equal to 1. (Hint: The area of a triangle is $\frac{1}{2}$(base) · (height).)
b. What is the probability that w is less than 20? Less than 10? More than 30?
c. What is the probability that w is between 10 and 30? (Hint: It might be easier first to find the probability that w is *not* between 10 and 30.)

6.7 In a certain coin-operated target game, the objective is to get as close as possible to the midpoint of a two-foot-long horizontal line segment. Identify the midpoint with the value 0 on a measurement scale, and let $x =$ the actual landing point. Any number between -1 and $+1$ is a possible value of x. The density curve for x is specified by the height function

$$f(x) = \begin{cases} .75(1 - x^2) & -1 \le x \le 1 \\ 0 & \text{otherwise} \end{cases}$$

The density curve follows. For any x between -1 and $+1$, the area under the curve to the left of x is $.5 + .75(x - \frac{1}{3}x^3)$.

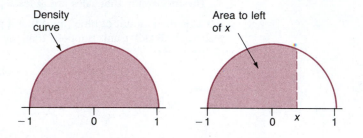

a. Verify that the total area under the density curve (the area to the left of 1) is 1.
b. Calculate $P(x < .5)$ (that is, the area under the density curve to the left of .5).
c. Calculate the probability that the landing point is within $\frac{1}{2}$ foot of the target.
d. What is the mean value of x? Hint: The density curve is symmetric.

6.8 An express mail service charges a special rate for any package that weighs less than one pound. Let x denote the weight of a randomly selected parcel that qualifies for this special rate. The probability distribution of x is specified by the accompanying density curve.

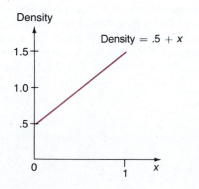

Use the fact that

area of a trapezoid = (base) · (average of two side lengths)

to answer each of the following questions.

a. What is the probability that a randomly selected package of this type is at most $\frac{1}{2}$ pound? Between $\frac{1}{4}$ pound and $\frac{1}{2}$ pound? At least $\frac{3}{4}$ pound?

b. It can be shown that $\mu_x = \frac{7}{12}$ and $\sigma_x^2 = \frac{11}{144}$. What is the probability that the value of x is more than one standard deviation from the mean value?

6.2   THE NORMAL DISTRIBUTION

A normal curve was first introduced in Chapter 2 as one with a shape that gives a very good approximation to histograms for many different data sets. There are actually many different normal curves rather than just one. Every one is bell-shaped, and a particular normal curve results from specifying where the curve is centered and how much it spreads out about its center. For a particular normal distribution, the density function, which gives the height of the curve above each value x on the measurement scale, is rather complicated. Fortunately we needn't deal with it explicitly. For our purposes, an acquaintance with some general properties and a table that gives certain normal curve areas will suffice.

▌ DEFINITION

A continuous random variable x is said to have a **normal distribution** if the density curve of x is a normal curve.* The mean value μ determines where the curve is centered on the measurement axis, and the standard deviation σ determines the extent to which the curve spreads out about μ.

Figure 7 illustrates normal density curves for several different values of μ and σ. As with all density curves, the total area under each curve is 1.

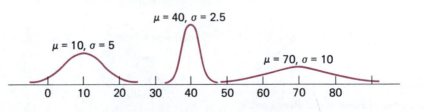

FIGURE 7

Several normal density curves.

---

* The density function for a normally distributed random variable is

$$f(x) = \frac{1}{\sqrt{2\pi\sigma^2}} \cdot e^{-(x-\mu)^2/2\sigma^2} \qquad \text{for } -\infty < x < \infty$$

where the symbols e and π represent numbers whose approximate values are

$$e \approx 2.7182818, \qquad \pi \approx 3.1415927$$

That is, for any number x, the value of $f(x)$ gives the height of the normal curve (with specified mean μ and standard deviation σ) above the value x on the number line.

The value of μ is the number on the measurement axis lying directly below the top of the bell. The value of σ can also be ascertained from a picture of the curve. Consider the normal curve pictured in Figure 8. Starting at the top of the bell (above 100) and moving to the right, the curve turns downward until it is above the value 110. After that point it continues to decrease in height but is turning up rather than down. Similarly, to the left of 100 it turns down until it reaches 90 and then begins to turn up. The curve changes from turning down to turning up at a distance of 10 on either side of μ, and thus $\sigma = 10$. In general, σ is the distance to either side of μ at which a normal curve changes from turning downward to turning upward.

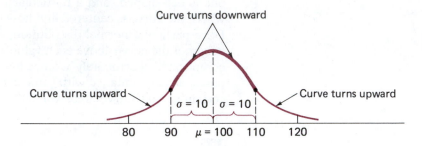

FIGURE 8

Pictorial identification of μ and σ.

When x has a normal distribution with specified values of μ and σ, the probability that an observed x value falls in some interval is the area under the corresponding normal density curve and above that interval. For example, let x denote the number of miles per gallon achieved by a particular type of car in a fuel efficiency test. Suppose that x is a normally distributed variable with $\mu = 27.0$ mpg and $\sigma = 1.5$ mpg. Figure 9 illustrates the curve areas that correspond to various probabilities. In Example 2, we showed how such areas could be evaluated given a formula for the cumulative area up to any value x. However, there is no nice formula of this type for a normal distribution. Instead we will utilize a single table of cumulative normal curve areas.

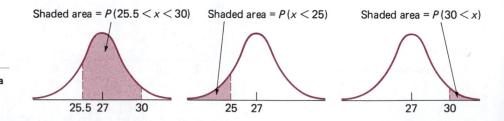

FIGURE 9

Probabilities as curve areas when x has a normal distribution with $\mu = 27.0$ and $\sigma = 1.5$.

THE STANDARD NORMAL DISTRIBUTION

Rather than tabulate normal curve areas separately for each different combination of μ and σ, statisticians have chosen a particular normal curve as a reference curve. Once we learn to use the table containing areas (i.e., probabilities) for this reference curve, it is easy to obtain areas under any other normal curve.

DEFINITION

A random variable z is said to have a **standard normal distribution** if it has a normal distribution with mean value $\mu = 0$ and standard deviation $\sigma = 1$. The corresponding normal curve is referred to as the **standard normal** or **z curve**.

The z curve is displayed in Figure 10(a). It is centered at $\mu = 0$ and the "turning points" are above $z = +1$ and $z = -1$. Appendix Table II, which also appears on the inside front cover, tabulates cumulative z curve areas of the sort shown in Figure 10(b) for many different values along the z axis. The smallest value for which the cumulative area is given is -3.49, a value quite far out in the lower tail of the z curve. The next smallest value for which the area appears is -3.48, then -3.47, then -3.46, and so on, in increments of .01, terminating with the cumulative area to the left of 3.49.

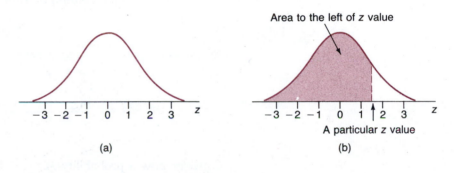

FIGURE 10

A standard normal (z) curve and a cumulative area.

USING THE TABLE OF STANDARD NORMAL CURVE AREAS

Let c denote a number between -3.49 and 3.49 that has two digits to the right of the decimal point (such as -1.76 or 0.58). For each such c, Appendix Table II gives

$$\left(\begin{array}{c}\text{area under } z \text{ curve}\\ \text{to the left of } c\end{array}\right) = P(z < c) = P(z \leq c)$$

To find this probability, locate

(i) the row labeled with the sign of c and the digits to either side of the decimal point (e.g., -1.7 or 0.5);
(ii) the column identified with the second digit to the right of the decimal point in c (e.g., .06 if $c = -1.76$).

The desired probability is the number at the intersection of this row and column.

EXAMPLE 4

The probability $P(z < -1.76)$ is found at the intersection of the -1.7 row and the .06 column of the z table. The result is

$$P(z < -1.76) = .0392$$

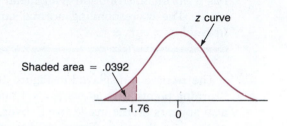

In other words, in a long sequence of observations selected from the standard normal distribution, roughly 3.9% of the observed values will be smaller than -1.76. Similarly,

$$P(z \leq 0.58) = \left(\begin{array}{c}\text{entry in 0.5 row and} \\ \text{.08 column of Table II}\end{array}\right) = .7190$$

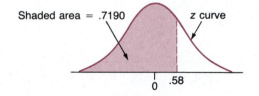

Consider now a probability $P(z < c)$ for which the number c is less than -3.49, for example, $P(z < -3.85)$. This probability does not appear in Appendix Table II, however, it must be the case that

$$P(z < -3.85) < P(z < -3.49)$$

Since $P(z < -3.49) = .0002$, it follows that $P(z < -3.85) < .0002$. In practice, we can say that

$$P(z < -3.85) \approx 0$$

Arguing in a similar fashion, $P(z \leq 4.18) > P(z \leq 3.49) = .9998$, from which we conclude that

$$P(z \leq 4.18) \approx 1$$

We can now use the probabilities tabulated in Appendix Table II to calculate other probabilities involving z. For example (see Figure 11), the probability that z falls in the interval between a lower limit a and an upper limit b is

$$P(a < z < b) = \left(\begin{array}{c}\text{the area under the } z \text{ curve} \\ \text{and above the interval} \\ \text{from } a \text{ to } b\end{array}\right) = P(z < b) - P(z < a)$$

That is, $P(a < z < b)$ is the difference between two cumulative areas. Similarly, the probability that z exceeds a value c is

$$P(c < z) = \left(\begin{array}{c}\text{the area under the } z \text{ curve} \\ \text{to the right of } c\end{array}\right) = 1 - P(z \le c)$$

In words, a "right-tail" area is 1 minus the corresponding cumulative area.

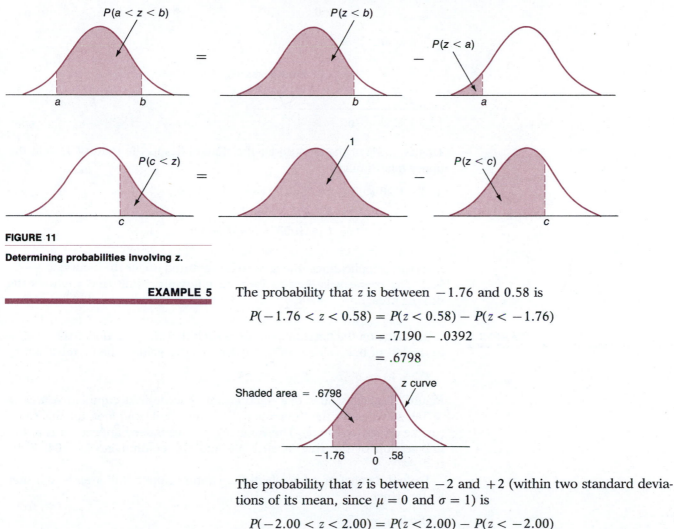

FIGURE 11

Determining probabilities involving z.

EXAMPLE 5

The probability that z is between -1.76 and 0.58 is

$$P(-1.76 < z < 0.58) = P(z < 0.58) - P(z < -1.76)$$
$$= .7190 - .0392$$
$$= .6798$$

The probability that z is between -2 and $+2$ (within two standard deviations of its mean, since $\mu = 0$ and $\sigma = 1$) is

$$P(-2.00 < z < 2.00) = P(z < 2.00) - P(z < -2.00)$$
$$= .9772 - .0228$$
$$= .9544 \approx .95$$

This last probability is the basis for one part of the Empirical Rule, which states that when a histogram is well approximated by a normal curve, roughly 95% of the values are within two standard deviations of the mean.

The probability that the value of z exceeds 1.96 is

$$P(1.96 < z) = -P(z \leq 1.96)$$
$$= 1 - .9750$$
$$= .0250$$

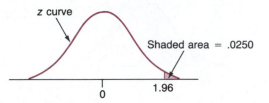

That is, 2.5% of the area under the z curve lies to the right of 1.96 in the upper tail. Similarly,

$$P(-1.28 < z) = \text{area to the right of } -1.28$$
$$= 1 - P(z \leq -1.28)$$
$$= 1 - .1003 = .8997 \approx .90$$

In some applications, the area to be captured under the z curve is specified. The problem is then to find the interval or limit that captures the desired area.

EXAMPLE 6

What value on the horizontal z axis is such that the area under the z curve to the left of that value is .67? If c denotes this values, then c must satisfy

$$P(z < c) = .67$$

Figure 12(a) illustrates this relationship. The desired cumulative area is .6700, so we enter the "main body" of the z table and look for this value (or the closest one to it). The value .6700 does indeed appear—it is at the intersection of the row labeled 0.4 and the column labeled .04. Thus $c = .44$.

To find a value of c for which the captured upper-tail area is .05, that is, for which

$$P(c < z) = .05$$

note that the cumulative area to the left of c must be .95 (see Figure 12(b)). A search for .9500 in Appendix Table II reveals the following:

.9495 is in the 1.6 row and the .04 column.

.9505 is in the 1.6 row and the .05 column.

Because .9500 is halfway between these two areas, we use $c = 1.645$.

What interval, symmetrically placed about zero, captures a central z curve area of .95? That is, for what value of c do we have

$$P(-c < z < c) = .95?$$

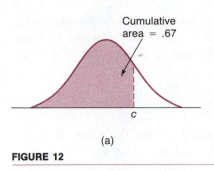

Cumulative area = .67

Upper-tail area = .05

Lower-tail area = .025

Central area = .95

c

c

$-c$ 0 c

(a)

(b)

(c)

FIGURE 12

Capturing specified areas under the z curve.

As Figure 12(c) shows, the cumulative area up to c must be .9750. The z table then reveals that $c = 1.96$. To check, note that

$$P(-1.96 < z < 1.96) = P(z < 1.96) - P(z < -1.96)$$
$$= .9750 - .0250$$
$$= .9500$$

Many statistical procedures utilize values that capture either particular tail areas or central areas under the z curve. Such values are often referred to as **z critical values**. Table 1 displays the most frequently used z critical values.

TABLE 1

The Most Useful z Critical Values

Critical Value, z	Area to the Right of z	Area to the Left of −z	Area between −z and z
1.28	.10	.10	.80
1.645	.05	.05	.90
1.96	.025	.025	.95
2.33	.01	.01	.98
2.58	.005	.005	.99
3.10	.001	.001	.998
3.30	.0005	.0005	.999

PROBABILITIES FOR AN ARBITRARY NORMAL DISTRIBUTION

Calculation of any probability involving a normal random variable can be reduced to a z curve calculation by means of the following result.

Let x be a normally distributed random variable with mean value μ and standard deviation σ. Then the *standardized variable*

$$z = \frac{x - \mu}{\sigma}$$

obtained by first subtracting the mean value and then dividing by the standard deviation, has a standard normal distribution. That is, the probability distribution of z is given by the standard normal curve.

To see how this result is used, suppose x has a normal distribution with $\mu = 30$ and $\sigma = 2$. Let's first consider $P(27 < x < 36)$. Subtracting 30 from each term inside the parentheses and dividing by 2 gives an equivalent event:

$$27 < x < 36 \quad \text{if and only if} \quad \frac{27 - 30}{2} < \frac{x - 30}{2} < \frac{36 - 30}{2}$$

Since the events are equivalent, the probabilities are the same. Because $z = (x - 30)/2$ has a standard normal distribution, we have

$$P(27 < x < 36) = P\left(\frac{27 - 30}{2} < \frac{x - 30}{2} < \frac{36 - 30}{2}\right)$$

$$= P(-1.50 < z < 3.00)$$

The desired probability is therefore the area under the z curve and above the interval from -1.50 to 3.00. Similarly,

$$x < 29 \quad \text{if and only if} \quad \frac{x - 30}{2} < \frac{29 - 30}{2}$$

so

$$P(x < 29) = P\left(\frac{x - 30}{2} < \frac{29 - 30}{2}\right) = P(z < -.50)$$

In general, **standardizing** the limit or limits on x—by subtracting the mean and dividing by the standard deviation—gives the corresponding limit or limits on a standard normal random variable.

When x has a normal distribution with mean μ and standard deviation σ,

$$P(a < x < b) = P\left(\frac{a - \mu}{\sigma} < z < \frac{b - \mu}{\sigma}\right) = \begin{array}{l} z \text{ curve area between} \\ \text{the standardized limits} \end{array}$$

$$P(x < a) = P\left(z < \frac{a - \mu}{\sigma}\right) = \begin{array}{l} z \text{ curve area to the left of the} \\ \text{standardized limit} \end{array}$$

$$P(b < x) = P\left(\frac{b - \mu}{\sigma} < z\right) = \begin{array}{l} z \text{ curve area to the right of the} \\ \text{standardized limit} \end{array}$$

EXAMPLE 7

Let x denote the systolic blood pressure (mm) of an individual selected at random from a certain population. Suppose that x has a normal distribution with mean $\mu = 120$ mm and standard deviation $\sigma = 10$ mm. (The article "Oral Contraceptives, Pregnancy, and Blood Pressure" (*J. Amer. Med. Assoc.* 222 (1972): 1507–10) reported on the results of a large study in which a sample histogram of blood pressures among women of similar ages was found to be well approximated by a normal curve.) To find the probability that x is between 110 and 140, we first standardize these limits:

$$\text{standardized lower limit} = \frac{110 - 120}{10} = -1.00$$

$$\text{standardized upper limit} = \frac{140 - 120}{10} = 2.00$$

Then

$$
\begin{aligned}
P(110 < x < 140) &= P(-1.00 < z < 2.00) \\
&= P(z < 2.00) - P(z < -1.00) \\
&= .9772 - .1587 \\
&= .8185
\end{aligned}
$$

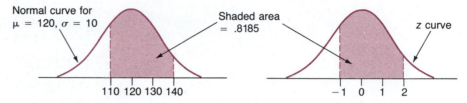

Thus if blood pressure values for many, many women from this population were determined, roughly 82% of the values would fall between 110 and 140 mm. Similarly,

$$\frac{125 - 120}{10} = \frac{5}{10} = .50$$

from which we conclude

$$
\begin{aligned}
P(125 < x) &= P(.50 < z) \\
&= 1 - P(z \leq .50) \\
&= 1 - .6915 \\
&= .3085.
\end{aligned}
$$

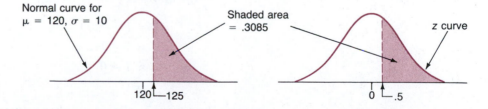

Since $\mu - 4\sigma = 120 - 40 = 80$, the probability that x falls more than four standard deviations below its mean is

$$P(x < 80) = P\left(z < \frac{80 - 120}{10}\right) = P(z < -4) \approx 0$$

If an individual's blood pressure is reported as less than 80, it is highly unlikely that the value was selected from the normal distribution specified here.

EXAMPLE 8   One method for determining the hardness of a metal involves impressing a hardened point into the surface of the metal and measuring the depth of penetration. Suppose that the hardness of a particular alloy is normally distributed with a mean value of 70 and a standard deviation of 4.

(a) What is the probability that the hardness of a randomly selected specimen is between 60 and 65? Is at least 50? To answer these questions, let

x = hardness of a randomly selected specimen

Then x has a normal distribution with $\mu = 70$ and $\sigma = 4$, so

$$P(60 < x < 65) = P\left(\frac{60 - 70}{4} < z < \frac{65 - 70}{4}\right)$$

$$= P(-2.50 < z < -1.25)$$

$$= \left(\begin{array}{c}\text{area to the left} \\ \text{of} -1.25\end{array}\right) - \left(\begin{array}{c}\text{area to the left} \\ \text{of} -2.50\end{array}\right)$$

$$= (-1.25 \ z \text{ table entry}) - (-2.50 \ z \text{ table entry})$$

$$= .1056 - .0062$$

$$= .0994$$

Note that $P(60 \leq x \leq 65) = .0994$ also, since x is a continuous random variable.

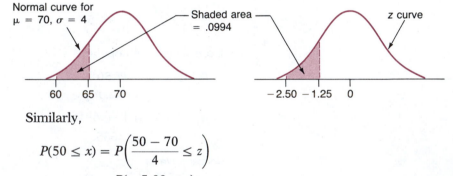

Similarly,

$$P(50 \leq x) = P\left(\frac{50 - 70}{4} \leq z\right)$$

$$= P(-5.00 \leq z)$$

$$= 1 - P(z < -5.00)$$

$$\approx 1 - 0$$

$$= 1$$

Thus there is almost no chance that hardness will be smaller than 50 (more than five standard deviations below the mean value).

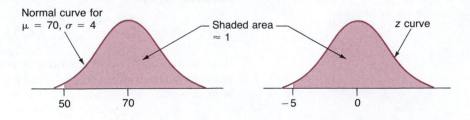

(b) Suppose that four specimens are independently selected. What is the probability that hardness exceeds 65 for exactly two of the four specimens? Let's call a specimen a success S if its hardness exceeds 65. Then

$$\pi = P(S) = P(65 < x) = P(-1.25 < z)$$
$$= 1 - (\text{area to the left of} -1.25) = .8944$$

The event of interest is that there are exactly two S's in a binomial experiment with $n = 4$ trials and success probability $\pi = .8944$. Using the binomial probability distribution formula gives us

$$p(2) = \left(\frac{4!}{2!2!}\right)(.8944)^2(.1056)^2$$
$$= (6)(.8944)^2(.1056)^2$$
$$= .0535$$

The process of standardizing a value is just a way of re-expressing a distance. If we standardize the value 34 using $\mu = 30$ and $\sigma = 2$, the result is

$$\frac{34 - 30}{2} = \frac{4}{2} = 2$$

This says that 34 is two standard deviations "above" (to the right of) the mean. Similarly,

$$\frac{27.5 - 30}{2} = \frac{-2.5}{2} = -1.25$$

so 27.5 is 1.25 standard deviations "below" (to the left of) the mean.

> Standardizing a value gives the distance from that value to the mean in units of standard deviation. A value above the mean yields a positive standardized value, while a value below the mean yields a negative standardized value.

EXAMPLE 9　If reaction time x to a certain stimulus is normally distributed, what is the probability that x is observed to be within one standard deviation of its mean? The value $\mu - \sigma$ is one standard deviation below the mean, and $\mu + \sigma$ is one standard deviation above the mean. Therefore

$$P\left(\begin{array}{l}x \text{ is within 1 standard} \\ \text{deviation of its mean}\end{array}\right) = P(\mu - \sigma < x < \mu + \sigma)$$

Standardizing limits gives us

$$\frac{(\mu + \sigma) - \mu}{\sigma} = \frac{\sigma}{\sigma} = 1.00, \qquad \frac{(\mu - \sigma) - \mu}{\sigma} = \frac{-\sigma}{\sigma} = -1.00$$

from which we obtain

$$
\begin{aligned}
P(\mu - \sigma < x < \mu + \sigma) &= P(-1.00 < z < 1.00) \\
&= .8413 - .1587 \\
&= .6826 \\
&\approx .68
\end{aligned}
$$

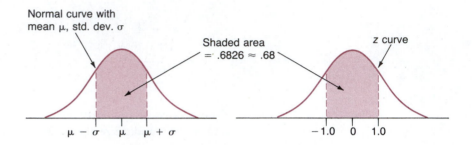

Replacing one standard deviation by two gives a probability of .9544 or roughly .95, and using three in place of one yields .9974 or roughly .997. Multiplication of these three probabilities by 100 gives the percentages quoted earlier in the Empirical Rule. Notice that these probabilities do not depend on μ or σ. More generally,

$$
P\left(\begin{array}{c} x \text{ is within } k \text{ standard} \\ \text{deviations of } \mu \end{array}\right) = P(\mu - k\sigma < x < \mu + k\sigma)
$$

$$
= P(-k < z < k)
$$

This probability depends only on k and not on μ or σ.

We saw that in some z problems, a probability (curve area) is specified and the corresponding limit is requested. This also occurs in other normal distribution settings.

EXAMPLE 10   The Environmental Protection Agency has in recent years developed a testing program to monitor vehicle emission levels of several pollutants. The article "Determining Statistical Characteristics of a Vehicle Emissions Audit Procedure" (*Technometrics* (1980): 483–93) describes the program, which involves using different vehicle configurations (combinations of weight, engine type, transmission, and axle ratios) and a fixed driving schedule (including cold- and hot-start phases, idling, accelerating, and decelerating). Data presented in the paper suggests that the normal distribution is a plausible model for the amount of oxides of nitrogen (g/mile) emitted. Let x denote the amount of this pollutant emitted by a randomly selected vehicle with a particular configuration. Suppose that x has a normal distribution with $\mu = 1.6$ and $\sigma = .4$. What pollution level c is such that 99% of all such vehicles emit pollution amounts less than c and only 1% exceed it? The distribution of x appears in Figure 13, with c identified on the measurement scale.

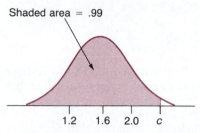

Shaded area = .99

FIGURE 13

The normal distribution for Example 10.

1.2   1.6   2.0 c

The value c satisfies $P(x < c) = .99$. By standardizing, this can be related to a z critical value:

$$P(x < c) = P\left(z < \frac{c - 1.6}{.4}\right) = .99$$

From Table 1 (given earlier in this section), the z critical value that captures a cumulative area of .99 is 2.33, i.e. $P(z < 2.33) = .99$. Thus

$$\frac{c - 1.6}{.4} = 2.33$$

from which

$$c = 1.6 + (.4)(2.33) = 2.532$$

EXERCISES   6.9–6.29   **SECTION 6.2**

6.9   Determine the following standard normal (z) curve areas.
  a. The area under the z curve to the left of 1.75
  b. The area under the z curve to the left of $-.68$
  c. The area under the z curve to the right of 1.20
  d. The area under the z curve to the right of -2.82
  e. The area under the z curve between -2.22 and .53
  f. The area under the z curve between -1 and 1
  g. The area under the z curve between -4 and 4

6.10   Determine each of the areas under the standard normal (z) curve.
  a. To the left of -1.28
  b. To the right of 1.28
  c. Between -1 and 2
  d. To the right of 0
  e. To the right of -5
  f. Between -1.6 and 2.5
  g. To the left of .23

6.11   Let z denote a random variable having a standard normal distribution. Determine each of the following probabilities.
   a. $P(z < 2.36)$
   b. $P(z \leq 2.36)$
   c. $P(z < -1.23)$
   d. $P(1.14 < z < 3.35)$
   e. $P(-.77 < z < -.55)$
   f. $P(-2.90 < z \leq 1.15)$
   g. $P(2 < z)$
   h. $P(-3.38 \leq z)$
   i. $P(z < 4.98)$

6.12   Let z denote a random variable having a normal distribution with $\mu = 0$ and $\sigma = 1$. Determine each of the following probabilities.
   a. $P(z < .10)$
   b. $P(z < -.10)$
   c. $P(.40 < z < .85)$
   d. $P(-.85 < z < -.40)$
   e. $P(-.40 < z < .85)$
   f. $P(-1.25 < z)$
   g. $P(z < -1.50 \text{ or } z > 2.50)$

6.13   Determine a number c to satisfy each of the following conditions (where z has a standard normal distribution).
   a. $P(z < c) = .5910$
   b. $P(z < c) = .4090$
   c. $P(c < z) = .0030$
   d. $P(-c < z < c) = .7540$

6.14   Determine the value c for which each of the following probabilities involving a standard normal random variable is as specified.
   a. $P(z < c) = .0075$
   b. $P(c < z) = .0040$
   c. $P(c < z) = .9830$
   d. $P(-c < z < c) = .9940$
   e. $P(z < -c \text{ or } c < z) = .0160$

6.15   Each of the following probabilities P involves a standard normal random variable z. Determine the value of c in each case so that P is as specified.
   a. $P(z < c) = .975$
   b. $P(z < c) = .90$
   c. $P(c < z) = .90$
   d. $P(c < z) = .005$
   e. $P(-c < z < c) = .99$
   f. $P(z > c \text{ or } z < -c) = .002$

6.16   Because $P(z < 0.44) = .67$, the value 0.44 is the 67th percentile of the standard normal distribution. Determine the value of each of the following percentiles for the standard normal distribution. (If the cumulative area that you must look for does not appear in the z table, use the closest entry.)
   a. The 91st percentile (Hint: Look for area .9100.)
   b. The 77th percentile
   c. The 50th percentile
   d. The 9th percentile
   e. The 23rd percentile

    f. How are the percentiles of (a) and (d) related? If p denotes a number between 0 and 1, what is the relationship between the $100p$th percentile and the $100(1 - p)$th percentile? What property of the z curve justifies your assertion?

6.17 Let x denote the amount of dye (ml) dispensed into one gallon of paint when mixing a certain shade. Suppose that x has a normal distribution with $\mu = 5$ ml and $\sigma = .2$ ml. Calculate the following probabilities.
    a. $P(x < 5.0)$
    b. $P(x < 5.4)$
    c. $P(x \le 5.4)$
    d. $P(4.6 < x < 5.2)$
    e. $P(4.5 < x)$
    f. $P(4.0 < x)$

6.18 Let x be the weight (lb) of a certain type of fish caught at a particular location. If x is normally distributed with a mean value of 3.8 lb and a standard deviation of 1.1 lb, calculate the following probabilities.
    a. $P(x < 4.0)$
    b. $P(3.25 < x)$
    c. $P(3 < x < 5)$
    d. $P(x < 8)$
    e. $P(8 < x)$
    f. $P(x < 6.363)$
    g. $P(1.644 < x < 5.956)$

6.19 A gasoline tank for a certain car is designed to hold 15 gallons. Suppose that the actual capacity x of a randomly chosen tank has a normal distribution with mean value 15.0 gal and standard deviation .10 gal.
    a. What is the probability that a randomly selected tank will hold at most 14.8 gal?
    b. What is the probability that a randomly selected tank will hold between 14.7 and 15.1 gal?

6.20 Refer back to Exercise 6.19, and suppose that the car on which the randomly selected tank is mounted gets exactly 25 mpg on a trip. What is the probability that the car can travel 370 miles without refueling? Hint: At 25 mpg, a trip of 370 miles requires how much gasoline?

6.21 Suppose that the force acting on a column that provides support for a building is normally distributed with mean 15.0 Kips and standard deviation 1.25 Kips. What is the probability that the force
    a. is at most 17 Kips?
    b. is less than 17 Kips?
    c. is between 12 and 17 Kips?
    d. differs from 15.0 Kips by more than two standard deviations?

6.22 Stress resistance x (psi) for a certain type of plastic sheet is normally distributed with mean value 30 and standard deviation .6.
    a. Calculate $P(29 < x < 32)$.
    b. Calculate $P(x < 28.2)$
    c. Calculate the probability that x is farther than 1.2 from its mean value.

6.23 Return to Exercise 6.22, involving stress resistance of a plastic sheet.
    a. Find the number c for which $P(x < c) = .01$. Hint: $P(z < -2.33) = .01$.
    b. Find the value of c for which 95% of all such sheets have stress resistances within c of the mean value, i.e., for which $P(30 - c < x < 30 + c) = .95$. Hint: $P(-1.96 < z < 1.96) = .95$

6.24 What is the probability that the value of a normally distributed random variable is observed to lie
a. within 1.5 standard deviations of its mean value?
b. farther than 2.5 standard deviations from its mean value?
c. between 1 and 2 standard deviations from its mean value?

6.25 The air pressure in a randomly selected tire put on a certain model new car is normally distributed with mean value 31 psi and standard deviation .2 psi.
a. What is the probability that the pressure for a randomly selected tire exceeds 30.5 psi?
b. What is the probability that the pressure for a randomly selected tire is between 30.5 and 31.5 psi? Is between 30 and 32 psi?
c. Suppose a tire is classed as underinflated if its pressure is less than 30.4 psi. What is the probability that at least one of the four tires on a car is underinflated? Hint: If A = *at least 1 tire is underinflated*, what is *not A*?

6.26 The time that it takes a randomly selected job applicant to perform a certain task is normally distributed with a mean value of 120 seconds and a standard deviation of 20 seconds. The fastest 10% are to be given advanced training. What task times qualify individuals for such training?

6.27 A machine that produces ball bearings has initially been set so that the true average diameter of the bearings it produces is .500 inch. A bearing is acceptable if its diameter is within .004 in. of this target value. Suppose, however, that the setting has changed during the course of production, so that the bearings have normally distributed diameters with mean value .499 in. and standard deviation .002 in. What percentage of the bearings produced will not be acceptable?

6.28 Suppose that net typing rate in words per minute (wpm) for experienced electric-typewriter touch typists is approximately normally distributed with a mean value of 60 wpm and a standard deviation of 15 wpm. (The paper "Effects of Age and Skill in Typing" (*J. of Exper. Psych.* (1984): 345–71) describes how net rate is obtained from gross rate by using a correction for errors.)
a. What is the probability that a randomly selected typist's net rate is at most 60 wpm? Less than 60 wpm?
b. What is the probability that a randomly selected typist's net rate is between 45 and 90 wpm?
c. Would you be surprised to find a typist in this population whose net rate exceeded 105 wpm? (Note: The largest net rate in a sample described in the paper cited is 104 wpm.)
d. Suppose that two typists are independently selected. What is the probability that both their typing rates exceed 75 wpm?

6.29 Because $P(z \leq 2.33) = .9901 \approx .99$, the value 2.33 is the 99th percentile of the standard normal distribution. That is, 99% of the area under the z curve lies to the left of 2.33. Let x have a normal distribution with mean value μ and standard deviation σ.
a. Show that $P(x \leq \mu + 2.33\sigma) = .99$, so that $\mu + 2.33\sigma$ is the 99th percentile of this normal distribution.
b. In Exercise 6.25, tire pressure x (psi) was assumed to be normally distributed with mean 31.0 and standard deviation .2. What is the 99th percentile of the tire pressure distribution (the value below which 99% of such tires have tire pressure)?
c. What is the 90th percentile of the tire pressure distribution? Hint: First find the 90th percentile for z.
d. What is the first percentile for the tire pressure distribution?

6.3 FURTHER APPLICATIONS OF THE NORMAL DISTRIBUTION

In this section we first show how probabilities for some discrete random variables can be approximated using a normal curve. The most important case of this concerns the calculation of binomial probabilities. Finally, as a prelude to statistical analysis, we present a method for judging whether it is plausible that a given sample has come from a normal population distribution.

THE NORMAL CURVE AND DISCRETE VARIABLES

The probability distribution of a discrete random variable x is represented pictorially by a probability histogram. The probability of a particular value is the area of the rectangle centered at that value. Possible values of x are isolated points on the number line, usually whole numbers. For example, if $x =$ the IQ of a randomly selected 8-year-old child, then x is a discrete random variable, since an IQ score must be a whole number.

Often a probability histogram can be very well approximated by a normal curve, as illustrated in Figure 14. In such cases, it is customary to say that x has *approximately* a normal distribution. The normal distribution can then be used to calculate approximate probabilities of events involving x.

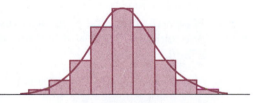

FIGURE 14

A normal curve approximation to a probability histogram.

EXAMPLE 11

The number of express mail packages mailed at a certain post office on a randomly selected day is approximately normally distributed with mean value 18 and standard deviation 6. Let's first calculate the approximate probability that $x = 20$. Figure 15(a) shows a portion of the probability histogram for x with the approximating normal curve superimposed. The area of the shaded rectangle is $P(x = 20)$. The left edge of this rectangle is at 19.5 on the horizontal scale, and the right edge is at 20.5. Therefore the desired probability is approximately the area under the normal curve between 19.5 and 20.5. Standardizing these limits gives us

$$\frac{20.5 - 18}{6} = .42, \qquad \frac{19.5 - 18}{6} = .25$$

from which we get

$$P(x = 20) \approx P(.25 < z < .42) = .6628 - .5987 = .0641$$

FIGURE 15

The normal approximation for Example 11.

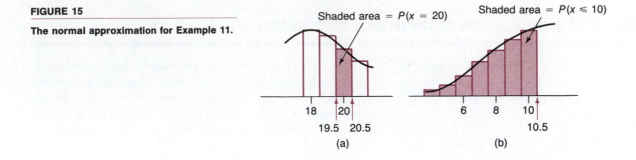

In a similar fashion, Figure 15(b) shows that $P(x \leq 10)$ is approximately the area under the normal curve to the left of 10.5. Thus,

$$P(x \leq 10) \approx P\left(z \leq \frac{10.5 - 18}{6}\right) = P(z \leq -1.25) = .1056$$

The calculation of probabilities in Example 11 illustrates the use of a **continuity correction**. Because the rectangle for $x = 10$ extends to 10.5 on the right, we use the normal curve area to the left of 10.5 rather than just 10. In general, if possible x values are consecutive whole numbers, then $P(a \leq x \leq b)$ will be approximately the normal curve area between limits $a - \frac{1}{2}$ and $b + \frac{1}{2}$.

THE NORMAL APPROXIMATION TO A BINOMIAL DISTRIBUTION

Figure 16 shows the probability histograms for two binomial distributions, one with $n = 25$, $\pi = .4$ and the other with $n = 25$, $\pi = .1$. For each distribution we computed

$$\mu = n\pi \quad \text{and} \quad \sigma = \sqrt{n\pi(1 - \pi)}$$

and then we superimposed a normal curve with this μ and σ on the corresponding probability histogram. A normal curve fits the probability histogram very well in the first case. When this happens, binomial probabilities

FIGURE 16

Normal approximations to binomial distributions.

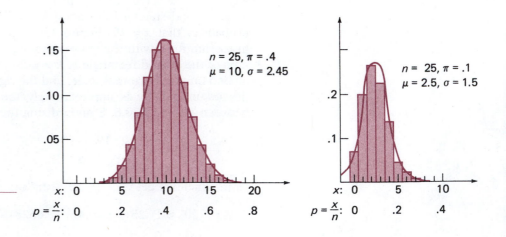

can be accurately approximated by areas under the normal curve. Because of this, statisticians say that both x (the number of S's) and x/n (the proportion of S's) are "approximately normally distributed". In the second case, the normal curve does not give a good approximation because the probability histogram is quite skewed, whereas the normal curve is symmetric. A normal curve would also give a poor approximation when $n = 25$, $\pi = .9$, since then the probability histogram would be centered at $\mu = 22.5$ and skewed in the opposite direction.

Let x be a binomial random variable based on n trials and success probability π, so that

$$\mu = n\pi, \qquad \sigma = \sqrt{n\pi(1 - \pi)}$$

If n and π are such that both

$$n\pi \geq 5 \qquad \text{and} \qquad n(1 - \pi) \geq 5$$

then x has approximately a normal distribution. This along with the continuity correction implies that

$$P(a \leq x \leq b) \approx P\left(\frac{a - \frac{1}{2} - \mu}{\sigma} \leq z \leq \frac{b + \frac{1}{2} - \mu}{\sigma}\right)$$

That is, the probability that x is between a and b inclusive is approximately the area under the approximating normal curve between $a - \frac{1}{2}$ and $b + \frac{1}{2}$. Similarly,

$$P(x \leq b) \approx P\left(z \leq \frac{b + \frac{1}{2} - \mu}{\sigma}\right)$$

$$P(a \leq x) \approx P\left(\frac{a - \frac{1}{2} - \mu}{\sigma} \leq z\right)$$

When either $n\pi < 5$ or $n(1 - \pi) < 5$, the binomial distribution is too skewed for the normal approximation to give accurate results.

EXAMPLE 12

Premature babies are those born more than three weeks early. *Newsweek* (May 16, 1988) reports that 10% of the live births in this country are premature. Suppose that 250 live births are randomly selected and the number x of "preemies" is determined. Since

$$n\pi = 250(.10) = 25 \geq 5$$
$$n(1 - \pi) = 250(.90) = 225 \geq 5$$

x has approximately a normal distribution, with

$$\mu = 250(.10) = 25, \qquad \sigma = \sqrt{250(.1)(.9)} = 4.743$$

The probability that x is between 15 and 30 (inclusive) is

$$P(15 \leq x \leq 30) = P\left(\frac{14.5 - 25}{4.743} < z < \frac{30.5 - 25}{4.743}\right)$$

$$= P(-2.21 < z < 1.16)$$

$$= .8770 - .0136$$

$$= .8634$$

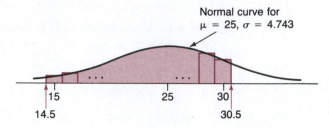

Normal curve for
$\mu = 25$, $\sigma = 4.743$

15
14.5

25

30
30.5

The event that *fewer than* 20 births are premature includes x values 19, 18, 17, . . . , 1, and 0, so

$$P(x < 20) = P(x \leq 19) \approx P\left(z < \frac{19.5 - 25}{4.743}\right)$$

$$= P(z < -1.16)$$

$$= .1230$$

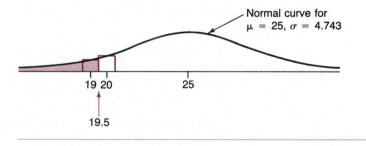

Normal curve for
$\mu = 25$, $\sigma = 4.743$

19 20

25

19.5

CHECKING FOR NORMALITY   Some of the most frequently used statistical methods are valid only when a sample $x_1, x_2, \ldots, x_n$ has come from a population distribution that is (at least approximately) normal. One way to see whether an assumption of population normality is plausible is to construct a **normal probability plot**. This plot utilizes certain quantities called **normal scores**. The values of the normal scores depend on the sample size n. For example, the normal scores when $n = 10$ are as follows:

$$-1.539, \quad -1.001, \quad -.656, \quad -.376, \quad -.123,$$
$$.123, \quad .376, \quad .656, \quad 1.001, \quad 1.539$$

To interpret these numbers, think of selecting sample after sample from a standard normal distribution, each one consisting of $n = 10$ observations. Then -1.539 is the long-run average of the smallest observation from each sample, -1.001 is the long-run average of the second smallest observation from each sample, and so on. Said another way, -1.539 is the mean value of the smallest observation in a sample of size 10 from the z distribution, -1.001 is the mean value of the second smallest observation, etc.

After ordering the sample observations from smallest to largest, the smallest normal score is paired with the smallest observation, the second smallest normal score with the second smallest observation, and so on. The first number in a pair is the normal score and the second is the observed x value. Each such pair can be represented as a point on a two-dimensional coordinate system. Consider, for example, the pair $(1.001, 35.0)$. The corresponding point lies above 1.001 on the horizontal axis and to the right of 35.0 on the vertical axis. This is illustrated in Figure 17.

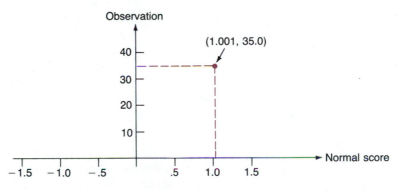

FIGURE 17

Representing a pair of numbers as a point.

Once all points are plotted, a substantial linear pattern in the plot suggests that population normality is plausible. On the other hand, a systematic departure from a straight-line pattern (such as curvature in the plot) casts doubt on the legitimacy of assuming a normal population distribution.

EXAMPLE 13

The following ten observations are widths of contact windows in integrated circuit chips:

3.21   2.49   2.94   4.38   4.02   3.62   3.30   2.85   3.34   3.81

The ten pairs for the normal probability plot are then

$(-1.539, 2.49)$	$(.123, 3.34)$
$(-1.001, 2.85)$	$(.376, 3.62)$
$(-.656, 2.94)$	$(.656, 3.81)$
$(-.376, 3.21)$	$(1.001, 4.02)$
$(-.123, 3.30)$	$(1.539, 4.38)$

The normal probability plot is shown in Figure 18. The linearity of the plot supports the assumption that the window width distribution from which these observations were drawn is normal.

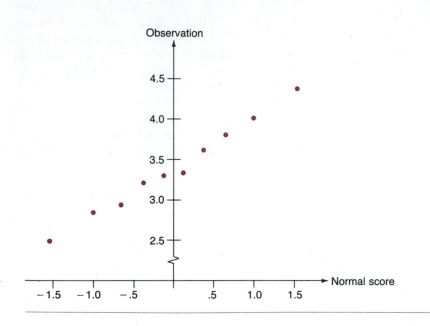

FIGURE 18

A normal probability plot for Example 13.

Extensive tabulations of normal scores for many different sample sizes are available. Alternatively, the better packages of statistical programs (such as MINITAB and SAS) can compute these scores on request and then construct a normal probability plot.

The judgment as to whether a plot does or doesn't show a substantial linear pattern is somewhat subjective. Particularly when n is small, normality should not be ruled out unless the departure from linearity is very clear cut. Figure 19 displays several plots that suggest a non-normal population distribution.

FIGURE 19

Plots suggesting nonnormality
(a) Indication that the population distribution is skewed;
(b) Indication that the population distribution has "heavier tails" than a normal curve;
(c) Presence of an outlier.

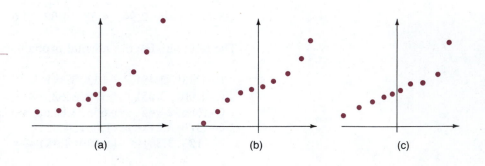

6.30   Let x denote the IQ score for an individual selected at random from a certain population. The value of x must be a whole number. Suppose that the distribution of x can be approximated by a normal distribution with mean value 100 and standard deviation 15. Calculate (approximations to) the following probabilities.
  a. $P(x = 100)$
  b. $P(x \leq 110)$
  c. $P(x < 110)$ Hint: $x < 110$ is the same as $x \leq 109$.
  d. $P(75 \leq x \leq 125)$

6.31   Suppose that the distribution of the number of items x produced by an assembly line during an eight-hour shift can be approximated by a normal distribution with mean value 150 and standard deviation 10.
  a. What is the probability that the number of items produced is at most 120?
  b. What is the probability that at least 125 items are produced?
  c. What is the probability that between 135 and 160 (inclusive) items are produced?

6.32   The number of vehicles leaving a turnpike at a certain exit during a particular time period has approximately a normal distribution with mean value 500 and standard deviation 75. What is the probability that the number of cars exiting during this period is
  a. at least 650?
  b. strictly between 400 and 550? ("Strictly" means that the values 400 and 550 are not included.)
  c. between 400 and 550 (inclusive)?

6.33   Let x have a binomial distribution with $n = 50$ and $\pi = .6$, so that $\mu = n\pi = 30$ and $\sigma = \sqrt{n\pi(1 - \pi)} = 3.4641$. Calculate the following probabilities using the normal approximation with the continuity correction.
  a. $P(x = 30)$
  b. $P(x = 25)$
  c. $P(x \leq 25)$
  d. $P(25 \leq x \leq 40)$
  e. $P(25 < x < 40)$ Hint: $25 < x < 40$ is the same as $26 \leq x \leq 39$.

6.34   Seventy percent of the bicycles sold by a certain store are mountain bikes. Among 100 randomly selected bike purchases, what is the approximate probability that
  a. at most 75 are mountain bikes?
  b. between 60 and 75 (inclusive) are mountain bikes?
  c. more than 80 are mountain bikes?
  d. at most 30 are *not* mountain bikes?

6.35   Suppose that 25% of the fire alarms in a large city are false alarms. Let x denote the number of false alarms in a random sample of 100 alarms. Give approximations to the following probabilities.
  a. $P(20 \leq x \leq 30)$
  b. $P(20 < x < 30)$
  c. $P(35 \leq x)$
  d. The probability that x is farther than two standard deviations from its mean value.

6.36   Suppose that 65% of all registered voters in a certain area favor a 7-day waiting period prior to purchase of a handgun. Among 225 randomly selected voters, what is the probability that
  a. at least 150 favor such a waiting period?

b. more than 150 favor such a waiting period?

c. fewer than 125 favor such a waiting period?

6.37 Flash bars manufacturered by a certain company are sometimes defective.

a. If 5% of all such bars are defective, could the techniques of this section be used to approximate the probability that at least five of the bars in a random sample of size 50 are defective? If so, calculate this probability; if not, explain why not.

b. Reconsider the question posed in part (a) for the probability that at least 20 bars in a random sample of size 500 are defective.

6.38 A company that manufactures mufflers for cars offers a lifetime warranty on its products provided that ownership of the car does not change. Suppose that only 20% of its mufflers are replaced under this warranty.

a. In a random sample of 400 purchases, what is the approximate probability that between 75 and 100 (inclusive) are replaced under warranty?

b. Among 400 randomly selected purchases, what is the probability that at most 70 are ultimately replaced under warranty?

c. If you were told that fewer than 50 among 400 randomly selected purchases were ever replaced under warranty, would you question the 20% figure? Explain.

6.39 Ten measurements of the steam rate (lb/hr) of a distillation tower were used to construct the given normal probability plot (Source: "A Self-Descaling Distillation Tower" (*Chem. Eng. Process* (1968): 79–84)). Based on this plot, do you think it is reasonable to assume that the normal distribution provides an adequate description of the steam rate distribution? Explain.

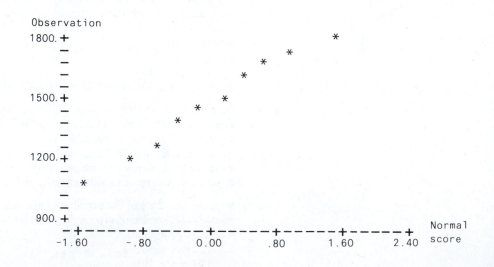

6.40 The accompanying normal probability plot was constructed using part of the data appearing in the paper "Trace Metals in Sea Scallops" (*Environ. Concentration and Toxicology* 19: 1326–34). The variable under study was the amount of cadmium in North Atlantic sea scallops. Does the sample data suggest that the cadmium concentration distribution is not normal? Explain.

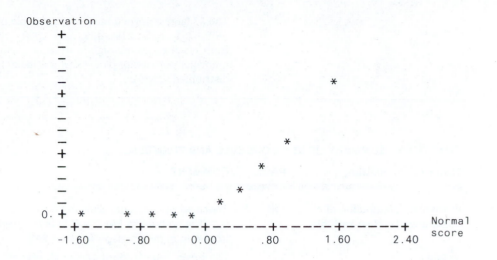

6.41 Consider the following ten observations on the lifetime (hr) for a certain type of component: 152.7, 172.0, 172.5, 173.3, 193.0, 204.7, 216.5, 234.9, 262.6, 422.6. Construct a normal probability plot, and comment on the plausibility of a normal distribution as a model for component lifetime.

6.42 The paper "The Load–Life Relationship for M50 Bearings with Silicon Nitride Ceramic Balls" (*Lubric. Eng.* (1984): 153–9) reported the accompanying data on bearing load life (million revs.); the corresponding normal scores are also given. Construct a normal probability plot. Is normality plausible?

x	Normal Score	x	Normal Score
47.1	−1.867	240.0	.062
68.1	−1.408	240.0	.187
68.1	−1.131	278.0	.315
90.8	−.921	278.0	.448
103.6	−.745	289.0	.590
106.0	−.590	289.0	.745
115.0	−.448	367.0	.921
126.0	−.315	385.9	1.131
146.6	−.187	392.0	1.408
229.0	−.062	505.0	1.867

6.43 The accompanying observations are DDT concentrations in the blood of 20 people

24	26	30	35	35	38	39	40	40	41
42	42	52	56	58	61	75	79	88	102

Use the normal scores from Problem 6.42 to construct a normal probability plot, and comment on the appropriateness of a normal probability model.

6.44 Consider the following sample of 25 observations on the diameter x (cm) of a disk used in a certain system:

16.01	16.08	16.13	15.94	16.05	16.27	15.89	15.84	16.15
16.19	16.22	15.95	16.10	15.92	16.04	15.82	16.15	16.06
15.66	16.07	16.13	16.11	15.78	15.99	16.29		

The 13 largest normal scores for a sample of size 25 are 1.965, 1.524, 1.263, 1.067, .905, .764, .637, .519, .409, .303, .200, .100, and 0. The 12 smallest scores result from placing a minus sign in front of each of the given nonzero scores. Construct a normal probability plot. Does it appear plausible that disk diameter is normally distributed? Explain.

CHAPTER 6 SUMMARY OF KEY CONCEPTS AND FORMULAS

TERM OR FORMULA	PAGE	COMMENT
Probability distribution of a continuous random variable x	159	Specified by a smooth (density) curve for which the total area under the curve is 1. The probability $P(a < x < b)$ is the area under the curve and above the interval from a to b; this is also $P(a \leq x \leq b)$.
μ_x and σ_x	162	The mean and standard deviation, respectively, of a continuous random variable x. These quantities describe the center and extent of spread about the center of the variable's probability distribution.
Normal distribution	167	A continuous probability distribution that has a bell-shaped density curve specified by a certain mathematical function. A particular normal distribution is determined by specifying values of μ and σ.
Standard normal distribution	169	This is the normal distribution with $\mu = 0$ and $\mu = 1$. The density curve is called the z curve, and z is the letter commonly used to denote a variable having this distribution. Areas under the z curve to the left of various values are given in Appendix Table II, which also appears on the inside front cover.
z critical value	173	A number on the z measurement scale that captures a specified tail area or central area
$z = \dfrac{x - \mu}{\sigma}$	173	z is obtained from x by "standardizing": subtracting the mean and then dividing by the standard deviation. When x has a normal distribution, z has a standard normal distribution. This fact implies that probabilities involving *any* normal random variable (any μ or σ) can be obtained from z curve areas.
Normal approximation to the binomial distribution	184	When both $n\pi \geq 5$ and $n(1 - \pi) \geq 5$, binomial probabilities are well-approximated by corresponding areas under a normal curve with $\mu = n\pi$ and $\sigma = \sqrt{n\pi(1 - \pi)}$.
Normal probability plot	186	A picture used to judge the plausibility of the assumption that a sample has been selected from a normal population distribution. If the plot is reasonably straight, this assumption is reasonable.

SUPPLEMENTARY EXERCISES   6.45–6.63

6.45 A pizza company advertises that it puts .5 lb of real mozzarella cheese on its medium pizzas. In fact, the amount of cheese on a randomly selected medium pizza is normally distributed with a mean value of .5 lb and a standard deviation of .025 lb.

a. What is the probability that the amount of cheese on a medium pizza is between .525 and .550 lb?

b. What is the probability that the amount of cheese on a medium pizza exceeds the mean value by more than two standard deviations?

c. What is the probability that three randomly selected medium pizzas all have at least .475 lb of cheese?

6.46 There are at least two things to keep in mind when interpreting EPA fuel efficiency ratings (mpg) for different automobiles. The first is that their values are determined under experimental conditions that are not necessarily representative of actual driving conditions. Second, each reported value is an average, so fuel efficiency for a randomly selected car may differ (considerably) from the average. Suppose, then, that fuel efficiency for a particular model car under specified conditions is normally distributed with mean 30.0 mpg and standard deviation 1.2 mpg.

   a. What is the probability that the fuel efficiency for a randomly selected car of this type is between 29 and 31 mpg?
   b. Would it surprise you to find that the efficiency of a randomly selected car of this model is less than 25 mpg?
   c. If three cars of this model are randomly selected, what is the probability that all three have efficiencies exceeding 32 mpg?
   d. Find a number c such that 95% of all cars of this model have efficiencies exceeding c (that is, $P(x > c) = .95$).

6.47 The amount of time spent by a statistical consultant with a client at their first meeting is a random variable (what else!) having a normal distribution with mean value 60 min and standard deviation 10 min.

   a. What is the probability that more than 45 minutes is spent at the first meeting?
   b. What amount of time is exceeded by only 10% of all clients?
   c. If the consultant assesses a fixed charge of $10 (for overhead) and then charges $50 per hour, what is the expected revenue from a client?

6.48 The lifetime of a certain brand of battery is normally distributed with mean value 6 hr and standard deviation .8 hr when it is used in a particular cassette player. Suppose two new batteries are independently selected and put into the player. The player will cease to function as soon as one of the batteries fails.

   a. What is the probability that the player functions for at least 4 hours?
   b. What is the probability that the cassette player works for at most 7 hours?
   c. Find a number c such that only 5% of all cassette players will function without battery replacement for more than c hours.

6.49 A machine producing vitamin E capsules operates so that the actual amount of vitamin E in each capsule is normally distributed with mean 5 mg and standard deviation .05 mg. What is the probability that a randomly selected capsule contains less than 4.9 mg of vitamin E? At least 5.2 mg?

6.50 Accurate labeling of packaged meat is difficult because of weight decrease due to moisture loss (defined as a percentage of the package's original net weight). Suppose that moisture loss for a package of chicken breasts is normally distributed with mean value 4.0% and standard deviation 1.0%. (This model is suggested in the paper "Drained Weight Labeling for Meat and Poultry: An Economic Analysis of a Regulatory Proposal" (*J. of Consumer Affairs* (1980): 307–25)). Let x denote the moisture loss for a randomly selected package.

   a. What is the probability that x is between 3.0% and 5.0%?
   b. What is the probability that x is at most 4.0%?
   c. What is the probability that x is at least 7.0%?
   d. Find a number c such that 90% of all packages have moisture losses below c%.
   e. What is the probability that moisture loss differs from the mean value by at least 1%?

6.51 The *Wall Street Journal* (Feb. 15, 1972) reported that General Electric was being sued in Texas for sex discrimination over a minimum height requirement of

5 ft 7 in. The suit claimed that this restriction eliminated more than 94% of adult females from consideration. Let x represent the height of a randomly selected adult woman. Suppose that x is approximately normally distributed with mean 66 in. (5 ft 6 in.) and standard deviation 2 in.

a. Is the claim that 94% of all women are shorter than 5 ft 7 in. correct?
b. What proportion of adult women would be excluded from employment due to the height restriction?

6.52 Suppose that your statistics professor tells you that the scores on a midterm exam were approximately normally distributed with a mean of 78 and a standard deviation of 7. The top 15% of all scores have been designated as A's. Your score is 89. Did you receive an A? Explain.

6.53 Suppose that the pH of soil samples taken from a certain geographic region is normally distributed with a mean pH of 6.00 and a standard deviation of .10. If the pH of a randomly selected soil sample from this region is determined, answer the following questions about it.

a. What is the probability that the resulting pH is between 5.90 and 6.15?
b. What is the probability that the resulting pH exceeds 6.10?
c. What is the probability that the resulting pH is at most 5.95?
d. What value will be exceeded by only 5% of all such pH values?

6.54 The light bulbs used to provide exterior lighting for a large office building have an average lifetime of 700 hr. If length of life is approximately normally distributed with a standard deviation of 50 hr, how often should all of the bulbs be replaced so that no more than 20% of the bulbs will have already burned out?

6.55 Soaring insurance rates have made it difficult for many people to afford automobile insurance. Suppose that 16% of all those driving in a certain city are uninsured. Consider a random sample of 200 drivers.

a. What is the mean value of the number who are uninsured, and what is the standard deviation of the number who are uninsured?
b. What is the (approximate) probability that between 25 and 40 (inclusive) drivers in the sample were uninsured?
c. If you learned that more than 50 among the 200 were uninsured, would you doubt the 16% figure? Explain.

6.56 Let x denote the duration of a randomly selected pregnancy (the time elapsed between conception and birth). Accepted values for the mean value and standard deviation of x are 266 days and 16 days, respectively. Suppose that the probability distribution of x is (approximately) normal.

a. What is the probability that the duration of pregnancy is between 250 and 300 days?
b. What is the probability that the duration of pregnancy is at most 240 days?
c. What is the probability that the duration of pregnancy is within 16 days of the mean duration?
d. A *Dear Abby* column dated January 20, 1973, contained a letter from a woman who stated that the duration of her pregnancy was exactly 310 days. (She wrote that the last visit with her husband, who was in the navy, occurred 310 days prior to birth.) What is the probability that the duration of pregnancy is at least 310 days? Does this probability make you a bit skeptical of the claim?
e. Some insurance companies will pay the medical expenses associated with childbirth only if the insurance has been in effect for more than 9 months (275 days). This restriction is designed to ensure that the insurance company has to pay benefits only for those pregnancies where conception occurred during coverage. Suppose that conception occurred two weeks after coverage began. What is the

probability that the insurance company will refuse to pay benefits because of the 275-day insurance requirement?

6.57   A machine that cuts corks for wine bottles operates so that the diameter of the corks produced is approximately normally distributed with mean 3 cm and standard deviation .1 cm. The specifications call for corks with diameters between 2.9 and 3.1 cm. A cork not meeting the specifications is considered defective. (A cork that is too small leaks and causes the wine to deteriorate, while a cork that is too large doesn't fit in the bottle.) What proportion of corks produced by this machine are defective?

6.58   Refer to Exercise 6.57. Suppose that there are two machines available for cutting corks. The one described in the preceding problem produces corks with diameters normally distributed with mean 3 cm and standard deviation .1 cm. The second machine produces corks with diameters normally distributed with mean 3.05 cm and standard deviation .01 cm. Which machine would you recommend? (Hint: Which machine would produce the fewest defective corks?)

6.59   Suppose that SAT math and verbal scores are approximately normally distributed. Both the math and verbal sections of the test have a distribution of scores with an average of 500 and a standard deviation of 100.
  a. If a student scored 620 on the verbal section, what percentile is associated with his or her score?
  b. If the same student scored 710 on the math section, what is the corresponding percentile?
  c. What score does a student have to achieve in order to be at the 90th percentile?

6.60   A certain bookstore deals exclusively in mysteries. Let x denote the number of pages in a book randomly selected from the store's inventory. Suppose that x has approximately a normal distribution with a mean value of 240 pages and a standard deviation of 40 pages.
  a. What is the probability that x is at least 250 pages?
  b. What is the probability that x is more than 250 pages?
  c. What is the probability that x is between 200 and 250 pages (inclusive)?
  d. If two books are independently selected, what is the probability that at least one has fewer than 200 pages?

6.61   The weight of parcels sent with a certain service has a normal distribution with mean value 10 lb and standard deviation 2 lb. This parcel service wishes to establish a weight limit w beyond which there will be a surcharge. What value w is such that exactly 99% of all parcels are at least one pound under the surchage weight w?

6.62   Suppose that only 40% of all drivers in a certain region regularly wear a seat belt. A random sample of 500 drivers is selected.
  a. What is the probability that at least half of those in the sample regularly wear seat belts?
  b. What is the probability that between 25% and 50% (inclusive) of those in the sample regularly wear a seat belt?

6.63   A probability distribution that has been very widely used to model lifetimes and various other phenomenon is the *exponential distribution*. The corresponding density curve has height

$$f(x) = \begin{cases} (1/\mu)e^{-x/\mu} & x \geq 0 \\ 0 & \text{otherwise} \end{cases}$$

where μ is the mean value of x and $e \approx 2.71828$. The corresponding density curve follows. For any value $x > 0$, the area under the curve to the left of x is $1 - e^{-x/\mu}$.

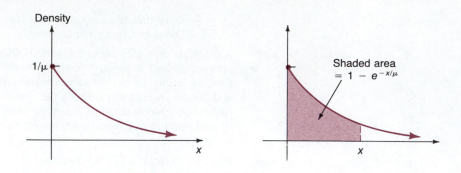

Suppose that the lifetime of a particular component has an exponential distribution with mean value 1000 hr.

a. What is the probability that the lifetime is at most 2000 hr?
b. What is the probability that the lifetime is at most 1000 hr? Why is this probability not .5?
c. What is the probability that the lifetime is between 500 and 2000 hr?

REFERENCES   The books referenced in Chapter 4 contain material on continuous probability distributions. The book by Chambers et al. listed among the Chapter 2 references is a good source for more information on probability plotting.

C H A P T E R 7

SAMPLING DISTRIBUTIONS

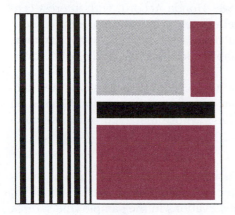

INTRODUCTION The inferential methods presented in subsequent chapters will use information contained in a sample to reach conclusions about one or more characteristics of the whole population. As an example, let μ denote the true mean nicotine content (in mg) for a particular brand of cigarette. To learn something about μ, a sample of $n = 4$ cigarettes might be obtained, resulting in $x_1 = 1.68$, $x_2 = 1.89$, $x_3 = 1.73$, and $x_4 = 1.95$, from which we calculate the sample mean, $\bar{x} = 1.813$. How close is this sample mean to the true population mean, μ? If another sample of 4 were selected and $\bar{x}$ computed, would this second $\bar{x}$ value be near 1.813 or might it be quite different?

These issues can be addressed by studying what is called the *sampling distribution* of $\bar{x}$. Just as the probability distribution of a numerical variable describes long-run behavior, the sampling distribution of $\bar{x}$ provides information about the long-run behavior of $\bar{x}$ based on many different samples, each of size n.

Properties of $\bar{x}$'s sampling distribution will then be used to develop inferential procedures for drawing conclusions about μ. In a similar manner, we will study the sampling distribution of a *sample proportion* (the fraction of individuals in a sample who have some specified property). This will enable us to describe procedures for making inferences about the corresponding population proportion π.

7.1 STATISTICS AND RANDOM SAMPLES

The objective of many investigations and research projects is to draw conclusions about how the values of some numerical variable x are distributed in a population. Although many aspects of the distribution might be of interest, attention is often focused on one particular population characteristic. Examples include:

1. x = fuel efficiency (mpg) for a 1989 Honda Accord, with interest centered on the mean value μ of fuel efficiency for all such cars;
2. x = thickness of a printed circuit board used in a personal computer, with interest focused on the variability in thickness as described by σ, the standard deviation of x;
3. x = time to first recurrence of skin cancer for a patient who was successfully treated using a particular therapy, with attention focused on the proportion π of such individuals whose first recurrence is within five years of original treatment.

STATISTICS Statistical methods for drawing conclusions about the distribution of a variable x are based on obtaining a sample of x values. As in earlier chapters, the letter n denotes the number of observations in the sample (the sample size). The observations themselves are denoted by $x_1, x_2, \ldots, x_n$. Once these values are available, an investigator must decide which computed quantities will be most informative in drawing the desired type of conclusion. For example, we shall see in the next few chapters that several standard statistical procedures for drawing conclusions about μ utilize calculated values of both $\bar{x}$ and s.

DEFINITION

Any quantity computed from values in a sample is called a **statistic**.

Several statistics other than $\bar{x}$ and s were introduced in Chapter 3. They included the *sample median* (the middle value in the ordered list of sample observations), the *sample range* (the difference between the largest and smallest sample values), a *trimmed mean*, and a *sample percentile*.

It is very important to appreciate the distinction between a population characteristic (i.e., a characteristic of the x distribution) and a statistic, which is a sample characteristic. The mean value of x, or the population mean, μ, is a population characteristic. The value of μ is a fixed number, such as 1.75 mg for nicotine content or 28.5 mpg for fuel efficiency. However, the value of μ is generally not known, which is why we take a sample of x values. The sample mean, $\bar{x}$, is a characteristic of the sample. Its value is a fixed number for any particular sample, but different samples typically result in different $\bar{x}$ values. It would be nice if the value of $\bar{x}$ turned out to be exactly μ; in practice, however, this almost never happens because of sampling variability. Some samples yield x values that are not very representative of the population distribution. Thus the value of μ might be 1.75, but one sample might yield $\bar{x} = 1.63$, whereas another might result in $\bar{x} = 2.04$.

SAMPLING DISTRIBUTIONS

The value of a statistic is subject to sampling variability. This variability may cause different samples to yield different conclusions about the population. Information about a statistic's long-run behavior is needed in order to attach a measure of reliability to conclusions. This information is obtained by using probability concepts to determine a statistic's sampling distribution.

▐▐▐ DEFINITION

The value of a statistic depends on the particular sample selected from the population and changes from sample to sample. A statistic is therefore a random variable and as such has a probability distribution. The probability distribution of a statistic is called its **sampling distribution**. The sampling distribution of a statistic describes the long-run behavior of the statistic's values when many different samples, each of size n, are obtained and the value of the statistic is computed for each one.

EXAMPLE 1

The record library of a particular classical music radio station contains five different recordings of Beethoven's Fifth Symphony. Because listener polls have identified this as the most popular classical work, the station manager has insisted that three different recordings of this symphony be played each month. The recordings all differ somewhat in playing time (due to differences in interpretations by conductors and orchestras). These times are 29.6, 29.9, 30.0, 30.2, and 30.8 minutes.

Suppose that the three records to be played in a given month are determined by writing each of the five playing times on a different slip of paper, mixing up the slips, and selecting three at random. Then any set of three times has the same chance of selection as any other set. Suppose that the statistic of interest here is the sample median playing time. We list each of the ten possible samples (ordered for convenience) and the corresponding value of this statistic.

Sample	Sample Median	Sample	Sample Median
29.6, 29.9, 30.0	29.9	29.6, 30.2, 30.8	30.2
29.6, 29.9, 30.2	29.9	29.9, 30.0, 30.2	30.0
29.6, 29.9, 30.8	29.9	29.9, 30.0, 30.8	30.0
29.6, 30.0, 30.2	30.0	29.9, 30.2, 30.8	30.2
29.6, 30.0, 30.8	30.0	30.0, 30.2, 30.8	30.2

Each sample has the same chance of occurring as any other sample, so the probability of any particular one is 1/10, or .10. That is, in a very long sequence of months, each sample would occur about 1/10 (or 10%) of the time. The only three possible values of the sample median are 29.9, 30.0, and 30.2. Since 30.0 is the sample median for four of the possible samples,

$$P(\text{sample median} = 30.0) = 4/10 = .40$$

Similarly,

$$P(\text{sample median} = 29.9) = .3$$

and

$$P(\text{sample median} = 30.2) = .3$$

The sampling distribution of the sample median can now be summarized in a probability distribution table:

Value of sample median	29.9	30.0	30.2
Probability of value	.3	.4	.3

Thus, in the long run, 40% of all months result in a sample median playing time of 30.0. Notice that the population median time is 30.0 (the middle value when all five possible times are ordered). In 40% of all months, the sample median equals the population median, but in 60% of all months the two differ. The population median remains fixed in value, but the sample median varies in value from sample to sample.

There are several interesting things to notice about Example 1. First of all, even though this experiment involved relatively few outcomes, obtaining the sampling distribution required some careful thought and calculation. Things would have been much worse if the radio station owned ten different recordings (a larger population) and five (a larger sample) were to be played each month. Just listing the possible outcomes would be extremely tedious. (There are 252 of them!) Second, the same reasoning could be used to obtain the sampling distribution of any other statistic. For example, replacing the value of the sample median by the value of $\bar{x}$ (29.83 for the first ordered sample, etc.) would lead to the sampling distribution of $\bar{x}$. Third, the calculations were done for sampling without replacement. If sampling had been with replacement, the same time might have been chosen twice or even on all three selections. Additional samples and corresponding values of the sample median would be possible, resulting in a more complicated sampling distribution. In conclusion, the sampling distribution depends not only on which statistic is under consideration but also on the sample size and the method of sampling.

EXAMPLE 2

A company maintains three offices in a certain area, each staffed by two employees. Information concerning yearly salaries (in thousands of dollars) is as follows:

Office	1		2		3	
Employee	1	2	3	4	5	6
Salary	14.7	18.6	15.2	18.6	10.8	14.7

A survey will be conducted to obtain information on average salary levels. Two of the six employees will be selected for inclusion in the survey. Suppose that six slips of paper numbered 1, 2, . . . , 6 are placed in a box and two are drawn without replacement. This ensures that each of the 15 possible employee pairs has probability 1/15 of being chosen. Computing the sample average salary $\bar{x}$ for each possible pair leads to the sampling distribution of $\bar{x}$ shown. For example, $\bar{x} = 16.65$ occurs for the four pairs (1, 2), (1, 4), (2, 6), and (4, 6), so

$$P(\bar{x} = 16.65) = 4/15 = .267.$$

$\bar{x}$ value	12.75	13.00	14.70	14.95	16.65	16.90	18.60
Probability	.133	.067	.200	.133	.267	.133	.067

Thus if this experiment is performed over and over again, each time with a new selection of two employees, in the long run the value $\bar{x} = 14.70$ will occur 20% of the time, the value $\bar{x} = 18.60$ only 6.7% of the time, and so on.

Now consider what happens if the method for obtaining a sample is changed. Suppose this time one of the three offices is selected at random (using just three slips of paper), and both employees from that office are included. This is called a *cluster sample* because a cluster (in this case an office) is selected at random and all individuals in the cluster are included in the sample. There are only three possible employee pairs, each one having probability 1/3 (or .333) of occurring. The resulting sampling distribution of $\bar{x}$ is now as follows:

$\bar{x}$ value	12.75	16.65	16.90
Probability	.333	.333	.333

If this experiment is repeated many times over, each of these three $\bar{x}$ values will occur approximately 33.33% of the time. These two sampling distributions of $\bar{x}$ are obviously quite different. Clearly the sampling distribution depends on the method of sampling used.

RANDOM SAMPLES   When an investigator takes a sample from a population of interest, the sampling is usually done without replacement. This implies that the results of successive selections are dependent. (See the discussion on sampling in Chapter 4.) The dependence will be quite negligible, however, when the sample size is small relative to the population size. Sampling without replacement is then almost like sampling with replacement, for which there is no dependence in successive draws and the distribution of possible values is identical on each selection.

▌▌▌ DEFINITION

A sample $x_1, x_2, \ldots, x_n$ of values of a numerical variable x is called a (**simple**) **random sample** if the sampled values are selected independently from the same population distribution.

In most applications, selection is random and without replacement, but the sample size is much smaller than the population size (at most 5% of the population is sampled). For practical purposes, the successive observations can be regarded as independent, so the sample can be considered a random sample as we have defined it. The x values in a random sample are usually obtained by first selecting n individuals or objects and then observing or determining the value of x for each one. *It is customary to refer to the selected individuals or objects, as well as to the x values themselves, as a random sample.* Thus we may speak of a "random sample of students" when the variable of interest is $x =$ grade point average, or a "random sample of houses" when a study is concerned with $x =$ January electricity usage.

Methods for analyzing both random samples and other types of samples are based on concepts of probability and results concerning the sampling distributions of various statistics. However, the statistics and methods employed are most easily understood in the case of random samples. **Throughout the remainder of this book, we deal with random samples.**

EXERCISES   7.1–7.13   **SECTION 7.1**

7.1   Explain the difference between a population characteristic and a statistic.

7.2   What is the difference between $\bar{x}$ and μ? Between σ and s?

7.3   For each of the following statements, identify the number that appears in bold face type as either the value of a population characteristic or a statistic.
  a. A department store reports that **84%** of all customers who use the store's credit plan pay their bills on time.
  b. A sample of 100 students at a large university had a mean age of **24.1** years.
  c. The Department of Motor Vehicles reports that **22%** of all vehicles registered in a particular state are imports.
  d. A hospital reports that, based on the ten most recent cases, the mean length of stay for surgical patients is **6.4** days.

e. A consumer group, after testing 100 batteries of a certain brand, reported an average life of **63** hours of use.

7.4   Assume that the 435 members of the U.S. House of Representatives are listed in alphabetical order. Explain how you would select a random sample of 20 members.

7.5   Describe how you might go about selecting a random sample of each of the following.
   a. Doctors practicing in Los Angeles County
   b. Students enrolled at a particular university
   c. Boxes in a warehouse
   d. Registered voters in your community
   e. Subscribers to a local newspaper
   f. Radios from a shipment of 1000

7.6   Consider the following "population": {1, 2, 3, 4}. Note that the population mean is

$$\mu = \frac{1 + 2 + 3 + 4}{4} = 2.5$$

   a. Suppose that a random sample of size two is to be selected without replacement from this population. There are twelve possible samples (provided that the order in which observations are selected is taken into account):

1, 2	1, 3	1, 4
2, 1	2, 3	2, 4
3, 1	3, 2	3, 4
4, 1	4, 2	4, 3

   Compute the sample mean for each of the twelve possible samples. Use this information to construct the sampling distribution of $\bar{x}$. (Display it in table form.)
   b. Suppose that a random sample of size two is to be selected, but this time sampling will be done *with* replacement. Using a method similar to that of part (a), construct the $\bar{x}$ sampling distribution. Hint: There are 16 different possible samples in this case.
   c. In what ways are the two sampling distributions of parts (a) and (b) similar? In what ways are they different?

7.7   Simulate sampling from the population of Exercise 7.6 by using four slips of paper individually marked 1, 2, 3, and 4. Select a sample of size 2 without replacement, and compute $\bar{x}$. Repeat this process 50 times, and construct a relative frequency distribution of the fifty $\bar{x}$ values. How does the relative frequency distribution compare to the sampling distribution of $\bar{x}$ derived in 7.6(a)?

7.8   Use the method of Exercise 7.6 to find the sampling distribution of the sample range when a random sample of size 2 is to be selected with replacement from the population {1, 2, 3, 4}. (Recall that sample range = largest observations − smallest observation.)

7.9   On four different occasions you have borrowed money from a friend in the amounts $1, $5, $10, and $20. The friend has four IOU slips with each of these amounts written on a different slip. He will randomly select two slips from among the four and ask that those IOU's be repaid immediately. Let t (the sample total) denote the amount that you must repay immediately.
   a. List all possible outcomes, the value of t for each, and then determine the sampling distribution of t.
   b. Calculate the mean value of t. How does μ_t relate to the "population" mean μ?

7.10 Refer back to Exercise 7.9. Suppose that instead of asking that the total amount on the two selected slips be repaid immediately, the maximum m of the two selected amounts is to be repaid immediately (e.g., $m = 10$ if the $5 and $10 slips are drawn). Obtain the sampling distribution of m if the two slips are selected
  a. without replacement;
  b. with replacement.

7.11 Consider the following "population": $\{2, 3, 3, 4, 4\}$. The value of μ is 3.2, but suppose this is not known to an investigator who therefore wants to estimate μ from sample data. Three possible statistics for estimating μ are:

  statistic #1: the sample mean, $\bar{x}$

  statistic #2: the sample median

  statistic #3: the average of the largest and smallest values in the sample

A random sample of size 3 will be selected without replacement. Provided that we disregard the order in which observations are selected, there are ten possible samples that might result (writing 3 and 3*, 4 and 4* to distinguish the two 3's and the two 4's):

2, 3, 3*	2, 3, 4	2, 3, 4*
2, 3*, 4	2, 3*, 4*	2, 4, 4*
3, 3*, 4	3, 3*, 4*	3, 4, 4*
3*, 4, 4*		

For each of these ten samples, compute statistics 1, 2, and 3. Construct the sampling distribution of each of these statistics. Which statistic would you recommend for estimating μ? Explain the reasons for your choice.

7.12 The three offices in Example 2 can be regarded as distinct segments, or *strata*, of the population. A *stratified sample* involves selecting a specified number of individuals from each stratum. When values within a stratum are similar but values in some strata differ substantially from those in other strata, stratified sampling can help ensure a representative sample. In the context of Example 2, suppose that a stratified sample consists of one of the two individuals from each stratum (each selected by tossing a fair coin). There are then eight possible samples. List them, and obtain the sampling distribution of $\bar{x}$.

7.13 Suppose that you have four books on your shelf that you are planning to read in the near future. Two are fictional works, containing 212 and 379 pages, respectively, and the other two are nonfiction, with 350 and 575 pages, respectively.
  a. Suppose that you randomly select two books from among these four to take on a one-week ski trip (in case you injure yourself). Let $\bar{x}$ denote the sample average number of pages for the two books selected. Obtain the sampling distribution of $\bar{x}$ and determine the mean of the $\bar{x}$ distribution.
  b. Suppose that you randomly select one of the two fiction books and also randomly select one of the two nonfiction books. Determine the sampling distribution of $\bar{x}$ and then calculate the mean of the $\bar{x}$ distribution.

7.2 THE SAMPLING DISTRIBUTION OF A SAMPLE MEAN

When the objective of a statistical investigation is to make an inference about the population mean, μ, it is natural to consider the sample mean, $\bar{x}$. In order to understand how inferential procedures based on $\bar{x}$ work, we must first study how sampling variability causes $\bar{x}$ to differ in value from one sample to another. The behavior of $\bar{x}$ is described by its sampling distribution. The sample size n and characteristics of the population—its shape, mean value μ, and standard deviation σ—are important in determining properties of the sampling distribution of $\bar{x}$.

It is helpful to begin by looking at the results of some sampling experiments. In the two examples that follow, we start with a specified x population distribution, fix a sample size n, and select 500 different random samples of this size. We then compute $\bar{x}$ for each sample and construct a sample histogram of these 500 $\bar{x}$ values. Because 500 is reasonably large (a reasonably long sequence of samples), the sample histogram should rather closely resemble the true sampling distribution of $\bar{x}$ (obtained from an unending sequence of $\bar{x}$ values). We repeat the experiment for several different values of n to see how the choice of sample size affects the sampling distribution. Careful examination of these sample histograms will aid in understanding the general results to be stated shortly.

EXAMPLE 3

The paper "Platelet Size in Myocardial Infarction" (*Brit. Med. J.* (1983): 449–51) presented evidence that suggests that the distribution of platelet volume was approximately normal in shape both for patients after acute myocardial infarction (a heart attack) and for control subjects who had no history of serious illness. The suggested values of μ and σ for the control-subject distribution were $\mu = 8.25$ and $\sigma = .75$. Figure 1 pictures the corresponding normal curve. The curve is centered at 8.25, the mean value of platelet volume. The value of the population standard deviation, .75, determines the extent to which the x distribution spreads out about its mean value.

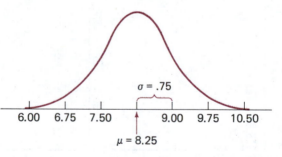

FIGURE 1

Normal distribution of platelet volume x with $\mu = 8.25$ and $\sigma = .75$.

We first used MINITAB to select 500 random samples from this normal distribution, each one consisting of $n = 5$ observations. A histogram of the resulting 500 $\bar{x}$ values appears in Figure 2(a). This procedure was repeated for samples of size $n = 10$, again for $n = 20$, and finally for $n = 30$. The resulting sample histograms of $\bar{x}$ values are displayed in Figures 2(b), (c), and (d).

The first thing to notice about the histograms is their shape. To a reasonable approximation, each of the four looks like a normal curve. The resemblance would be even more striking if each histogram had been based on many more than 500 $\bar{x}$ values. Second, each histogram is centered approximately at 8.25, the mean of the population being sampled. Had the histograms been based on an unending sequence of $\bar{x}$ values, their centers would have been exactly the population mean, 8.25.

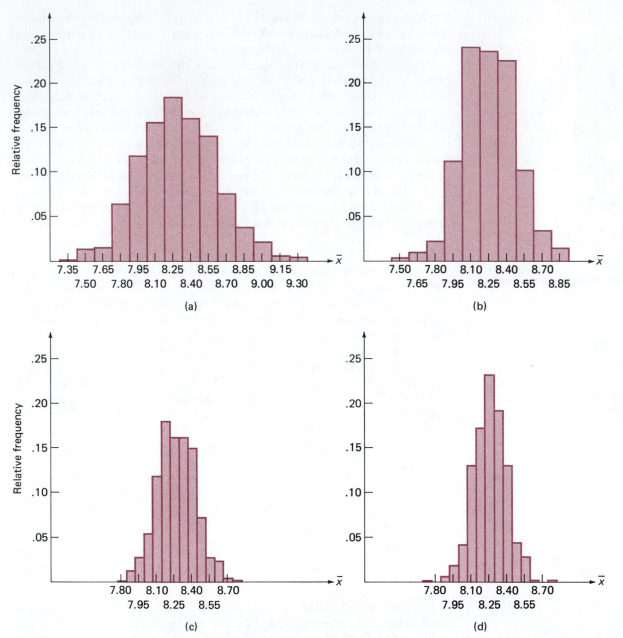

FIGURE 2

Sample histograms for x̄ based on 500 samples, each consisting of n observations.

(a) n = 5
(b) n = 10
(c) n = 20
(d) n = 30

The final aspect of the histograms to note is their spread relative to one another. The smaller the value of n, the greater the extent to which the sampling distribution spreads out about the mean value. This is why the histograms for n = 20 and n = 30 are based on narrower class intervals than those for the two smaller sample sizes. For the larger sample sizes, most of the x̄ values are quite close to 8.25. This is the effect of averaging. When n is small, a single unusual x value can result in an x̄ value far from the center. With a larger sample size, any unusual x values, when averaged in with the other sample values, still tend to yield an x̄ value close to μ.

Combining these insights yields a result that should appeal to your intuition: $\bar{x}$ **based on a large n tends to be closer to μ than does $\bar{x}$ based on a small n.**

EXAMPLE 4

Now consider properties of the $\bar{x}$ sampling distribution when the population distribution is quite skewed (and thus very unlike a normal distribution). The May 8, 1983, issue of the *Los Angeles Times* contained data on the amount spent per pupil by each of 254 school districts located in Southern California. Figure 3 displays a MINITAB histogram of the data. The histogram has several peaks and a very long upper tail.

Let's now regard the 254 values as comprising a population, so that the histogram of Figure 3 shows the distribution of values in the population.

```
MIDDLE OF    NUMBER OF
INTERVAL     OBSERVATIONS
   1625.0        6    ******
   1675.0       49    **************************************************
   1725.0       24    ************************
   1775.0       30    ******************************
   1825.0       48    ************************************************
   1875.0       29    *****************************
   1925.0       15    ***************
   1975.0        8    ********
   2025.0       11    ***********
   2075.0        9    *********
   2125.0        6    ******
   2175.0        6    ******
   2225.0        3    ***
   2275.0        2    **
   2325.0        0
   2375.0        0
   2425.0        1    *
   2475.0        1    *
   2525.0        1    *
   2575.0        0
   2625.0        1    *
   2675.0        0
   2725.0        1    *
   2775.0        1    *
   2825.0        0
   2875.0        0
   2925.0        1    *
   2975.0        0
   3025.0        0
   3075.0        0
   3125.0        0
   3175.0        0
   3225.0        0
   3275.0        0
   3325.0        0
   3375.0        0
   3425.0        0
   3475.0        0
   3525.0        0
   3575.0        0
   3625.0        0
   3675.0        0
   3725.0        1    *
```

FIGURE 3

The population distribution for Example 4 ($\mu = 1864$).

The skewed shape makes identification of the mean value from the picture more difficult than for a normal distribution. We found the average of these 254 values to be $\mu = 1864$, so that is the balance point for the population histogram. The median population value is 1818, less than μ and reflective of the distribution's positively skewed nature.

For each of the sample sizes $n = 5, 10, 20,$ and 30, as well as for $n = 50$, we selected 500 random samples of size n. This was done with replacement to approximate more nearly the usual situation in which the sample size n is only a small fraction of the population size. (Without replacement, $n = 30$ results in more than 10% of the population being sampled; see the discussion at the end of Section 7.1.) We then constructed a histogram of the 500 $\bar{x}$ values for each of the five sample sizes. These histograms are displayed in Figures 4(a) through (e).

```
MIDDLE OF    NUMBER OF
INTERVAL     OBSERVATIONS
 1685.0       0
 1700.0       3    ***
 1715.0       5    *****
 1730.0      17    *****************
 1745.0      24    ************************
 1760.0      29    *****************************
 1775.0      31    *******************************
 1790.0      22    **********************
 1805.0      41    *****************************************
 1820.0      38    **************************************
 1835.0      27    ***************************
 1850.0      32    ********************************
 1865.0      34    **********************************
 1880.0      34    **********************************
 1895.0      28    ****************************
 1910.0      20    ********************
 1925.0      22    **********************
 1940.0      15    ***************
 1955.0      15    ***************
 1970.0      10    **********
 1985.0       8    ********
 2000.0       8    ********
 2015.0       4    ****
 2030.0       6    ******
 2045.0       4    ****
 2060.0       2    **
 2075.0       0
 2090.0       4    ****
 2105.0       2    **
 2120.0       2    **
 2135.0       4    ****
 2150.0       0
 2165.0       0
 2180.0       0
 2195.0       0
 2210.0       0
 2225.0       1    *
 2240.0       1    *
 2255.0       3    ***
 2270.0       2    **
 2285.0       0
 2300.0       0
 2315.0       1    *
 2330.0       0
```

(a)   (continued)

FIGURE 4

Minitab histograms of 500 $\bar{x}$ values, each based on n observations.

(a) $n = 5$ (average $\bar{x} = 1863.5$)

Unlike the normal population case, these histograms all differ in shape. In particular, the histograms become progressively less skewed as the sample size, n, increases. The histograms for $n = 10$ and $n = 20$ are substantially more symmetric than is the population distribution, and those for $n = 30$ and $n = 50$ are very much like normal curves. This is the effect of averaging. Even when n is large, one of the few large x values in the population appears only infrequently in the sample. When one does appear, its contribution to $\bar{x}$ is swamped by the contributions of more typical sample values. The normal curve shape of the histograms for $n = 30$ and $n = 50$ are exactly what is predicted by the Central Limit Theorem, to be introduced shortly. According to this theorem, even if the population distribution bears no resemblance whatsoever to a normal curve, the $\bar{x}$ sampling distribution is approximately normal when the sample size n is reasonably large.

```
EACH * REPRESENTS 2 OBSERVATIONS

MIDDLE OF   NUMBER OF
INTERVAL    OBSERVATIONS
  1685.0        0
  1700.0        1    *
  1715.0        1    *
  1730.0        3    **
  1745.0        4    **
  1760.0        8    ****
  1775.0       18    *********
  1790.0       31    ***************
  1805.0       49    *************************
  1820.0       50    *************************
  1835.0       53    ***************************
  1850.0       43    **********************
  1865.0       49    *************************
  1880.0       44    **********************
  1895.0       25    *************
  1910.0       25    *************
  1925.0       20    **********
  1940.0       20    **********
  1955.0        9    *****
  1970.0        8    ****
  1985.0        8    ****
  2000.0        7    ****
  2015.0        6    ***
  2030.0        2    *
  2045.0        6    ***
  2060.0        1    *
  2075.0        2    *
  2090.0        2    *
  2105.0        1    *
  2120.0        1    *
  2135.0        1    *
  2150.0        0
  2165.0        0
  2180.0        1    *
  2195.0        0
  2210.0        0
  2225.0        0
  2240.0        0
  2255.0        1    *
```

FIGURE 4   (continued)

Minitab histograms of 500 $\bar{x}$ values, each based on n observations.
(b) $n = 10$ (average $\bar{x} = 1865.4$)

(b)   (continued)

EACH * REPRESENTS 2 OBSERVATIONS

MIDDLE OF   NUMBER OF
INTERVAL   OBSERVATIONS
1685.0   0
1700.0   0
1715.0   0
1730.0   0
1745.0   1   *
1760.0   3   **
1775.0   12   ******
1790.0   22   **********
1805.0   30   ***************
1820.0   52   **************************
1835.0   67   **********************************
1850.0   65   *********************************
1865.0   58   *****************************
1880.0   38   *******************
1895.0   34   *****************
1910.0   44   **********************
1925.0   20   **********
1940.0   14   *******
1955.0   15   ********
1970.0   13   *******
1985.0   3   **
2000.0   3   **
2015.0   2   *
2030.0   2   *
2045.0   1   *
2060.0   1   *

(c)

EACH * REPRESENTS 2 OBSERVATIONS

MIDDLE OF   NUMBER OF
INTERVAL   OBSERVATIONS
1685.0   0
1700.0   0
1715.0   0
1730.0   0
1745.0   0
1760.0   1   *
1775.0   1   *
1780.0   12   ******
1805.0   33   ****************
1820.0   44   **********************
1835.0   77   ***
1850.0   77   ***
1865.0   80   **
1880.0   53   **************************
1895.0   45   ***********************
1910.0   30   ***************
1925.0   22   ***********
1940.0   13   *******
1955.0   3   **
1970.0   4   **
1985.0   4   **
2000.0   1   *

(d)

FIGURE 4 (continued)

Minitab histograms of 500 $\bar{x}$ values, each based on n observations.
(c) $n = 20$ (average $\bar{x} = 1865.0$)
(d) $n = 30$ (average $\bar{x} = 1862.1$)

(continued)

FIGURE 4 (continued)

Minitab histograms of 500 $\bar{x}$ values, each based on n observations.
(e) $n = 50$ (average $\bar{x} = 1864.5$)

```
EACH * REPRESENTS 2 OBSERVATIONS

MIDDLE OF   NUMBER OF
INTERVAL    OBSERVATIONS
1775.0         0
1790.0         4    **
1805.0        10    *****
1820.0        28    **************
1835.0        72    ************************************
1850.0        91    *********************************************
1865.0       102    ***************************************************
1880.0        93    **********************************************
1895.0        68    **********************************
1910.0        22    ***********
1925.0         8    ****
1940.0         2    *
```

(e)

The averages of the 500 $\bar{x}$ values for the five different sample sizes are all quite close to the population mean $\mu = 1864$. If each histogram had been based on an unending sequence of $\bar{x}$ values rather than just 500 values, each histogram would have been centered at exactly 1864. Thus different values of n change the shape but not the center of $\bar{x}$'s sampling distribution. Comparison of the five $\bar{x}$ histograms in Figure 4 also shows that as n increases, the extent to which the histogram spreads out about its center decreases markedly. Increasing n both changes the shape of the distribution and squeezes it in toward the center, so that $\bar{x}$ based on a large n is a less variable quantity than is $\bar{x}$ based on a small value of n.

GENERAL RULES CONCERNING THE SAMPLING DISTRIBUTION OF $\bar{x}$

Examples 3 and 4 suggest that for any n, the center of the $\bar{x}$ distribution (the mean value of $\bar{x}$) coincides with the mean of the population being sampled, but the spread of the $\bar{x}$ distribution decreases as n increases. Additionally, the histograms then indicate that the standard deviation of $\bar{x}$ is smaller for large n than for small n. The sample histograms also suggest that in some cases, the $\bar{x}$ distribution is approximately normal in shape. These observations are stated more formally in the following general rules.

GENERAL RULES CONCERNING THE $\bar{x}$ SAMPLING DISTRIBUTION

Let $\bar{x}$ denote the mean of the observations in a random sample of size n from a population having mean μ and standard deviation σ. Denote the mean value of the $\bar{x}$ distribution by $\mu_{\bar{x}}$ and the standard deviation of the $\bar{x}$ distribution by $\sigma_{\bar{x}}$. Then the following rules hold:

Rule 1. $\mu_{\bar{x}} = \mu$

Rule 2. $\sigma_{\bar{x}} = \sigma/\sqrt{n}$

Rule 3. When the population distribution is normal, the sampling distribution of $\bar{x}$ is also normal for any sample size n.

Rule 4. When n is sufficiently large, the sampling distribution of $\bar{x}$ is well approximated by a normal curve, even when the population distribution is not itself normal.

Rule 1, $\mu_{\bar{x}} = \mu$, says that the $\bar{x}$ distribution is always centered at the mean of the population sampled. Rule 2, $\sigma_{\bar{x}} = \sigma/\sqrt{n}$, not only says that the spread of the $\bar{x}$ distribution decreases as n increases but also it gives a precise relationship between the standard deviation of the $\bar{x}$ distribution and the population standard deviation. When $n = 4$, for example,

$$\sigma_{\bar{x}} = \sigma/\sqrt{n} = \sigma/\sqrt{4} = \sigma/2$$

so that the $\bar{x}$ distribution has a standard deviation only half as large as the population standard deviation. Rules 3 and 4 specify circumstances under which the $\bar{x}$ distribution is normal or approximately normal: when the population is normal or when the sample size is large. Figure 5 illustrates these rules by showing several $\bar{x}$ distributions superimposed over a graph of the population distribution.

FIGURE 5

Population distribution and sampling distributions of $\bar{x}$.
(a) A symmetric population distribution
(b) A skewed population distribution

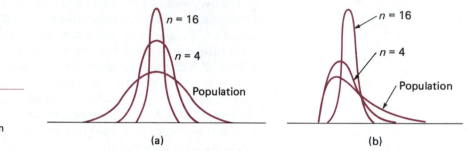

(a)　　　　　　　(b)

Rule 4 is the statement of a very important result called the **Central Limit Theorem**. This theorem says that when n is sufficiently large, the $\bar{x}$ distribution is approximately normal, no matter what the population distribution looks like. This result has enabled statisticians to develop some large-sample procedures for making inferences about a population mean μ even when the shape of the population distribution is unknown.

Recall that a variable is standardized by subtracting the mean value and then dividing by its standard deviation. Using rules 1 and 2 to standardize $\bar{x}$ gives an important consequence of the last two rules.

If n is large or the population distribution is normal, the standardized variable

$$z = \frac{\bar{x} - \mu_{\bar{x}}}{\sigma_{\bar{x}}} = \frac{\bar{x} - \mu}{\sigma/\sqrt{n}}$$

has (approximately) a standard normal (z) distribution.

Application of the Central Limit Theorem in specific problem situations requires a rule of thumb for deciding whether n is indeed sufficiently large. Such a rule is not as easy to come by as you might think. Look back at Figure 4, which shows the approximate sampling distribution of $\bar{x}$ for $n = 5$,

10, 20, 30, and 50 when the population distribution is quite skewed. Certainly the histogram for $n = 10$ is not well described by a normal curve, and this is still true of the histogram for $n = 20$, particularly in the tails of the histogram (far away from the mean value). Among the five histograms, only the ones for $n = 30$ and $n = 50$ have reasonably normal shapes.

On the other hand, when the population distribution is normal, the sampling distribution of $\bar{x}$ is normal for any n. If the population distribution is somewhat skewed but not to the extent of Figure 3, we might expect the $\bar{x}$ sampling distribution to be a bit skewed for $n = 5$ but quite well fit by a normal curve for n as small as 10 or 15. What value of n it will take for a normal curve to give a good approximation to $\bar{x}$'s sampling distribution will depend on how much the population distribution differs from a normal distribution. The closer the population distribution is to being normal, the smaller the value of n necessary for the Central Limit Theorem approximation to be accurate.

The rule that many statisticians recommend is conservative.

> The Central Limit Theorem can safely be applied if n exceeds 30.

If the population distribution is believed to be reasonably close to a normal distribution, an n of 15 or 20 is often large enough for $\bar{x}$ to have approximately a normal distribution. At the other extreme, we can imagine a distribution with a much longer tail than that of Figure 3, in which case even $n = 40$ or 50 would not suffice for approximate normality of $\bar{x}$. In practice, however, very few population distributions are likely to be this badly behaved.

To see how information regarding the sampling distribution of $\bar{x}$ can be used, consider the following three examples.

EXAMPLE 5   Let x denote the duration of a randomly selected song (in minutes) for a certain type of songbird. Suppose that the mean value of song duration is $\mu = 1.5$ min and that the standard deviation of song duration is $\sigma = .9$ min. (These values are suggested by the results of a large random sample of house finch songs, as reported in "Song Dialects and Colonization in the House Finch" (*Condor* (1975): 407–22.) The sampling distribution of $\bar{x}$ based on a random sample of $n = 25$ song durations then also has mean value

$$\mu_{\bar{x}} = 1.5 \text{ min}$$

That is, the sampling distribution of $\bar{x}$ is centered at 1.5. The standard deviation of $\bar{x}$ is

$$\sigma_{\bar{x}} = \frac{\sigma}{\sqrt{n}} = \frac{.9}{\sqrt{25}} = .18$$

only one-fifth as large as the population standard deviation, σ.

EXAMPLE 6   A meat market claims that the average fat content of its ground chuck is 12%. Let x denote the fat content of a randomly selected package of ground chuck. Suppose that x is normally distributed, with $\sigma = 1.6\%$. Consider selecting $n = 16$ packages and determining the fat content of each one (observations $x_1, x_2, \ldots, x_{16}$), and let $\bar{x}$ denote the resulting sample average fat content. Because the x distribution is normal, $\bar{x}$ is also normally distributed. If the market's claim is correct, the $\bar{x}$ sampling distribution has mean value

$$\mu_{\bar{x}} = \mu = 12$$

and standard deviation

$$\sigma_{\bar{x}} = \frac{\sigma}{\sqrt{n}} = \frac{1.6}{\sqrt{16}} = .40$$

To calculate a probability involving $\bar{x}$, we simply standardize by subtracting the mean value, 12, and dividing by the standard deviation (of $\bar{x}$), which is .40. For example, the probability that the sample average fat content is between 11.6% and 12.8% is calculated by first standardizing the interval limits:

lower limit: $\dfrac{11.6 - 12}{.40} = -1.0$

upper limit: $\dfrac{12.8 - 12}{.40} = 2.0$

Then

$$P(11.6 \le \bar{x} \le 12.8) = \left(\begin{array}{c}\text{area under the } z \text{ curve}\\ \text{between } -1.0 \text{ and } 2.0\end{array}\right)$$

$$= \left(\begin{array}{c}\text{area to the}\\ \text{left of } 2.0\end{array}\right) - \left(\begin{array}{c}\text{area to the}\\ \text{left of } -1.0\end{array}\right)$$

$$= .9772 - .1587$$

$$= .8185$$

The probability that the sample average fat content is at least 13% is, since $(13 - 12)/.40 = 2.5$,

$$P(13 \le \bar{x}) = \left(\begin{array}{c}\text{area under the } z \text{ curve}\\ \text{to the right of } 2.5\end{array}\right) = .0062$$

If the x distribution is as described and the claim is correct, a sample average fat content based on 16 observations exceeds 13% for less than 1% of all such samples. Thus, observation of an $\bar{x}$ value that exceeds 13% casts doubt on the market's claim that the average fat content is 12%.

EXAMPLE 7   A cigarette manufacturer asserts that one of its brands of cigarettes has an average nicotine content of $\mu = 1.8$ mg per cigarette. Smokers of this brand would probably not be disturbed if the mean were less than 1.8 but would be unhappy if it exceeded 1.8. Let x denote the nicotine content of a randomly selected cigarette and suppose that σ, the standard deviation of the x distribution, is .4.

An independent testing organization is asked to analyze a random sample of 36 cigarettes. Let $\bar{x}$ be the average nicotine content for this sample. The sample size, $n = 36$, is large enough to invoke the Central Limit Theorem and regard the $\bar{x}$ distribution as being approximately normal. The standard deviation of the $\bar{x}$ distribution is

$$\sigma_{\bar{x}} = \frac{\sigma}{\sqrt{n}} = \frac{.4}{\sqrt{36}} = .0667 \text{ mg}$$

If the manufacturer's claim is correct, we know that

$$\mu_{\bar{x}} = \mu = 1.8 \text{ mg}$$

Suppose the sample resulted in a mean of $\bar{x} = 1.84$. Does this result indicate that the manufacturer's claim is incorrect?

We can answer this question by looking at the sampling distribution of $\bar{x}$. Due to sampling variability, even if $\mu = 1.8$ we know that $\bar{x}$ will deviate somewhat from this value. Is it likely that we would see a sample mean as large as 1.84 when the true population mean is 1.8? If the company's claim is correct,

$$P(\bar{x} \geq 1.84) \approx P\left(z > \frac{1.84 - 1.8}{.0667}\right)$$

$$= P(z > .6)$$

$$= \left(\begin{array}{c}\text{area under the } z \text{ curve} \\ \text{to the right of } .6\end{array}\right)$$

$$= .2743$$

The value $\bar{x} = 1.84$ does not exceed 1.8 by enough to cast substantial doubt on the manufacturer's claim. Values of $\bar{x}$ as large as 1.84 will be observed about 27.43% of the time when a random sample of size 36 is taken from a population with mean 1.8 and standard deviation .4.

OTHER CASES

We now know a great deal about the sampling distribution of $\bar{x}$ in two cases, that of a normal population distribution and that of a large sample size. What happens when the population distribution is not normal and n is small? Unfortunately, while it is still true that $\mu_{\bar{x}} = \mu$ and $\sigma_{\bar{x}} = \sigma/\sqrt{n}$, there is no general result about the shape of the $\bar{x}$ distribution. When the objective is to make an inference about the center of such a population, one way to proceed is to replace the normality assumption with some other distributional model for the population. Statisticians have proposed and studied a number of such models. Then theoretical methods or simulation can be used to describe the $\bar{x}$ distribution corresponding to the assumed model. An alternative path is to use an inferential procedure based on statistics other than $\bar{x}$.

EXERCISES 7.14–7.25   **SECTION 7.2**

7.14 A random sample is to be selected from a population with mean $\mu = 100$ and standard deviation $\sigma = 10$. Determine the mean and standard deviation of the $\bar{x}$ sampling distribution for each of the following sample sizes.
a. $n = 9$ **b.** $n = 15$ **c.** $n = 36$
d. $n = 50$ **e.** $n = 100$ **f.** $n = 400$

7.15 For which of the sample sizes given in Exercise 7.14 would it be reasonable to think that the $\bar{x}$ sampling distribution will be approximately normal in shape?

7.16 Explain the difference between σ and $\sigma_{\bar{x}}$. Between μ and $\mu_{\bar{x}}$.

7.17 Suppose that a random sample of size 64 is to be selected from a population with mean 40 and standard deviation 5.
a. What are the mean and standard deviation of the $\bar{x}$ sampling distribution? Describe the shape of the $\bar{x}$ sampling distribution.
b. What is the approximate probability that $\bar{x}$ will be within .5 of the population mean μ?
c. What is the approximate probability that $\bar{x}$ will differ from μ by more than .7?

7.18 The time that a randomly selected individual waits for an elevator in an office building has a uniform distribution over the interval from 0 to 1 min. (The uniform distribution is discussed briefly in Example 1 in Section 6.1.) It can be shown that, for this distribution, $\mu = .5$ and $\sigma = .289$.
a. Let $\bar{x}$ be the sample average waiting time for a random sample of 16 individuals. What are the mean value and standard deviation of $\bar{x}$'s sampling distribution?
b. Answer part (a) for a random sample of 50 individuals. In this case, sketch a picture of a good approximation to the actual $\bar{x}$ distribution.

7.19 Let x denote the time (min) that it takes a fifth grade student to read a certain passage. Suppose that the mean value and standard deviation of x are $\mu = 2$ min and $\sigma = .8$ min, respectively.
a. If $\bar{x}$ is the sample average time for a random sample of $n = 9$ students, where is the $\bar{x}$ distribution centered and how much does it spread out about the center (as described by its standard deviation)?
b. Repeat part (a) for a sample of size $n = 20$ and again for a sample of size $n = 100$. How do the centers and spreads of the three $\bar{x}$ distributions compare to one another? Which sample size would be most likely to result in an $\bar{x}$ value close to μ, and why?

7.20 Suppose that the mean value of interpupillary distance for all adult males is 65 mm and the population standard deviation is 5 mm.
a. If the distribution of interpupillary distance is normal and a sample of $n = 25$ adult males is selected, what is the probability that the sample average distance $\bar{x}$ for these 25 will be between 64 and 67 mm? At least 68 mm?
b. Suppose that a sample of 100 adult males is obtained. Without assuming that interpupillary distance is normally distributed, what is the approximate probability that the sample average distance is between 64 and 67 mm? At least 68 mm?

7.21 Suppose that a sample of size 100 is to be drawn from a population with standard deviation 10.
a. What is the probability that the sample mean will be within 2 of the value of μ?
b. For this example ($n = 100$, $\sigma = 10$), complete each statement by computing the appropriate value.
i. Approximately 95% of the time, $\bar{x}$ is within _____ of μ.
ii. Approximately .3% of the time, $\bar{x}$ is farther than _____ from μ.

7.22 A manufacturing process is designed to produce bolts with a .5″ diameter. Once each day, a random sample of 36 bolts is selected and the diameters recorded. If the resulting sample mean is less than .49″ or greater than .51″, the process is shut down for adjustment. The standard deviation for diameter is .02″. What is the probability that the manufacturing line will be shut down unnecessarily? (Hint: Find the probability of observing an $\bar{x}$ in the shut-down range when the true process mean really is .5″.)

7.23 College students with a checking account typically write relatively few checks in any given month, while full-time residents typically write many more checks during a month. Suppose that 50% of a bank's accounts are held by students and 50% by full-time residents. Let x denote the number of checks written in a given month by a randomly selected bank customer.

   a. Give a sketch of what the probability distribution of x might look like.
   b. Let the mean value of x be 22.0 and the standard deviation 16.5. If a random sample of $n = 100$ customers is selected and $\bar{x}$ denotes the sample average number of checks written during a particular month, where is the sampling distribution of $\bar{x}$ centered, and what is the standard deviation of the $\bar{x}$ distribution? Sketch a rough picture of the sampling distribution.
   c. Referring to part (b), what is the approximate probability that $\bar{x}$ is at most 20? At least 25? What result are you using to justify your computations?

7.24 An airplane with room for 100 passengers has a total baggage limit of 6000 lb. Suppose that the total weight of the baggage checked by an individual passenger is a random variable x with mean value 50 lb and standard deviation 20 lb. If 100 passengers board a flight, what is the approximate probability that the total weight of their baggage will exceed the limit? (Hint: With $n = 100$, the total weight exceeds the limit precisely when the average weight $\bar{x}$ exceeds 6000/100.)

7.25 Manufacturing processes are often monitored in order to determine whether unusual circumstances (such as inferior raw materials or tool wear) are influencing the quality of the manufactured product. The thickness of the coating (in mm) applied to disk drives is a characteristic that determines the usefulness of the product. When no unusual circumstances are present, the thickness (x) has a normal distribution with mean 3 mm and standard deviation .05 mm. Suppose the process will be monitored by selecting a random sample of 16 drives from each shift's production and determining $\bar{x}$, the mean coating thickness for the sample.

   a. Describe the sampling distribution of $\bar{x}$ (for a sample of size 16).
   b. When no unusual circumstances are present, we expect $\bar{x}$ to be within $3\sigma_{\bar{x}}$ of 3 mm, the desired value. An $\bar{x}$ value farther from 3 than $3\sigma_{\bar{x}}$ is interpreted as an indication of a problem that needs attention. Compute $3 \pm 3\sigma_{\bar{x}}$. (A plot over time of $\bar{x}$ values with horizontal lines drawn at the limits $\mu \pm 3\sigma_{\bar{x}}$ is called a *process control chart*.)
   c. Referring to part (b), what is the probability that a sample mean will be outside $3 \pm 3\sigma_{\bar{x}}$ just by chance (that is, when there are no unusual circumstances)?
   d. Suppose that a machine used to apply the coating is out of adjustment, resulting in a mean coating thickness of 3.05 mm. What is the probability that a problem will be detected when the next sample is taken? (Hint: This will occur if $\bar{x} > 3 + 3\sigma_{\bar{x}}$ or $\bar{x} < 3 - 3\sigma_{\bar{x}}$ when $\mu = 3.05$.)

7.3 THE SAMPLING DISTRIBUTION OF A SAMPLE PROPORTION

The objective of many statistical investigations is to draw a conclusion about the proportion of individuals or objects in a population that possess a specified property; for example, Maytag washers that don't require service during the warranty period, Europeans who favor the deployment of a certain type of missile, or smokers who regularly smoke nonfilter cigarettes. In such situations, any individual or object that possesses the property of interest is labeled a success (S), and one that does not possess the property is termed a failure (F). The letter π denotes the proportion of S's in the population. The value of π is a number between 0 and 1, and 100π is the percentage of S's in the population. Thus $\pi = .75$ means that 75% of the population members are S's, while $\pi = .01$ identifies a population containing only 1% S's and 99% F's.

The value of π is usually unknown to an investigator. When a random sample of size n is selected from this type of population, some of the individuals in the sample are S's and the remaining individuals in the sample are F's. The statistic that will provide a basis for making inferences about π is p, the **sample proportion of S's**:

$$p = \frac{\text{the number of S's in the sample}}{n}$$

For example, if $n = 5$ and three S's result, then $p = 3/5 = .6$.

Just as making inferences about μ requires knowing something about the sampling distribution of the statistic $\bar{x}$, making inferences about π requires first learning about properties of the sampling distribution of the statistic p. For example, when $n = 5$, possible values of p are 0, .2 (from 1/5), .4, .6, .8, and 1. The sampling distribution of p gives the probability of each of these six possible values (the long-run proportion of the time that each value would occur if samples with $n = 5$ were selected over and over again). Similarly, when $n = 100$, the 101 possible values of p are 0, .01, .02, . . . , .98, .99, and 1. The sampling distribution of p can then be used to calculate probabilities such as $P(.3 \le p \le .7)$ and $P(.75 \le p)$.

Before stating some general rules concerning the sampling distribution of p, we present the results of two sampling experiments as aids to developing an intuitive understanding of the rules. In each example, we selected a population having a specified value of π and obtained 500 random samples, each of size n, from the population. We then computed p for each sample and constructed a sample histogram of these 500 values. As with $\bar{x}$, this was repeated for several different values of n to show how the sampling distribution changes with increasing sample size.

EXAMPLE 8

The percentage of females in the European labor force varies widely from country to country. The publication *European Marketing Data and Statistics* (17th ed., 1981) reports that Ireland's work force had the lowest percentage of females, 26.5%, while the U.S.S.R. had the highest, with 50.4%. We decided to simulate sampling from Ireland's labor force with S denoting a female worker and F a male worker, so $\pi = .265$. A computer was used to select 500 samples of size $n = 5$, then 500 samples of size $n = 10$, then 500 samples with $n = 25$, and finally 500 samples with $n = 50$. The sample histograms of the 500 values of p for the four sample sizes are displayed in Figure 6.

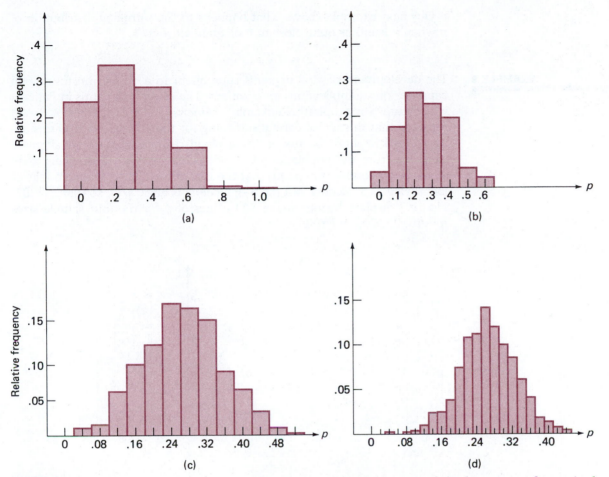

FIGURE 6

Histograms of 500 values of p, each based on a random sample of size n (π = .265).

(a) n = 5
(b) n = 10
(c) n = 25
(d) n = 50

The most noticeable feature of the histogram shapes is the progression toward the shape of a normal curve as n increases. The histogram for n = 5 is definitely skewed. The two smallest possible values of p, which are 0 and .2, occurred 122 and 174 times, respectively, in the 500 samples, while the two largest values, .8 and 1, occurred only 3 and 1 times, respectively. The histogram for n = 10 is considerably more bell-shaped, although it still has a slight positive skew. The histogram for n = 25 exhibits very little skewness, and the histogram for n = 50 looks like a normal curve.

Although the skewness of the first two histograms makes the location of their centers (balance points) a bit difficult, all four histograms appear to be centered at roughly .265, the value of π for the population being sampled. Had the histograms been based on an unending sequence of samples (or, equivalently, on theoretical calculations using the binomial probability distribution) instead of just 500 samples, each histogram would have been centered at exactly .265. Finally, as was the case with the sampling distribution of $\bar{x}$, the histograms spread out more for small n than for large n. The value of p based on a large sample size tends to be closer to π, the population proportion of S's, than does p from a small sample.

Our next example shows what happens to the sampling distribution of p when π is either quite close to 0 or quite close to 1.

EXAMPLE 9

The development of viral hepatitis subsequent to a blood transfusion can cause serious complications for a patient. The paper "Hepatitis in Patients with Acute Nonlymphatic Leukemia" (*Amer. J. of Med.* (1983): 413–21) reported that in spite of careful screening for those having a hepatitis antigen, viral hepatitis occurs in 7% of blood recipients. Here we simulate sampling from the population of blood recipients, with S denoting a recipient who contracts hepatitis (not the sort of characteristic one thinks of identifying as a success, but the S–F labeling is arbitrary), so that $\pi = .07$. Figure 7 displays histograms of 500 values of p for the four sample sizes $n = 10$, 25, 50, and 100.

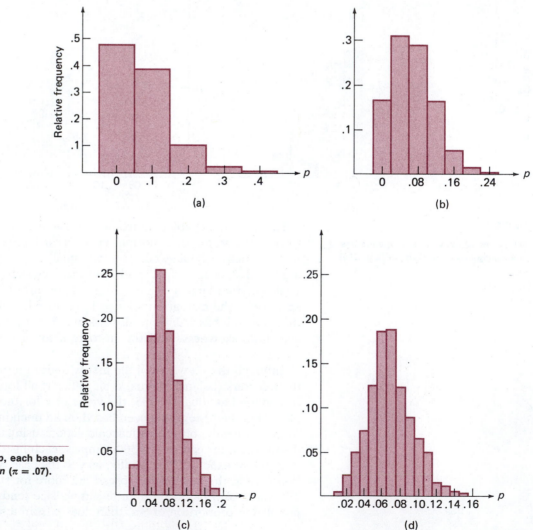

FIGURE 7

Histograms of 500 values of p, each based on a random sample of size n ($\pi = .07$).

(a) $n = 10$
(b) $n = 25$
(c) $n = 50$
(d) $n = 100$

As was the case in Example 8, all four histograms are centered at approximately the value of π for the population being sampled. (The average values of p are .0690, .0677, .0707, and .0694.) If the histograms had been based on an unending sequence of samples, they would all have been centered at exactly $\pi = .07$. Again the spread of a histogram based on a large n is smaller than the spread of a histogram resulting from a small sample size. The larger the value of n, the closer the sample proportion, p, tends to be to the value of the population proportion, π.

Furthermore, there is a progression toward the shape of a normal curve as n increases. However, the progression is much slower here than in the previous example because the value of π is so extreme. (The same thing would happen for $\pi = .93$, except that the histograms would be negatively rather than positively skewed.) The histograms for $n = 10$ and $n = 25$ exhibit substantial skew, and the skew of the histogram for $n = 50$ is still moderate (compare Figure 7(c) to Figure 6(d)). Only the histogram for $n = 100$ is reasonably well fit by a normal curve. It appears that whether a normal curve provides a good approximation to the sampling distribution of p depends on the values of both n and π. Knowing just that $n = 50$ is not enough to guarantee that the shape of the histogram is approximately normal.

The previous two examples suggest that the sampling distribution of p depends both on n, the sample size, and on π, the proportion of S's in the population. These results are stated more formally in the following general rules.

GENERAL RULES CONCERNING THE p SAMPLING DISTRIBUTION

Let p be the proportion of successes in a random sample of size n from a population whose proportion of S's is π. Denote the mean value of p by μ_p and the standard deviation by σ_p. Then the following rules hold.

Rule 1. $\mu_p = \pi$

Rule 2. $\sigma_p = \sqrt{\pi(1 - \pi)/n}$

Rule 3. When n is sufficiently large, the sampling distribution of p is approximately normal.

Thus the sampling distribution of p is always centered at the value of the population success proportion π, and the extent to which the distribution spreads out about π decreases as the sample size n increases.

Examples 8 and 9 indicate that both π and n must be considered in judging whether p's sampling distribution is approximately normal.

> The farther the value of π is from .5, the larger the value of n must be in order for the normal approximation to the sampling distribution of p to be accurate. A conservative rule of thumb is that if both $n\pi \geq 5$ and $n(1 - \pi) \geq 5$, then it is safe to use the normal approximation.

A sample size of $n = 100$ is not by itself sufficient to justify the use of the normal approximation. If $\pi = .01$, the distribution of p will be very positively skewed, so a bell-shaped curve will not give a good approximation. Similarly, if $n = 100$ and $\pi = .99$ (so $n(1 - \pi) = 1 < 5$), the distribution of p will have a substantial negative skew. The conditions $n\pi \geq 5$ and $n(1 - \pi) \geq 5$ ensure that the sampling distribution of p is not too skewed. If $\pi = .5$, the normal approximation can be used for n as small as 10, while for $\pi = .05$ or $.95$, n should be at least 100.

EXAMPLE 10

The proportion of all blood recipients stricken with viral hepatitis was given as .07 in the paper referenced in Example 9. Suppose that a new treatment is developed that is believed to reduce the incidence rate of viral hepatitis. This treatment is given to $n = 200$ blood recipients. Only six of the 200 patients contract hepatitis. This appears to be a favorable result, since $p = 6/200 = .03$. The question of interest to medical researchers is: Does this result indicate that the true (long-run) proportion of patients who contract hepatitis after the experimental treatment will be less than .07, or could this result be plausibly attributed to sampling variability (i.e., the fact that p will typically deviate from its mean value, π)? If the treatment is ineffective,

$$\mu_p = \pi = .07$$

$$\sigma_p = \sqrt{\pi(1 - \pi)/200} = \sqrt{(.07)(.93)/200} = .018$$

Furthermore, since

$$n\pi = 200(.07) = 14 \geq 5$$

and

$$n(1 - \pi) = 200(.93) = 186 \geq 5,$$

the sampling distribution of p is approximately normal. Then

$$P(p < .03) = P\left(z < \frac{.03 - .07}{.018}\right)$$

$$= P(z < -2.2)$$

$$= \text{area to the left of } -2.2 \text{ under the } z \text{ curve}$$

$$= .0139$$

Thus, it is unlikely that a sample proportion .03 or smaller would be observed if the treatment really is ineffective. The new treatment appears to yield a smaller incidence rate for the disease than occurs without any treatment.

EXERCISES 7.26–7.32   **SECTION 7.3**

7.26   A random sample is to be selected from a population that has a proportion of successes $\pi = .65$. Determine the mean and standard deviation of the sampling distribution of p for each of the following sample sizes.
   a. $n = 10$ **b.** $n = 20$ **c.** $n = 30$
   d. $n = 50$ **e.** $n = 100$ **f.** $n = 200$

7.27   For which of the sample sizes given in Exercise 7.26 would the sampling distribution of p be approximately normal if $\pi = .65$? If $\pi = .2$?

7.28   A certain chromosome defect occurs in only one out of 200 Caucasian adult males. A random sample of $n = 100$ males is obtained.
   a. What is the mean value of the sample proportion p (the number of defects divided by 100), and what is the standard deviation of the sample proportion?
   b. Does p have approximately a normal distribution in this case? Explain.
   c. What is the smallest value of n for which the sampling distribution of p is approximately normal?

7.29   A column in the April 22, 1985, issue of *Newsweek* reported that 55% of all women in the American work force regard themselves as underpaid. Although this conclusion was based on sample data, suppose that, in fact, $\pi = .55$, where π represents the true proportion of working women who believe that they are underpaid.
   a. Would p based on a random sample of only ten working women have approximately a normal distribution? Explain why or why not.
   b. What are the mean value and standard deviation of p based on a random sample of size 400?
   c. When $n = 400$, what is $P(.5 \le p \le .6)$?
   d. Suppose now that $\pi = .4$. For a random sample of $n = 400$ working women, what is $P(.5 < p)$?

7.30   The article "Thrillers" (*Newsweek*, April 22, 1985) states, "Surveys tell us that more than half of America's college graduates are avid readers of mystery novels." Let π denote the actual proportion of college graduates who are avid readers of mystery novels. Consider p to be based on a random sample of 225 college graduates.
   a. If $\pi = .5$, what are the mean value and standard deviation of p? Answer this question when $\pi = .6$. Does p have approximately a normal distribution in both cases? Explain.
   b. Calculate $P(p \ge .6)$ both when $\pi = .5$ and when $\pi = .6$.
   c. Without doing any calculations, how do you think the probabilities in part (b) would change if n were 400 rather than 225?

7.31   Suppose that a particular candidate for public office is in fact favored by 48% of all registered voters in the district. A polling organization takes a random sample of 500 voters and will use p, the sample proportion, to estimate π. What is the approximate probability that p will be greater than .5, causing the polling organization to incorrectly predict the result of the upcoming election?

7.32   A manufacturer of electric typewriters purchases plastic print wheels from a vendor. When a large shipment is received, a random sample of 200 print wheels is selected and each is inspected. If the sample proportion of defectives is more than .02, the entire shipment will be returned to the vendor.
   a. What is the approximate probability that the shipment will be returned if the true proportion of defectives in the shipment is .05?
   b. What is the approximate probability that the shipment will not be returned when the true proportion of defectives in the shipment is .10?

CHAPTER 7 SUMMARY OF KEY CONCEPTS AND FORMULAS

TERM OR FORMULA	PAGE	COMMENT
Statistic	198	Any quantity whose value is computed from sample data
Sampling distribution	199	The probability distribution of a statistic: the sampling distribution describes the long-run behavior of the statistic
Random sample	202	A sample for which values are selected independently from the same population distribution
Sampling distribution of $\bar{x}$	204	The probability distribution of the sample mean $\bar{x}$, based on a random sample of size n. Key properties of the $\bar{x}$ sampling distribution are $\mu_{\bar{x}} = \mu$ and $\sigma_{\bar{x}} = \sigma/\sqrt{n}$ (where μ and σ are the population mean and standard deviation, respectively). In addition, when the population distribution is normal or the sample size is large, the sampling distribution of $\bar{x}$ is (approximately) normal.
Central Limit Theorem	212	This important theorem states that when n is sufficiently large, the $\bar{x}$ distribution will be approximately normal. The standard rule of thumb is that the theorem can safely be applied when n exceeds 30.
Sampling distribution of p	217	The probability distribution of the sample proportion p, based on a random sample of size n. When the sample size is sufficiently large, the sampling distribution of p is approximately normal, with $\mu_p = \pi$ and $\sigma_p = \sqrt{\pi(1-\pi)/n}$ (where π is the true population proportion).

SUPPLEMENTARY EXERCISES 7.33–7.39

7.33 The nicotine content in a single cigarette of a particular brand is a random variable with mean .8 mg and standard deviation .1 mg. If 100 of these cigarettes are analyzed, what is the probability that the resulting sample mean nicotine content will be less than .79? Less that .77?

7.34 Let $x_1, x_2, \ldots, x_{100}$ denote the actual net weights (in lb) of 100 randomly selected bags of fertilizer. Suppose that the weight of a randomly selected bag is a random variable with mean 50 lb and variance 1 lb². Let $\bar{x}$ be the sample mean weight ($n = 100$).
a. Describe the sampling distribution of $\bar{x}$.
b. What is the probability that the sample mean is between 49.75 lb and 50.25 lb?
c. What is the probability that the sample mean is less than 50 lb?

7.35 Suppose that 20% of the subscribers of a cable television company watch the shopping channel at least once a week. The cable company is trying to decide whether to replace this channel with a new local station. A survey of 100 subscribers will be undertaken. The cable company has decided to keep the shopping channel if the sample proportion is greater than .25. What is the approximate probability that the cable company will keep the shopping channel, even though the true proportion who watch it is only .20?

7.36 Although a lecture period at a certain university lasts exactly 50 minutes, the actual lecture time of a statistics instructor on any particular day is a random variable with mean value 52 min and standard deviation 2 min. Suppose that times of different lectures are independent of one another. Let $\bar{x}$ represent the mean of 36 randomly selected lecture times.

    a. What are the mean value and standard deviation of the sampling distribution of $\bar{x}$?

    b. What is the probability that the sample mean exceeds 50 min? 55 min?

7.37 Water permeability of concrete is an important characteristic in assessing suitability for various applications. Permeability can be measured by letting water flow across the surface and determining the amount lost (in./hr). Suppose that the permeability index x for a randomly selected concrete specimen of a particular type is normally distributed with mean value 1000 and standard deviation 150.

    a. How likely is it that a single specimen will have a permeability index between 850 and 1300?

    b. If the permeability index is determined for each specimen in a random sample of size 10, how likely is it that the sample average permeability index will be between 950 and 1100? Between 850 and 1300?

7.38 Suppose that 60% of all students taking Elementary Statistics write their names in their textbooks. A random sample of 100 students is selected.

    a. What is the mean value of the proportion among the 100 sampled who have their names in their texts? What is the standard deviation of the sample proportion?

    b. What is the chance that at most 50% of those sampled have their names in their texts?

    c. Answer part (b) if there are 400 students in the random sample.

7.39 The amount of money spent by a customer at a discount store has a mean of $100 and a standard deviation of $30. What is the probability that a randomly selected group of 50 shoppers will spend a total of more than $5300? (Hint: The total will be more than $5300 when the sample average exceeds what value?)

REFERENCES The books by Freedman et al. and by Moore, both listed in the Chapter 1 References, give excellent informal discussions of sampling distributions at a very elementary level.

ESTIMATION USING A SINGLE SAMPLE

INTRODUCTION The objective of inferential statistics is to use sample data to increase our knowledge about the corresponding population. Often, data is collected to obtain information that allows the investigator to estimate the value of some population characteristic, such as population mean, μ, or a population proportion π. This could be accomplished by using the sample data to arrive at a single number that represents a plausible value for the characteristic of interest. Alternately, one could construct an entire interval of plausible values for the characteristic. These two estimation techniques, *point estimation* and *interval estimation*, are introduced in this chapter.

8.1 POINT ESTIMATION

The simplest approach to estimating a population characteristic involves using sample data to compute a single number that can be regarded as a plausible value of the characteristic. For example, sample data might suggest that 1.1 mg is a plausible value for μ, the true mean nicotine content of Players 100 mm cigarettes. (This is the value stated on the package.) In a different setting, a sample survey of students at a particular university might lead to the statement that .41 is a plausible value for π, the true proportion of students who favor a fee for recreational activities.

DEFINITION

A **point estimate** of a population characteristic is a single number that is based on sample data and represents a plausible value of the characteristic.

In the examples just given, 1.1 is a point estimate of μ and .41 is a point estimate of π. The adjective *point* reflects the fact that the estimate is a single point on the number line.

A point estimate is obtained by first selecting an appropriate statistic. The estimate is then the value of the statistic for the given sample. Thus the computed value of the sample mean $\bar{x}$ could provide a point estimate of a population mean μ.

EXAMPLE 1

The sale of human organs for transplantation raises some difficult ethical issues both inside and outside the medical community. Let π denote the proportion of all U.S. doctors who oppose such activity. (With success identified as opposition, π is then the population proportion of successes.) The very large number of doctors makes it impractical to determine the exact value of π by soliciting an opinion from every doctor. Suppose that a random sample of n doctors is obtained. Then the statistic

$$p = \frac{\text{number of successes in the sample}}{n}$$

which is the sample proportion of successes, is an obvious candidate for use in obtaining a point estimate of π. An article in the Nov. 24, 1978, *Los Angeles Times* reported that in a random sample of $n = 244$ doctors, the number opposed to the sale of organs was 184. Thus, based on this sample, the point estimate of π, the proportion of all doctors opposed to the sale of organs, is

$$p = \frac{184}{244} = .754$$

For purposes of estimating a population proportion π, it would be difficult to suggest a statistic other than p that could reasonably be used. In other situations there are several statistics that can be used to obtain an estimate. The following example presents such a case.

EXAMPLE 2

The paper "Effects of Roadside Conditions on Plants and Insects" (*J. Appl. Ecology* (1988): 709–15) reported the results of an experiment to evaluate the effect of nitrous oxide from automobile exhaust on roadside soil. The concentration of soluble nitrogen (in mg/g) was recorded at twenty roadside locations. Suppose that the observed readings are as given, along with a dot diagram of the data. (This data is compatible with summary values given in the paper.)

1.7	1.9	2.0	1.7	1.6	1.7	1.8	1.7	1.8	1.8
1.6	1.7	2.1	1.7	1.8	1.7	1.8	1.4	1.9	1.7

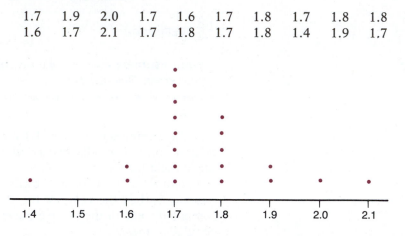

Suppose that a point estimate of μ, the true mean soluble nitrogen concentration, is desired. An obvious choice of a statistic for estimating μ is the sample mean, $\bar{x}$. However, there are other possibilities. We might consider using a trimmed mean or even the sample median, since the data set exhibits some symmetry. (If the corresponding population distribution is symmetric, μ is also the population median.)

The three statistics and the resulting estimates of μ are

$$\text{sample mean} = \bar{x} = \frac{\sum x}{n} = \frac{35.1}{20} = 1.76$$

$$\text{sample median} = \frac{1.7 + 1.7}{2} = 1.70$$

$$10\% \text{ trimmed mean} = \left(\begin{array}{c} \text{average of middle} \\ \text{16 observations} \end{array} \right) = \frac{28.0}{16} = 1.75$$

The estimates differ somewhat from one another. The choice among them should depend on which statistic tends to produce an estimate closest to the true value. The following subsections discuss criteria for choosing among competing statistics.

**CHOOSING A STATISTIC FOR
COMPUTING AN ESTIMATE**

The point of the previous example is that there may be more than one statistic that can reasonably be used to obtain a point estimate of a specified population characteristic. Loosely speaking, the statistic used should be one that tends to yield an accurate estimate, i.e., an estimate close to the value of the population characteristic. Information on the accuracy of estimation for a particular statistic is provided by the statistic's sampling distribution. Figure 1 pictures the sampling distributions of three different statistics. The value of the population characteristic, which we refer to as the *true value*, is marked on the measurement axis.

The distribution pictured in Figure 1(a) is that of a statistic unlikely to yield an estimate close to the true value. The distribution is centered to the right of the true value, making it very likely that an estimate (a value of the statistic for a particular sample) will be substantially larger than the true value. If this statistic is used to compute an estimate based on a first sample, then another estimate based on a second sample, and another estimate based on a third sample, and so on, the long-run average value of these estimates will considerably exceed the true value.

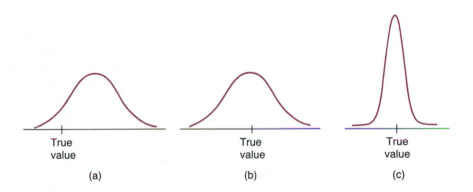

FIGURE 1

Sampling distributions of three different statistics for estimating a population characteristic.

The sampling distribution of Figure 1(b) is centered at the true value. Thus, while one estimate may be smaller than the true value and another may be larger, when this statistic is used many times over with different samples, there will be no long-run tendency to over- or underestimate the true value. However, while this sampling distribution is correctly centered it spreads out quite a bit about the true value. That is, the statistic's standard deviation is relatively large. Because of this, some estimates resulting from the use of this statistic will be far above or below the true value— even though there is no systematic tendency to underestimate or overestimate the true value. In contrast, the mean value of the statistic with the distribution appearing in Figure 1(c) is exactly the true value of the population characteristic (implying no systematic estimation error), and the statistic's standard deviation is relatively small. Estimates using this third statistic will almost always be quite close to the true value, certainly more often than estimates resulting from the statistic with the sampling distribution shown in Figure 1(b).

▌▌▌ DEFINITION

A statistic with mean value equal to the value of the population character-istic being estimated is said to be an **unbiased statistic**. A statistic that is not unbiased is said to be **biased**.

Let $x_1, x_2, \ldots, x_n$ be a random sample, as discussed in Chapter 7. One of the general results concerning the sampling distribution of $\bar{x}$ is that $\mu_{\bar{x}} = \mu$. This result says that no matter what the value of μ, the $\bar{x}$ distribution is centered at that value. For example, if $\mu = 100$, that is where the $\bar{x}$ distribution is centered, whereas if $\mu = 5200$, then the $\bar{x}$ distribution is centered at 5200. Therefore, $\bar{x}$ is an unbiased statistic for estimating μ. Similarly, since the sampling distribution of p is centered at π ($\mu_p = \pi$), it follows that p is an unbiased statistic for estimating a population proportion.

Our definition of the sample variance is

$$s^2 = \frac{\sum (x - \bar{x})^2}{n - 1}$$

This statistic is a good choice for obtaining a point estimate of the population variance σ^2. It can be shown that s^2 is an unbiased statistic for estimating σ^2. That is, whatever the value of σ^2, the sampling distribution of s^2 is centered at that value. It is precisely for this reason—to obtain an unbiased statistic—that the divisor $n - 1$ is used. The statistic

$$\frac{\sum (x - \bar{x})^2}{n}$$

which has a more natural divisor than s^2, *is* biased. Estimates using this statistic tend to be smaller than σ^2.

EXAMPLE 3　The paper "Diallel Analysis of Pod Length and Shelling Percent of Winged Beans" (*Field Crops Research*, (1987): 209–216) reported the following lengths for nine winged bean pods:

12.1　18.5　17.2　27.7　17.9　12.9　17.0　15.0　14.5

Then

$$\sum x = 152.8 \qquad \sum x^2 = 2762.86$$
$$\bar{x} = 16.978 \qquad \sum (x - \bar{x})^2 = \sum x^2 - n\bar{x}^2 = 168.588$$

Let σ^2 denote the variance of pod length for the population of all such pods. Using s^2 to estimate σ^2 yields

$$s^2 = \frac{\sum (x - \bar{x})^2}{n - 1} = \frac{168.588}{8} = 21.074$$

Using the average squared deviation (with divisor $n = 9$), the resulting estimate is

$$\binom{\text{average squared}}{\text{deviation}} = \frac{\sum (x - \bar{x})^2}{n} = \frac{168.588}{9} = 18.732$$

Because s^2 is an unbiased statistic for estimating σ^2, most statisticians would recommend using the estimate 21.074.

An obvious choice of a statistic for estimating the population standard deviation σ is the sample standard deviation, s. For the data given in Example 3, $s = 4.591$, so our point estimate for the population standard deviation σ is 4.591. Unfortunately, the fact that s^2 is an unbiased statistic for estimating σ^2 does not imply that s is an unbiased statistic for estimating σ. The sample standard deviation tends to underestimate slightly the true value of σ. However, unbiasedness is not the only criterion by which a statistic can be judged, and there are other good reasons for using s to estimate σ. In what follows, whenever we need to estimate σ based on a single random sample, we use the statistic s to obtain a point estimate.

Using an unbiased statistic that has a small standard deviation guarantees that there will be no systematic tendency to under- or overestimate the value of the population characteristic and that estimates will almost always be relatively close to the true value.

> Given a choice between several unbiased statistics that could be used for estimating a population characteristic, the best statistic to use is the one with the smallest standard deviation.

Let us now return to the problem of estimating a population mean, μ. The obvious choice of statistic for obtaining a point estimate of μ is the sample mean, $\bar{x}$, an unbiased statistic for this purpose. However, when the population distribution is symmetric, $\bar{x}$ is not the only choice. Other unbiased statistics for estimating μ in this case include the sample median and any trimmed mean (with the same number of observations trimmed from each end of the ordered sample). Which statistic should be used? The following result is helpful in making a choice.

1. If the population distribution is normal, then $\bar{x}$ has a smaller standard deviation than any other unbiased statistic for estimating μ. However, in this case, a trimmed mean with a small trimming percentage (such as 10%) performs almost as well as $\bar{x}$.
2. When the population distribution is symmetric, with heavy tails compared to the normal curve, a trimmed mean is a better statistic than $\bar{x}$ for estimating μ.

When the population distribution is unquestionably normal, the choice is clear: use $\bar{x}$ to estimate μ. But with a heavy-tailed distribution, a trimmed mean gives protection against one or two outliers in the sample that might otherwise drastically affect the value of the estimate.

EXERCISES 8.1–8.8   **SECTION 8.1**

8.1 Three different statistics are being considered for estimating a population characteristic. The sampling distributions of the three statistics are shown below.

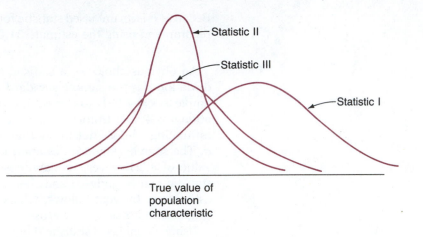

Which statistic would you recommend? Explain your choice.

8.2 Why is an unbiased statistic generally preferred over a biased statistic for estimating
a population characteristic? Does unbiasedness by itself guarantee that the estimate will be close to the true value? Explain. Under what circumstances might
you choose a biased statistic over an unbiased statistic if two statistics are available
for estimating a population characteristic?

8.3 The accompanying radiation readings (in mR/hr) were obtained from television
display areas in a sample of ten department stores ("Many Color TV Set Lounges
Show Highest Radiation," *J. of Environ. Health* (1969): 359–60).

.40   .48   .60   .15   .50   .80   .50   .36   .16   .89

a. Assuming that the distribution of radiation readings is approximately normal,
give a point estimate of μ, the true mean radiation level for TV lounges.
b. The recommended limit for this type of radiation is .5 mR/hr. Use the given
data to estimate the proportion of TV lounges with radiation levels exceeding
the recommended limit.

8.4 A random sample of $n = 12$ four-year-old red pine trees was selected, and the diameter (in.) of each tree's main stem was measured. The resulting observations were
as follows.

11.3   10.7   12.4   15.2   10.1   12.1   16.2   10.5   11.4   11.0   10.7   12.0

a. Compute a point estimate of σ, the population standard deviation of main stem
diameter. What statistic did you use to obtain your estimate?
b. Making no assumption whatsoever about the shape of the population distribution of diameter, give a point estimate for the population median diameter (that
is, for the middle diameter value in the entire population of four-year-old red
pine trees). What statistic did you use to obtain the estimate?
c. Suppose that the population distribution of diameter is symmetric but with heavier tails than the normal distribution. Give a point estimate of the population
mean diameter based on a statistic that gives some protection against the presence of outliers in the sample. What statistic did you use?

d. Suppose that the diameter distribution is normal. Then the 90th percentile of the diameter distribution is $\mu + 1.28\sigma$ (so 90% of all trees have diameters below this value). Compute a point estimate for this percentile. (Hint: First compute an estimate of μ in this case; then use it along with your estimate of σ from part (a).)

8.5 Each person in a random sample of 20 students at a particular university was asked whether they were registered to vote. The responses were (with R = registered, N = not registered)

R	R	N	R	N	N	R	R	R	N
N	R	R	R	R	R	N	R	R	R

Use this data to estimate π, the true proportion of all students at the university who are registered to vote.

8.6 **a.** A random sample of ten houses in a particular area, each of which is heated with natural gas, is selected and the amount of gas (therms) used during the month of January is determined for each house. The resulting observations are

    103   156   118   89   125   147   122   109   138   99

Let μ_J denote the average gas usage during January by all houses in this area. Compute a point estimate of μ_J.

b. Suppose that there are 10,000 houses in this area that use natural gas for heating. Let τ denote the total amount of gas used by all of these houses during January. Estimate τ using the data of part (a). What statistic (sample-based quantity) did you use in computing your estimate?

c. Use the data in part (a) to estimate π, the proportion of all houses that used at least 100 therms.

d. Give a point estimate of the population median usage (the middle value in the population of all houses) based on the sample of part (a). What statistic did you use?

8.7 Referring to Exercise 8.6, suppose that August gas usage is determined for these same ten houses, yielding the observations (in the same order)

    42   57   50   26   43   62   68   50   47   29

Let μ_d denote the average difference between January and August usage for all houses in this area. Use this data along with that of Exercise 8.6(a) to compute a point estimate of μ_d. What statistic did you use?

8.8 After taking a random sample of 80 components of a certain type, 12 are found to be defective.

a. Give a point estimate of the proportion of all such components that are *not* defective.

b. A system is to be constructed by randomly selecting two of these components and connecting them in series, as shown here.

The series connection implies that the system will function if and only if neither component is defective (i.e., both components work properly). Estimate the proportion of all such systems that work properly. Hint: If π denotes the probability that a component works properly, how can P(system works) be expressed in terms of π?

8.2   A LARGE-SAMPLE CONFIDENCE INTERVAL FOR A POPULATION MEAN

Section 8.1 discussed how to estimate a population characteristic using a point estimate (a single number). Recall that a point estimate results from selecting an appropriate statistic and computing its value for the given sample. It would be nice if a statistic could be found for which the resulting point estimate was exactly the value of the characteristic being estimated. However, the estimate (the value of the statistic) depends on which sample is selected. Different samples generally yield different estimates, due to sampling variability. In practice, only rarely is a sample selected for which the estimate is exactly equal to the value of the population characteristic. We can only hope that the chosen statistic produces an estimate close to the value of the population characteristic. These considerations suggest the desirability of indicating how precisely the population characteristic has been estimated. The point estimate by itself conveys no information about its closeness to the value of the population characteristic. While an estimate of 157.6 may represent our best guess for the value of μ, it is not the only plausible value.

Suppose that, instead of reporting a point estimate as the single most credible value for the population characteristic, we report an entire interval of reasonable values based on the sample data. We might, for example, be confident that for some population, the value of the average serum cholesterol level, μ, is in the interval from 156.4 to 158.8. The narrowness of this interval implies that we have rather precise information about the value of μ. If, with the same high degree of confidence, we could state only that μ was between 145.3 and 169.9, it would be clear that our knowledge concerning the value of μ is relatively imprecise.

DEFINITION

A **confidence interval** for a population characteristic is an interval of plausible values for the characteristic. It is constructed so that, with a chosen degree of confidence, the value of the characteristic will be captured inside the interval.

Associated with each confidence interval is a **confidence level**. The confidence level provides information on how much "confidence" we can have in the method used to construct the interval estimate. Usual choices for confidence levels are 90%, 95%, and 99%, although other levels are also possible. If we were to construct a 95% confidence interval, using the technique to be described shortly, we would be using a method that is "successful" 95% of the time. That is, if this method is used to generate an interval estimate over and over again with different samples, in the long run, 95% of the resulting intervals would capture the true value of the characteristic being estimated. Similarly, a 99% confidence interval is one that is constructed using a method that is, in the long run, successful in capturing the true value of the population characteristic 99% of the time.

We will discuss factors that affect the choice of confidence level after we examine the method for constructing confidence intervals. We begin by considering a large sample confidence interval for a population mean μ. Let $x_1, x_2, \ldots, x_n$ denote a random sample from a population with mean μ and standard deviation σ.

> In this section, *we assume that n is large enough for the Central Limit Theorem to apply*. Then the sampling distribution of $\bar{x}$ is described approximately by a normal curve with mean $\mu_{\bar{x}} = \mu$ and standard deviation $\sigma_{\bar{x}} = \sigma/\sqrt{n}$.

The development of a confidence interval for μ is easiest to understand if we select a particular confidence level, say 95%. Appendix Table II, the table of standard normal (z) curve areas, can be used to determine a value of z such that a central area of .95 falls between $-z$ and z. In this case, the remaining area of .05 is divided equally between the two tails, as shown in Figure 2.

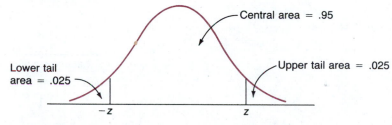

FIGURE 2

A normal curve such that the area to the left of z is .975.

Locating .975 in the body of Appendix Table II, we find that the corresponding z is 1.96. The area under the standard normal curve between -1.96 and 1.96 is .95. This implies the following result.

> For *any* normal curve, an area of .95 is associated with values that are within 1.96 standard deviations of the mean.

Since, for large samples, the $\bar{x}$ distribution is approximately normal, approximately 95% of all samples result in $\bar{x}$ values that are within 1.96 standard deviations ($1.96\sigma_{\bar{x}}$) of $\mu_{\bar{x}}$. Since $\mu_{\bar{x}} = \mu$ and $\sigma_{\bar{x}} = \sigma/\sqrt{n}$, an equivalent statement is this:

Approximately 95% of all samples result in an $\bar{x}$ value that is within $1.96\sigma/\sqrt{n}$ of μ.

If $\bar{x}$ is within $1.96\sigma/\sqrt{n}$ of μ, the interval

$$\bar{x} - 1.96 \cdot \frac{\sigma}{\sqrt{n}} \quad \text{to} \quad \bar{x} + 1.96 \cdot \frac{\sigma}{\sqrt{n}}$$

will capture μ (and this will happen for 95% of all possible samples). However, if $\bar{x}$ is farther away from μ than $1.96\sigma/\sqrt{n}$ (which will happen for about 5% of all possible samples), this interval will not include the true value of μ. This is pictured in Figure 3.

When n is large, a **95% confidence interval for μ** is

$$\left(\bar{x} - 1.96 \cdot \frac{\sigma}{\sqrt{n}}, \ \bar{x} + 1.96 \cdot \frac{\sigma}{\sqrt{n}}\right)$$

An abbreviated formula for the interval is

$$\bar{x} \pm 1.96 \cdot \frac{\sigma}{\sqrt{n}}$$

where $+$ gives the upper limit and $-$ the lower limit of the interval.

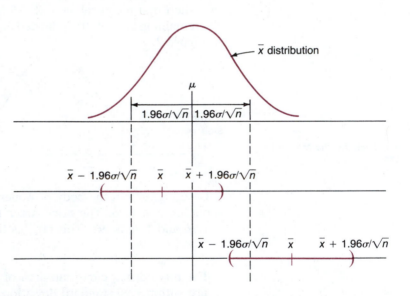

FIGURE 3

The mean μ is captured in the interval from $\bar{x} - 1.96\sigma/\sqrt{n}$ to $\bar{x} + 1.96\sigma/\sqrt{n}$ when $\bar{x}$ is within $1.96\sigma/\sqrt{n}$ of μ.

EXAMPLE 4

Many of us who watch television think that far too much time is devoted to commercials. People involved in various aspects of television advertising obviously have a different perspective. The paper "The Impact of Infomercials: Perspectives of Advertisers and Advertising Agencies" (*J. of Ad. Research* (1983): 25–32) reported on a survey of $n = 62$ such individuals. Each person was asked what he or she believed to be the optimum amount of allocated time per hour for commercials during prime time. The resulting sample average was $\bar{x} = 8.20$ min. Let μ denote the average allocation of time believed optimal for the population of all individuals involved in television advertising, and let σ denote the population standard deviation. Although σ is not usually known, for illustrative purposes suppose that

$\sigma = 4.5$. Then a 95% confidence interval for μ is

$$\bar{x} \pm 1.96 \cdot \frac{\sigma}{\sqrt{n}} = 8.20 \pm \frac{(1.96)(4.5)}{\sqrt{62}}$$

$$= 8.20 \pm 1.12$$

$$= (7.08, 9.32)$$

That is, based on sample data, plausible values of μ are those between 7.08 min and 9.32 min. A 95% confidence level is associated with the method used to obtain this interval.

The 95% confidence interval for μ in Example 4 is (7.08, 9.32). It is tempting to say that there is a 95% chance of μ being between 7.08 and 9.32. Do not yield to this temptation! The 95% refers to the percentage of *all* possible samples that result in an interval that includes μ. Said another way, if we take sample after sample from the population and use each one separately to compute a 95% confidence interval, in the long run roughly 95% of these intervals will capture μ. Figure 4 illustrates this for 100 intervals: 93 of the intervals include μ, whereas 7 do not. Our interval (7.08, 9.32) either includes μ or it does not (remember, the value of μ is fixed but not known to us). We cannot make a chance (probability) statement concerning this particular interval. *The confidence level 95% refers to the method used to construct the interval rather than to any particular interval, such as the one we obtained.*

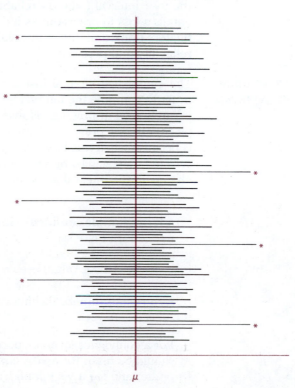

FIGURE 4

One hundred 95% confidence intervals for μ computed from 100 different samples (* identifies intervals that don't include μ).

The formula given for a 95% confidence interval can easily be adapted to other confidence levels. The choice of a 95% confidence level led to the use of the z value 1.96 (chosen to achieve a central area of .95) in the formula. Any other confidence level can be obtained by using an appropriate z critical value in place of 1.96. For example, Appendix Table II shows that a z curve area of .99 lies between -2.58 and 2.58. A 99% confidence interval for μ is obtained by using 2.58 in place of 1.96 in the formula for the 95% interval.

> The general formula for a **large-sample confidence interval for μ** (when the value of σ is known) is
>
> $$\bar{x} \pm (z \text{ critical value}) \cdot \frac{\sigma}{\sqrt{n}}$$
>
> The desired confidence level determines which critical value is used.
>
> The three most commonly used confidence levels, 90%, 95%, and 99%, require the critical values 1.645, 1.96, and 2.58, respectively.

Why settle for 95% confidence when 99% confidence is possible? The higher confidence level comes with a price tag: the resulting interval is wider than the 95% interval. The width of the 95% interval is $2(1.96\sigma/\sqrt{n})$, whereas the 99% interval has width $2(2.58\sigma/\sqrt{n})$. The higher *reliability* of the 99% interval (where "reliability" is specified by the confidence level) entails a loss in precision (as indicated by the wider interval). Many investigators think that a 95% interval gives a reasonable compromise between reliability and precision.

A CONFIDENCE INTERVAL WHEN σ IS UNKNOWN

The confidence interval just developed still has one major drawback: in order to compute the interval limits, σ must be known. Unfortunately, this is rarely true in practice. However, when n is large, it is legitimate to replace σ in our previous formula by the sample standard deviation s. This is because when n is large, $s \approx \sigma$ for almost any sample that is likely to be selected. Replacing σ by s introduces very little extra variability, so no other change in the interval is required.

> **A large-sample confidence interval for a population mean μ** is given by the formula
>
> $$\bar{x} \pm (z \text{ critical value}) \cdot \frac{s}{\sqrt{n}}$$
>
> As a rule of thumb, this interval is appropriate when the sample size exceeds 30.
>
> For a confidence level of approximately 95%, the z critical value 1.96 should be used; for a 90% confidence level, the value 1.645 is appropriate; and for a 99% confidence level, the value 2.58 is used.

EXAMPLE 5

A primary cause of hearing loss in many individuals is exposure to noise. The paper "Effects of Steady State Noise upon Human Hearing Sensitivity from 8,000 to 20,000 Hz" (*Amer. Indus. Hygiene Assoc. J.* (1980): 427–32) reported on a study involving a sample of 44 individuals who had substantial exposure to industrial noise (jet engines, turbines, and the like). Each individual was asked to identify the loudness level at which signals at various frequencies became audible. The sample mean loudness level for detecting a 10-kHz frequency signal was $\bar{x} = 32.0$ dB, and the sample standard deviation was $s = 22.0$ dB. A 90% confidence interval for μ, the mean detection level for the population of all individuals exposed to noise of this type, is

$$\bar{x} \pm (1.645) \cdot \frac{s}{\sqrt{n}} = 32.0 \pm (1.645) \cdot \frac{22.0}{\sqrt{44}}$$
$$= 32.0 \pm 5.5$$
$$= (26.5, 37.5)$$

Even though the confidence level is only 90% rather than 95%, the resulting interval is rather wide because s is quite large—there is much variability in detection level—and n is not terribly large. However, individuals with minimal noise exposure have an average detection level on the order of 20 dB. So even though our estimate of μ is relatively imprecise, it is clear that noise exposure can have negative effects.

The variable of interest in each of these examples (time in Example 4 and loudness level in Example 5) is continuous. The large-sample confidence interval formulas are valid not only in this case but also when μ is the mean value of a discrete variable. The following example illustrates its use with discrete data.

EXAMPLE 6

The behavior of bats has long been a fascinating subject to many people. One aspect of particular interest has been roosting behavior. Individuals of most species do not forage continuously at night. Instead, they commonly return to the roost for varying amounts of time between foraging periods. The paper "Night Roosting Behavior of the Little Brown Bat" (*J. of Mammology* (1982): 464–74) reported that when bats of this species return to the roost, they often have trouble gaining a foothold to hang in their usual upside-down position. Let x denote the number of attempts made by a randomly selected little brown bat before gaining a foothold, and let μ denote the mean value of x. (If you picture a population of bats returning to roost, μ is the mean value of x in the population.) For a sample of $n = 48$ bats attempting to roost, the sample mean number of attempts to succeed was $\bar{x} = 13.7$ and the sample standard deviation was $s = 14.1$. A 95% confidence interval for μ is then

$$\bar{x} \pm (1.96) \cdot \frac{s}{\sqrt{n}} = 13.7 \pm (1.96) \frac{(14.1)}{\sqrt{48}}$$
$$= 13.7 \pm 4.0 = (9.7, 17.7)$$

While this interval estimate is not terribly precise, one would presumably go batty acquiring enough data to increase the precision substantially.

THE GENERAL FORM OF A CONFIDENCE INTERVAL

Many confidence intervals have the same general form as the large-sample intervals for μ that we just considered. We started with a statistic $\bar{x}$ from which a point estimate for μ was obtained. The standard deviation of this statistic is $\sigma/\sqrt{n}$, which can be computed when the value of σ is known. This resulted in a confidence interval of the form

$$\left(\begin{array}{c}\text{point estimate using}\\\text{a specified statistic}\end{array}\right) \pm \left(\begin{array}{c}\text{critical}\\\text{value}\end{array}\right) \cdot \left(\begin{array}{c}\text{standard deviation}\\\text{of the statistic}\end{array}\right)$$

When σ was unknown, we estimated the standard deviation of the statistic by $s/\sqrt{n}$, yielding the interval

$$\left(\begin{array}{c}\text{point estimate using}\\\text{a specified statistic}\end{array}\right) \pm \left(\begin{array}{c}\text{critical}\\\text{value}\end{array}\right) \cdot \left(\begin{array}{c}\text{estimated standard}\\\text{deviation of the}\\\text{statistic}\end{array}\right)$$

For a population characteristic other than μ, a statistic for estimating the characteristic will be selected. Then (drawing on statistical theory) a formula for the standard deviation of the statistic will be given. In practice it will almost always be necessary to estimate this standard deviation (using something analogous to $s/\sqrt{n}$ rather than $\sigma/\sqrt{n}$), so the second interval will be the prototype confidence interval. The estimated standard deviation of the statistic is sometimes referred to as the **standard error** in the statistical literature.

CHOOSING THE SAMPLE SIZE

When σ is known, the 95% confidence interval for μ is based on the fact that, for approximately 95% of all random samples, $\bar{x}$ will be within $1.96\sigma/\sqrt{n}$ of μ. The quantity $1.96\sigma/\sqrt{n}$ is sometimes called the **bound on the error of estimation** associated with a 95% confidence level—with 95% confidence, the point estimate $\bar{x}$ will be no further than this from μ. Before collecting any data, an investigator may wish to determine a sample size for which a particular value of the bound is achieved. For example, with μ representing the average fuel efficiency in mpg for all cars of a certain type, the objective of an investigation may be to estimate μ to within 1 mpg with 95% confidence. The value of n necessary to achieve this is obtained by equating 1 to $1.96\sigma/\sqrt{n}$ and solving for n.

In general, suppose it is desired to estimate μ to within an amount B (the specified error of estimation) with 95% confidence. Finding the necessary sample size requires solving the equation $B = 1.96\sigma/\sqrt{n}$ for n. The result is

$$n = \left[\frac{1.96\sigma}{B}\right]^2$$

Notice that a large value of σ forces n to be large, as does a small value of B.

EXAMPLE 7

The current version of a certain bias-ply tire is known to give an average of 20,000 miles of tread wear with a standard deviation of 2000 miles. A change in the manufacture of the tire has been proposed that, it is hoped, will increase the average tread wear without changing the standard devia-

tion. To find how many prototype tires of the new type should be manufactured and tested in order to estimate μ, the true average tread life, to within
500 miles with 95% confidence, we use $\sigma = 2000$ and $B = 500$ in the formula for n:

$$n = \left[\frac{(1.96)(2000)}{500}\right]^2 = (7.84)^2 = 61.5$$

A sample size of $n = 62$ will suffice to achieve the desired precision.

Use of the sample size formula requires that σ be known. Of course, this
is rarely the case. One possibility is to carry out a preliminary study and
use the resulting sample standard deviation (or a somewhat larger value,
to be conservative) to determine n for the main part of the study. Another
possibility is simply to make an educated guess about what value of σ might
result and use it in calculating n. For a population distribution that is not
too skewed, dividing the range (the difference between the largest and
smallest values) by 4 often gives a rough idea of the standard deviation. In
the tire-wear problem of Example 7, suppose that σ is unknown but tread
life is believed to be between 28,000 and 36,000 miles. Then a reasonable
value to use as the standard deviation in the formula for n would be
$(36,000 - 28,000)/4 = 2000$.

EXERCISES 8.9–8.22   SECTION 8.2

8.9 Discuss how each of the following factors affects the width of the large-sample confidence interval for μ when σ is known.
   a. Confidence level
   b. Sample size
   c. Population standard deviation

8.10 The formula used to compute a confidence interval for μ when n is large and σ is
known is

$$\bar{x} \pm (z \text{ critical value}) \cdot \frac{\sigma}{\sqrt{n}}$$

What is the appropriate z critical value for each of the following confidence levels?
   a. 95%   **b.** 90%   **c.** 99%   **d.** 80%   **e.** 85%

8.11 Suppose that a random sample of 50 bottles of a particular brand of cough medicine
is selected and the alcohol content of each bottle is determined. Let μ denote the
average alcohol content for the population of all bottles of the brand under study.
Suppose that the sample of 50 results in a 95% confidence interval for μ of (7.8, 9.4).
   a. Would a 90% confidence interval have been narrower or wider than the given
   interval? Explain your answer.
   b. Consider the following statement: There is a 95% chance that μ is between 7.8
   and 9.4. Is this statement correct? Why or why not?
   c. Consider the following statement: If the process of selecting a sample of size 50
   and then computing the corresponding 95% confidence interval is repeated 100
   times, 95 of the resulting intervals will include μ. Is this statement correct? Why
   or why not?

8.12 *U.S.A. Today* (Jan. 15, 1986) reported on a study of medical costs incurred by automobile accident victims. A survey of 135 accident victims resulted in an average medical cost of $565 for motorists wearing seat belts and an average of $1200 for motorists who were not wearing seat belts. Suppose that the sample sizes and standard deviations for the two groups were as given in the accompanying table.

	Sample Size n	Medical Cost	
		$\bar{x}$	s
Wearing seat belts	90	565	268
No seat belts	45	1200	506

a. Construct a 90% confidence interval for the mean medical cost for accident victims who were wearing seat belts.
b. Construct a 90% confidence interval for the mean medical cost for accident victims who were not wearing seat belts.
c. What degree of confidence do you have that *both* intervals of parts (a) and (b) simultaneously include the corresponding true mean values?

8.13 A manufacturer of video recorder tapes sells tapes labeled as giving 6 hours of playing time. Sixty-four of these tapes are selected and the actual playing time for each is determined. If the mean and standard deviation of the 64 observed playing times are 352 min and 8 min, respectively, construct a 99% confidence interval for μ, the true average playing time for 6-hr tapes made by this manufacturer. Based on your interval, do you think that the manufacturer could be accused of false advertising? Explain.

8.14 Computer equipment can be very sensitive to high temperatures. As a result, when testing computer components, the temperature at which a malfunction occurs is of interest. Let μ denote the average temperature at which components of a certain type fail. A random sample of 49 components was tested by exposing the components to increasing temperatures and recording the temperature at which each component first failed. The resulting observations were used to compute $\bar{x} = 89°F$ and $s = 14°F$. Find a 95% confidence interval for μ.

8.15 The amount of time spent on housework by women who work outside the home was examined in the paper "The Effects of Wife's Employment Time on Her Household Work Time" (*Home Econ. Research J.* (1983): 260–65). Each person in a sample of 362 working women was asked to indicate how much time she spent each day on certain household activities. Some of the results are summarized in the accompanying table.

Activity	Average (min/day)	Standard Deviation
All housework	348.95	176.9
Food preparation	74.43	50.4
Cleaning	72.09	72.6

a. Construct a 90% confidence interval for the average amount of time that all working women spend on housework.
b. Construct a 95% confidence interval for the mean time that working women spend on food preparation.
c. Would a 95% confidence interval for the average amount of time spent on cleaning be wider or narrower than the 95% interval for the average amount of time spent in food preparation? Explain.

d. How would you explain the very large standard deviation for the amount of time spent on cleaning?

8.16 The paper "External Search Effort: An Investigation Across Several Product Categories" (*J. of Consumer Research* (1987): 83–91) reported on the results of a survey on buying habits.

a. For a sample of buyers of small TVs, it was reported that the mean number of visits to retail stores prior to making a purchase was 2.29. Suppose that this mean was based on a sample of size 100 and that the sample standard deviation was 1.6. Construct a 95% confidence interval for the true mean number of retail visits made by buyers of small TVs.

b. For a sample of buyers of large TVs, the mean number of retail visits was reported to be 3.25. Assuming the sample size was also 100 and that the sample standard deviation was 1.7, obtain a 95% confidence interval for the true mean number of retail visits made by buyers of large TVs.

c. Do the intervals constructed in parts (a) and (b) overlap? Based on these two intervals, would you conclude that buyers of large TVs make more retail visits, on the average, than buyers of small TVs? Explain your answer.

8.17 Anyone who has owned a dog or cat knows that caring for a pet can be expensive. The paper "Veterinary Health Care Market for Dogs" (*J. of Amer. Vet. Med. Assoc.* (1984): 207–8) studied annual veterinary expenditures for households owning dogs. The average veterinary expenditure was reported as $74 per year, and the corresponding standard deviation was (approximately) $40. Suppose that these statistics had been calculated based on a random sample of size 144 (the actual sample was much larger). Construct a 90% confidence interval for the average yearly veterinary expenditure for dog-owning households.

8.18 A manufacturer of college textbooks is interested in estimating the strength of the bindings produced by a particular binding machine. Strength can be measured by recording the force required to pull the pages from the binding. If this force is measured in pounds, how many books should be tested in order to estimate with 95% confidence the average force required to break the binding to within .1 lb? Assume that σ is known to be .8 lb.

8.19 The paper "The Variability of Blood Pressure Measurements in Children" (*Amer. J. of Public Health* (1983): 1207–11) reported the results of a study of 99 third-grade children. The sample consisted of 53 boys and 46 girls. The values of several variables, including systolic blood pressure, diastolic blood pressure, and pulse, were recorded for each child. Means and standard deviations are given in the accompanying table.

	Boys		Girls	
	Average	Standard Deviation	Average	Standard Deviation
Systolic blood pressure	101.7	9.8	101.8	9.8
Diastolic blood pressure	59.2	9.8	60.3	10.1
Pulse	86.7	10.3	86.7	10.2

a. Construct 90% confidence intervals for the mean systolic blood pressure for boys and for girls. Are the intervals similar? Based on your intervals, do you think that there is a difference in mean systolic blood pressure for boys and girls? Explain.

b. Construct a 99% confidence interval for boys' average pulse rate.

8.20 Suppose that the study discussed in Exercise 8.19 is to be considered a pilot study and that a large-scale study will follow.

a. The researchers would like to estimate the true average pulse rate for girls to within 1 beat/minute with 95% confidence. How many girls should be included in the study?

b. Suppose that in addition to estimating the average pulse rate for girls to within 1 beat/minute, the researchers also want to be able to estimate both the average systolic blood pressure and the average diastolic blood pressure for girls to within 1.5 with 95% confidence. How many girls should be studied in order to achieve all of these goals?

8.21 The paper "National Geographic, the Doomsday Machine," which appeared in the *J. of Irreproducible Results* (yes, there really is a journal by that name—it's a spoof of technical journals!) predicted dire consequences resulting from a nationwide buildup of *National Geographic*. The author's predictions are based on the observation that the number of subscriptions for *National Geographic* is on the rise and that no one ever throws away a copy of the *National Geographic*. A key to the analysis presented in the paper is the weight of an issue of the magazine. Suppose that you were assigned the task of estimating the average weight of a particular issue of the *National Geographic*. How many copies should you sample in order to estimate the average weight to within .1 oz with 95% confidence? Assume that σ is known to be 1 oz.

8.22 The formula described in this section for determining sample size corresponds to a confidence level of 95%. What would be the appropriate formula for determining sample size when the desired confidence level is 90%? 99%?

8.3 A LARGE-SAMPLE CONFIDENCE INTERVAL FOR A POPULATION PROPORTION

Often an investigator wishes to make inferences about the proportion of individuals or objects in a population that possess a particular property. For example, a university administrator might be interested in the proportion of students who favor on-line computer registration. In a different setting, a quality control engineer would be concerned about the proportion of manufactured parts that are classified as defective. Let

$$\pi = \begin{pmatrix} \text{proportion of population that possess} \\ \text{the property of interest} \end{pmatrix}$$

In this section we consider estimating π using information in a random sample of size n from the population. In Section 7.1, we used the sample proportion

$$p = \frac{\text{number in sample that possess property of interest}}{n}$$

to calculate a point estimate of π. We can also use p to form a confidence interval for π.

Although a confidence interval for π can be obtained when n is small, our focus will be on the large-sample case. The justification for the large-sample interval rests on properties of the sampling distribution of p:

1. The sampling distribution of p is centered at π, that is, $\mu_p = \pi$. Therefore p is an unbiased statistic for estimating π.

2. The standard deviation of p is $\sigma_p = \sqrt{\pi(1 - \pi)/n}$.

3. As long as n is large ($n\pi \geq 5$ and $n(1 - \pi) \geq 5$), the sampling distribution of p is well approximated by a normal curve.

The accompanying box summarizes these properties.

> When n is large, p has a distribution that is approximately normal with mean π and standard deviation $\sqrt{\pi(1 - \pi)/n}$.

Applying the same line of reasoning that we used to develop the large-sample confidence interval for μ gives us the interval estimate

$$p \pm (z \text{ critical value})\sqrt{\pi(1 - \pi)/n}$$

Since π is unknown, $\sigma_p = \sqrt{\pi(1 - \pi)/n}$ must be estimated. As long as the sample size is large, the value of $\sqrt{p(1 - p)/n}$ should be close to $\sqrt{\pi(1 - \pi)/n}$ and can be used in its place.

> **A large-sample confidence interval for π is**
>
> $$p \pm (z \text{ critical value})\sqrt{p(1 - p)/n}$$
>
> This interval can be used as long as $np \geq 5$ and $n(1 - p) \geq 5$. The appropriate z critical value depends on the confidence level specified.

EXAMPLE 8

The 1983 Tylenol poisoning episode focused attention on the desirability of packaging various commodities in a tamper-resistant manner. The article "Tamper-Resistant Packaging: Is It Really?" (*Package Engr.* (June 1983): 96–104) reported the results of a survey dealing with consumer attitudes toward such packaging. One question asked of the sample of 270 consumers was, "Would you be willing to pay extra for tamper-resistant packages?" The number of yes responses was 189. Let π denote the proportion of all consumers who would pay extra for such packaging. A point estimate of π is

$$p = \frac{189}{270} = .700$$

Since

$$np = (270)(.700) = 189 \geq 5$$
$$n(1 - p) = (270)(.300) = 81 \geq 5,$$

the large-sample interval can be used. A 95% confidence interval requires the z critical value 1.96. The required computations follow.

$$.700 \pm 1.96\sqrt{(.700)(.300)/270} = .700 \pm (1.96)(.028)$$
$$= .700 \pm .055$$
$$= (.645, .755)$$

We can be 95% confident that π is between .645 and .755. This result should be encouraging to those manufacturers who want to pass on extra packaging costs to the consumer.

CHOOSING THE SAMPLE SIZE

When estimating μ using a large sample, $1.96\sigma_{\bar{x}} = 1.96\sigma/\sqrt{n}$ is called the *bound on error of estimation* because, for 95% of all samples, $\bar{x}$ is within this distance of μ. Similarly, when p is used to estimate π,

$$\text{bound on error} = 1.96\sigma_p = 1.96\sqrt{\pi(1-\pi)/n}$$

For 95% of all samples, p is within $1.96\sqrt{\pi(1-\pi)/n}$ of π.

Suppose that an investigator wishes to estimate π to within B (a specified error bound) with 95% confidence. Then the sample size n should be chosen to satisfy

$$B = 1.96\sqrt{\pi(1-\pi)/n}$$

Solving this equation for n results in

$$n = \pi(1-\pi)\left(\frac{1.96}{B}\right)^2$$

Unfortunately, the use of this formula requires the value of π, which is unknown. A conservative solution follows from the observation that $\pi(1-\pi)$ is never larger than .25 (its value when $\pi = .5$). Replacing $\pi(1-\pi)$ by .25, the maximum value, yields

$$n = .25\left(\frac{1.96}{B}\right)^2$$

Using this formula to obtain n gives us a sample size with which we can be 95% confident that p will be within B of π, no matter what the value of π.

EXAMPLE 9

The article "What Kinds of People Do Not Use Seat Belts" (*Amer. J. of Public Health* (1977): 1043–49) reported on a survey with the objective of studying characteristics of drivers and seat belt usage. Let π denote the proportion of all drivers of cars with seat belts who use them. Suppose that at the outset of the study, the investigators wished to estimate π to within an amount .02 with 95% confidence. The required sample size is

$$n = (.25)(1.96/.02)^2$$
$$= (.25)(9604)$$
$$= 2401$$

A very large sample would be required to achieve the desired accuracy. If the researchers were willing to settle for $B = .05$, the required sample size would then be

$$n = (.25)(1.96/.05)^2 = 384.16$$

Thus using $n = 385$, a much smaller sample size, satisfies the requirement.

8.23   The use of the interval

$$p \pm (z \text{ critical value}) \cdot \sqrt{\frac{p(1-p)}{n}}$$

requires a large sample. For each of the following combinations of n and p, indicate whether the given interval would be appropriate.

  a. $n = 50$ and $p = .30$ **e.** $n = 100$ and $p = .70$
  b. $n = 50$ and $p = .05$ **f.** $n = 40$ and $p = .25$
  c. $n = 15$ and $p = .45$ **g.** $n = 60$ and $p = .25$
  d. $n = 100$ and $p = .01$ **h.** $n = 80$ and $p = .10$

8.24   Discuss how each of the following factors affects the width of the confidence interval for π.
  a. The confidence level
  b. The sample size
  c. The value of p

8.25   In recent years, a number of student deaths have been attributed to participation in high school sports. The paper "Concussion Incidences and Severity in Secondary School Varsity Football Players" (*Amer. J. of Public Health* (1983): 1370–75) reported the results of a survey of 3,063 high school varsity football players. Each participant was asked to provide information on injuries and illnesses incurred as a result of participation in the 1977 football season. Loss of consciousness due to concussion was reported by 528 players. Use this information to construct a 90% confidence interval for π, the proportion of all high school football players who suffer loss of consciousness due to concussion.

8.26   *Los Angeles Times* (Jan. 6, 1987) reported on a survey of 200 patients at Boston's Brigham Hospital and San Francisco's H. C. Moffitt Hospital. Eighty of the patients surveyed said they prefer that doctors call them by their first name. Use this information to construct a 95% confidence interval for π, the proportion of all patients at these two hospitals who prefer doctors to address them by their first names.

8.27   The article referenced in Problem 8.26 reported that of the 200 patients surveyed, 110 thought a doctor should wear a white coat when seeing patients and 96 said it was acceptable for a doctor to wear tennis shoes.
  a. Construct a 90% confidence interval for the proportion of all patients at these two hospitals who believe that a doctor should wear a white coat.
  b. Construct a 95% confidence interval for the proportion of all patients at these two hospitals who think it is acceptable for a doctor to wear tennis shoes.

8.28   The attitudes of classroom teachers are important when introducing a new technology into the classroom. The paper "Attitudes towards Microcomputers in Learning" (*Educ. Research.* (1987): 137–45) reported that 67% of teachers surveyed think that computers are now essential tools in the classroom. Suppose that this information was based on a random sample of $n = 200$ elementary school teachers. Calculate a 90% confidence interval for π, the true proportion of elementary school teachers who think the computer is an essential classroom tool.

8.29   Many universities allow courses to be taken on a credit/no credit basis. This encourages students to take courses outside their areas of emphasis without having to worry about the effect on their grade point averages. The paper "Effects of Transition from Pass/No Credit to Traditional Letter Grade System" (*J. of Exper. Educ.* (Winter 1981/82): 88–90) compared student performance under the pass/no credit and letter-grade systems. Sixty-three university professors who had taught under both grading systems participated in the study. Thirty-five responded "yes" when asked whether they felt that students under the letter-grade system performed better on tests and assignments. Construct a 99% confidence interval for π, the proportion of university professors who have taught under both systems and feel that student performance is better under the letter-grade system.

8.30 The paper "Worksite Smoking Cessation Programs: A Potential for National Impact" (*Amer. J. of Public Health* (1983): 1395–96) investigated the effectiveness of smoking cessation programs that appeal to all smokers at a particular worksite and not just those who have expressed an interest in giving up smoking. The program tested involved group meetings and monetary incentives for attending meetings and for not smoking. Of those who chose to participate in the experiment, 91% successfully stopped smoking and were still abstinent six months later. Suppose that 70 people were involved in the experiment and that these 70 are considered to be a sample of all people who participate in such a program. Let π denote the success rate (the proportion of participants who are still nonsmokers six months after completing the program) for this program. Find a 99% confidence interval for π.

8.31 In order to estimate the proportion of students at a particular university who favor the sale of beer on campus, a random sample of 100 students is selected. Of the selected students, 43 support the sale of beer. Let π denote the true proportion of the university's students who favor the sale of beer on campus. Estimate π using a 90% confidence interval. Does the width of the interval suggest that precise information about π is available? Why or why not?

8.32 Steroid use among athletes is currently a controversial issue. The paper "Anabolic–Androgenic Steroid Use among 1,010 College Men" (*The Physician and Sports Medicine* (1988): 75–9) reported that 17 of 1,010 college men surveyed used steroids. Assuming the 1,010 men surveyed were selected randomly, estimate the true proportion of college men who use steroids by obtaining a 99% confidence interval.

8.33 The paper "Television and Human Values: A Case for Cooperation" (*J. of Home Econ.* (Summer 1982): 18–23) reported that 48% of U.S. homes had more than one television set.

a. Suppose that the statistic reported in this paper was based on a random sample of 200 homes. Construct a 95% confidence interval for π, the proportion of all U.S. homes that have more than one television set.

b. The same paper reported that the average viewing time per household was about six hours per day. If you wanted to construct an interval estimate for the average television viewing time for all U.S. households, would you use the following interval?

$$p \pm (z \text{ critical value}) \cdot \sqrt{\frac{p(1 - p)}{n}}$$

Explain your answer.

8.34 A consumer group is interested in estimating the proportion of packages of ground beef sold at a particular store that have an actual fat content exceeding the fat content stated on the label. How many packages of ground beef should be tested in order to estimate this proportion to within .05 with 95% confidence?

8.35 A manufacturer of small appliances purchases plastic handles for coffee pots from an outside vendor. If a handle is cracked, it is considered defective and must be discarded. A very large shipment of plastic handles is received. The proportion of defective handles, π, is of interest. How many handles from the shipment should be inspected in order to estimate π to within .1 with 95% confidence?

8.36 Cornell University's Cooperative Education Department conducted a study of soda consumption among children (*Consumer's Research* (Nov. 1983)). An estimate of the proportion of children under age 3 who drink soda at least once every three days was desired. How large a sample should be selected in order to obtain an estimate within .05 of the true value with 95% confidence?

8.37 The formula given in the text for computing the sample size necessary to estimate π to within an amount B has an associated confidence level of 95%. How would you modify the formula to obtain the sample size required for 99% confidence? Will the value of n for 99% confidence be larger than that for 95% confidence (based on the same B)? Explain.

8.4 A SMALL-SAMPLE CONFIDENCE INTERVAL FOR THE MEAN OF A NORMAL POPULATION

The large-sample confidence interval for μ that we discussed in Section 8.2 is appropriate whatever the shape of the population distribution. This is because it is based on the Central Limit Theorem, which says that when n is sufficiently large, the $\bar{x}$ sampling distribution is approximately normal for any population distribution. When n is small, the Central Limit Theorem does not apply and the interval we developed earlier should not be used. One way to proceed in the small-sample case is to make a specific assumption about the shape of the population distribution and then to use an interval that is valid only under this assumption. The confidence interval we will develop in this section is based on the assumption that the population distribution is normal.

In order to understand the derivation of this confidence interval, it is instructive to begin by taking another look at the large-sample 95% confidence interval. We know that $\mu_{\bar{x}} = \mu$ and that $\sigma_{\bar{x}} = \sigma/\sqrt{n}$; also, when n is large, the $\bar{x}$ distribution is approximately normal. These facts imply that the standardized variable,

$$z = \frac{\bar{x} - \mu}{\sigma/\sqrt{n}}$$

has approximately a standard normal distribution. Since the interval from -1.96 to 1.96 captures an area of .95 under the z curve, approximately 95% of all samples result in an $\bar{x}$ value satisfying

$$-1.96 < \frac{\bar{x} - \mu}{\sigma/\sqrt{n}} < 1.96$$

Manipulating these inequalities to isolate μ in the middle results in the equivalent inequalities

$$\bar{x} - 1.96 \cdot \frac{\sigma}{\sqrt{n}} < \mu < \bar{x} + 1.96\frac{\sigma}{\sqrt{n}}$$

The term $\bar{x} - (1.96\sigma/\sqrt{n})$ is the lower limit of the 95% large-sample confidence interval for μ, and $\bar{x} + (1.96\sigma/\sqrt{n})$ is the upper limit. If σ is unknown, the sample standard deviation, s, is used. When n is large, the substitution of s for σ has little effect.

When n is small, the shape of the $\bar{x}$ distribution may not be approximately normal and depends on the shape of the population distribution. However, when the population distribution is itself normal, the $\bar{x}$ distribution is normal even for small sample sizes. It follows that $(\bar{x} - \mu)/(\sigma/\sqrt{n})$ has a standard normal distribution. But, since σ will usually be unknown, we must estimate σ with the sample standard deviation, resulting in the standardized variable

$$t = \frac{\bar{x} - \mu}{s/\sqrt{n}}$$

The value of s may not be all that close to σ when n is small. As a consequence, the use of s in place of σ introduces extra variability, so the distribution of t is more spread out than the z curve. (The value of $(\bar{x} - \mu)/(\sigma/\sqrt{n})$ will vary from sample to sample because different samples generally result in different $\bar{x}$ values. There is even more variability in $(\bar{x} - \mu)/(s/\sqrt{n})$, because different samples may result in different values of both $\bar{x}$ and s.)

We need to know the probability distribution of the standardized variable t for a small sample from a normal population in order to develop an appropriate confidence interval. This requires that we first learn about probability distributions called t distributions.

t DISTRIBUTIONS

Just as there are many different normal distributions, there are also many t distributions. While normal distributions are distinguished from one another by their mean μ and standard deviation σ, the t distributions are distinguished by a positive whole number called *degrees of freedom (df)*. There is a t distribution with 1 df, another with 2 df, and so on.

IMPORTANT PROPERTIES OF *t* DISTRIBUTIONS:

1. The t curve corresponding to any fixed number of degrees of freedom is bell-shaped and centered at zero, just as is the standard normal (z) curve.
2. Any t curve is more spread out than the z curve.
3. As the number of degrees of freedom increases, the spread of the corresponding t curve decreases.
4. As the number of degrees of freedom increases, the corresponding sequence of t curves approaches the z curve.

These properties are illustrated in Figure 5, which shows several t curves along with the z curve.

Appendix Table III (which also appears inside the back cover) gives selected upper-tail critical values for various t distributions. The central areas for which values are tabulated are .80, .90, .95, .98, .99, .998, and .999. To find a particular critical value, go down the left margin of the table to the row labeled with the desired number of degrees of freedom. Then move over in that row to the column headed by the desired central area. For ex-

FIGURE 5

Comparison of the z curve and t curves for 12 df and 4 df.

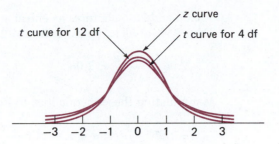

ample, the value in the 12-df row under the column corresponding to central area .95 is 2.18, so 95% of the area under the t curve with 12 degrees of freedom lies between -2.18 and 2.18. Moving over two columns, the critical value for central area .99 (still with 12 df) is 3.06. Moving down the .99 column to the 20-df row, the critical value is 2.85, so the area between -2.85 and 2.85 is .99 under the t curve with 20 degrees of freedom. Notice that the critical values increase as you move to the right in each row. This is necessary in order to capture a larger central area and a smaller tail area. In each column, the critical values decrease as you move downward, reflecting decreasing spread for t distributions with larger degrees of freedom.

The larger the number of degrees of freedom, the more closely the t curve resembles the z curve. To emphasize this, we have included the z critical values as the last row of the t table. Furthermore, once the number of degrees of freedom exceeds 30, the critical values change very little as the number of degrees of freedom increases. For this reason, Table III jumps from 30 df to 40 df, then to 60 df, then to 120 df, and finally to the row of z critical values. If we need a critical value for a number of degrees of freedom between those tabulated, we just use the critical value for the closest df. For df > 120, we use the z critical values.

THE t CONFIDENCE INTERVAL

The fact that the sampling distribution of $(\bar{x} - \mu)/(\sigma/\sqrt{n})$ is approximately the z (standard normal) distribution when n is large is what led to the large-sample z interval. In the same way, the following proposition provides the key to obtaining a small-sample confidence interval when the population distribution is normal.

> Let $x_1, x_2, \ldots, x_n$ constitute a random sample from a normal population distribution. Then the sampling distribution of the standardized variable
>
> $$t = \frac{\bar{x} - \mu}{s/\sqrt{n}}$$
>
> is the t distribution with $n - 1$ degrees of freedom.

To see how this result leads to the desired confidence interval, consider the case $n = 25$, so that df $= 24$. From Table III, the interval between

-2.06 and 2.06 captures a central area of $.95$ under the t curve with 24 df. Then 95% of all samples (with $n = 25$) from a normal population result in

$$-2.06 < \frac{\bar{x} - \mu}{s/\sqrt{n}} < 2.06$$

Manipulating these inequalities to isolate μ yields

$$\bar{x} - 2.06 \cdot \frac{s}{\sqrt{25}} < \mu < \bar{x} + 2.06 \cdot \frac{s}{\sqrt{25}}$$

The 95% confidence interval for μ in this case extends from the lower limit $\bar{x} - (2.06s/\sqrt{25})$ to the upper limit $\bar{x} + (2.06s/\sqrt{25})$. This interval can be written

$$\bar{x} \pm 2.06 \cdot \frac{s}{\sqrt{25}}$$

The major difference between this interval and the large-sample interval is the use of the t critical value 2.06 rather than the z critical value 1.96. The extra uncertainty that results from estimating σ using a small sample causes the t interval to be wider than the z interval.

If the sample size is something other than 25 or if the desired confidence level is something other than 95%, a different t critical value (obtained from Table III) is used in place of 2.06.

Let $x_1, x_2, \ldots, x_n$ be a random sample from a normal population distribution with mean value μ. Then a **small-sample confidence interval for μ** has the form

$$\bar{x} \pm (t \text{ critical value}) \cdot \frac{s}{\sqrt{n}}$$

where the critical value is based on $(n - 1)$ df. Appendix Table III gives critical values appropriate for each of the confidence levels 90%, 95%, and 99%, as well as several other less frequently used confidence levels.

This confidence interval is appropriate for small n only when the population distribution is (at least approximately) normal. If this is not the case, as might be suggested by a normal probability plot or box plot, another method should be used.

EXAMPLE 10

The use of synthetic male hormones (technically: anabolic steroids) is widespread in sports that require great muscular strength. The article "Side Effects of Anabolic Steroids in Weight Trained Men" (*The Physician and Sports Med.* (Dec. 1983): 87–98) reported on a study of a sample of 20 bodybuilders who were current users of such steroids. The sample average weekly dose for oral agents was $\bar{x} = 173$ mg, and the sample standard deviation was $s = 45$ mg. Suppose that, in the population of all body builders who use oral steroids, the distribution of the weekly dose is normal with mean value μ. To compute a 95% confidence interval for μ, the t critical value for $n - 1 = 19$ df is needed. From Table III, this value is 2.09. The 95% confidence interval is then

$$\bar{x} \pm (t \text{ critical value}) \cdot \frac{s}{\sqrt{n}} = 173 \pm (2.09) \cdot \frac{45}{\sqrt{20}}$$

$$= 173 \pm 21.0$$

$$= (152, 194)$$

We can be highly confident that the true average weekly dose is somewhere between 152 mg and 194 mg. The article reports that a manufacturer's recommended dose for a certain oral steroid is between 35 mg and 70 mg per week, so our analysis certainly suggests excessive steroid use among bodybuilders.

EXAMPLE 11

Chronic exposure to asbestos fiber is a well known health hazard. The paper "The Acute Effects of Chrysotile Asbestos Exposure on Lung Function" (*Environ. Research* (1978): 360–72) reported results of a study based on a sample of construction workers who had been exposed to asbestos over a prolonged period. Among the data given in the article were the following (ordered) values of pulmonary compliance (cm^3/cm H_2O) for each of 16 subjects eight months after the exposure period. (Pulmonary compliance is a measure of lung elasticity, or how effectively the lungs are able to inhale and exhale):

| 167.9 | 180.8 | 184.8 | 189.8 | 194.8 | 200.2 | 201.9 | 206.9 |
| 207.2 | 208.4 | 226.3 | 227.7 | 228.5 | 232.4 | 239.8 | 258.6 |

A normal probability plot of this data is quite straight, so it seems plausible that the population pulmonary compliance distribution is approximately normal. For this data set, we have

$$n = 16 \qquad \sum x = 3356.0 \qquad \sum x^2 = 712673.82$$

$$\bar{x} = \frac{3356.0}{16} = 209.75$$

$$s^2 = \frac{\sum x^2 - n\bar{x}^2}{n - 1}$$

$$= \frac{712673.82 - 16(209.75)^2}{15}$$

$$= 583.52$$

$$s = \sqrt{583.52} = 24.16$$

Let μ denote the mean pulmonary compliance for the population of all men who have extensive exposure to asbestos. The t critical value for a 90% confidence interval when df $= 16 - 1 = 15$ is 1.75. The confidence interval is then obtained as follows:

$$\bar{x} \pm (1.75) \cdot \frac{s}{\sqrt{n}} = 209.75 \pm (1.75)\frac{(24.16)}{\sqrt{16}}$$

$$= 209.75 \pm 10.57$$

$$= (199.18, 220.32)$$

With 90% confidence, the mean pulmonary compliance μ is between 199.18 and 220.32 cm³/cm H_2O. Remember that the 90% confidence level implies that 90% of the time, the method used to construct this interval successfully captures μ.

EXERCISES   8.38–8.47   **SECTION 8.4**

8.38 Given a variable that has a t distribution with the specified degrees of freedom, what percentage of the time will its value fall in the indicated region?
   a. 10 df, between -1.81 and 1.81
   b. 10 df, between -2.23 and 2.23
   c. 24 df, between -2.06 and 2.06
   d. 24 df, between -2.80 and 2.80
   e. 24 df, outside the interval from -2.80 and 2.80
   f. 24 df, to the right of 2.80
   g. 10 df, to the left of -1.81

8.39 The formula used to compute a confidence interval for the mean of a normal population when n is small is

$$\bar{x} \pm (t \text{ critical value}) \cdot \frac{s}{\sqrt{n}}$$

What is the appropriate t critical value for each of the following confidence levels and sample sizes?
   a. 95% confidence, $n = 17$ **d.** 90% confidence, $n = 25$
   b. 90% confidence, $n = 12$ **e.** 90% confidence, $n = 13$
   c. 99% confidence, $n = 24$ **f.** 95% confidence, $n = 10$

8.40 Family food expenditures were investigated in the paper "Household Production of Food: Expenditures, Norms, and Satisfaction" (*Home Econ. Research J.* (March 1983): 27). A sample of Iowa homes resulted in an average weekly food expenditure of $164 and a standard deviation of $85. Assuming these results were based on a random sample of size 25, construct a 95% confidence interval for μ, the mean weekly food expenditure for Iowa families. Interpret your interval.

8.41 Blood concentrations of growth hormone and glucose levels were measured for 16 low–birth weight infants four days after birth ("Serum Concentrations of Growth Hormone, Insulin, Free Thyroxine, Thyrotropin, and Cortisol in Very Low Birth Weight Infants Receiving Total Parenteral Nutrition," *Amer. J. of Diseases of Children* (1988): 993–5).

Growth Hormone (μg/l)		Glucose (mmol/l)	
$\bar{x}$	s	$\bar{x}$	s
40.3	16.0	6.6	3.2

a. Assuming that the glucose level distribution is approximately normal, construct a 90% confidence interval for the true mean glucose level of low–birth weight infants at four days after birth.

b. Calculate a 95% confidence interval for the true mean growth hormone level of low–birth weight infants at four days after birth. What assumptions about the growth hormone distribution are required for the validity of the interval computed?

8.42   A triathlon consisting of swimming, cycling, and running is one of the more strenuous amateur sporting events. The paper "Cardiovascular and Thermal Response of Triathlon Performance" (*Medicine and Science in Sports and Exercise* (1988): 385–9) reported on a research study involving nine male triathletes. Maximum heartrate (beats/min) was recorded while performing each of the three events.

	$\bar{x}$	s
Swimming	188	7.2
Biking	186	8.5
Running	194	7.8

a. Assuming that the heartrate distribution for each event is approximately normal, construct 95% confidence intervals for the true mean heartrate of triathletes for each event.

b. Do the intervals in part (a) overlap? Based on the computed intervals, do you think there is evidence that the mean maximum heartrate is higher for running than for the other two events? Explain.

8.43   Authors of the paper "Quality of Carrots Dehydrated by Three Home Methods" (*Home Econ. Research J.* (Sept. 1983): 81) examined the time and energy required to dehydrate carrots. Five one-quart containers of sliced carrots were dehydrated by each method, and the accompanying data was reported.

	Time Required for Dehydration (min)		Energy Required for Dehydration (W · hr)	
Method	Mean	(Approximate) Standard Deviation	Mean	(Approximate) Standard Deviation
Convection oven	428	35	2199	100
Food dehydrator	513	20	3920	170
Microwave oven	135	4	1431	30

a. Compute a 90% confidence interval for the average time required to dehydrate one quart of carrots using a convection oven.

b. Given that the method of dehydration is by food dehydrator, would a 90% confidence interval for the average time required be wider or narrower than the 90% interval for the convection oven method computed in part (a)? Explain without actually computing the interval.

c. Construct a 95% confidence interval for the average energy required to dehydrate one quart of carrots with each of the three methods. Based on these intervals, what conclusions can you draw about the relative efficiency of the three methods?

8.44 The effectiveness of various drugs used to treat horses is discussed in the paper "Factors Involved in the Choice of Routes of Administration of Antimicrobial Drugs" (*J. of Amer. Vet. Assoc.* (1984): 1076–82). One characteristic of interest is the *half-life* of a drug (the length of time until the concentration of the drug in the blood is one-half of the initial value). Given below are the reported values of the sample size and the half-life sample mean and standard deviation for three drugs under study.

Drug	n	Sample Mean	Sample Standard Deviation
Gentamicin	7	1.85 hr	.231 hr
Trimethoprin	6	3.16 hr	.845 hr
Sulfadimethoxine	6	10.62 hr	2.560 hr

a. Construct a 90% confidence interval for the mean half-life of gentamicin. Is this confidence interval valid whatever the distribution of half-lives? Explain.
b. Construct individual 90% confidence intervals for the average half-lives of trimethoprin and sulfadimethoxine.
c. Interpret each of the intervals in parts (a) and (b). Do any of the intervals overlap? If a shorter half-life is desirable (since it would indicate quicker absorption of the drug), based on your confidence intervals would you be able to recommend one of the three drugs over the others? Explain.

8.45 The paper "Surgeons and Operating Rooms: Underutilized Resources" (*Amer. J. of Public Health* (1983): 1361–65) investigated the number of operations per year performed by doctors in various medical specialties.
a. Nine plastic surgeons were asked to indicate the number of operations performed in the previous year. The resulting sample mean was 263.7 operations. If the sample standard deviation was 50.4, find a 90% confidence interval for the average number of operations per year for the population of all plastic surgeons.
b. Twenty-two neurosurgeons were also surveyed concerning the number of operations performed during the previous year. The sample mean was 58.5 operations. If the sample standard deviation was 12.1, find a 99% confidence interval for μ, the mean number of operations performed per year by neurosurgeons.
c. What assumptions are required about the distribution of the number of operations performed in one year for the two populations of doctors in order for the intervals in parts (a) and (b) to be appropriate?

8.46 A wine manufacturer sells a Cabernet with a label that asserts an alcohol content of 11%. Sixteen bottles of this Cabernet are randomly selected and analyzed for alcohol content. The resulting observations are

10.8	9.6	9.5	11.4	9.8	9.1	10.4	10.7
10.2	9.8	10.4	11.1	10.3	9.8	9.0	9.8

a. Construct a 95% confidence interval for μ, the average alcohol content of the bottles of Cabernet produced by this manufacturer. (Assume that alcohol content is normally distributed.)
b. Based on your interval in part (a), do you think that the manufacturer is incorrect in its label claim? Explain.

8.47 Five students visiting the student health center for a free dental examination during National Dental Hygiene Month were asked how many months had passed since their last visit to a dentist. Their responses were as follows:

6   17   11   22   29

If these five students can be considered to be a random sample of all students participating in the free check-up program, construct a 95% confidence interval for the mean number of months elapsed since the last visit to a dentist for the population of students participating in the program.

CHAPTER 8   SUMMARY OF KEY CONCEPTS AND FORMULAS

TERM OR FORMULA	PAGE	COMMENT
Point estimate	227	A single number, based on sample data, that represents a plausible value of a population characteristic.
Unbiased statistic	230	A statistic that has a sampling distribution with a mean equal to the value of the population characteristic to be estimated.
Confidence interval	234	An interval that is computed from sample data and provides plausible values for a population characteristic.
Confidence level	234	A number that provides information on how much "confidence" we can have in the method used to construct a confidence interval estimate. The confidence level specifies the percentage of all possible samples that will produce an interval containing the true value of the population characteristic.
$\bar{x} \pm (z \text{ critical value}) \cdot \dfrac{s}{\sqrt{n}}$	238	A formula used to construct a confidence interval for μ when the sample size is large.
$n = \left[\dfrac{1.96\sigma}{B} \right]^2$	240	A formula used to compute the sample size necessary for estimating μ to within an amount B with 95% confidence.
$p \pm (z \text{ critical value}) \cdot \sqrt{\dfrac{p(1-p)}{n}}$	245	A formula used to construct a confidence interval for π when the sample size is large.
$n = \pi(1-\pi) \left[\dfrac{1.96}{B} \right]^2$	246	A formula used to compute the sample size necessary for estimating π to within an amount B with 95% confidence.
$\bar{x} \pm (t \text{ critical value}) \cdot \dfrac{s}{\sqrt{n}}$	252	A formula for constructing a confidence interval for μ when the sample size is small and when it is reasonable to assume that the population distribution is normal.

SUPPLEMENTARY EXERCISES   8.48–8.60

8.48   Television advertisers are becoming concerned about the use of video cassette recorders (VCRs) to tape television shows, since many viewers fast-forward through the commercials when viewing taped shows. A survey conducted by A. C. Nielsen Co. of 1100 VCR owners found that 715 used the fast-forward feature to avoid commercials on taped programs (the *Los Angeles Times*, Sept. 2, 1984). Construct and interpret a 95% confidence interval for the proportion of all VCR owners who use the fast-forward feature to avoid advertisements.

8.49   Stock researcher Norman Fosback published a study assessing the effect of stock tips that appeared in the *Wall Street Journal*. A sample of companies receiving favorable mention showed a mean one-day price increase of 5.5 points (the *Los*

Angeles Times, April 30, 1984). Suppose the sample consisted of 100 observations and that the sample standard deviation was 3.6. Construct a 95% confidence interval for μ, the true mean one-day increase of companies receiving positive mention.

8.50 The paper "The Market for Generic Brand Products" (*J. of Marketing* (1984): 75–83) reported that in a random sample of 1442 shoppers, 727 purchased generic brands. Estimate the true proportion of all shoppers who purchase generic brands by using a 99% confidence interval.

8.51 Each year as Thanksgiving draws near, inspectors from the Department of Weights and Measures weigh turkeys randomly selected from grocery store freezers to see if the marked weight is accurate. The *San Luis Obispo Telegram-Tribune* (Nov. 22, 1984) reported that of 1000 birds weighed, 486 required remarking. Use a 95% confidence interval to estimate π, the true proportion of all frozen turkeys with an incorrectly marked weight. Based on your interval, do you think that it is plausible that more than half of all frozen turkeys are incorrectly marked? Explain.

8.52 About 14% of all dogs suffer from an infection of the urinary tract sometime during their lifetime. The preferred treatment for an infection caused by the urinary bacteria *pseudomonas* is the antibiotic tetracycline. In a study of healthy adult dogs who received a daily dose of tetracycline of 55 mg/kg body weight, the mean concentration of tetracycline in the urine was 138 μg/mL and the sample standard deviation was 65 μg/mL. (*Source:* "Therapeutic Strategies Involving Antimicrobial Treatment of the Canine Urinary Tract," *J. of Amer. Vet Assoc.* (1984): 1162–64). Suppose the mean and standard deviation given had been computed using a sample of $n = 10$ observations. Construct and interpret a 95% confidence interval for the mean concentration of tetracycline in the urine of dogs receiving daily tetracycline (55 mg/kg body weight).

8.53 The paper "Chlorinated Pesticide Residues in the Body Fat of People in Iran" (*Environ. Research* (1978): 419–22) summarized the results of an Iranian study of a sample of $n = 170$ tissue specimens. It was found that the sample mean DDT concentration and sample standard deviation were 8.13 ppm and 8.34 ppm, respectively.
a. Construct and interpret a 95% confidence interval for μ, the true mean DDT concentration.
b. If the above summary data had resulted from a sample of only $n = 15$ specimens, do you think a t interval would have been appropriate? Explain.

8.54 In a survey of 1,515 people, 606 said they thought that autoworkers were overpaid (Associated Press, Aug. 15, 1984). Treating the 1,515 people as a random sample of the American public, use a 90% confidence interval to estimate the true proportion of Americans who think autoworkers are overpaid.

8.55 The effect of anaesthetic on the flow of aqueous humour (a fluid of the eye) was investigated in the paper "A Method for Near-Continuous Determination of Aqueous Humour Flow: Effects of Anaesthetics, Temperature and Indomethacin" (*Exper. Eye Research* (1984): 435–53). Summary quantities for aqueous flow rate (μl/min) observed under three different anaesthetics are given.

Anaesthetic	n	Mean Flow Rate	Standard Deviation
Pentobarbitol	191	.99	.235
Urethane	13	1.47	.314
Ketamine	16	.99	.164

a. Construct a 95% confidence interval for the true mean flow rate when under the effects of pentobarbitol.

b. Construct a 95% confidence interval for the true mean flow rate under urethane. Give two reasons why this interval is wider than that in part (a).

c. Construct a 95% confidence interval for the true mean flow rate under the anaesthetic ketamine. Note that the sample mean was the same for the pentobarbitol and the ketamine samples and yet the corresponding 95% confidence intervals are different. What factors contribute to this difference? Explain.

8.56 In a random sample of 31 inmates selected from residents of the prison in Angola, Louisiana ("The Effects of Education on Self-Esteem of Male Prison Inmates," *J. of Correctional Educ.* (1982): 12–18), 25 were Caucasian. Use a 90% confidence interval to estimate the true proportion of inmates (at this prison) who are Caucasian.

8.57 When n is large, the statistic s is approximately unbiased for estimating σ and has approximately a normal distribution. The standard deviation of this statistic when the population distribution is normal is $\sigma_s \approx \sigma/\sqrt{2n}$, which can be estimated by $s/\sqrt{2n}$. A large-sample confidence interval for the population standard deviation σ is then

$$s \pm (z \text{ critical value}) \cdot \frac{s}{\sqrt{2n}}$$

Use the data of Exercise 8.55 to obtain a 95% confidence interval for the true standard deviation of flow rate under pentobarbitol.

8.58 The interval from -2.33 to 1.75 captures an area of .95 under the z curve. This implies that another large-sample 95% confidence interval for μ has lower limit $\bar{x} - (2.33)s/\sqrt{n}$ and upper limit $\bar{x} + (1.75)s/\sqrt{n}$. Would you recommend using this 95% interval over the 95% interval $\bar{x} \pm (1.96)s/\sqrt{n}$ discussed in the text? Explain. (*Hint:* Look at the width of each interval.)

8.59 Suppose that an individual's morning waiting time for a certain bus is known to have a uniform distribution on the interval from 0 min to an unknown upper limit θ min. A 95% confidence interval for θ based on a random sample of n waiting times can be shown to have lower limit $\max(x)$ and upper limit $\max(x)/(.05)^{1/n}$. If $n = 5$ and the resulting waiting times are 4.2, 3.5, 1.7, 1.2, and 2.4, obtain the confidence interval. (*Hint:* $(0.5)^{1/5} = .5493$.) Notice that the confidence interval here is not of the form (estimate) $\pm$ (critical value)(standard deviation).

8.60 As an example of a situation in which several different statistics could reasonably be used to calculate a point estimate, consider a population of N invoices. Associated with each invoice is its "book value", the recorded amount of that invoice. Let T denote the total book value, a known amount. Some of these book values are erroneous. An audit will be carried out by randomly selecting n invoices and determining the audited (correct) value for each one. Suppose that the sample gives the following results (in dollars).

Invoice:	1	2	3	4	5
Book value:	300	720	526	200	127
Audited value:	300	520	526	200	157
Error:	0	200	0	0	−30

Let

$\bar{y} = $ sample mean book value

$\bar{x} = $ sample mean audited value

$\bar{d} = $ sample mean error

Several different statistics for estimating the total audited (correct) value have been proposed (c.f. "Statistical Models and Analysis in Auditing", *Statistical Science* (1989: 2–33). These include

$$\text{mean per unit statistic} = N\bar{x}$$
$$\text{difference statistic} = T - N\bar{d}$$
$$\text{ratio statistic} = T \cdot (\bar{x}/\bar{y})$$

If $N = 5000$ and $T = 1,761,300$, calculate the three corresponding point estimates. (The cited paper also discusses confidence intervals.)

REFERENCES Again, the books by Freedman et al. and by Moore listed in the Chapter 1 References contain very informal and lucid discussions of confidence intervals at a level comparable to that of this text. The Devore book referenced in Chapter 4 gives a somewhat more general introduction to confidence intervals.

C H A P T E R 9

HYPOTHESIS TESTING USING A SINGLE SAMPLE

INTRODUCTION Chapter 8 considered problems in which the primary goal was to estimate the unknown value of some population characteristic. Alternatively, sample data may be collected in order to decide whether some claim or statement about a population characteristic is plausible. For example, a consumer group investigating claims made by the pharmaceutical company that makes a prescription pain reliever might want to determine if the mean dosage per capsule is in fact 5 mg, as asserted by the manufacturer.

A *hypothesis* is a claim or a statement about one or more population characteristics. The hypothesis $\mu = 5$ mg, where μ is the true mean dosage, corresponds to the manufacturer's claim in the preceding example. The consumer group would like to use data resulting from a sample of capsules to make a decision between the two rival hypotheses $\mu = 5$ mg (the assertion that the manufacturer's claim is correct) and $\mu \neq 5$ mg (which says that the manufacturer's claim is incorrect). This chapter presents hypothesis-testing procedures that facilitate making such decisions.

9.1 HYPOTHESES AND TEST PROCEDURES

A **hypothesis** is a claim or statement either about the value of a single population characteristic or about the values of several characteristics. One example of a hypothesis is the claim $\mu = 100$, where μ is the average IQ for all first-born children. Another example is the statement $\pi > .1$, where π is the proportion of all television sets of a certain brand that need repair while under warranty. The statements $\bar{x} > 110$ and $p = .15$ are not hypotheses because neither $\bar{x}$ nor p is a *population* characteristic.

In any hypothesis-testing problem, there are two contradictory hypotheses under consideration. One hypothesis might be $\mu = 100$ and the other $\mu \neq 100$, or one might be $\pi = .1$ and the other $\pi > .1$. The objective is to decide, based on sample information, which of the two hypotheses is correct. A criminal trial is a familiar situation in which a choice between two contradictory claims must be made. The person accused of the crime must be judged either innocent or guilty. Under the American system of justice, the individual on trial is initially presumed innocent. Only strong evidence to the contrary will cause the innocence claim to be rejected in favor of a guilty verdict. The burden is thus put on the prosecution to prove the guilty claim. The French perspective in criminal proceedings is the opposite of ours. There, once enough evidence has been presented to justify bringing an individual to trial, the initial assumption is that the accused is guilty. The burden of proof then falls on the accused to establish innocence.

A **test of hypotheses** is a method for deciding which of the two contradictory claims (hypotheses) is the correct one. As in a judicial proceeding, we shall initially assume that a particular one of the two hypotheses is the correct one. In carrying out a test, this claim will be rejected in favor of the second (alternative) claim if sample evidence is incompatible with the initial assumption.

▌▌▌ DEFINITION

The **null hypothesis**, denoted by H_0, is the claim that is initially assumed to be true. The other hypothesis is referred to as the **alternative hypothesis** and is denoted by H_a.

In carrying out a test of H_0 versus H_a, the hypothesis H_0 will be rejected in favor of H_a only if sample evidence strongly suggests that H_0 is false. If the sample does not contain such evidence, H_0 will not be rejected. The two possible conclusions are then *reject H_0* or *fail to reject H_0*.

EXAMPLE 1

Consider a machine that produces ball bearings. Because of variation in the machining process, bearings produced by this machine do not have identical diameters. Let μ denote the true average diameter for bearings currently being produced. Suppose that the machine was initially calibrated to achieve the design specification $\mu = .5$ in. However, the manufacturer is now concerned that the diameters no longer conform to this specification. That is, the hypothesis $\mu \neq .5$ in. must now be considered a possibility. If sample evidence suggests that $\mu \neq .5$ in, the production process will have to be halted while recalibration takes place. Because this is costly, the manufacturer wants to be quite sure that $\mu \neq .5$ in. before undertaking recalibration. Under these circumstances, it is sensible to select the null hypothesis as $H_0: \mu = .5$ (the specification is being met, so recalibration is unnecessary) and the alternative hypothesis as $H_a: \mu \neq .5$. Only compelling sample evidence would then result in H_0 being rejected in favor of H_a.

EXAMPLE 2

A pack of a certain brand of cigarettes displays the statement, "1.5 mg nicotine average per cigarette by FTC method." Let μ denote the mean nicotine content per cigarette for all cigarettes of this brand. Then the advertised claim is that $\mu = 1.5$ mg. People who smoke this brand would probably be unhappy if it turned out that μ exceeded the advertised value. Suppose a sample of cigarettes of this brand is selected, and the nicotine content of each cigarette is determined. The sample results can then be used to test the hypothesis $\mu = 1.5$ mg against the hypothesis $\mu > 1.5$ mg. The accusation that the company is understating mean nicotine content is a serious one, and it is reasonable to require compelling sample evidence before concluding that $\mu > 1.5$. This suggests that the claim $\mu = 1.5$ should be selected as the null hypothesis and $\mu > 1.5$ as the alternative hypothesis. Then $H_0: \mu = 1.5$ would be rejected in favor of $H_a: \mu > 1.5$ only when sample evidence strongly suggests that the initial assumption $\mu = 1.5$ mg is no longer tenable.

Because the alternative of interest in Example 1 was $\mu \neq .5$, it was natural to state H_0 as the equality claim $\mu = .5$. However, the alternative hypothesis in Example 2 was stated as $\mu > 1.5$ (true average nicotine content exceeds the advertised level), from which it might seem more reasonable to state H_0 as $\mu \leq 1.5$ rather than $\mu = 1.5$. After all, the average level might actually be less than what the company advertises! Suppose, though, that sample evidence leads to the rejection of $\mu = 1.5$ in favor of the claim $\mu > 1.5$. Then, intuitively, the sample would offer even less support to values of μ smaller than 1.5 when compared to the claim $\mu > 1.5$. Thus, explicitly testing $H_0: \mu = 1.5$ is equivalent to implicitly testing the null hypothesis $\mu \leq 1.5$. We have chosen to state a null hypothesis as an equality claim.

The form of a null hypothesis is

H_0: population characteristic = hypothesized value

where the hypothesized value is a specific number determined by the problem context. The alternative hypothesis then has one of the following three forms:

H_a: population characteristic > hypothesized value

H_a: population characteristic < hypothesized value

H_a: population characteristic ≠ hypothesized value

Thus we might test H_0: $\pi = .1$ versus H_a: $\pi < .1$; but we will not consider testing H_0: $\mu = 50$ versus H_a: $\mu > 100$. The number appearing in the alternative hypothesis must be identical to the hypothesized value in H_0.

We previously noted that the American and French judicial systems operate from different perspectives when it comes to the initial presumption of innocence or guilt. Similarly, the selection of a null hypothesis—the claim that is initially assumed true—sometimes depends on the viewpoint of the investigator.

EXAMPLE 3

A customer is considering the purchase of many components of a certain type from a particular manufacturer and so is concerned about the long-run percentage of defective components. After reflection, the customer decides that 10% is the dividing line between acceptable and unacceptable defective rates. Let π denote the true proportion of this manufacturer's components that are defective. The manufacturer may be the only one currently making this component or may have offered the customer favorable purchase terms. In that case, the customer would want to purchase from this manufacturer unless sample evidence strongly suggests an unacceptable defective rate. It is then sensible to test

H_0: $\pi = .1$ versus H_a: $\pi > .1$,

so that the alternative hypothesis is identified with an unacceptable defective percentage.

On the other hand, the customer might wish to place the burden of proof on the manufacturer and require them to show that its defective rate is acceptable. This suggests testing

H_0: $\pi = .1$ versus H_a: $\pi < .1$,

with the alternative now stating that the defective rate is acceptable. The purchase from this manufacturer would then not be made unless sample evidence strongly suggested rejecting H_0 in favor of H_a.

Once H_0 and H_a have been formulated, we need a method for using sample data to determine whether H_0 should be rejected. This is accomplished through the use of a **test procedure**. Just as an incorrect verdict may be handed down in a criminal trial, there is some chance that when a test procedure is applied to sample data, the wrong conclusion may be drawn. Before looking formally at test procedures, we will examine the issue of errors in hypothesis testing.

EXERCISES 9.1–9.10   **SECTION 9.1**

9.1   Explain why the statement $\bar{x} = 50$ is not a legitimate hypothesis.

9.2   For the following pairs, indicate which don't comply with our rules for setting up hypotheses and explain why.
   a. $H_0: \mu = 15$, $H_a: \mu = 15$ **d.** $H_0: \mu = 123$, $H_a: \mu = 125$
   b. $H_0: \pi = .4$, $H_a: \pi > .6$ **e.** $H_0: p = .1$, $H_a: p \neq .1$
   c. $H_0: \mu = 123$, $H_a: \mu < 123$

9.3   In order to determine whether the pipe welds in a nuclear power plant meet specifications, a random sample of welds is selected, and tests are conducted on each weld in the sample. Weld strength is measured as the force required to break the weld. Suppose that the specifications state that the mean strength of welds should exceed 100 lb/in? The inspection team decides to test $H_0: \mu = 100$ versus $H_a: \mu > 100$. Explain why it might be preferable to use this H_a rather than $\mu < 100$.

9.4   Researchers have postulated that, due to differences in diet, Japanese children have a lower mean blood cholesterol level than American children. Suppose that the mean level for American children is known to be 170. Let μ represent the true mean blood cholesterol level for Japanese children. What hypotheses should the researchers test?

9.5   A certain university has decided to implement plus/minus grading as long as there is evidence that more than 60% of the faculty favor the change. A random sample of faculty will be selected and the resulting data used to test the relevant hypotheses. If π represents the true proportion of all faculty that favor a change to +/− grading, which of the following pair of hypotheses should the administration test?

   $H_0: \pi = .6$ versus $H_a: \pi < .6$

or

   $H_0: \pi = .6$ versus $H_a: \pi > .6$

Explain your choice.

9.6   A large manufacturing plant monitors the proportion of defective items produced. In an effort to reduce this proportion, the plant has implemented a program of statistical quality control (SQC). Assuming that the proportion of defectives before SQC was .01, what hypotheses should the plant managers test to determine if the program has been effective?

9.7   A water-quality control board reports that the water is unsafe for drinking if the mean nitrate concentration exceeds 30 ppm. Water specimens from a well will be analyzed and appropriate hypotheses tested to determine if the well should be closed. If your drinking water comes from this well, what hypotheses would you want the water-quality board to test? Explain.

9.8   A county commissioner must make a decision on a resolution that would commit substantial resources to the construction of a sewer in an outlying residential area. His fiscal decisions have been criticized in the past and so he decides to take a survey of constituents to find out whether they favor spending money for a sewer system. He will vote to appropriate funds only if he can be fairly certain that a majority of the people in his district favor the measure. What hypotheses should he test?

9.9 Many older homes have electrical systems that use fuses rather than circuit breakers. A manufacturer of 40-A fuses wants to make sure that the mean amperage at which its fuses burn out is in fact 40. If the mean amperage is lower than 40, customers will complain because the fuses require replacement too often. If the mean amperage is higher than 40, the manufacturer might be liable for damage to an electrical system due to fuse malfunction. In order to verify the mean amperage of the fuses, a sample of fuses is to be selected and inspected. If a hypothesis test is to be performed on the resulting data, what null and alternative hypotheses would be of interest to the manufacturer?

9.10 An automobile manufacturer offers a 50,000-mile extended warranty on its new cars. You plan on buying one of these cars and must decide whether to purchase the extended warranty. (The ordinary warranty is for 12,000 miles.) Suppose that a recent magazine article has reported the number of miles at which 30 of this manufacturer's cars first needed repair. This information can be used to conduct a test of hypotheses that will aid you in your decision.
a. What hypotheses would you test in each of the following two cases?
  i. The extended warranty is very expensive.
  ii. The extended warranty is not very expensive.
b. Explain your choice of hypotheses in part (a).

9.2 ERRORS IN HYPOTHESIS TESTING

Hypothesis testing, like estimation, involves making an inference about a population characteristic. Since inferences are generally based on information from a sample rather than a census, they are subject to error. Before an inferential procedure is employed, it is important to understand the potential errors that might occur and to know something about the likelihood of making such errors.

In reaching a judgment on the innocence or guilt of a defendant in a criminal trial, two different types of errors must be considered. The defendant may be found guilty when in fact he or she is innocent. Alternatively, the defendant may be found innocent even though guilty. Similarly, there are two different types of errors that might be made when making a decision in a hypothesis-testing problem. One type of error involves rejecting H_0 even though H_0 is true. The second type of error results from failing to reject H_0 when it is false.

DEFINITION

The error of rejecting H_0 when H_0 is true is called a **type I error**.

The error in which H_0 is not rejected when it is false is called a **type II error**.

No reasonable test procedure comes with a guarantee that neither type of error will be made: this is the price paid for basing an inference on a sample. With any procedure, there is some chance that a type I error will be made, and there is also some chance that a type II error will result.

EXAMPLE 4

In the fall of 1987, the U.S. Department of Transportation released information on the proportion of airline flights that were "on time." An on-time arrival is one that arrives within 15 minutes of its scheduled arrival time (Associated Press, Nov. 11, 1987). Overall, 77% of all September (1987) flights arrived on time. Suppose an airline that performed poorly during September were to offer its employees incentives for improved performance. Monetary rewards will be paid if, in an upcoming month, the airline's proportion of on-time flights exceeds the overall industry rate of .77. Let π be the true proportion of the airline's flights that are on time during the month of interest. A random sample of flights might be selected and used as a basis for choosing between

$$H_0: \pi = .77 \quad \text{and} \quad H_a: \pi > .77$$

In this context, a type I error (rejecting a true H_0) would result in the airline's rewarding its employees when in fact their true proportion of on-time flights did not exceed .77. A type II error (not rejecting a false H_0) would result in the airline employees *not* receiving a reward which in fact they deserved.

EXAMPLE 5

(*Example 2 continued*) With μ denoting the true mean nicotine content (in mg) for cigarettes of the brand under study, the hypotheses to be tested are

$$H_0: \mu = 1.5 \quad \text{versus} \quad H_a: \mu > 1.5$$

The null hypothesis states that the manufacturer's advertised claim of 1.5 mg is correct. The alternative hypothesis states that the true mean nicotine content exceeds the value claimed by the manufacturer.

In this context, a type I error would mean concluding that μ is greater than 1.5 when in fact it is not. A possible consequence of this type of error is that the manufacturer would be falsely accused of misleading advertising. A type II error would result when the true mean nicotine content in fact exceeds 1.5 but this is not detected, so we fail to reject H_0. A consequence of a type II error is that the manufacturer is allowed to continue advertising a value that understates the true mean nicotine content.

Examples 4 and 5 illustrate the two different types of error that might occur when testing hypotheses. Note that type I and type II errors, and the associated consequences of making such errors, are quite different.

In choosing a test procedure (a method for deciding whether or not to reject H_0), we would like a small probability of drawing an incorrect conclusion. Some commonly used notation will facilitate a discussion of error probabilities.

▐▐▐ **DEFINITION**

The probability of a type I error is denoted by α and is called the **level of significance** of the test. Thus a test with $\alpha = .01$ is said to have a level of significance of .01, or to be a level .01 test. The probability of a type II error is denoted by β.

The ideal test procedure has both $\alpha = 0$ and $\beta = 0$. However, since we must base our decision on incomplete information—a sample rather than a census—it is impossible to achieve this ideal. The standard test procedures do allow the user to control α, but provide no direct control over β.

Since α represents the probability of rejecting a true null hypothesis, selecting $\alpha = .05$ results in a test procedure that, used over and over with different samples, rejects a true H_0 about five times in a hundred. Selecting $\alpha = .01$ results in a test procedure with a type I error rate of 1% in long-term repeated use. Choosing a small value for α implies that the user wants to employ a procedure for which the risk of a type I error is quite small.

A natural question that arises at this point is the following: If the user can select α, the probability of making a type I error, why would anyone ever select $\alpha = .05$ rather than $\alpha = .01$? Why not always select a very small value for α? In order to achieve a small probability of making a type I error, the corresponding test procedure will require the evidence against H_0 to be very strong before the null hypothesis can be rejected. While this makes a type I error unlikely, it increases the risk of a type II error (*not* rejecting H_0 when it should have been rejected). Therefore, if a type II error has serious consequences, it may be a good idea to select a somewhat larger value for α.

In general, there is a compromise between small α and small β, leading to the following widely accepted principle for specifying a test procedure.

> After thinking about the consequences of type I and type II errors, identify the largest α that is tolerable for the problem. Then employ a test procedure that uses this maximum acceptable value—rather than anything smaller—as the level of significance (because using a smaller α increases β).

Thus if you decide that $\alpha = .05$ is tolerable, you should not use a test with $\alpha = .01$, because the smaller α inevitably results in larger β. The values of α most frequently used in practical problems are .05 and .01 (a 1-in-20 or a 1-in-100 chance of rejecting H_0 when it is actually true), but the choice in any given problem depends on the seriousness of a type I error relative to a type II error in that context.

EXAMPLE 6

A television manufacturer claims that (at least) 90% of its sets will need no service during the first three years of operation. A consumer agency wishes to check this claim, so it obtains a random sample of $n = 100$ purchasers and asks each whether or not the set purchased needed repair during the first three years. Let p be the sample proportion of responses indicating no repair (so that no repair is identified with a success). Let π denote the true proportion of successes. The agency does not want to claim false advertising unless sample evidence strongly suggests that $\pi < .9$. The appropriate hypotheses are then

$$H_0: \pi = .9 \quad \text{versus} \quad H_a: \pi < .9$$

In this context, a type I error consists of saying that the manufacturer's claim is fallacious ($\pi < .9$) when in fact the manufacturer is correct in its claim. A type II error occurs if the manufacturer's claim is incorrect but the consumer agency fails to detect it. Since a type I error has quite serious consequences, the consumer agency may decide that a type I error probability of .01, but no larger, can be tolerated. They would then use a test procedure with $\alpha = .01$.

We are now ready to take a formal look at some general hypothesis-testing procedures. In the remainder of this chapter, we will consider testing hypotheses about either a population mean μ or a population proportion π.

EXERCISES 9.11–9.17

SECTION 9.2

9.11　A manufacturer of handheld calculators receives very large shipments of printed circuits from a supplier. It is too costly and time-consuming to inspect all incoming circuits, so when each shipment arrives, a sample is selected for inspection. Information from the sample is then used to test $H_0: \pi = .05$ versus $H_a: \pi < .05$, where π is the true proportion of defectives in the shipment. If the null hypothesis is rejected, the shipment is accepted and the circuits are used in the production of calculators. If the null hypothesis cannot be rejected, the entire shipment is returned to the supplier due to inferior quality. (A shipment is defined to be of inferior quality if it contains 5% or more defectives.)
a. In this context, define type I and type II errors.
b. From the calculator manufacturer's point of view, which type of error would be considered more serious?
c. From the printed circuit supplier's point of view, which type of error would be considered more serious?

9.12　Water samples are taken from water used for cooling as it is being discharged from a power plant into a river. It has been determined that as long as the mean temperature of the discharged water is at most 150°F, there will be no negative effects on the river's ecosystem. To investigate whether the plant is in compliance with regulations that prohibit a mean discharge water temperature above 150°, 50 water samples will be taken at randomly selected times, and the temperature of each sample recorded. The resulting data will be used to test the hypotheses

$$H_0: \mu = 150° \quad \text{versus} \quad H_a: \mu > 150°$$

In the context of this example, describe type I and type II errors. Which type of error would you consider to be more serious? Explain.

9.13 The purchasing manager for a large office complex must decide between two competing brands of fluorescent light bulbs. One is the brand currently used in the building and has a mean life of 900 hours. The new brand is more expensive, but the manufacturer claims that the mean life exceeds that of the current brand.
a. What hypotheses about μ, the true mean life for the new brand, should the purchasing manager test? Explain your choice.
b. For the hypotheses of part (a), describe type I and type II errors.

9.14 The marketing department for a computer company must determine the selling price for a new model of personal computer. In order to make a reasonable profit, the company would like the computer to sell for $3200. If more than 30% of the potential customers would be willing to pay this price, the company will adopt it. A survey of potential customers is to be carried out; it will include a question asking the maximum amount that the respondent would be willing to pay for a computer with the features of the new model. Let π denote the proportion of all potential customers who would be willing to pay $3200 or more. Then the hypotheses to be tested are

$$H_0: \pi = .3 \quad \text{versus} \quad H_a: \pi > .3$$

In the context of this example, describe type I and type II errors. Discuss the possible consequences of each type of error.

9.15 Occasionally, warning flares of the type contained in most automobile emergency kits fail to ignite. A consumer advocacy group is to investigate a claim against a manufacturer of flares brought by a person who claims that the proportion of defectives is much higher than the value of .1 claimed by the manufacturer. A large number of flares will be tested and the results used to decide between $H_0: \pi = .1$ and $H_a: \pi > .1$, where π represents the true proportion of defectives for flares made by this manufacturer. If H_0 is rejected, charges of false advertising will be filed against the manufacturer.
a. Explain why the alternative hypothesis was chosen to be $H_a: \pi > .1$.
b. In this context, describe type I and type II errors and discuss the consequences of each.

9.16 Suppose that you are an inspector for the Fish and Game Department, and you are given the task of determining whether to prohibit fishing along part of the California coast. You will close an area to fishing if it is determined that fish in that region have an unacceptably high mercury content.
a. If a mercury concentration of 5 ppm is the maximum considered safe, which pair of hypotheses would you test?

$$H_0: \mu = 5 \quad \text{versus} \quad H_a: \mu > 5$$

or

$$H_0: \mu = 5 \quad \text{versus} \quad H_a: \mu < 5$$

Give the reasons for your choice.
b. Would you prefer a significance level of .1 or .01 for your test? Explain.

9.17 An automobile manufacturer is considering using robots for part of its assembly process. Converting to robots is an expensive process, so it will be undertaken only if there is strong evidence that the proportion of defective installations is lower for the robots than for human assemblers. Let π denote the true proportion of defective installations for the robots. It is known that human assemblers have a defect proportion of .02.

a. Which of the following pairs of hypotheses should the manufacturer test?

$H_0: \pi = .02$ versus $H_a: \pi < .02$

or

$H_0: \pi = .02$ versus $H_a: \pi > .02$

Explain.

b. In the context of this example, describe type I and type II errors.

c. Would you prefer a test with $\alpha = .01$ or $\alpha = .1$? Explain your reasoning.

9.3 LARGE-SAMPLE HYPOTHESIS TESTS FOR A POPULATION MEAN

Now that some general concepts of hypothesis testing have been introduced, we are ready to turn our attention to the development of test procedures. A preliminary example will aid in understanding the general method.

EXAMPLE 7

The Food and Nutrition Board of the National Academy of Sciences reports that the mean daily sodium intake should not exceed 3300 mg. In a study of sodium intake (*Consumer Reports* (1984): 17–22), a sample of U.S. residents was found to have a mean daily sodium intake of 4600 mg. Suppose that this result was based on a sample of size 100 and that the population standard deviation is $\sigma = 1100$ mg. The researchers were interested in determining whether the mean daily sodium intake for U.S. residents exceeded the maximum recommended level.

Let μ represent the true mean sodium intake for all U.S. residents. The hypotheses to be tested are

$H_0: \mu = 3300$ versus $H_a: \mu > 3300$

The null hypothesis will be rejected only if there is strong evidence indicating that $\mu > 3300$ mg.

Suppose that H_0 is in fact true, so that $\mu = 3300$ mg. Even when this is the case, the sample mean $\bar{x}$ might differ somewhat from 3300 due simply to sampling variability. The researcher's sample resulted in $\bar{x} = 4600$ mg, a value larger than 3300. Is the difference between $\bar{x}$ and 3300 reasonably attributable to chance variation from sample to sample, or is $\bar{x}$ so much larger than 3300 that the only plausible conclusion is that $\mu > 3300$?

This question can be answered by examining the sampling distribution of $\bar{x}$. Since the sample size is large ($n = 100$), if H_0 is true the $\bar{x}$ sampling distribution is approximately normal, with

$$\mu_{\bar{x}} = \mu = 3300$$

and

$$\sigma_{\bar{x}} = \frac{\sigma}{\sqrt{n}} = \frac{1100}{\sqrt{100}}$$

It follows that when H_0 is true, the standardized variable

$$z = \frac{\bar{x} - 3300}{1100/\sqrt{100}}$$

has approximately a standard normal distribution (one described by the standard normal curve). For our sample,

$$z = \frac{4600 - 3300}{1100/\sqrt{100}} = \frac{1300}{110} = 11.82$$

Based on what we know about the standard normal distribution, 11.82 is an extremely unusual value. There are two possible explanations for this. Either H_0 is true and we have seen something incredibly rare (an $\bar{x}$ that is more that 11 standard deviations above what would be expected were H_0 true), or H_0 is false and the reason the observed sample mean is so large is because μ actually exceeds 3300. Because it would be so surprising to see an $\bar{x}$ value as large as 4600 when $\mu = 3300$, the sample provides quite convincing evidence against H_0. The null hypothesis should therefore be rejected.

Suppose that the sample had instead resulted in an $\bar{x}$ value of 3375 mg. Then

$$z = \frac{\bar{x} - 3300}{1100/\sqrt{100}} = \frac{3375 - 3300}{110} = .68$$

If $\mu = 3300$ mg, the observed $\bar{x}$ is less than one standard deviation above what would be expected. This is not an unusual occurrence for a normal random variable, and so the sample results are compatible with H_0. The null hypothesis would not be rejected in this case.

The preceding example illustrates the rationale behind large-sample procedures for testing hypotheses about μ. We begin by assuming that the null hypothesis is true. The sample is then examined in light of this assumption. Either the sample mean has a value that is compatible with H_0 or it has a value much different from what would be expected were H_0 true. Based on this assessment, a decision to reject or fail to reject H_0 can be made.

In the previous example, we calculated

$$z = \frac{\bar{x} - 3300}{1100/\sqrt{n}}$$

and used its value to decide between H_0 and H_a. This z is an example of a **test statistic**. Notice that the value of σ was assumed known. In practice, this is seldom the case. When the sample size is large, the statistic obtained by using s in place of σ still has approximately a standard normal distribution.

A **test statistic** is a quantity calculated from sample data that is used as a basis for reaching a decision in a hypothesis test.

When using a large sample ($n > 30$) to test a null hypothesis of the form

$$H_0: \mu = \text{hypothesized value}$$

the appropriate test statistic is

$$z = \frac{\bar{x} - \text{hypothesized value}}{s/\sqrt{n}}$$

This test statistic has an approximately standard normal distribution when H_0 is true.

As you probably noticed, the two cases examined in Example 7 ($z = 11.82$ and $z = .68$) were such that the decision between rejecting or not rejecting H_0 was very clear-cut. A decision in other cases may not be so obvious. For example, what if the sample had resulted in $z = 2.0$? Because the area under the standard normal curve to the right of 2.0 is .0228, a value as large or larger than 2.0 would occur for only about 2.3% of all samples when H_0 is true. Is this unusual enough to warrant rejection of H_0? How extreme must the value of z be before H_0 should be rejected?

This question is answered by specifying what is called a **rejection region**. Then H_0 is rejected if the computed value of the test statistic falls within the rejection region.

The appropriate rejection region for a hypothesis test is determined both by the form of the alternative hypothesis H_a and by the value of α (the probability of a type I error) selected by the person conducting the test. Since the test statistic

$$z = \frac{\bar{x} - \text{hypothesized value}}{s/\sqrt{n}}$$

has (approximately) a standard normal distribution when H_0 is true and n is large, we know how the z statistic will behave in this case. We can then determine which values are consistent with H_0 and which values are quite unusual. This is precisely what is needed for selecting a rejection region.

The alternative hypothesis will contain one of the three inequalities $>$, $<$, or $\neq$. If the alternative is

$$H_a: \mu > \text{hypothesized value}$$

then H_0 should be rejected in favor of H_a if $\bar{x}$ considerably exceeds the hypothesized value. This is equivalent to rejecting H_0 if the value of z is quite large and positive (a z value far out in the upper tail of the z curve). The desired α is then achieved by using the z critical value that captures an upper-tail z curve area α as the cutoff value for the rejection region. Similarly, if

$$H_a: \mu < \text{hypothesized value}$$

then H_0 should be rejected if $\bar{x}$ is considerably less than the hypothesized value or, equivalently, if z is too far out in the lower tail of the z curve to be consistent with H_0. In the case

H_a: $\mu \neq$ hypothesized value

rejection of H_0 is appropriate if the computed value of z is too far out in either tail of the z curve. Figure 1 illustrates the choice of rejection region for a given α in each case.

Once the inequality in H_a is identified and a value of α selected, the appropriate rejection region can be specified. Procedures for large-sample tests of hypotheses about a population mean are summarized in the accompanying box.

SUMMARY OF LARGE-SAMPLE z TESTS FOR μ

Null hypothesis: H_0: $\mu =$ hypothesized value

Test statistic: $z = \dfrac{\bar{x} - \text{hypothesized value}}{s/\sqrt{n}}$

Alternative Hypothesis	Rejection Region
H_a: $\mu >$ hypothesized value	Reject H_0 if $z > z$ critical value (upper-tailed test)
H_a: $\mu <$ hypothesized value	Reject H_0 if $z < -z$ critical value (lower-tailed test)
H_a: $\mu \neq$ hypothesized value	Reject H_0 if either $z > z$ critical value or $z < -z$ critical value (two-tailed test)

The z critical value in the rejection region is determined by the desired level of significance α. The bottom row of Appendix Table III contains critical values corresponding to the most frequently used significance levels.

Suppose, for example, that the null hypothesis is H_0: $\mu = 25$ and that $\alpha = .05$ is specified. If the alternative is H_a: $\mu > 25$, the appropriate test procedure is upper-tailed. The level $\alpha = .05$ is then located in the "level of significance for a one-tailed test" row along the bottom margin of Table III. The desired critical value, 1.645, appears directly above $\alpha = .05$ in the "z critical value" row. In this case, H_0 will be rejected if $z > 1.645$ but not otherwise. Similarly, if the alternative is H_a: $\mu < 25$, the test is lower-tailed and, therefore, one-tailed, so for $\alpha = .05$ the same critical value, 1.645, is used. Here H_0 will be rejected if $z < -1.645$. Finally, H_a: $\mu \neq 25$ requires a two-tailed test. The correct critical value for this test when $\alpha = .05$ is not 1.645 because, as illustrated in Figure 1(c), the value .05 must be divided equally between the two tails, giving an area of .025 for each tail. Then

Alternative Hypothesis   Rejection Region and $\alpha = P(\text{type I error})$

(a) H_a: $\mu >$ hypothesized value

(b) H_a: $\mu <$ hypothesized value

(c) H_a: $\mu \neq$ hypothesized value

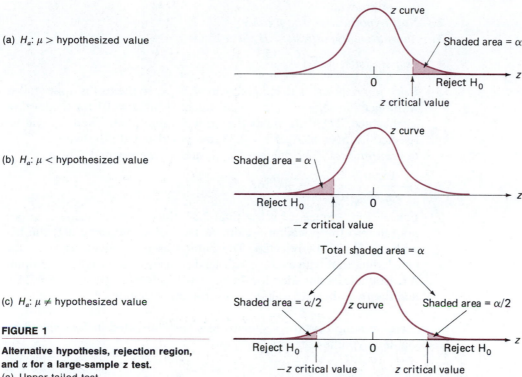

FIGURE 1

Alternative hypothesis, rejection region, and α for a large-sample z test.
(a) Upper-tailed test
(b) Lower-tailed test
(c) Two-tailed test

the critical value capturing an upper-tail z curve area of .025 is identified. This is accomplished by entering the "level of significance for a two-tailed test" row along the bottom margin of Table III, moving over to $\alpha = .05$, and looking directly above to the corresponding entry in the "z critical value" row. The resulting z critical value is 1.96, so H_0 will be rejected in favor of H_a: $\mu \neq 25$ either if $z > 1.96$ or if $z < -1.96$.

EXAMPLE 8

In recent years, a substantial amount of research has focused on possible relationships between chemical contamination of various sorts and mental retardation. The article "Increased Lead Burdens and Trace-Mineral Status in Mentally Retarded Children" (*J. of Special Educ.* (1982): 87–89) reported data on hair-lead concentration for a sample of mentally retarded children for which the cause of retardation was unknown. Summary quantities were

$$n = 40 \qquad \bar{x} = 15.90 \text{ ppm} \qquad s = 8.40 \text{ ppm}$$

The paper states that 15 ppm is considered the acceptable upper limit of hair-lead concentration. Does the given sample data support the research hypothesis that true average hair-lead concentration for all such mentally retarded children exceeds the acceptable upper limit?

In answering this type of question, it is helpful to proceed in an organized manner by following a fixed sequence of steps.

1. *Population characteristic of interest:* μ = the average hair-lead concentration for all mentally retarded children with the cause of retardation unknown.
2. *Null hypothesis:* H_0: $\mu = 15$
3. *Alternative hypothesis:* H_a: $\mu > 15$
4. *Test statistic:* $z = \dfrac{\bar{x} - 15}{s/\sqrt{n}}$
5. *Rejection region:* The inequality in H_a is $>$, so the test is upper-tailed with rejection region $z > z$ critical value. Using $\alpha = .01$ for the level of significance, Table III yields the critical value 2.33. Then H_0 will be rejected in favor of H_a if $z > 2.33$ and not rejected otherwise.
6. *Computations:* The values of n, $\bar{x}$, and s are provided. Thus

$$z = \frac{15.90 - 15}{8.40/\sqrt{40}} = \frac{.90}{1.33} = .68$$

7. *Conclusion:* Since .68 is less than 2.33, the computed value of z does not fall into the rejection region. At level of significance .01, the hypothesis H_0 is not rejected. The data does not provide support for concluding that true average hair-lead concentration exceeds 15 ppm. A reasonable explanation for the observed difference between $\bar{x} = 15.9$ and the hypothesized value 15 is sampling variation.

We recommend that the sequence of steps illustrated in Example 8 be used in any hypothesis-testing analysis.

STEPS IN A HYPOTHESIS-TESTING ANALYSIS

1. Describe the population characteristic about which hypotheses are to be tested.
2. State the null hypothesis, H_0.
3. State the alternative hypothesis, H_a.
4. Display the test statistic to be used, with substitution of the hypothesized value identified in step 2 but *without* any computation at this point.
5. Identify the rejection region. This is accomplished by first using the inequality in H_a to determine whether an upper, lower, or two-tailed test is appropriate and then going to an appropriate table to obtain the critical value corresponding to the selected level of significance α.
6. Compute all quantities appearing in the test statistic; then compute the value of the test statistic itself.
7. State the conclusion (which will be to reject H_0 if the value of the test statistic falls in the rejection region and not to reject H_0 otherwise). The conclusion should be stated in the context of the problem, and the level of significance used should be included.

Steps 1–3 constitute a statement of the problem, steps 4 and 5 state how the conclusion will be reached, and steps 6 and 7 give the analysis and conclusion.

EXAMPLE 9

A certain type of brick is being considered for use in a particular construction project. It is decided that the brick will be used unless sample evidence strongly suggests that the true average compressive strength is below 3200 lb/in.^{2}. A random sample of 36 bricks is selected and each is subjected to a compressive strength test. The resulting sample average compressive strength and the sample standard deviation of compressive strength are 3109 lb/in.^{2} and 156 lb/in.^{2}, respectively. State the relevant hypotheses and carry out a test to reach a decision using level of significance .05.

1. μ = true average compressive strength for this type of brick
2. $H_0: \mu = 3200$
3. $H_a: \mu < 3200$ (so that the brick will be used unless H_0 is rejected in favor of H_a)
4. Test statistic:

$$z = \frac{\bar{x} - 3200}{s/\sqrt{n}}$$

5. Because < appears in H_a, a lower-tailed test is used. For $\alpha = .05$, the appropriate one-tailed critical value is obtained from Table III as 1.645. Thus H_0 will be rejected if $z < -1.645$.
6. $n = 36$, $\bar{x} = 3109$, and $s = 156$, so

$$z = \frac{3109 - 3200}{156/\sqrt{36}} = \frac{-91}{26} = -3.50$$

That is, $\bar{x}$ has been observed to fall 3.5 estimated standard deviations (of $\bar{x}$) below what would have been expected were H_0 true.
7. Since $-3.50 < -1.645$, H_0 is rejected at the 5% level of significance. We conclude that true average compressive strength is below 3200, so the brick should not be used. In reaching this conclusion, we may have made a type I error (rejecting H_0 when it is true), but the low level of significance, .05, has made that unlikely.

EXAMPLE 10

An automobile manufacturer recommends that any purchaser of one of its new cars bring it in to a dealer for a 3000-mile checkup. The company wishes to know whether the true average mileage at this initial servicing differs from 3000. A random sample of 50 recent purchasers resulted in a sample average mileage of 3208 miles and a sample standard deviation of 273 miles. Does the data strongly suggest that the true average mileage for this checkup is something other than the value recommended by the company? State and test the relevant hypotheses, using a level of significance of .01.

1. μ = true average mileage of cars brought to the dealer for 3000-mile checkup
2. $H_0: \mu = 3000$
3. $H_a: \mu \neq 3000$ (which says that μ differs from what the manufacturer recommends)

4. Test statistic

$$z = \frac{\bar{x} - 3000}{s/\sqrt{n}}$$

5. The alternative hypothesis uses $\neq$, so a two-tailed test should be used. Table III gives the $\alpha = .01$ critical value as 2.58 (*not* 2.33, the *one*-tailed critical value for $\alpha = .01$). The hypothesis H_0 will now be rejected if either $z > 2.58$ or $z < -2.58$.

6. $n = 50$, $\bar{x} = 3208$, and $s = 273$, for which

$$z = \frac{3208 - 3000}{273/\sqrt{50}} = \frac{208}{38.61} = 5.39$$

7. Since 5.39 is in the upper tail of the two-tailed rejection region (5.39 > 2.58), H_0 is rejected using $\alpha = .01$. The data does strongly suggest that the true average initial checkup mileage differs from the manufacturer's recommended value.

STATISTICAL VERSUS PRACTICAL SIGNIFICANCE

Carrying out a test amounts to deciding whether the value obtained for the test statistic could plausibly have resulted when H_0 is true. If the value doesn't deviate too much from what is expected when H_0 is true, there is no compelling reason for rejecting H_0 in favor of H_a. But suppose that the observed value is quite far out in the appropriate tail of the test statistic's sampling distribution when H_0 is true (e.g., a large positive value of z when H_a contains the inequality $>$). One could continue to believe that H_0 is true and that such a value arose just through chance variation (a very unusual and "unrepresentative" sample). However, in this case a more plausible explanation for what was observed is that H_0 is false and H_a is true.

When the value of the test statistic falls in the rejection region, it is customary to say that the result is **statistically significant** at the chosen level α. The finding of statistical significance means that, in the investigator's opinion, the observed deviation from what was expected under H_0 cannot plausibly be attributed just to chance variation. Unfortunately, though, statistical significance cannot be equated with the conclusion that the true situation differs from what H_0 states in any practical sense. That is, even after H_0 has been rejected, the data may suggest that there is no *practical* difference between the true value of the population characteristic and what the null hypothesis states that value to be.

EXAMPLE 11

Let μ denote the true average IQ for children in a certain region of the U.S. The average IQ for all children in the U.S. is 100. Education authorities are interested in testing

$$H_0: \mu = 100 \qquad \text{versus} \qquad H_a: \mu > 100$$

Using a significance level of $\alpha = .001$, the hypothesis H_0 will be rejected if $z > 3.09$. A sample of 2500 children resulted in the values

$$n = 2500 \qquad \bar{x} = 101.0 \qquad s = 15.0$$

Then

$$z = \frac{101.0 - 100}{15/\sqrt{2500}} = 3.33$$

Because $3.33 > 3.09$, we reject H_0. The true average IQ for this region does appear to exceed 100.

However, with $n = 2500$, the point estimate $\bar{x} = 101.0$ is almost surely very close to the true value of μ. So, it looks as though H_0 was rejected because $\mu \approx 101$ rather than 100. And, from a practical point of view, a one-point IQ difference has no significance. So the statistically significant result does not have any practical consequences.

EXERCISES   9.18–9.31   **SECTION 9.3**

9.18   Let μ denote the true average amount of surface area in square feet covered by 1 gal of a particular oil-based paint. A researcher wishes to test the hypotheses $H_0: \mu = 400$ versus $H_a: \mu > 400$ using a sample size of 50. Give the appropriate test statistic and rejection region for each of the given significance levels.
a. .01   **b.** .05   **c.** .10   **d.** .13

9.19   A sample of size 75 is to be used to decide between the hypotheses $H_0: \mu = 14$ and $H_a: \mu \neq 14$, where μ is the true average filled weight in ounces for containers coming off a certain production line. Give the test statistic and rejection region associated with each of the given significance levels.
a. .05   **b.** .01   **c.** .10   **d.** .24

9.20   A scale for rating a politician's image solicits responses from -5 (very negative) to 5 (very positive). Let μ denote the true average image rating of a particular politician. A large-sample z statistic is to be used to test the hypotheses $H_0: \mu = 0$ versus $H_a: \mu < 0$. Determine the appropriate rejection region for each of the given significance levels.
a. .01   **b.** .05   **c.** .10

9.21   The desired percentage of silicon dioxide in a certain type of cement is 5.5. A random sample of $n = 36$ specimens gave a sample average percentage of $\bar{x} = 5.21$ and a sample standard deviation of $s = .38$. Let $\mu =$ the true average percentage of silicon dioxide in this type of cement. Test $H_0: \mu = 5$ versus $H_a: \mu \neq 5$ using a significance level of .01.

9.22   A coating designed to retard corrosion is applied to 35 metal bars. These bars are buried in soil for a specified time, and the maximum penetration (in mils) is then determined. The resulting summary data is $\bar{x} = 52.7$ and $s = 4.8$. Let μ denote the true average maximum penetration, and test $H_0: \mu = 50$ versus $H_a: \mu > 50$, using $\alpha = .05$.

9.23   The Environmental Protection Agency sets limits on the maximum allowable concentration of certain chemicals in drinking water. For the substance PCB, the limit has been set at 5 ppm. A random sample of 36 water specimens from the same well results in a sample mean PCB concentration of 4.82 ppm and sample standard deviation of .6 ppm.
a. Is there sufficient evidence to substantiate the claim that the well water is safe? Use a .01 level of significance.
b. Would you recommend using a significance level greater than .01? Why or why not?

9.24 The paper "Undergraduate Marijuana Use and Anger" (*J. of Psych.* (1988): 343–7) reported that the mean and standard deviation on an anger expression scale were $\bar{x} = 42.72$ and $s = 6.05$, respectively, for a sample of $n = 47$ frequent marijuana users. Suppose that the population mean score for nonusers is 41.5.

a. Does the data indicate that frequent users have a mean anger expression score that is higher than that for nonusers? State and test the relevant hypotheses using $\alpha = .05$.

b. Describe type I and type II errors for the hypotheses tested in part (a).

9.25 Data resulting from a study on white rabbit fertility appears in the paper "Genetic Analysis of Litter Traits in Bauscat and Giza White Rabbits" (*Animal Production* (1987): 123–34). For the Giza white rabbit, a sample of 59 litters resulted in a mean litter weight of 1888 g and a standard deviation of 99 g. Does this data provide sufficient evidence to conclude that the true mean litter weight is less than 2000 g? Use a test with $\alpha = .05$.

9.26 Forty-eight seven-year-old children who were reading at below grade level were asked to read a passage aloud. The number of corrections and omissions were noted. ("Strategies Used in the Early Stages of Learning to Read: A Comparison of Children and Adults," *Educ. Research* (1987): 83–93). The sample mean number of corrections and omissions was 17.68. Suppose that the sample also resulted in a standard deviation of 4.2.

a. If the population mean number of corrections and omissions for children who are reading at grade level is 19.9, does the data provide sufficient evidence that the mean number of corrections and omissions for children who are reading below grade level is different from 19.9? Carry out a test using a significance level of .01.

b. Would your conclusion have been different if a significance level of .05 had been used?

9.27 Each plant in a sample of 79 soybean plants grown in a particular soil type was analyzed to determine an iron-deficiency score ("Evaluation of Soybean Genotypes for Iron-Deficiency Chlorosis in Potted Calcareous Soil," *Crop Sci.* (1987): 953–7). The resulting sample mean and standard deviation were 4.2 and .6, respectively. Is there sufficient evidence to indicate that the iron-deficiency score for all soybean plants grown in this soil type exceeds 4.0? Use $\alpha = .01$.

9.28 The chemical Permethrin has been recommended as a way of controlling insects on plant leaves. One study on the effectiveness of Permethrin was summarized in the paper "Evaluation of Nonlethal Side Effects of Permethrin Used in Predator Exclusion Experiments" (*Environ. Entomol.* (1987): 1012–18). An inspection of 96 plants resulted in a mean of 7.2 insects per plant, with a standard deviation of 6.3.

a. Is there sufficient evidence to indicate that the true mean number of insects per plant is less than 8? Use $\alpha = .05$.

b. Would your conclusion in part (a) have been different if a significance level of .01 had been used?

9.29 Minor surgery on horses under field conditions requires a reliable short-term anesthetic producing good muscle relaxation, minimal cardiovascular and respiratory changes, and a quick, smooth recovery with minimal aftereffects so that horses can be left unattended. The article "A Field Trial of Ketamine Anesthesia in the Horse" (*Equine Vet. J.* (1984): 176–9) reported that for a sample of $n = 73$ horses to which ketamine was administered under certain conditions, the sample average lateral recumbency (lying-down) time was 18.86 min and the standard deviation was 8.6 min. Does this data suggest that true average lateral recumbency time under these conditions is less than 20 min? Use the seven-step procedure to test the appropriate hypotheses at level of significance .10.

9.30 One of the biggest problems facing researchers who must solicit survey data by mail is that of nonresponse. One method that has been proposed as a way of increasing both the response rate and the quality of responses is to offer a monetary incentive for returning a questionnaire. The paper "The Effect of Monetary Inducement on Mailed Questionnaire Response Quality" (*J. of Marketing Research* (1980): 265–8) examines some of these issues. One hypothesis of interest to the researchers was that providing a cash incentive for completing a survey would lead to fewer questions left unanswered. An appliance warranty questionnaire containing 17 questions that had been used extensively was mailed to people who had purchased a major appliance during the previous year. A \$.25 payment was included with each questionnaire. In the past, when no money was included with the survey, the mean number of questions left unanswered on returned forms was 1.38. In this experiment, 174 surveys were returned and the number of unanswered questions was determined for each one. The resulting sample mean was .81. Suppose that the sample standard deviation was 1.8. Using a level .01 test, can you conclude that including \$.25 with each survey results in a mean number of unanswered questions that is smaller than 1.38?

9.31 To check a manufacturer's claim that its audio tapes have an average playing time of at least 90 min, 900 tapes are randomly selected and timed. These yield a sample average playing time of 89.95 min and a sample standard deviation of .3 min. Does the data refute the manufacturer's claim? Comment on the statistical and practical significance of this result.

9.4 LARGE-SAMPLE HYPOTHESIS TESTS FOR A POPULATION PROPORTION

Let π denote the proportion of individuals or objects in a specified population that possess a certain property. A random sample of n individuals or objects is to be selected from the population. The sample proportion

$$p = \frac{\text{number that possess property}}{n}$$

is the natural statistic for making inferences about π.

Consider testing hypotheses about π. The null hypothesis will be of the form

$H_0: \pi = \text{hypothesized value}$

and the alternative hypothesis will be similar, but with the $=$ in H_0 replaced by $>$, $<$, or $\neq$. For example, we might want to test

$H_0: \pi = .25$ versus $H_a: \pi < .25$.

The rationale used to construct a test procedure is similar to that used in the previous section. If H_0 is true and the sample size is large enough, the sampling distribution of p is approximately normal, with mean

$\mu_p = \pi = \text{hypothesized value}$

and standard deviation

$\sigma_p = \sqrt{\pi(1 - \pi)/n}$

$\quad = \sqrt{(\text{hypothesized value})(1 - \text{hypothesized value})/n}$

In this case,

$$z = \frac{p - \text{hypothesized value}}{\sqrt{(\text{hypothesized value})(1 - \text{hypothesized value})/n}}$$

has appoximately a standard normal distribution.

This z variable serves as the test statistic. Once a significance level α is selected, a rejection region can be determined. The nature of the rejection region—upper-tailed, lower-tailed, or two-tailed—depends on which of the three inequalities $>$, $<$, or $\neq$ (respectively) appears in H_a.

SUMMARY OF LARGE-SAMPLE z TESTS FOR π

Null hypothesis: H_0: $\pi =$ hypothesized value

Test statistic: $z = \dfrac{p - \text{hypothesized value}}{\sqrt{(\text{hypoth. value})(1 - \text{hypoth. value})/n}}$

Alternative Hypothesis	Rejection Region
H_a: $\pi >$ hypothesized value	Reject H_0 if $z > z$ critical value (upper-tailed test)
H_a: $\pi <$ hypothesized value	Reject H_0 if $z < -z$ critical value (lower-tailed test)
H_a: $\pi \neq$ hypothesized value	Reject H_0 if either $z > z$ critical value or $z < -z$ critical value (two-tailed test)

The z critical values corresponding to the most frequently used levels of significance appear in the bottom row of Appendix Table III.

This test can be used if n satisfies both

$n(\text{hypothesized value}) \geq 5$ and

$n(1 - \text{hypothesized value}) \geq 5$.

EXAMPLE 12

The article "Statistical Evidence of Discrimination" (*J. of Amer. Stat. Assoc.* (1982): 773–83) discussed the court case *Swain v. Alabama* (1965), in which it was alleged that there was discrimination against blacks in grand jury selection. Census data suggested that 25% of those eligible for grand jury service were black, yet a random sample of 1050 individuals called to appear for possible duty yielded only 177 blacks. Using a level .01 test, does this data support a conclusion of discrimination?

1. The population characteristic of interest here is $\pi =$ the true proportion of all those called for possible grand jury service who are black.
2. H_0: $\pi = .25$

3. H_a: $\pi < .25$ (discrimination exists)
4. Since

   $n \cdot$ (hypothesized value) $= 1050(.25) \geq 5$

   $n \cdot (1 -$ hypothesized value) $= 1050(.75) \geq 5$,

   the large-sample test is appropriate. The test statistic is

   $$z = \frac{(p - .25)}{\sqrt{(.25)(.75)/n}}$$

5. The inequality in H_a implies the use of a lower-tailed test, with H_0 rejected if $z < -z$ critical value. From the bottom row of Table III, z critical value $= 2.33$ for a one-tailed, level .01 test. The rejection region is then $z < -2.33$.
6. The denominator of z is

   $$\sqrt{(.25)(.75)/1050} = .0134$$

   and

   $$p = \frac{177}{1050} = .169$$

   so

   $$z = \frac{.169 - .250}{.0134} = \frac{-.081}{.0134} = -6.04$$

7. Since $-6.04 < -2.33$, the hypothesis H_0 is rejected at level .01. Evidence of discrimination seems very clear. Unfortunately, the court looked only at the numerator difference, $-.081$, rather than at z itself. In the court's view, the difference was not large enough to establish a *prima facie* (without further examination) case.

EXAMPLE 13

Environmental problems associated with leaded gasolines are well known. Many motorists have tampered with their cars' emission-control devices in order to save money by purchasing leaded rather than unleaded gas. A *Los Angeles Times* article (March 17, 1984) reported that 15% of all California motorists have engaged in such tampering. Suppose that after obtaining a random sample of 200 cars from one particular California county, the emission-control devices of 21 of them are found to have been modified. Does this suggest that the proportion of cars in this county with tampered devices differs from the statewide proportion? We will use a test with a significance level of .05.

1. $\pi =$ the proportion of cars in this county with modified emission-control devices.
2. H_0: $\pi = .15$
3. H_a: $\pi \neq .15$

4. Since $(200)(.15) \geq 5$ and $(200)(.85) \geq 5$, the z test can be used. The test statistic is

$$z = \frac{(p - .15)}{\sqrt{(.15)(.85)/n}}$$

5. Table III shows that the critical value for a two-tailed level .05 test is 1.96, so H_0 will be rejected if either $z > 1.96$ or $z < -1.96$.

6. $\sqrt{(.15)(.85)/200} = .0252$ and $p = 21/200 = .105$, so

$$z = \frac{(.105 - .15)}{.0252} = \frac{-.045}{.0252} = -1.79$$

7. Since -1.79 is neither greater than 1.96 nor less than -1.96, the hypothesis H_0 cannot be rejected at level .05. The data does not suggest that the proportion of cars in this county having modified devices differs from the statewide proportion.

EXERCISES 9.32–9.45 **SECTION 9.4**

9.32 For which of the following null hypotheses H_0 and sample sizes n is the large-sample z test appropriate?
a. $H_0: \pi = .2$, $n = 25$ c. $H_0: \pi = .9$, $n = 100$
b. $H_0: \pi = .6$, $n = 10$ d. $H_0: \pi = .05$, $n = 75$

9.33 A pizza parlor is considering replacing its oven with a new one. The new oven is particularly suited to baking large (16″) pizzas. Let π denote the proportion of all pizzas ordered that are large. A random sample of $n = 150$ pizza orders yielded 120 that were large. Test $H_0: \pi = .75$ versus $H_a: \pi > .75$ using a significance level of .10.

9.34 A new edition of a certain textbook has been sent to 8000 faculty across the country for examination. Let π denote the proportion of these books that are sold to a book-buyer. (Such sales are a serious problem for publishers and authors.) If a random sample of 100 faculty members yields 28 who sold their examination copies, test $H_0: \pi = .40$ versus $H_a: \pi < .40$, using $\alpha = .05$.

9.35 A telephone company is trying to decide whether some new lines in a large community should be installed underground. Because a small surcharge will have to be added to telephone bills to pay for the extra installation costs, the company has decided to survey customers and proceed only if the survey strongly indicates that more than 60% of all customers favor underground installation. If 118 of 160 customers surveyed favor underground installation in spite of the surcharge, what should the company do? Test using a significance level of .05.

9.36 The incidence of a certain type of chromosome defect in the U.S. adult male population is believed to be 1 in 80. A random sample of 600 individuals in U.S. penal institutions revealed 12 men who have such defects. Can it be concluded that the incidence rate of this defect among prisoners differs from the presumed rate for the entire adult male population? State and test the relevant hypotheses using $\alpha = .05$. What type of error might you have made in reaching a conclusion?

9.37 Airport security checks were the subject of a recent Associated Press report (*San Luis Obispo Telegram–Tribune*, Oct. 23, 1987). Tests were conducted in 1986–87. FAA officials tried to sneak concealed weapons through airport security checks at 136 different airports. Of 2419 attempts, 496 resulted in nondetection. Is there sufficient evidence to indicate that the proportion of weapons that would not be detected by airport security is greater than .15? Use a significance level of .01.

9.38 Many consumers are turning to generics as a way of reducing the cost of prescription medications. The paper "Commercial Information on Drugs: Confusing to the Physician?" (*J. of Drug Issues* (1988): 245–57) gave the results of a survey of 102 doctors. Only 47 of those surveyed knew the generic name for the drug methadone. Does this data support the hypothesis that fewer than half of all physicians know the generic name for methadone? Test the appropriate hypotheses using a .01 significance level.

9.39 The *Los Angeles Times* (Dec. 11, 1984) reported that 32% of all adult Americans have attended at least one year of college. Suppose that a random sample of 200 adults in the western U.S. included 82 with one or more years of college. (This is consistent with summary values reported.) Does this data support the assertion that a higher proportion of westerners (as compared to the U.S. as a whole) have attended at least one year of college? Test the appropriate hypotheses using a significance level of .01.

9.40 How do young people make decisions? This question was examined in the paper "Decision Making and Young People" (*J. of Drug Educ.* (1988): 109–13). Each person in a random sample of 216 seventh graders was asked whether they agreed with the statement, "It is best for me to do the first thing that comes into my mind." A total of 107 agreed with that statement. Based on this information, would you conclude that the true proportion of seventh graders who think it best to do the first thing that comes to mind is different from .5? Test the relevant hypotheses, using $\alpha = .05$.

9.41 A statewide health-care poll revealed that 484 of 1008 Californians surveyed felt that their lives contain a great deal of stress ("Poll: Laid-Back Californians Also Get Stressed, Depressed," Associated Press, March 29, 1988). Is there sufficient evidence to conclude that fewer than half of all Californians feel that their life is stressful? Use $\alpha = .05$.

9.42 The psychological impact of subliminal advertising has been the subject of much speculation. To decide whether people believe that the use of subliminal advertising is ethical, the authors of the paper "Public Perceptions of Subliminal Advertising" (*J. of Adver.* (Jan. 1984): 40–44) conducted a survey of 145 residents of Washington, D.C. Of those surveyed, 58 felt that the use of subliminal advertising was acceptable. Does this data provide sufficient evidence to conclude that fewer than half of Washington's residents find subliminal advertising acceptable? Use a level .05 test.

9.43 A U.S. House of Representatives subcommittee has been hearing testimony on a possible link between problem pregnancies and working with video display terminals (VDTs). A survey of United Airlines employees who work full-time on VDTs found that of 48 pregnancies, 15 resulted in miscarriage (*Los Angeles Times*, March 11, 1984). According to the March of Dimes, there is a 10% miscarriage rate for the general population.
a. Does the data strongly indicate that the miscarriage rate of women who work full-time on VDTs is higher than that of the general female population? Use a level .01 test.
b. On the basis of your work in part (a), can it be concluded that full-time work on VDTs tends to *cause* miscarriages? Explain.

9.44 A woman who smokes during pregnancy increases health risks to the infant. ("Understanding the Intentions of Pregnant Nullipara to Not Smoke Cigarettes after Childbirth", *J. of Drug Educ.* (1988): 115–20). Suppose that a sample of 300 pregnant women who smoked prior to pregnancy contained 51 who quit smoking during pregnancy. (This is consistent with summary data given in the paper.) Does this data support the theory that fewer than 25% of female smokers quit smoking during pregnancy? Use $\alpha = .05$.

9.45 Is it appropriate for a physician to help a gravely ill person die? In a survey of 588 doctors, 365 responded that it was sometimes right to agree to hasten a patient's death ("Hemlock Poll: Doctors Favor Mercy Killing," Associated Press, Feb. 19, 1988). Based on this information, would you conclude that more than 60% of all doctors feel it is sometimes appropriate to help a seriously ill person die?
a. Test the relevant hypotheses using a .01 level of significance.
b. Would your conclusion have been different if a significance level of .05 had been employed?

9.5 SMALL-SAMPLE HYPOTHESIS TESTS FOR THE MEAN OF A NORMAL POPULATION

The large-sample hypothesis testing procedures for μ discussed in Section 9.3 can be used without having to make any specific assumptions about the population distribution. The justification for these procedures is invalid, though, when n is small because the Central Limit Theorem can no longer be used. As with confidence intervals, one way to proceed is to make a specific assumption about the nature of the population distribution and then to develop testing procedures that are valid in this more specialized situation. Here we shall restrict consideration to the case of a normal population distribution. The result on which a test procedure is based is then the same as the one we used in Chapter 8 to obtain a t confidence interval.

> When $x_1, x_2, \ldots, x_n$ constitute a random sample of size n from a normal distribution, the probability distribution of the standardized variable
>
> $$t = \frac{\bar{x} - \mu}{s/\sqrt{n}}$$
>
> is the t distribution with $n - 1$ degrees of freedom.

The null hypothesis is stated just as it was for the large-sample test concerning μ:

$H_0: \mu =$ hypothesized value

When H_0 is true, replacing μ in t by the hypothesized value gives a test statistic with a sampling distribution that is known (t with $n - 1$ df). A rejection region giving the desired significance level is then obtained by using the appropriate t critical value from the $n - 1$ df row of Appendix Table III.

SUMMARY OF t TEST FOR THE MEAN OF A NORMAL POPULATION

Null hypothesis: H_0: μ = hypothesized value

Test statistic: $t = \dfrac{\bar{x} - \text{hypothesized value}}{s/\sqrt{n}}$

Alternative Hypothesis	Rejection Region
H_a: μ > hypothesized value	Reject H_0 if t > t critical value (upper-tailed test)
H_a: μ < hypothesized value	Reject H_0 if t < $-t$ critical value (lower-tailed test)
H_a: $\mu \neq$ hypothesized value	Reject H_0 if either t > t critical value or t < $-t$ critical value (two-tailed test)

The t critical value in the rejection region is based on $n - 1$ degrees of freedom and is determined by the desired level of significance. Table III contains one-tailed and two-tailed critical values corresponding to the most frequently used significance levels.

The test statistic here is the same as the large-sample z statistic of Section 9.3. It is labeled t to emphasize that it has a t distribution when H_0 is true, rather than the z distribution.

EXAMPLE 14

The low population density of the Amazon region has long puzzled geographers and other social scientists. Some have suggested that environmental conditions are inimical to support of a large population. The paper "Anthrosols and Human Carrying Capacity in Amazonia" (*Annals of the Assoc. of Amer. Geog.* (1980): 553–66) suggests otherwise. The author's case for this viewpoint rests largely on an analysis of black-earth soil samples, which gives evidence of the presence of large and sedentary Indian populations prior to the European influx. The accompanying stem-and-leaf display gives pH values for the 29 black-earth soil samples discussed in the paper. (pH is a numerical measure of acidity and is related to availability of soil nutrients.) Does this data indicate that true average black-earth pH differs from 5.0, the pH value for many other types of soil in the region? The display gives evidence of a somewhat skewed distribution (confirmed by a normal probability plot) and the largest value, 7.9, is a mild outlier. However, $n = 29$ is close to the sample size required for the large-sample z test, which requires no specific assumption about the pH distribution, so it seems safe to use the t test here. Let's state and test the relevant hypotheses at a level of significance of .05.

4l	.2, .3
4h	.5, .6, .6, .6, .6, .6, .6, .9
5l	.3, .3, .3, .4, .4, .4
5h	.5, .5, .5, .6, .6, .6, .7, .9, .9
6l	.1, .2
6h	
7l	.0
7h	.9

1. μ = the true average black-earth soil pH in Amazonia
2. H_0: $\mu = 5.0$
3. H_a: $\mu \neq 5.0$ ($\neq$ because of the phrase *differs from*, which indicates a departure from H_0 in either direction)
4. Test statistic:

$$t = \frac{\bar{x} - 5.0}{s/\sqrt{n}}$$

5. Rejection region: reject H_0 if either $t > t$ critical value or $t < -t$ critical value. Because the test is two-tailed, first locate .05 in the two-tailed significance level row along the bottom margin of Table III. Moving up that column to the $n - 1 = 28$ df row gives critical value 2.05, so H_0 will be rejected if either $t > 2.05$ or $t < -2.05$.

6. $\sum x = 155.6$ and $\sum (x - \bar{x})^2 = 18.23$ (either from the deviations or by using the computational formula $\sum x^2 - n\bar{x}^2$). Thus

$$\bar{x} = 155.6/29 = 5.37$$

$$s^2 = 18.23/28 = .651,$$

so $s = .807$. The computed value of t is

$$t = \frac{5.37 - 5.0}{.807/\sqrt{29}} = \frac{.37}{.150} = 2.47$$

7. The value $t = 2.47$ is in the upper tail of the rejection region ($2.47 > 2.05$), so H_0 is rejected at level .05. The true average pH of black-earth soil does appear to be something other than 5.0.

EXAMPLE 15

Extensive data collected during the first half of this century showed clearly that in those years, American-born Japanese children grew faster than did Japanese-born Japanese children. A research paper ("Do American-Born Japanese Children Still Grow Faster than Native Japanese?" *Amer. J. of Phys. Anthro.* (1975): 187–94) conjectured that improved economic and environmental conditions in postwar Japan had greatly narrowed this gap. To investigate the validity of this conjecture, a large sample of Hawaiian-born Japanese children was obtained, and the children were categorized with respect to age. There were thirteen 11-year-old boys in the sample (most of the children sampled were older). The sample average height of these 13 boys was 146.3 cm and the sample standard deviation was 6.92 cm. The average height of native-born 11-year-old Japanese children at that time was known to be 139.7 cm. Does this data suggest that the true average height for Hawaiian-born male 11-year-olds exceeds that for their native-born counterparts? The investigators were willing to assume that the population height distribution was normal. The relevant hypotheses can then be tested using a t test with level of significance .01. The steps in the analysis are as follows.

1. μ = true average height for all Hawaiian-born 11-year-old male Japanese children.
2. H_0: $\mu = 139.7$
3. H_a: $\mu > 139.7$
4. Test statistic:

$$t = \frac{\bar{x} - 139.7}{s/\sqrt{n}}$$

5. Rejection region: because the inequality in H_a is $>$, an upper-tailed test is appropriate. The hypothesis H_0 should be rejected in favor of H_a if $t > t$ critical value. With df $= n - 1 = 13 - 1 = 12$, moving along the bottom margin of Table III to one-tailed level of significance .01 and up to the 12 df row gives a t critical value of 2.68. The rejection region is therefore $t > 2.68$.
6. With $n = 13$, $\bar{x} = 146.3$, and $s = 6.92$, the computed value of t is

$$t = \frac{146.3 - 139.7}{6.92/\sqrt{13}} = \frac{6.60}{1.92} = 3.44$$

7. Since 3.44 is in the rejection region (3.44 > 2.68), at level of significance .01 we reject H_0 in favor of H_a. It seems clear that at this age, the average height of Hawaiian-born Japanese children exceeds that of their native-born counterparts.

While the growth differential at age 11 appears to be substantial, analysis of data on older children suggests that the gap narrows considerably.

Strictly speaking, the validity of the one-sample t test requires that the population distribution be normal. In practice, the test can be used even when the population distribution is somewhat nonnormal as long as n is not too small. Statisticians say that the test is "robust" to mild departures from normality, by which they mean that error probabilities are still approximately what they would be in the case of a normal population distribution.

EXERCISES 9.46–9.55   **SECTION 9.5**

9.46   Let μ denote the true average surface area covered by 1 gal of a certain paint. The hypothesis H_0: $\mu = 400$ is to be tested against H_a: $\mu > 400$. Assuming that paint coverage is normally distributed, give the appropriate test statistic and rejection region for each given sample size and significance level.
  a. $n = 10$, $\alpha = .05$ **c.** $n = 25$, $\alpha = .001$
  b. $n = 18$, $\alpha = .01$ **d.** $n = 50$, $\alpha = .10$

9.47 Suppose that, at a certain store, the amount of air pressure (psi) in new tires of a particular type is normally distributed with mean value μ. For testing $H_0: \mu = 30$ versus $H_a: \mu \neq 30$, give the test statistic and rejection region for each sample size and significance level listed in Exercise 9.46.

9.48 The National Bureau of Standards previously reported the value of selenium content in orchard leaves to be .08 ppm. The paper entitled "A Neutron Activation Method for Determining Submicrogram Selenium in Forage Grasses" (*Soil Sci. Amer. J.* (1975): 57–60) reported the following selenium content for five determinations:

   .072   .073   .080   .078   .088

   a. What assumption about the selenium content distribution must you be willing to make in order to use a t test for testing $H_0: \mu = .08$ versus $H_a: \mu \neq .08$?

   b. Use a t test at level .01 to test the hypotheses stated in part (a).

9.49 The times of first sprinkler activation (in sec) for a series of tests of fire-prevention sprinkler systems that use an aqueous film-forming foam were

   27   41   22   27   23   35   30   33   24   27   28   22   24

   ("Use of AFFF in Sprinkler Systems," *Fire Tech.* (1976): 5). The system has been designed so that the true average activation time is supposed to be at most 25 sec. Does the data strongly indicate that the design specifications have not been met? Test the relevant hypotheses using a significance level of .05. What assumptions are you making about the distribution of activation times?

9.50 The paper "Distinguishing the Dimensions of Valence and Belief Consistency in Depressive and Nondepressive Information Processing" (*Cognitive Therapy and Research* (1988): 391–407) reported that the mean and standard deviation of scores on a self-image scale for 13 clinically depressed women were 28.63 and 13.61, respectively.

   a. Suppose that the mean score for nondepressed women is 28.0. What hypotheses would you test to determine whether the mean score for depressed women differs from 28.0?

   b. Assuming that the self-image score distribution is approximately normal, use a significance level of .01 to test the hypotheses in part (a).

   c. In the context of this problem, describe type I and type II errors.

9.51 The paper "Orchard Floor Management Utilizing Soil-Applied Coal Dust for Frost Protection" (*Agric. and Forest Meteorology* (1988): 71–82) reported the following values for soil heat flux of eight plots covered with coal dust.

   34.7   35.4   34.7   37.7   32.5   28.0   18.4   24.9

   The mean soil heat flux for plots covered only with grass is 29.0 Assuming that the heat flux distribution is approximately normal, does the data suggest that the coal dust is effective in increasing the mean heat flux over that for grass? Test the appropriate hypotheses using a .05 significance level.

9.52 A new method for measuring phosphorus levels in soil is described in the paper "A Rapid Method to Determine Total Phosphorus in Soils" (*Soil Sci. Amer. J.* (1988): 1301–4). Suppose a sample of 11 soil specimens, each with a true phosphorus content of 548 mg/kg, is analyzed using the new method. The resulting sample mean and standard deviation for phosphorus level are 587 and 10, respectively.

   a. Is there evidence that the mean phosphorus level reported by the new method differs significantly from the true value of 548 mg/kg? Use $\alpha = .05$.

   b. What assumptions must you make in order for the test in part (a) to be appropriate?

9.53 A number of veterinary procedures on pigs require the use of a general anesthetic. To evaluate the effects of a certain anesthetic, it was administered to four pigs and various bodily functions were measured. ("Xylazine-Ketamine-Oxymorphone: An Injectable Anesthetic Combination in Swine," *J. of Amer. Vet. Med. Assoc.* (1984): 182–4). Average normal heartrate for pigs is considered to be 114 beats per minute. The heartrates for the four pigs under anesthesia were 116, 85, 118, and 118 beats/ min. Use a level .10 test to determine whether the anesthetic results in a mean heartrate that differs significantly from the mean normal heartrate.

9.54 A certain type of soil was determined to have a natural mean pH of 8.75. The authors of the paper "Effects of Brewery Effluent on Agricultural Soil and Crop Plants" (*Environ. Pollution* (1984): 341–51) treated soil samples with various dilutions of an acidic effluent. Five soil samples were treated with a solution of 25% water and 75% effluent. The mean and standard deviation of the five pH measurements were 8.00 and .05, respectively. Does this data indicate that at this concentration, the effluent results in a mean pH that is lower than the natural pH of the soil? Use a level .01 test.

9.55 Much concern has been expressed in recent years regarding the practice of using nitrates as meat preservatives. In one study involving possible effects of these chemicals, bacteria cultures were grown in a medium containing nitrates. The rate of uptake of radio-labeled amino acid was then determined for each one, yielding the accompanying observations.

7251	6871	9632	6866	9094	5849	8957	7978
7468	7064	7494	7883	8178	7523	8724	

Suppose it is known that true average uptake for cultures without nitrates is 8000. Does the data suggest that the addition of nitrates results in a decrease in true average uptake? Test the appropriate hypotheses using a significance level of .10. Be sure to state any assumptions that are necessary to validate the use of your test.

9.6 *P*-VALUES

One way to report the result of a hypothesis-testing analysis is simply to say whether or not the null hypothesis was rejected at a specified level of significance. Thus an investigator might state that H_0 was rejected at level of significance .05 or that use of a level .01 test resulted in nonrejection of H_0. This type of statement is somewhat inadequate because nothing is said about whether the computed value of the test statistic just barely fell into the rejection region or whether it exceeded the critical value by a very large amount. A related difficulty is that such a report imposes the specified significance level on other decision makers. There are many decision situations in which individuals might have different views concerning the consequences of type I and type II errors. Each individual would then want to select his or her own personal significance level—some selecting $\alpha = .05$, others .01, and so on—and reach a conclusion accordingly. This could result in some individuals rejecting H_0, whereas others might conclude that the data does not show a strong enough contradiction of H_0 to justify its rejection.

EXAMPLE 16

The true average time to the initial relief of pain for the current best-selling pain reliever is known to be 10 min. Let μ denote the true average time to relief for a company's newly developed pain reliever. The company wishes to produce and market this product only if it provides quicker relief than does the current best seller, so it wishes to test

$H_0: \mu = 10$ versus $H_a: \mu < 10$.

Only if experimental evidence leads to rejection of H_0 will the new pain reliever be introduced. After weighing the relative seriousness of the two types of errors, a single level of significance must be agreed upon and a decision—to reject H_0 and introduce the pain reliever or not to do so—must be made at that level.

Now suppose that the new product has been introduced. The company supports its claim of quicker relief by stating, based on an analysis of experimental data, that $H_0: \mu = 10$ was rejected in favor of $H_a: \mu < 10$, using a level of significance of $\alpha = .10$. Any particular individual contemplating a switch to this new pain reliever would naturally want to reach his or her own conclusion concerning the validity of the claim. Individuals who are satisfied with the current best seller would view a type I error (concluding that the new product provides quicker relief when it actually doesn't) as serious, so they might wish to use $\alpha = .05$, .01, or an even smaller level. Unfortunately, the nature of the company's statement prevents an individual decision maker from reaching a conclusion at such a level. The company has imposed its own choice of significance level on others. The report could have been done in a manner that allowed each individual flexibility in drawing a conclusion at a personally selected α.

A more informative way to report the result of a test is to give the *P*-value associated with the test. A *P*-value conveys a great deal of information about the strength of the evidence against H_0 and allows an individual decision maker to draw a conclusion at any specified level α. Before giving a general definition, consider how the conclusion in a hypothesis-testing problem depends on the selected α.

EXAMPLE 17

The nicotine-content problem discussed in Example 2 involved testing

$H_0: \mu = 1.5$ versus $H_a: \mu > 1.5$

The rejection region was upper-tailed, with H_0 rejected if $z > z$ critical value. Suppose that $z = 2.10$. The accompanying table displays the rejection region for each of four different values of α along with the resulting conclusion.

Level of Significance α	Rejection Region	Conclusion for $z = 2.10$
.05	$z > 1.645$	Reject H_0
.025	$z > 1.96$	Reject H_0
.01	$z > 2.33$	Don't reject H_0
.005	$z > 2.58$	Don't reject H_0

For α relatively large, the z critical value is not very far out in the upper tail; 2.10 exceeds the critical value, and so H_0 is rejected. However, as α decreases, the critical value increases. For small α the z critical value is large; 2.10 does not exceed it, and H_0 is not rejected.

Recall that for an upper-tailed z test, α is just the area under the z curve to the right of the critical value. That is, once α is specified, the critical value is chosen to capture upper-tail area α. Table II shows that the area to the right of 2.10 is .0179. Using an α larger than .0179 corresponds to z critical value <2.10. An α less than .0179 necessitates using a z critical value that exceeds 2.10. The decision at a particular level α thus depends on how the selected α compares to the tail area captured by the computed z. This is illustrated in Figure 2. Notice in particular that .0179, the captured tail area, is the smallest level α at which H_0 can be rejected, because using any smaller α results in a z critical value that exceeds 2.10, so that 2.10 is then not in the rejection region.

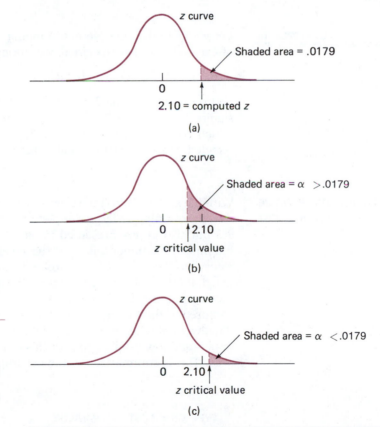

FIGURE 2

Relationship between α and tail area captured by computed z.
(a) Tail area captured by computed z
(b) When $\alpha > .0179$, z critical value < 2.10 and H_0 is rejected
(c) When $\alpha < .0179$, z critical value > 2.10 and H_0 is not rejected

In general, suppose that the sampling distribution of a test statistic has been determined when H_0 is true. Then for a specified α, the rejection region is determined by finding a critical value or values that capture a tail area of α (upper-, lower-, or two-tailed, whichever is appropriate) under the sampling distribution curve. The smallest α for which H_0 could be rejected is the tail area captured by the computed value of the test statistic. This smallest α is the P-value.

▊▊▊ **DEFINITION**

The **P-value** is the smallest level of significance at which H_0 can be rejected.

Once the P-value has been determined, the conclusion at any particular level α results from comparing the P-value to α:

1. If P-value $\leq \alpha$, then reject H_0 at level α
2. If P-value $> \alpha$, then do not reject H_0 at level α

EXAMPLE 18

Let μ denote the true average stopping distance in feet of a particular type of car under certain specified conditions. An investigator tested

$$H_0: \mu = 125 \quad \text{versus} \quad H_a: \mu < 125$$

and reported that the P-value = .0372. What conclusion is appropriate at significance level .01? By the definition of the P-value, the *smallest* α at which H_0 can be rejected is .0372; any smaller α results in H_0 not being rejected. Since $\alpha = .01$ is smaller that the P-value, H_0 shouldn't be rejected.

CALCULATING THE *P*-VALUE FOR A *z* TEST

With the aid of Table II the (approximate) P-value can be calculated for any z test (not only the large-sample z test concerning μ discussed in Section 9.3 but also others presented in later sections and chapters). The test of Example 17 is upper-tailed, so the P-value is the area captured under the z curve to the right of the computed value of z. If the test had been lower-tailed, the P-value would have been the area under the z curve to the left of the computed z (the lower-tail area). For a two-tailed test, finding the area captured in the tail in which z falls—e.g., the upper tail when $z = 2.43$ or the lower tail when $z = -.92$—determines $\alpha/2$ (because α is the sum of the upper- and lower-tail areas). In this case the P-value is twice the captured area. This is summarized in the accompanying box.

TYPE OF z TEST	P-VALUE
Upper-tailed	Area under the z curve to the right of computed z
Lower-tailed	Area under the z curve to the left of computed z
Two-tailed	Twice the area captured in the tail determined by the computed z

EXAMPLE 19   Suppose that a *z* test for testing

$$H_0: \mu = 10 \quad \text{versus} \quad H_a: \mu < 10$$

results in the value $z = -1.49$. Since this is a lower-tail test, the associated *P*-value is the area under the standard normal curve and to the left of

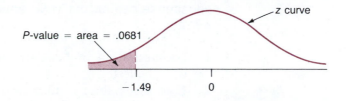

P-value = area = .0681

z curve

−1.49   0

−1.49. From Table II, the desired area is .0681. This is illustrated in the accompanying picture. Therefore the *P*-value = .0681. At significance level .10, the hypothesis H_0 will be rejected (since .0681 < .10), but rejection of H_0 is not justified at significance level .05.

EXAMPLE 20   The mean systolic blood pressure for white males aged 35–44 in the United States is 127.2. The paper "Blood Pressure in a Population of Diabetic Persons Diagnosed after 30 Years of Age" (*Amer. J. of Public Health* (1984): 336–9) reports that the mean blood pressure and standard deviation of a sample of 101 diabetic males aged 35–44 are 130 and 8, respectively. The researchers were interested in determining whether the mean systolic blood pressure of 35–44-year-old diabetic males differed from that of 35–44-year-old males in the general population. A significance level of .01 was selected. The analysis again proceeds via a sequence of steps.

1. Let μ denote the true mean systolic blood pressure for diabetic males aged 35–44.
2. $H_0: \mu = 127.2$
3. $H_a: \mu \neq 127.2$
4. Test statistic: $z = \dfrac{\bar{x} - 127.2}{s/\sqrt{n}}$
5. Calculations:

$$z = \frac{130 - 127.2}{8/\sqrt{101}} = \frac{2.80}{.796} = 3.52$$

6. Determination of the *P*-value: From Table II, the area under the *z* curve and to the right of 3.52 is .0002. Since this is a two-tailed test,

   P-value = 2(.0002) = .0004.

7. **Conclusion**   The null hypothesis will be rejected at significance level .01 (since the *P*-value is smaller than the significance level).

P-VALUES FOR t TESTS

Table II can be used to compute the *P*-value for *z* tests. However, because Table III gives only limited information on *t* distributions, the best that we can do is to establish an upper and/or lower bound on the *P*-value for a *t* test. The procedure for doing this is summarized in the accompanying box.

PROCEDURE FOR CALCULATING BOUNDS ON THE *P*-VALUE FOR A *t* TEST

1. Locate the row of Table III corresponding to the df associated with the test.
2. Look across this row to determine where the computed value of *t* (or $|t|$ if *t* is negative) falls relative to the critical values.
3. Determine the bounds for the *P*-value by referring to one of the two *level of significance* rows at the bottom of the table, depending on whether the test is one- or two-tailed.
   a. If the computed *t* falls between two critical values in the table, the *P*-value will be between the corresponding levels of significance.
   b. If the computed *t* exceeds the largest critical value in the row, then the *P*-value is smaller than the level of significance for the largest critical value in the row.
   c. If the computed *t* is smaller than the smallest critical value in the row, then the *P*-value is larger than the significance level associated with the smallest critical value in the row.

EXAMPLE 21

Federal officials have investigated the problems associated with disposal of hazardous wastes. One disposal site is the abandoned Stringfellow acid pits in Riverside County, Calif. The EPA sampled water from 11 wells in nearby Glen Avon (*Los Angeles Times*, May 31, 1984). The EPA standard for maximum allowable radiation level of drinking water is 15 pCi/l. Suppose that the sample of 11 water specimens resulted in a sample mean radiation level of 22.5 pCi/l and a sample standard deviation of 8. Is this sufficient evidence to indicate that the water in Glen Avon has a mean radiation level that exceeds the EPA standard?

1. Let μ denote the true mean radiation level for well water in Glen Avon.
2. $H_0: \mu = 15$
3. $H_a: \mu > 15$
4. Test statistic: $t = \dfrac{\bar{x} - 15}{s/\sqrt{n}}$
5. Computations:

$$t = \frac{22.5 - 15}{8/\sqrt{11}} = \frac{7.5}{2.4} = 3.125$$

6. Determination of the *P*-value: Since the sample size is 11, df = 10. Looking in the 10 df row of Table III, we find that

$$2.76 < 3.125 < 3.17$$

where 2.76 is the critical value associated with a one-tailed significance level of .01 and 3.17 is the critical value associated with significance level .005. Therefore,

$$.005 < P\text{-value} < .01$$

7. Conclusion: The null hypothesis will be rejected if $\alpha = .01$ or larger. It will not be rejected if $\alpha = .005$ or smaller.

Although our *t* table, like *t* tables in all other texts, limits us to bounds on the *P*-value, almost any good statistical computer package is programmed to calculate the *P*-value resulting from a *t* test. With this information, the *t* table is unnecessary—a conclusion can be drawn simply by comparing the *P*-value to the selected level of significance.

EXAMPLE 22   An automobile manufacturer who wishes to advertise that one of its models achieves 30 mpg decides to carry out a fuel efficiency test. Six nonprofessional drivers are selected and each one drives a car from Phoenix to Los Angeles. The resulting miles-per-gallon figures are

$$x_1 = 27.2, \ x_2 = 29.3, \ x_3 = 31.2$$
$$x_4 = 28.4, \ x_5 = 30.3, \ x_6 = 29.6$$

Assuming that fuel efficiency (mpg) is normally distributed under these circumstances, does the data contradict the claim that true average fuel efficiency is (at least) 30?

With μ denoting true average fuel efficiency, the hypotheses of interest are

$$H_0\text{: } \mu = 30 \text{ versus } H_a\text{: } \mu < 30$$

The alternative statement is the contradiction of prior belief. We used MINITAB to perform a *t* test; the output is given here. ("SE MEAN" denotes the standard error of the mean, which is $s/\sqrt{n} = .57$.)

```
TEST OF MU = 30.0 VS MU L.T. 30.0
 N     MEAN     STDEV    SE MEAN         T    P VALUE
 6    29.33      1.41      0.57     - 1.16      0.15
```

The *P*-value is .15, the smallest level at which H_0 can be rejected. Thus, even at level of significance .10, the hypothesis H_0 cannot be rejected ($\alpha <$ *P*-value). The data does not contradict the prior belief.

SECTION 9.6

9.56 For which of the given P-values will the null hypothesis be rejected when performing a level .05 test?
  a. .001   **b.** .021   **c.** .078   **d.** .047   **e.** .148

9.57 Pairs of P-values and significance levels, α, are given. For each pair, state whether the observed P-value will lead to rejection of H_0 at the given significance level.
  a. P-value = .084, α = .05   **d.** P-value = .084, α = .10
  b. P-value = .003, α = .001   **e.** P-value = .039, α = .01
  c. P-value = .498, α = .05   **f.** P-value = .218, α = .10

9.58 Let μ denote the true average reaction time to a certain stimulus. For a large-sample z test of $H_0: \mu = 5$ versus $H_a: \mu > 5$, find the P-value associated with each of the given values of the test statistic.
  a. 1.4   **b.** .9   **c.** 1.9   **d.** 2.4   **e.** $-.1$

9.59 Newly purchased automobile tires of a certain type are supposed to be filled to a pressure of 30 lb/in². Let μ denote the true average pressure. Find the P-value associated with each given z statistic value for testing $H_0: \mu = 30$ versus $H_a: \mu \neq 30$.
  a. 2.1   **b.** -1.7 **c.** $-.5$ **d.** 1.4   **e.** -5

9.60 To assess the impact of quality circles (groups of employees who meet to discuss issues related to product quality) on employee job satisfaction, the authors of the paper "Consequences of Quality Circles in an Industrial Setting: A Longitudinal Assessment" (*Academy Mgmt. J.* (1988): 338–58) studied 73 employees who were participating in quality circles. The mean job satisfaction score for this sample was 3.18 and the standard deviation was .99. Suppose that the mean score on this scale for all workers is known to be 3.12. (This is consistent with values given in the paper.)
  a. What hypotheses should be tested if the researcher wants to determine whether the mean job satisfaction score for employees participating in quality circles is higher than that of the general working population?
  b. Would a z or t test statistic be used to test the hypotheses in part (a)?
  c. What is the value of the test statistic for this data?
  d. Find the P-value associated with the value of the test statistic computed in part (c).
  e. Will H_0 be rejected if a significance level of .05 is selected?
  f. Would the conclusion in part (e) have been different for a significance level of .01? Of .10?

9.61 A soda manufacturer is interested in determining whether its bottling machine tends to overfill. Each bottle is supposed to contain 12 oz of fluid. A random sample of size 36 is taken from bottles coming off the production line, and the contents of each bottle are carefully measured. It is found that the mean amount of soda for the sample of bottles is 12.1 oz and the sample standard deviation is .2 oz. The manufacturer will use this information to test $H_0: \mu = 12$ versus $H_a: \mu > 12$.
  a. What value does the z test statistic take for this data?
  b. Find the P-value associated with the value of z computed in part (a).
  c. If the manufacturer decides on a level .05 test, should H_0 be rejected in favor of the conclusion that the machine is overfilling?

9.62 A sample of 40 speedometers of a particular brand is obtained and each is checked for accuracy at 55 mph. The resulting sample average and sample standard deviation are 53.8 and 1.3, respectively. Let μ denote the true average reading when the actual speed is 55 mph. Compute a P-value and use it and a significance level of .01 to decide whether the sample evidence strongly suggests that μ is not 55.

9.63 The drying time of a particular brand and type of paint is known to have a mean of 75 min. In an attempt to improve drying time, a new additive has been developed. Use of the additive in 100 test samples of the paint yields an observed mean drying time of 68.5 min and a standard deviation of 9.4 min. Using a significance level of .01, does the experimental evidence indicate that the additive improves (shortens) drying time? Use a *P*-value to conduct your test.

9.64 The national mean cholesterol level is approximately 210 (*Science 84* (April 1984): 16). Each person in a group of men with unusually high cholesterol levels (over 265) was treated with a new drug, cholestyramine. After taking the drug for a given length of time, cholesterol determinations were made. Suppose that 100 men participated in the study. After treatment with the drug, the mean cholesterol level for the 100 men was 228 and the sample standard deviation was 12. Let μ denote the average cholesterol level for all men taking this drug. One question of interest is whether men taking the drug still have a mean cholesterol level that exceeds the national average. Compute the *P*-value for this data. If a .05 significance level is chosen, what conclusion would you draw?

9.65 Student reading skills have long been a major focus of concern in the education community. The article "Can an Advance Organizer Technique Compensate for Poor Reading Performance?" (*J. of Exptl. Educ.* (Summer 1986): 217–22) reported on one study of reading comprehension. A random sample of 100 students was selected, and each was shown an "organizer" (an introduction to orient students to the nature of the test material) prior to taking a test. The resulting sample mean test score was 32.96 and the sample standard deviation was 9.24. Suppose that through long experience with this test, it is known that the true average score without an organizer is 32. Does this data suggest that preliminary exposure to an organizer increases the true average score on this test?
  a. Use a significance level of .05 to test the appropriate hypotheses.
  b. Determine the *P*-value for the test of part (a), and then indicate for which significance levels the null hypothesis would be rejected.

9.66 A researcher collected data in order to test $H_0: \mu = 17$ versus $H_a: \mu > 17$. Place bounds on the *P*-value for each of the given t test statistic values and associated degrees of freedom.
  a. $t = 1.84$, df $= 14$ **d.** $t = 1.32$, df $= 8$
  b. $t = 3.74$, df $= 25$ **e.** $t = 2.67$, df $= 45$
  c. $t = 2.42$, df $= 13$

9.67 Place bounds on the *P*-value for a two-tailed t test for each case.
  a. $t = 2.3$, df $= 6$ **c.** $t = 4.2$, df $= 24$
  b. $t = -3.0$, df $= 14$ **d.** $t = -1.3$, df $= 17$

9.68 One method for straightening wire prior to coiling it to make a spring is called "roller straightening." The paper "The Effect of Roller and Spinner Wire Straightening on Coiling Performance and Wire Properties" (*Springs* (1987): 27–8) reported on the tensile properties of wire. Suppose that a sample of 16 wires is selected and each is tested to determine tensile strength (N/mm²). The resulting sample mean and standard deviation are 2160 and 30, respectively.
  a. The mean tensile strength for springs made using spinner straightening is 2150 N/mm². What hypotheses should be tested to determine if the mean tensile strength for the roller method exceeds 2150?
  b. Assuming that the tensile strength distribution is approximately normal, what test statistic would you use to test the hypotheses in part (a)?
  c. What is the value of the test statistic for this data?
  d. Place bounds on the *P*-value associated with the value of the test statistic computed in part (c).
  e. For a level .05 test, what conclusion would you reach?

9.69 The accompanying times (in sec) to first sprinkler activation were first presented in Exercise 9.49.

27   41   22   27   23   35   30   33   24   27   28   22   24

(See "Use of AFFF in Sprinkler Systems," *Fire Tech.* (1976): 5). The system has been designed so that the true average activation time is supposed to be at most 25 sec. Does the data strongly indicate that the design specifications have not been met? Answer by computing bounds on the *P*-value. Use $\alpha = .01$.

CHAPTER 9   SUMMARY OF KEY CONCEPTS AND FORMULAS

TERM OR FORMULA	PAGE	COMMENT
Hypothesis	262	A statement about the value of a population characteristic
Null hypothesis, H_0	262	The hypothesis initially assumed to be true. It has the form $H_0: \left(\begin{array}{c}\text{population}\\\text{characteristic}\end{array}\right) = \left(\begin{array}{c}\text{hypothesized}\\\text{value}\end{array}\right)$
Alternative hypothesis, H_a	262	A hypothesis that specifies a claim that is contradictory to H_0 and is judged the more plausible claim when H_0 is rejected
Type I error	266	Rejection of H_0 when H_0 is true; the probability of a type I error is denoted by α
Type II error	266	Nonrejection of H_0 when H_0 is false; the probability of a type II error is denoted by β
Test statistic	273	The quantity computed from sample data for making a decision between H_0 and H_a
Rejection region	273	Specifies values of the test statistic for which H_0 should be rejected
P-value	294	The smallest significance level for which H_0 can be rejected
$z = \dfrac{\bar{x} - \text{hypothesized value}}{s/\sqrt{n}}$	274	A test statistic for testing $H_0: \mu = $ hypothesized value when the sample size is large
$z = \dfrac{p - \text{hypothesized value}}{\sqrt{\dfrac{(\text{hyp. val.})(1 - \text{hyp. val.})}{n}}}$	282	A test statistic for testing $H_0: \pi = $ hypothesized value when the sample size is large
$t = \dfrac{\bar{x} - \text{hypothesized value}}{s/\sqrt{n}}$	287	A test statistic for testing $H_0: \mu = $ hypothesized value when the sample size is small and it is reasonable to assume that the population distribution is normal

SUPPLEMENTARY EXERCISES   9.70–9.82

9.70 Sixty-five clerical workers at a large financial service organization participated on a health risk analysis, and the resulting data was summarized in the paper "Workplace Stress and Indicators of Coronary-Disease Risk" (*Acad. Mgmt. J.* (1988): 686–98). For systolic blood pressure, the sample mean and standard deviation were 111.63 and 11.94, respectively. Is there sufficient evidence to conclude that the mean systolic blood pressure for all clerical workers at this business exceeds 110? Test the relevant hypotheses using a .01 significance level.

9.71 A Norwegian study of 105 males born in 1962 with birth weights of 2500 g or less was described in the article "Males with Low Birth Weight Examined at 18 Years of Age'' (*J. of Amer. Med. Assoc.* (1984): 3248). When examined in 1981 by the Norwegian military draft board, 7 of the 105 were declared unfit for military service. The Norwegian draft board declared 6.2% (a proportion of .062) of all 18-year-olds examined in 1981 unfit for military service. Does the data provide sufficient evidence to indicate that the true proportion of males with birth weights of 2500 g or less who are unfit is higher than that of the general population? Use a .01 significance level.

9.72 A standard method for recovering minerals and metals from biological materials results in a mean copper recovery of 63 ppm when used to treat oyster tissue. A new treatment method was described in the paper ''Simple Sample Digestion of Sewage and Sludge for Multi-Element Analysis'' (*J. Environ. Sci. and Health* (1984): 959–72). Suppose this new treatment is used to treat $n = 40$ bits of oyster tissue, resulting in a sample mean copper recovery and a sample standard deviation of 62.6 ppm and 3.7 ppm, respectively. Is there evidence to suggest that the mean copper recovery is lower for the new method than for the standard one? Use a .01 significance level.

9.73 Past experience has indicated that the true response rate is 40% when individuals are approached with a request to fill out and return a particular questionnaire in a stamped and addressed envelope. An investigator believes that if the person distributing the questionnaire is stigmatized in some obvious way, potential respondents would feel sorry for the distributor and thus tend to respond at a rate higher than 40%. To investigate this theory, a distributor is fitted with an eyepatch. Of the 200 questionnaires distributed by this individual, 109 were returned. Does this strongly suggest that the response rate in this situation does exceed the rate in the past?
a. State and test the appropriate hypotheses at significance level .05.
b. Compute the *P*-value for this data and then use it to carry out a test.

9.74 The drug cibenzoline is currently being investigated for possible use in controlling cardiac arrhythmia. The paper ''Quantification of Cibenzoline in Human Plasma by Gas Chromatography–Negative Ion Chemical–Ionization Mass Spectrometry'' (*J. of Chromatography* (1984): 403–9) describes a new method of determining the concentration of cibenzoline in a solution. After 5 ng of cibenzoline was added to a solution, the concentration was measured by the new method. This process was repeated three times, resulting in $n = 3$ concentration readings. The sample mean and standard deviation were reported to be 4.59 ng and .08 ng, respectively. Does this data suggest that the new method produces a mean concentration reading that is too small (less than 5 ng)? Use a .05 significance level and test the appropriate hypotheses.

9.75 The increasing number of senior citizens has made this group an attractive target market for retailers. An understanding of how the elderly feel about various consumer problems is, therefore, important to retailers. The paper ''Consumer Problems and Complaint Actions of Older Americans'' (*J. of Retailing* (1981): 107–23) reported that in a sample of 404 elderly individuals who shop for grocery items, 270 were satisfied with their purchases, whereas 134 were dissatisfied. Suppose that the proportion of all individuals who are satisfied with grocery items is .8 (a value suggested in the paper). Does the sample data suggest that the proportion of elderly people who are satisfied is smaller than the proportion of all individuals who are satisfied? State the relevant hypotheses and carry out the appropriate test using a .01 significance level.

9.76 Police departments across the country have recently voiced concerned that too many calls to the 911 emergency telephone number are not true emergencies. Suppose that the police chief in a particular city is contemplating an advertising campaign to warn of the consequences of abusing the 911 number. Because of the cost, the campaign can be justified only if more than 25% of all 911 calls are not emergencies. A random sample of 200 recent calls to the 911 number is selected, and it is determined that 56 were nonemergency calls. Does this sample data support going ahead with the ad campaign? Test the relevant hypotheses using significance level .10.

9.77 To investigate whether sudden infant death syndrome (SIDS) might be related to an imbalance between peptides affecting respiration, the authors of the paper "Postmortem Analysis of Neuropeptides in Brains from Sudden Infant Death Victims" (*Brain Research* (1984): 279–85) measured cortex met-enkephalin levels (pmol/g wet weight) in brain tissue of 12 SIDS victims. The resulting sample mean and standard deviation were 7.66 and 3.78, respectively. The mean level for children who are not victims of SIDS was reported to be 7.48. Using a .05 significance level, test to determine if the true mean met-enkephalin level of SIDS victims is higher than that of children who are not victims of SIDS.

9.78 The effect of discharging wastewater from a dairy processing plant into groundwater used to irrigate kidney beans was examined in the paper "Effect of Industrial Dairy Processing Effluent on Soil and Crop Plants" (*Environ. Pollution* (1984): 97–106). The wastewater was rich in bicarbonates and calcium, so it was thought that irrigating with a 50% solution of wastewater might promote growth. Suppose that 40 kidney bean plants were irrigated with this mixture, resulting in a sample mean root length of 5.46 cm and a sample standard deviation of .55 cm. The mean root length for kidney bean plants irrigated with uncontaminated water is known to be 5.20 cm. Does this data support the hypothesis that irrigation with the 50% wastewater solution results in a mean root length that is greater than 5.20? Use a .05 significance level.

9.79 The paper referenced in Exercise 9.78 also gave information on root length for pearl millet. When irrigated with uncontaminated water, the mean root length is 6.40 cm. A sample of 40 plants irrigated with a 50% wastewater solution resulted in a sample mean root length of 4.76 cm and a sample standard deviation of .48 cm. Does the data strongly suggest that irrigation with the wastewater mixture results in a mean root length that differs from 6.40? Use a .05 level test.

9.80 A student organization uses the proceeds from a particular soft-drink dispensing machine to finance its activities. The price per can had been $.40 for a long time, and the average daily revenue during that period had been $50.00. The price was recently increased to $.45 per can. A random sample of $n = 20$ days subsequent to the price increase yielded a sample average revenue and sample standard deviation of $47.30 and $4.20, respectively. Does this data suggest that the true average daily revenue has decreased from its value prior to the price increase? Test the appropriate hypotheses using $\alpha = .05$.

9.81 A hot-tub manufacturer advertises that with its heating equipment, a temperature of 100°F can be achieved in at most 15 min. A random sample of 32 tubs is selected and the time necessary to achieve a 100°F temperature is determined for each tub. The sample average time and sample standard deviation are 17.5 min and 2.2 min, respectively. Does this data cast doubt on the company's claim?
a. Carry out a test of hypotheses using significance level .05.
b. Compute the *P*-value, and use it to reach a conclusion at level .05.

9.82 Let π denote the probability that a coin will land heads side up. The coin is tossed 10,000 times, and 5100 heads result. Using a significance level of .05, would you reject the assertion that the coin is fair? From a practical point of view, does the unfairness of the coin bother you?

REFERENCES The books by Freedman et al. and by Moore, listed in earlier chapter references, are excellent sources. Their orientation is primarily conceptual with a minimum of mathematical development, and both sources offer many valuable insights. The introduction to hypothesis testing in the Devore book (Chapter 4 References) is more comprehensive than the present text, especially with respect to type II error probabilities and determination of sample size.

COMPARING TWO POPULATIONS OR TREATMENTS

INTRODUCTION Many investigations are carried out for the purpose of comparing two populations. For example, hospital administrators might want to compare the average length of hospitalization for those patients having private health insurance with that for those covered by Medicare. This would involve a comparison of two population means. A consumer organization might wish to determine whether there is a difference between two manufacturers of personal computers with respect to the proportion of machines that need repair during the first two years of use. This would require comparison of two population proportions. This chapter introduces hypothesis tests and confidence intervals that can be used when comparing two populations on the basis of either means or proportions.

10.1 INFERENCES CONCERNING THE DIFFERENCE BETWEEN TWO POPULATION MEANS BASED ON LARGE INDEPENDENT SAMPLES

An investigator who wishes to compare two populations is often interested either in estimating the difference between the two population means or in testing hypotheses about this difference. In order to accomplish either task, information (in the form of a sample) must be obtained from each population. The sample information is then used to make inferences about the difference between the population means.

In earlier chapters, μ was used to denote the mean of the single population under study. When comparing two populations, it is necessary to use notation that distinguishes between the characteristics of the first population and those of the second. This is accomplished by using subscripts on quantities such as μ and σ^2. Similarly, subscripts on sample statistics such as $\bar{x}$ indicate to which sample these quantities refer.

NOTATION

	Mean Value	Variance	Standard Deviation
Population 1	μ_1	σ_1^2	σ_1
Population 2	μ_2	σ_2^2	σ_2

	Sample Size	Mean	Variance	Standard Deviation
Sample from population 1	n_1	$\bar{x}_1$	s_1^2	s_1
Sample from population 2	n_2	$\bar{x}_2$	s_2^2	s_2

We will compare the means of two populations by focusing on their difference, $\mu_1 - \mu_2$. Note that when $\mu_1 - \mu_2 = 0$, the two population means are identical. That is,

$\mu_1 - \mu_2 = 0$ is equivalent to $\mu_1 = \mu_2$

Similarly,

$\mu_1 - \mu_2 > 0$ is equivalent to $\mu_1 > \mu_2$

and

$\mu_1 - \mu_2 < 0$ is equivalent to $\mu_1 < \mu_2$

Before developing inferential procedures concerning $\mu_1 - \mu_2$, we need to consider how the two samples, one from each population, are selected. Two samples are said to be **independent** if the selection of the individuals or objects that make up one sample has no bearing on the selection of those in the other sample. However, when observations from the first sample are paired in some meaningful way with observations in the second sample, the data is said to be **paired**. For example, to study the effectiveness of a speed-reading course, the reading speed of subjects could be measured prior to taking the class and again after completion of the course. This gives rise to two related samples—one from the population of individuals who have not taken this particular course (the "before" measurements) and one from the population of individuals who have had such a course (the "after" measurements). These samples are paired. The two samples are not independently chosen, since the selection of individuals from the first (before) population completely determines which individuals make up the sample from the second (after) population. In this section and the next, we consider procedures based on independent samples. Procedures for paired data will be presented in Section 3.

Because $\bar{x}_1$ provides an estimate of μ_1 and $\bar{x}_2$ gives an estimate of μ_2, it is natural to use $\bar{x}_1 - \bar{x}_2$ as a point estimate of $\mu_1 - \mu_2$. The value of $\bar{x}_1$ varies from sample to sample (it is a *statistic*), as does the value of $\bar{x}_2$. Thus $\bar{x}_1 - \bar{x}_2$ is itself a statistic and therefore has a sampling distribution. Our inferential methods will be based on information about the sampling distribution of $\bar{x}_1 - \bar{x}_2$. The next box summarizes what is known about the $\bar{x}_1 - \bar{x}_2$ distribution if the two samples are independently chosen.

PROPERTIES OF THE SAMPLING DISTRIBUTION OF $\bar{x}_1 - \bar{x}_2$

If the samples on which $\bar{x}_1$ and $\bar{x}_2$ are based are selected independently of one another, then

1. $\mu_{x_1 - \bar{x}_2} = \begin{pmatrix} \text{the mean value} \\ \text{of } \bar{x}_1 - \bar{x}_2 \end{pmatrix} = \mu_{\bar{x}_1} - \mu_{\bar{x}_2} = \mu_1 - \mu_2$

   Thus the $\bar{x}_1 - \bar{x}_2$ sampling distribution is always centered at the value of $\mu_1 - \mu_2$, so $\bar{x}_1 - \bar{x}_2$ is an unbiased statistic for estimating $\mu_1 - \mu_2$.

2. $\sigma^2_{\bar{x}_1 - \bar{x}_2} = \begin{pmatrix} \text{variance of} \\ \bar{x}_1 - \bar{x}_2 \end{pmatrix} = \sigma^2_{\bar{x}_1} + \sigma^2_{\bar{x}_2} = \dfrac{\sigma_1^2}{n_1} + \dfrac{\sigma_2^2}{n_2}$

   and

   $\sigma_{\bar{x}_1 - \bar{x}_2} = \begin{pmatrix} \text{standard deviation} \\ \text{of } \bar{x}_1 - \bar{x}_2 \end{pmatrix} = \sqrt{\dfrac{\sigma_1^2}{n_1} + \dfrac{\sigma_2^2}{n_2}}$

3. When n_1 and n_2 are both large, $\bar{x}_1$ and $\bar{x}_2$ each have approximately normal distributions (the Central Limit Theorem). This implies that the sampling distribution of $\bar{x}_1 - \bar{x}_2$ is also approximately normal (even if the two population distributions themselves are not normal).

When the sample sizes are large, the properties of the $\bar{x}_1 - \bar{x}_2$ distribution imply that $\bar{x}_1 - \bar{x}_2$ can be standardized to obtain a variable with a sampling distribution that is approximately the standard normal (z) curve. Unfortunately, the values of σ_1^2 and σ_2^2 will rarely be known, but if n_1 and n_2 are both large (typically at least 30), s_1^2 and s_2^2 can be used in their places. This gives the following key result, on which large-sample tests and confidence intervals are based.

> When n_1 and n_2 are both large, the distribution of
>
> $$z = \frac{\bar{x}_1 - \bar{x}_2 - (\mu_1 - \mu_2)}{\sqrt{\dfrac{s_1^2}{n_1} + \dfrac{s_2^2}{n_2}}}$$
>
> is described approximately by the standard normal (z) distribution.

TEST PROCEDURES

In a test designed to compare two population means, the null hypothesis will be of the form

H_0: $\mu_1 - \mu_2$ = hypothesized value

Often the hypothesized value will be zero, with the null hypothesis then saying that there is no difference between the population means. The alternative hypothesis will involve the same hypothesized value, but will use one of the three inequalities >, <, or ≠. As an example, let μ_1 and μ_2 denote the average fuel efficiencies (mpg) for two models of a certain type of car, equipped with 4-cylinder and 6-cylinder engines, respectively. The hypotheses under consideration might be

H_0: $\mu_1 - \mu_2 = 5$

versus

H_a: $\mu_1 - \mu_2 > 5$

This null hypothesis claims that average efficiency for the 4-cylinder engine exceeds the average efficiency for the 6-cylinder engine by exactly 5 mpg. The alternative hypothesis states that the difference between the true average efficiencies is more than 5 mpg.

A test statistic is obtained by replacing $\mu_1 - \mu_2$ in the standardized z variable (given in the previous box) by the hypothesized value, which appears in H_0. Thus the z statistic for testing H_0: $\mu_1 - \mu_2 = 5$ is

$$z = \frac{\bar{x}_1 - \bar{x}_2 - 5}{\sqrt{\dfrac{s_1^2}{n_1} + \dfrac{s_2^2}{n_2}}}$$

When H_0 is true, the sampling distribution of this z statistic is approximately the standard normal (z) curve. The type I error probability can now be controlled by using an appropriate z critical value to determine the rejection region for the test.

LARGE-SAMPLE z TESTS FOR $\mu_1 - \mu_2$

Null hypothesis: $H_0: \mu_1 - \mu_2 =$ hypothesized value

Test statistic: $z = \dfrac{\bar{x}_1 - \bar{x}_2 - \text{hypothesized value}}{\sqrt{\dfrac{s_1^2}{n_1} + \dfrac{s_2^2}{n_2}}}$

ALTERNATIVE HYPOTHESIS	REJECTION REGION
$H_a: \mu_1 - \mu_2 >$ hypothesized value	$z > z$ critical value (upper-tailed test)
$H_a: \mu_1 - \mu_2 <$ hypothesized value	$z < -z$ critical value (lower-tailed test)
$H_a: \mu_1 - \mu_2 \neq$ hypothesized value	Either $z > z$ critical value or $z < -z$ critical value (two-tailed test)

The appropriate z critical value is determined by the choice of significance level. Critical values for the most frequently used values of α appear in the bottom row of Table III.

EXAMPLE 1

A number of studies have focused on the question of whether children born to women smokers differ physiologically from children born to nonsmokers. The paper "Placental Transfer of Lead, Mercury, Cadmium, and Carbon Monoxide in Women" (*Environ. Research* (1978): 494–503) reported on results from one such investigation. Does the accompanying data on blood–lead concentration (μg/l) suggest that the true mean concentration for smokers' newborn children exceeds that for nonsmokers? Test the appropriate hypotheses at level of significance .05.

Sample	Sample Size	Sample Mean	Sample SD
Mothers who smoke	109	8.9	3.3
Mothers who don't smoke	333	8.1	3.5

1. $\mu_1 =$ average lead concentration for all newborns born to smoking mothers

   $\mu_2 =$ average lead concentration for all newborns born to nonsmoking mothers

   $\mu_1 - \mu_2 =$ difference in average concentrations.

2. $H_0: \mu_1 - \mu_2 = 0$

3. $H_a: \mu_1 - \mu_2 > 0$

4. Test statistic:

$$z = \frac{\bar{x}_1 - \bar{x}_2 - 0}{\sqrt{\dfrac{s_1^2}{n_1} + \dfrac{s_2^2}{n_2}}}$$

5. Using level of significance .05, the upper-tailed z critical value is 1.645. The hypothesis H_0 will be rejected if $z > 1.645$.

6. Computations:

$$z = \frac{8.9 - 8.1 - 0}{\sqrt{\dfrac{(3.3)^2}{109} + \dfrac{(3.5)^2}{333}}}$$

$$= .8/.37$$

$$= 2.16$$

7. Since 2.16 is greater than 1.645, we reject H_0 at level .05. The data supports the claim that the average lead concentration is higher for the children of smokers than for those of nonsmokers.

COMPARING TREATMENTS

Often an experiment is carried out in order to compare two different treatments or to compare the effect of a treatment with the effect of *no* treatment ("treatment versus control"). For example, an agricultural experimenter might wish to compare weight gains for animals on two different diets. Let μ_1 denote the expected weight gain (expected response) for an animal on diet (treatment) 1. That is, if the population of all animals were placed on diet 1, then μ_1 would be the population average weight gain. This population does not actually exist, but we can conceptualize it— and the observed weight gains constitute a random sample from this conceptual population. Similarly, μ_2 can be viewed either as the expected weight gain for an animal fed diet 2 or as the average weight gain for the conceptual population consisting of all animals that could receive diet 2. Again, the observed weight gains represent a random sample from this conceptual population. The important point is that our two-sample z test (as well as other two-sample procedures) can be applied to compare conceptual populations.

EXAMPLE 2

The paper "Testing vs. Review: Effects on Retention" (*J. of Educ. Psych.* (1982): 18–22) reported on an experiment designed to compare several different methods ("treatments") for enhancing retention of school material just studied. After high school students studied a brief history text, each one either took a test (method 1) or spent equivalent time reviewing selected passages (method 2). Two weeks afterward, each student took a retention test. Summary data from the experiment follows. Does retention appear to be better with one method than with the other?

Treatment	Sample Size	Sample Mean	Sample SD
Method 1 (test)	31	12.4	4.5
Method 2 (review)	34	11.0	3.1

1. Let μ_1 denote the true average retention score for all high school students who might be assigned to method 1 (the mean of a conceptual population) and define μ_2 analogously for method 2. Then $\mu_1 - \mu_2$ is the difference between the true average retention scores for the two methods.
2. $H_0: \mu_1 - \mu_2 = 0$ (no difference between methods)
3. $H_a: \mu_1 - \mu_2 \neq 0$ (the methods differ)
4. Test statistic:

$$z = \frac{\bar{x}_1 - \bar{x}_2 - 0}{\sqrt{\dfrac{s_1^2}{n_1} + \dfrac{s_2^2}{n_2}}}$$

5. Using $\alpha = .05$, H_0 will be rejected if $z > 1.96$ or if $z < -1.96$
6. Computations:

$$z = \frac{12.4 - 11.0 - 0}{\sqrt{\dfrac{(4.5)^2}{31} + \dfrac{(3.1)^2}{34}}}$$
$$= 1.4/.967$$
$$= 1.45$$

7. Since 1.45 is between -1.96 and 1.96 we do not reject H_0. Retention does not seem to depend on which method is used.

Because the test procedure presented in this section uses a z test statistic, P-values can be computed using the same method as for the one-sample z tests. For the two-tailed test of Example 2, the P-value is twice the area to the right of 1.45 under the z curve. Since the area to the right of 1.45 is .0735 (from Appendix Table II),

P-value $= 2(.0735) = .1470$

Since the P-value is larger than the significance level selected for the test ($\alpha = .05$), H_0 is not rejected.

Comparisons and Causation   If the assignment of treatments to the individuals or objects used in a comparison of treatments is not made by the investigators, the study is said to be **observational**. As an example, the article "Lead and Cadmium Absorption among Children near a Nonferrous Metal Plant" (*Environ. Research* (1978): 290–308) reported data on blood–lead concentrations for two different samples of children. The first sample was drawn from a population residing within 1 km of a lead smelter, while those in the second sample were selected from a rural area much farther from the smelter. It was the parents of children, rather than the investigators, who determined whether the children would be in the close-to-smelter group or the far-from-smelter group. As a second example, a letter in the *J. of the Amer. Med. Assoc.* (May 19, 1978) reported on a comparison of doctors' longevity after medical school graduation for those with an academic affiliation and those in private practice. (The letter writer's stated objective was to see whether "publish or perish" really meant "publish *and* perish.") Here again, an investigator did not start out with a group of doctors and assign some to academic and others to nonacademic careers. The doctors themselves selected their groups.

The difficulty with drawing conclusions based on an observational study is that a statistically significant difference may be due to some underlying factors that have not been controlled, rather than to conditions that define the groups. Does the type of medical practice itself have an effect on longevity, or is the observed difference in lifetimes caused by other factors, which themselves led graduates to choose academic or nonacademic careers? Similarly, is the observed difference in blood–lead concentration levels due to proximity to the smelter? Perhaps there are other physical and socioeconomic factors related both to choice of living area and to concentration.

In general, rejection of H_0: $\mu_1 - \mu_2 = 0$ in favor of H_a: $\mu_1 - \mu_2 > 0$ suggests that, on the average, higher values of the variable are *associated* with individuals in the first population or receiving the first treatment than with those in the second population or receiving the second treatment. But *association does not imply causation*. Strong statistical evidence for a causal relationship can be built up over time through many different comparative studies that point to the same conclusions (as in the many investigations linking smoking to lung cancer). A **randomized controlled experiment,** in which investigators assign subjects in some prescribed random fashion to the treatments or conditions being compared, is particularly effective in suggesting causality. With such random assignment, the investigator and other interested parties will have more confidence in the conclusion that an observed difference was caused by the difference in treatments or conditions. Such carefully controlled studies are more easily carried out in the hard sciences than in social-science contexts, which may explain why the use of statistical methods is less controversial in the former than the latter disciplines.

A CONFIDENCE INTERVAL

A large-sample confidence interval for $\mu_1 - \mu_2$ can be obtained from the same z variable on which the test procedures were based. When n_1 and n_2 are both large, approximately 95% of all samples from the two populations will be such that

$$-1.96 < \frac{\bar{x}_1 - \bar{x}_2 - (\mu_1 - \mu_2)}{\sqrt{\dfrac{s_1^2}{n_1} + \dfrac{s_2^2}{n_2}}} < 1.96$$

The 95% confidence interval results from manipulations that isolate $\mu_1 - \mu_2$ in the middle (just as isolating μ in $-1.96 < (\bar{x} - \mu)/(s/\sqrt{n}) < 1.96$ led to the large-sample interval for μ in Chapter 8):

$$\bar{x}_1 - \bar{x}_2 - 1.96\sqrt{\frac{s_1^2}{n_1} + \frac{s_2^2}{n_2}} < \mu_1 - \mu_2 < \bar{x}_1 - \bar{x}_2 + 1.96\sqrt{\frac{s_1^2}{n_1} + \frac{s_2^2}{n_2}}$$

A confidence level other than 95% is achieved by using an appropriate z critical value in place of 1.96.

The **large-sample confidence interval for $\mu_1 - \mu_2$** is

$$\bar{x}_1 - \bar{x}_2 \pm (z \text{ critical value})\sqrt{\frac{s_1^2}{n_1} + \frac{s_2^2}{n_2}}$$

The z critical values associated with the most frequently used confidence levels appear in the bottom row of Table III.

EXAMPLE 3

Much attention has been focused in recent years on merger activity among business firms. Many business analysts are interested in knowing how various characteristics of merged firms compare to those of nonmerged firms. The article "Abnormal Returns from Merger Profiles" (*J. of Finan. Quant. Analysis* (1983): 149–62) reported the accompanying sample data on price–earnings ratios for two samples of firms.

Type of Firm	Sample Size	Sample Mean	Sample SD
Merged	44	7.295	7.374
Nonmerged	44	14.666	16.089

Let μ_1 and μ_2 denote the true average price–earnings ratios for all merged and nonmerged firms, respectively. Then a 99% confidence interval for the difference $\mu_1 - \mu_2$ between true average price–earnings ratios is

$$7.295 - 14.666 \pm (2.58)\sqrt{\frac{(7.374)^2}{44} + \frac{(16.089)^2}{44}}$$

$$= -7.371 \pm (2.58)(2.668)$$

$$= -7.371 \pm 6.883$$

$$= (-14.254, -.488)$$

Thus we are highly confident that $\mu_1 - \mu_2$ is between -14.254 and $-.488$. (That is, other factors being equal, the true average price–earnings ratio is less for merged firms than for nonmerged firms by between roughly .5 and 14.3.)

EXERCISES 10.1–10.17   **SECTION 10.1**

10.1 Consider two populations for which $\mu_1 = 30$, $\sigma_1 = 2$, $\mu_2 = 25$, and $\sigma_2 = 3$. Suppose that two independent random samples of sizes $n_1 = 40$ and $n_2 = 50$ are selected. Describe the approximate sampling distribution of $\bar{x}_1 - \bar{x}_2$ (center, spread, and shape).

10.2 An article in the Nov. 1983 *Consumer Reports* compared various types of batteries. The average lifetimes of Duracell Alkaline AA batteries and Eveready Energizer Alkaline AA batteries were given as 4.1 hr, and 4.5 hr, respectively. Suppose that these are the population average lifetimes.

a. Let $\bar{x}_1$ be the sample average lifetime of 100 Duracell batteries and $\bar{x}_2$ the sample average lifetime of 100 Eveready batteries. What is the mean value of $\bar{x}_1 - \bar{x}_2$ (that is, where is the sampling distribution of $\bar{x}_1 - \bar{x}_2$ centered)? How does your answer depend on the specified sample sizes?

b. Suppose that population standard deviations of lifetime are 1.8 hr for Duracell batteries and 2.0 hr for Eveready batteries. With the sample sizes as given in part (a), what is the variance of the statistic $\bar{x}_1 - \bar{x}_2$, and what is its standard deviation?

c. For the sample sizes as given in part (a), draw a picture of the approximate sampling distribution curve of $\bar{x}_1 - \bar{x}_2$ (include a measurement scale on the horizontal axis). Would the shape of the curve necessarily be the same for sample sizes of 10 batteries of each type? Explain.

10.3 Let μ_1 and μ_2 denote the true average tread life for two different brands of size FR78-15 radial tires. Test

$$H_0: \mu_1 - \mu_2 = 0 \text{ versus } H_a: \mu_1 - \mu_2 \neq 0$$

at level .05 using the following data: $n_1 = 40$, $\bar{x}_1 = 36{,}500$, $s_1 = 2200$, $n_2 = 40$, $\bar{x}_2 = 33{,}400$, and $s_2 = 1900$.

10.4 Let μ_1 denote the true average life for a certain brand of FR78-15 radial tire, and let μ_2 denote the true average tread life for bias-ply tires of the same brand and sizes. Test

$$H_0: \mu_1 - \mu_2 = 10{,}000 \text{ versus } H_a: \mu_1 - \mu_2 > 10{,}000$$

at level .01 using the following data: $n_1 = 40$, $\bar{x}_1 = 36{,}500$, $s_1 = 2200$, $n_2 = 40$, $\bar{x}_2 = 23{,}800$, and $s_2 = 1500$.

10.5 Leucocyte (white blood cell) counts in thoroughbred horses have recently been studied as a possible aid to the diagnosis of respiratory viral infections. The accompanying data on neutrophils (the most numerous kind of leucocyte) was reported in a comparative study of counts in horses of different ages ("Leucocyte Counts in the Healthy English Thoroughbred in Training" *Equine Vet. J.* (1984): 207–9).

Age	Sample Size	Sample Mean	Sample Standard Deviation
2-year-olds	197	51	5.6
4-year-olds	77	56	4.3

Does this data suggest that the true average neutrophil count for 4-year-olds exceeds that for 2-year-olds? Let μ_1 and μ_2 denote the true average counts for 2- and 4-year-old horses, respectively. Carry out a test of

$$H_0: \mu_1 - \mu_2 = 0 \text{ versus } H_a: \mu_1 - \mu_2 < 0$$

using level of significance .001. Be sure to give the test statistic, rejection region, computations, and conclusion (stated in the problem context).

10.6 The Associated Press reported on the spending patterns of single Americans (*San Luis Obispo Telegram–Tribune*, April 12, 1988). The mean amount of money spent each year on clothing was $657 for women and $735 for men. Suppose that these means had been computed based on information obtained from independently chosen random samples of 500 women and 300 men, and that the two sample standard deviations were $260 (for women) and $320 (for men). Is there sufficient evidence to conclude that the average amount of money spent on clothes differs for single men and women? Test the appropriate hypotheses using a significance level of .01.

10.7 The number of friends consulted for advice before purchasing a large TV or VCR was examined by the authors of the paper "External Search Effort: An Investigation across Several Product Categories" (*J. of Consumer Research* (1987): 83–93). Summary statistics consistent with information in the paper are given in the accompanying table.

Type of Purchase	Number of Purchasers (n)	Mean Number of Friends Consulted	Standard Deviation
Large TV	50	1.65	.4
VCR	50	3.26	.6

Does this data indicate that the average number of friends consulted prior to purchase is higher for those purchasing a VCR than for those purchasing a large TV? Test the relevant hypotheses using a level .01 test.

10.8 *U.S.A. Today* (Jan. 15, 1986) described a study of automobile accident victims at Newton–Wellesley Hospital in Massachusetts. The mean cost for medical treatment for a motorist wearing a seat belt at the time of the accident was $565, while the mean cost for a motorist who was not wearing a seat belt was $1200. Suppose that sample sizes and standard deviations were as given in the accompanying table.

	n	$\bar{x}$	s
Seat belts	45	565	220
No seat belts	90	1200	540

Estimate the true mean difference in the cost of medical treatment for belted and unbelted accident victims, using a 95% confidence interval.

10.9 The paper "Blood Pressure in Japanese Children during the First Three Years of Life" (*Amer. J. of Diseases of Children* (1988): 875–81) gave the accompanying data on systolic blood pressure (mm Hg).

Age	Sex	n	Sample Mean	Sample Standard Deviation
1 year	M	74	93	9
	F	72	92	8
3 years	M	86	96	10
	F	66	96	10

a. Construct a 90% confidence interval estimate for the true difference in mean systolic blood pressure for 1-year-old males and 1-year-old females.
b. Using a significance level of .05, test to determine if the mean systolic blood pressure of 1-year-old males differs from that of 3-year-old males.
c. Repeat part (b) for females.

10.10 Do teenage boys worry more than teenage girls? This is one of the questions addressed by the authors of the paper "The Relationship of Self-Esteem and Attributional Style to Young People's Worries" (*J. of Psych.* (1987): 207–15). A scale called the "Worries Scale" was administered to a group of teenagers, and the results are summarized in the accompanying table.

	n	Sample Mean Score	Sample Standard Deviation
Girls	108	62.05	9.5
Boys	78	67.59	9.7

Is there sufficient evidence to conclude that teenage boys score higher on the Worries Scale than teenage girls? Use a significance level of $\alpha = .05$.

10.11 The accompanying data on water salinity (%) was obtained during a study of seasonal influence of Amazon River water on biological production in the western tropical Atlantic ("Influence of Amazon River Discharge on the Marine Production System off Barbados, West Indies" *J. of Marine Research* (1979): 669–81). Let μ_1 denote the true average salinity level of water samples collected during the summer, and define μ_2 analogously for samples collected during the winter. Compute a 99% confidence interval for $\mu_1 - \mu_2$. Is it necessary to make any assumptions about the two salinity distributions? Explain.

Period	Sample Size	Sample Mean	Sample Standard Deviation
Summer	51	33.40	.428
Winter	54	35.39	.294

10.12 Some astrologists have speculated that people born under certain sun-signs are more extroverted than people born under other signs. The accompanying data was taken from the paper "Self-Attribution Theory and the Sun-Sign" (*J. of Social Psych.* (1984): 121–26). The Eysenck Personality Inventory (EPI) was used to measure extroversion and neuroticism.

Sun-sign	Sample size n	Extroversion $\bar{x}$	s	Neuroticism $\bar{x}$	s
Water signs	59	11.71	3.69	12.32	4.15
Other signs	186	12.53	4.14	12.23	4.11
Winter signs	73	11.49	4.28	11.96	4.22
Summer signs	49	13.57	3.71	13.27	4.04

a. Is there sufficient evidence to indicate that those born under water signs have a lower mean extroversion score than those born under other (nonwater) signs? Use a level .01 test.

b. Does the data strongly suggest that those born under winter signs have a lower mean extroversion score than those born under summer signs? Use a level .05 test.

c. Does the data indicate that those born under water signs differ significantly from those born under other signs with respect to mean neuroticism score? Use $\alpha = .05$.

d. Do those born under winter signs differ from those born under summer signs with respect to mean neuroticism score? Compute the P-value associated with the test statistic and use it to state a conclusion for significance level .01.

10.13 Recorded speech can be compressed and played back at a faster rate. The paper "Comprehension by College Students of Time-Compressed Lectures" (*J. of Exper. Educ.* (Fall 1975): 53–6) gave the results of a study designed to test comprehension of time-compressed speech. Fifty students listened to a 60-min lecture and took a comprehension test. Another 50 students heard the same lecture time-compressed to 40 min. The sample mean and standard deviation of comprehension scores for the normal speed group were 9.18 and 4.59, respectively, and those for the time-compressed group were 6.34 and 4.93, respectively.

a. Use a level .01 test to determine if the true mean comprehension score for students hearing a time-compressed lecture is significantly lower than the true mean score for students who hear a lecture at normal speed.

b. Estimate the difference in the true mean comprehension scores for normal and time-compressed lectures using a 95% confidence interval.

10.14 In a study of attrition among college students, 587 students were followed through their college years ("The Prediction of Voluntary Withdrawal from College: An Unsolved Problem" *J. of Exper. Educ.* (Fall 1980): 29–45). Of the 587 students in the sample, 87 withdrew from college for various reasons (although none were dismissed for academic reasons). The accompanying table gives summary statistics on SAT scores for the "persisters" and the "withdrawers."

	SAT Verbal Score	
	Persisters	Withdrawers
n	500.0	87.0
$\bar{x}$	491.0	503.0
s	80.6	78.8

Is there sufficient evidence to indicate a difference in true mean SAT verbal scores for students who withdraw from college and those who graduate? Perform the relevant hypothesis test using a significance level of .10.

10.15 The earning gap between men and women has been the subject of many investigations. One such study ("Sex, Salary, and Achievement: Reward-Dualism in Academia" *Sociology of Educ.* (1981): 71–85) gave the accompanying information for samples of male and female college professors. Suppose that these statistics were based on samples of size 50 chosen from the populations of all male and all female college professors.

	Males		Females	
	Mean	Standard Deviation	Mean	Standard Deviation
Salary	1634.10	715.00	1091.80	418.80
Years at univ.	7.93	8.04	6.25	7.65

a. Does the data strongly suggest that the mean salary for women is lower than the mean salary for men? Use a level .05 test. Does the resulting conclusion by itself point to discrimination against female professors, or are there some possible nondiscriminating explanations for the observed difference?

b. If the results of the test in part (a) were to be summarized using a *P*-value, what value should be reported?

c. Estimate the difference in the mean number of years at the university for men and women using a 90% confidence interval. Based on your interval, do you think that there is a significant difference between the true mean number of years at the university for men and women? Explain.

10.16 Should quizzes be given at the beginning or the end of a lecture period? The paper "On Positioning the Quiz: An Empirical Analysis" (*Accounting Review* (1980): 664–70) provides some insight. Two sections of an introductory accounting class were given identical instructions to read and study assigned text materials. Three quizzes and a final exam were given during the term, with one section taking quizzes at the beginning of the lecture and the other section taking quizzes at the end. Final exam scores for the two groups are summarized. Does the accompanying data indicate that there is a significant difference in the true mean final exam score for students who take quizzes at the beginning of class and those who take quizzes at the end?

	Quiz at Beginning	Quiz at End
Sample size	40	40
Mean	143.7	131.7
Standard deviation	21.2	20.9

10.17 Celebrity endorsement of products is a common advertising technique. A group of 98 people was shown an ad containing a celebrity endorsement, and a second group of 98 was shown the same ad but using an unknown actor ("Effectiveness of Celebrity Endorsees" *J. of Ad. Research* (1983): 57–62). Each participant rated the commercial's believability on a scale of 0 (not believable) to 10 (very believable). Results were as follows.

Type of Ad	Mean	Standard Deviation
Celebrity ad	3.82	2.63
Noncelebrity ad	3.97	2.51

Is there sufficient evidence to indicate that use of a celebrity endorsement results in a true mean believability rating that differs from the true mean rating for noncelebrity endorsements? Use a level .05 test.

10.2 SMALL-SAMPLE INFERENCES CONCERNING A DIFFERENCE BETWEEN TWO NORMAL POPULATION MEANS WHEN SAMPLES ARE INDEPENDENTLY CHOSEN

The z test and confidence interval discussed in Section 10.1 were large-sample procedures. Their use required no specific assumptions about the population distributions, as long as n_1 and n_2 were large enough to ensure that the sampling distributions of $\bar{x}_1$ and $\bar{x}_2$ were approximately normal. When at least one of the sample sizes is small, the z procedures are not appropriate.

In order to proceed in the small-sample case, we make some assumptions about the population distributions and then develop procedures that are known to perform well when these assumptions are met.

BASIC ASSUMPTIONS IN THIS SECTION

1. Both population distributions are normal.
2. The two population standard deviations are identical ($\sigma_1 = \sigma_2$), with σ denoting the common value.

Figure 1 pictures four different pairs of population distributions. Only Figure 1(a) is consistent with the basic assumptions. If you believe that the relevant picture for your problem is given in Figure 1(b)—normal population distributions but σ_1 substantially different from σ_2—then the methods of this section are inappropriate. There are several possible inferential procedures in this case, but (unlike the procedures of this section) there is still controversy among statisticians about which one should be used. The distributions in Figure 1(c) have exactly the same shape and spread, but one is shifted so that it is centered to the right of the other one. Appropriate inferential procedures in this case can be found in more

FIGURE 1

Four possible population distribution pairs.
(a) Both normal, $\sigma_1 = \sigma_2$
(b) Both normal, $\sigma_1 \neq \sigma_2$
(c) Same shape and spread, differing only in location
(d) Very different shapes and spreads

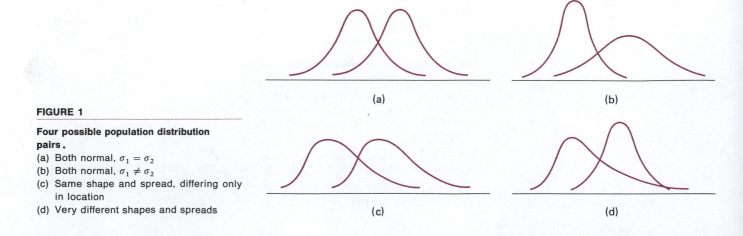

advanced texts. Finally, inference involving population distributions with very different shapes and spreads, as in Figure 1(d), can be very complicated. Good advice from a statistician is particularly important here.

In Chapter 7 we saw that if the population distribution is normal, the sampling distribution of $\bar{x}$ is also normal for *any* sample size. This fact was important in the development of the one-sample t test and confidence interval. In this section, both population distributions are assumed normal, so the sampling distributions of both $\bar{x}_1$ and $\bar{x}_2$ are normal, even for small samples. This in turn implies that the sampling distribution of $\bar{x}_1 - \bar{x}_2$ is also normal for any n_1 and n_2.

The facts concerning the mean value and standard deviation of $\bar{x}_1 - \bar{x}_2$ given in Section 10.1 remain valid here. We know that

$$\mu_{\bar{x}_1 - \bar{x}_2} = \mu_1 - \mu_2$$

which says that the $\bar{x}_1 - \bar{x}_2$ distribution is centered at $\mu_1 - \mu_2$. We are still assuming that the two random samples are obtained independently of one another and, in addition, that $\sigma_1 = \sigma_2 = \sigma$. The common standard deviation σ is then used in place of σ_1 and σ_2 in the formula for the standard deviation of $\bar{x}_1 - \bar{x}_2$, resulting in

$$\sigma_{\bar{x}_1 - \bar{x}_2} = \sqrt{\frac{\sigma_1^2}{n_1} + \frac{\sigma_2^2}{n_2}}$$

$$= \sqrt{\frac{\sigma^2}{n_1} + \frac{\sigma^2}{n_2}}$$

$$= \sqrt{\sigma^2 \left(\frac{1}{n_1} + \frac{1}{n_2} \right)}$$

If $\bar{x}_1 - \bar{x}_2$ is now standardized by subtracting $\mu_{\bar{x}_1 - \bar{x}_2}$ and dividing by $\sigma_{\bar{x}_1 - \bar{x}_2}$, the result is a variable with a standard normal distribution:

$$z = \frac{\bar{x}_1 - \bar{x}_2 - (\mu_1 - \mu_2)}{\sqrt{\sigma^2 \left(\dfrac{1}{n_1} + \dfrac{1}{n_2} \right)}}$$

ESTIMATING σ^2   Before this new standardized variable can be used as a test statistic or manipulated to yield a confidence interval, it is necessary to estimate σ^2. An obvious way to estimate σ^2 from the first sample alone is to use s_1^2, the sample variance of the n_1 observations in that sample. Similarly, the sample variance for the n_2 observations in the second sample also gives an estimate of σ^2. Intuitively, combining s_1^2 and s_2^2 in some fashion should lead to an estimate that is better than either one individually. A first thought might be to use $(s_1^2 + s_2^2)/2$, the ordinary average of the two variances. However, if n_1 is larger than n_2, the estimate s_1^2 will tend to be closer to σ^2 than will s_2^2. That is, the sample variance associated with the larger sample size should be more accurate than the one based on the smaller sample size. This suggests that a weighted average should be used, with the variance from a larger sample receiving heavier weight.

▌▌▌ **DEFINITION**

The statistic for estimating the common variance σ^2 is

$$s_p^2 = \left(\frac{n_1 - 1}{n_1 + n_2 - 2}\right) s_1^2 + \left(\frac{n_2 - 1}{n_1 + n_2 - 2}\right) s_2^2$$

The computed value of s_p^2 is called the **pooled estimate of σ^2** (*pooled* is synonymous with *combined*).

The two quantities in parentheses in the formula for s_p^2 always have a sum of 1. If $n_1 = n_2$, then $s_p^2 = (\frac{1}{2})s_1^2 + (\frac{1}{2})s_2^2$, the ordinary average of the two sample variances. But whenever $n_1 \neq n_2$, the s^2 based on the larger sample size will receive more weight (be multiplied by a larger number) than will the other s^2. The multipliers of s_1^2 and s_2^2 in s_p^2 might at first seem unnatural. The reason for using them has to do with degrees of freedom. The first sample contributes $n_1 - 1$ df to the estimation of σ^2, and the second contributes $n_2 - 1$ df. The two samples are independent, so the number of degrees of freedom associated with s_p^2 is the sum $(n_1 - 1) + (n_2 - 1) = n_1 + n_2 - 2$.

EXAMPLE 4

The paper "Anthropometric and Physical Performance Characteristics of Male Volleyball Players" (*Canadian J. of Appl. Sports Sci.* (1982): 182–8) reported on a comparison of Russian and Finnish volleyball players. One of the variables studied was the height increase in body center of gravity (cm) during a vertical jumping test. Summary data appears in the accompanying table.

Sample	Sample Size	Sample Mean	Sample SD
Finns	$n_1 = 14$	$\bar{x}_1 = 46.0$	$s_1 = 3.2$
Russians	$n_2 = 10$	$\bar{x}_2 = 49.4$	$s_2 = 4.3$

The authors of the paper assumed for purposes of analysis that both height-increase population distributions (Finns and Russians) were normal with the same unknown standard deviation σ. The pooled estimate of σ^2 is then

$$s_p^2 = \left(\frac{14 - 1}{14 + 10 - 2}\right)(3.2)^2 + \left(\frac{10 - 1}{14 + 10 - 2}\right)(4.3)^2$$

$$= \left(\frac{13}{22}\right)(10.24) + \left(\frac{9}{22}\right)(18.49)$$

$$= 6.05 + 7.56$$

$$= 13.61$$

This gives us

$$s_p = \sqrt{13.61} = 3.69$$

as the pooled estimate of σ. Notice that $s_p^2 = 13.61$ is between $s_1^2 = 10.24$ and $s_2^2 = 18.49$. It is closer to s_1^2 because s_1^2 is based on a larger sample size than s_2^2.

When σ^2 is replaced by its estimate, s_p^2, the result is a standardized value having a t distribution.

The standardized variable

$$t = \frac{\bar{x}_1 - \bar{x}_2 - (\mu_1 - \mu_2)}{\sqrt{s_p^2 \left(\dfrac{1}{n_1} + \dfrac{1}{n_2} \right)}}$$

has t distribution with df $= n_1 + n_2 - 2$ (provided the basic assumptions of this section are met).

TESTING HYPOTHESES ABOUT $\mu_1 - \mu_2$

As in the previous section, we consider testing a null hypothesis of the form

H_0: $\mu_1 - \mu_2$ = hypothesized value

The test statistic results from replacing $\mu_1 - \mu_2$ in t by the hypothesized value (which often is zero). When H_0 is true, the sampling distribution of the test statistic is the t distribution with $n_1 + n_2 - 2$ df. The type I error probability is then controlled by using an appropriate t critical value to specify the rejection region.

SUMMARY OF THE POOLED t TEST

Null hypothesis: H_0: $\mu_1 - \mu_2$ = hypothesized value

Test statistic: $t = \dfrac{\bar{x}_1 - \bar{x}_2 - \text{hypothesized value}}{\sqrt{s_p^2 \left(\dfrac{1}{n_1} + \dfrac{1}{n_2} \right)}}$

ALTERNATIVE HYPOTHESIS	REJECTION REGION
H_a: $\mu_1 - \mu_2 >$ hyp. value	$t > t$ critical value
H_a: $\mu_1 - \mu_2 <$ hyp. value	$t < -t$ critical value
H_a: $\mu_1 - \mu_2 \neq$ hyp. value	either $t > t$ critical value or $t < -t$ critical value

The test is based on $n_1 + n_2 - 2$ df, so the t critical value is obtained from that row and the appropriate column (depending on choice of α and whether the test is one-tailed or two-tailed) of Appendix Table III.

EXAMPLE 5

British health officials have recently expressed concern about problems associated with vitamin D deficiency among certain immigrants. Doctors have conjectured that such a deficiency could be related to the amount of fiber in a person's diet. The chemical compound H-labeled 25-hydroxy-vitamin D_3 ($25(OH)D_3$) is the major circulating form of vitamin D, and its plasma half-life is intimately related to the body's vitamin D level. An experiment to compare plasma half-lives for two groups of healthy individuals, one placed on a normal diet and the other on a high-fiber diet, resulted in the following data (from "Reduced Plasma Half-Lives of Radio-Labelled $25(OH)D_3$ in Subjects Receiving a High-Fibre Diet" *Brit. J. of Nutrit.* (1983): 213–16).

Normal diet	19.1	24.0	28.6	29.7	30.0	34.8	
High-fiber diet	12.0	13.0	13.6	20.5	22.7	23.7	24.8

Does this data suggest that true average plasma half-life for persons placed on a normal diet exceeds true average half-life for those on a high-fiber diet? Assuming that the two plasma half-life distributions are normal with equal variances, let's test the appropriate hypotheses at level of significance .01 using the pooled t test.

1. $\mu_1 - \mu_2 =$ difference between true average plasma half-lives for those on normal diets (population 1) and those on high-fiber diets (population 2).
2. $H_0: \mu_1 - \mu_2 = 0$
3. $H_a: \mu_1 - \mu_2 > 0$ (equivalently, $\mu_1 > \mu_2$)
4. Test statistic:

$$t = \frac{\bar{x}_1 - \bar{x}_2 - 0}{\sqrt{s_p^2\left(\frac{1}{n_1} + \frac{1}{n_2}\right)}}$$

5. The form of H_a ($>$) implies that an upper-tailed test is appropriate. With $\alpha = .01$ and

    $$df = n_1 + n_2 - 2 = 6 + 7 - 2 = 11,$$

    Table III gives the t critical value 2.72, so H_0 will be rejected in favor of H_a if $t > 2.72$.
6. The first sample (normal diet) yields $\bar{x}_1 = 27.70$ and $s_1^2 = 29.63$. The summary quantities for the second sample are $\bar{x}_2 = 18.61$ and $s_2^2 = 30.80$. Then

    $$s_p^2 = \frac{5}{11}(29.63) + \frac{6}{11}(30.80) = 30.27$$

    so

    $$t = \frac{27.70 - 18.61 - 0}{\sqrt{(30.27)\left(\frac{1}{6} + \frac{1}{7}\right)}} = \frac{9.09}{3.06} = 2.97$$

7. The value $t = 2.97$ is in the rejection region for a level .01 test $(2.97 > 2.72)$, so we reject H_0 in favor of H_a. The sample data suggests that true average plasma half-life is higher for individuals on a normal diet than for those on a high-fiber diet.

Rather than specifying a rejection region corresponding to a fixed significance level, we can compute the value of t and then obtain bounds on the associated P-value. If t is between two consecutive critical values in the $(n_1 + n_2 - 2)$-df row of the t table, the associated levels of significance (along the bottom margin) give upper and lower bounds. Only an upper bound is given if the computed t exceeds the largest critical value in the row, and only a lower bound is given when t is smaller than the smallest tabled value in the row.

In Example 5, the test was based on $n_1 + n_2 - 2 = 6 + 7 - 2 = 11$ degrees of freedom, and the computed value of t was 2.97. Then, from the 11-df row of the t table, we find that

$$2.72 < \text{computed value of } t < 3.11$$

Since the test was a one-tailed test,

$$.005 < P\text{-value} < .01$$

Thus H_0 would be rejected at the .01 level of significance.

The pooled t test is highly recommended by virtually all statisticians when the two basic asssumptions of this section are at least approximately satisfied. Normal probability plots can be used to check the plausibility of the normality assumptions. There is a formal test procedure, called an F test, for testing $H_0: \sigma_1^2 = \sigma_2^2$ versus $H_a: \sigma_1^2 \neq \sigma_2^2$. Some statisticians suggest that this test be carried out as a preliminary to the pooled t test, with the latter test used only if $H_0: \sigma_1^2 = \sigma_2^2$ is not rejected. But for technical reasons, we (along with many other statisticians) do not recommend this approach.*

Suppose that an investigator believes, prior to collecting data, that variability is roughly the same in the two populations or for the two treatments. Then, if calculated values of s_1^2 and s_2^2 are of roughly the same order of magnitude, use of the pooled t test is reasonable (particularly if the two sample sizes are not too different). This was certainly the case in Example 5.

A SMALL-SAMPLE CONFIDENCE INTERVAL

A small-sample confidence interval for $\mu_1 - \mu_2$ is easily obtained from the basic t variable of this section. Both the derivation of and formula for the interval are very similar to that of the large-sample z interval discussed in the previous section.

---

* Both the pooled t test and the F test are based on the assumption of normal population distributions. But the F test is much more sensitive to departures from this assumption than is the t test. A significant value of the F statistic might result not because σ_1^2 and σ_2^2 differ greatly but because the population distributions are slightly nonnormal.

> **The pooled t confidence interval for $\mu_1 - \mu_2$**, which is valid when both population distributions are normal and $\sigma_1 = \sigma_2$, is
>
> $$\bar{x}_1 - \bar{x}_2 \pm (t \text{ critical value}) \sqrt{s_p^2 \left(\frac{1}{n_1} + \frac{1}{n_2} \right)}$$
>
> The t critical value is based on $n_1 + n_2 - 2$ degrees of freedom. Critical values associated with the most frequently used confidence levels appear in Table III.

EXAMPLE 6 Much research effort has been expended in studying possible causes of the pharmacological and behavioral effects resulting from smoking marijuana. The article "Intravenous Injection in Man of Δ^9THC and $11-OH-\Delta^9$THC" (*Science* (1982): 633) reported on a study of two chemical substances thought to be instrumental in marijuana's effects. Subjects were given one of the two substances in increasing amounts and asked to say when the effect was first perceived. Data values are necessary dose to perception per kilogram of body weight.

Δ^9THC   19.54   14.47   16.00   24.83   26.39   11.49

$(\bar{x}_1 = 18.79, \quad s_1 = 5.91)$

$11-OH-\Delta^9$THC   15.95   25.89   20.53   15.52   14.18   16.00

$(\bar{x}_2 = 18.01, \quad s_2 = 4.42)$

From these values, $s_p^2 = 27.23$. Assuming normality of the two distributions and $\sigma_1 = \sigma_2$, we will calculate a 95% confidence interval for $\mu_1 - \mu_2$, the difference in mean dose to perception for the two substances. From the 10-df row of the t table, the t critical value is 2.23. The interval is therefore

$$(18.79 - 18.01) \pm 2.23 \sqrt{27.23 \left(\frac{1}{6} + \frac{1}{6} \right)} = .78 \pm 6.72$$

$$= (-5.94, 7.50)$$

This interval is rather wide because s_p^2 is large and the two sample sizes are small. Notice that the interval includes 0, so 0 is one of the many plausible values for $\mu_1 - \mu_2$. That is, it is plausible that there is no difference in the mean dose necessary for perception for the two substances.

EXERCISES   10.18–10.35   **SECTION 10.2**

10.18 Let μ_1 and μ_2 denote true average stopping distances (feet) for two different types of cars traveling at 50 mph. Assuming normality and $\sigma_1 = \sigma_2$, test

$H_0: \mu_1 - \mu_2 = 0$ versus $H_a: \mu_1 - \mu_2 \neq 0$

at level .05 using the following data: $n_1 = 6$, $\bar{x}_1 = 122.7$, $s_1 = 5.59$, $n_2 = 5$, $\bar{x}_2 = 129.3$, and $s_2 = 5.25$.

10.19 Suppose that μ_1 and μ_2 are true mean stopping distances (feet) at 50 mph for cars of two certain types equipped with disk brakes and with pneumatic brakes, respectively. Use the pooled t test at significance level .01 to test

$$H_0: \mu_1 - \mu_2 = -10 \text{ versus } H_a: \mu_1 - \mu_2 < -10$$

for the following data: $n_1 = 6$, $\bar{x}_1 = 115.7$, $s_1 = 5.03$, $n_2 = 6$, $\bar{x}_2 = 129.3$, and $s_2 = 5.38$.

10.20 In an experiment to study the effects of exposure to ozone, 20 rats were exposed to ozone in the amount of 2 ppm for a period of 30 days. The average lung volume for these rats after exposure was determined to be 9.28 ml with a standard deviation of .37, whereas the average lung volume for a control group of 17 rats with similar initial characteristics was 7.97 ml with a standard deviation of .41 ("Effect of Chronic Ozone Exposure on Lung Elasticity in Young Rats" *J. Appl. Physiology* (1974): 92–7). Does this data indicate that there is an increase in true average lung volume associated with ozone? Letting μ_1 and μ_2 denote the true average lung volumes for the exposed and unexposed conditions, test

$$H_0: \mu_1 - \mu_2 = 0 \text{ versus } H_a: \mu_1 - \mu_2 > 0.$$

Use a level .01 test.

10.21 A paper in the *J. of Nervous and Mental Disorders* ((1968): 136–46) reported the following data on the amount of dextroamphetamine excreted by a sample of children having organically related disorders and a sample of children with nonorganic disorders. (Dextroamphetamine is a drug commonly used to treat hyperkinetic children.) Use a level .05 test to decide whether the data suggests a difference in mean dextroamphetamine excretion for children with organic and nonorganic disorders.

Organic	17.53	20.60	17.62	28.93	27.10
Nonorganic	15.59	14.76	13.32	12.45	12.79

10.22 An experiment to assess the effects of automobile pollution was described in the paper "Effects of Roadside Conditions on Plants and Insects" (*J. of Appl. Ecology* (1988): 709–15). Twenty soil specimens taken .6 m from the roadside resulted in a mean nitrogen concentration of 1.70 mg/g. Twenty soil specimens taken 6 m from the roadside resulted in a mean nitrogen concentration of 1.35 mg/g. Suppose that the sample standard deviations were .4 (at .6 m) and .3 (at 6 m). Is there sufficient evidence to conclude that the mean nitrogen concentration in soil .6 m from the roadside is higher than that at 6 m? Test the relevant hypotheses using $\alpha = .01$.

10.23 The paper "Effect of Carbohydrate and Vitamin B_6 on Fuel Substrates during Exercise in Women" (*Medicine and Science in Sports and Exercise* (1988): 223–39) compared a group of women who participated in regular aerobic exercise with a group who did not. Data on percent body fat was obtained by skinfolds measured at seven sites and is summarized in the following table.

	Sample Size	Mean Percent Body Fat	Sample Standard Deviation
Aerobic exercise	5	20.1	4.2
No aerobic exercise	5	20.4	3.4

a. Estimate the true mean difference in percent body fat between the two groups using a 90% confidence interval.

b. Does your interval in part (a) contain zero? How would you interpret the interval?

c. What assumptions about the two populations are necessary in order for the interval of part (a) to be valid?

10.24 The accompanying table gives data that appeared in the paper "Development and Clinical Trial of a Minimal-Contact, Cognitive-Behavioral Treatment for Tension Headache" (*Cognitive Therapy and Research* (1988): 325–39). Two different treatments for headache (a relaxation therapy and a cognitive-behavioral therapy) were compared. Is there sufficient evidence to support the researchers' claim that the cognitive-behavioral therapy is more effective than relaxation in increasing the mean number of headache-free days (for a one-week period)? Test the relevant hypotheses using a level .05 test.

Therapy	Sample Size	Mean Number of Headache-Free Days	Sample Standard Deviation
Relaxation	24	3.82	1.75
Cognitive-behavioral	24	5.71	1.43

10.25 Toxaphene is an insecticide that has been identified as a pollutant in the Great Lakes ecosystem. To investigate the effect of toxaphene exposure on animals, groups of rats were given toxaphene in their diet. The paper "Reproduction Study of Toxaphene in the Rat" (*J. of Environ. Sci. Health* (1988): 101–126) reported weight gains (in grams) for rats given a low dose (4 ppm) and for control rats whose diet did not include the insecticide. The accompanying table gives data for both male and female rats.

	Males			Females		
	n	$\bar{x}$	s	n	$\bar{x}$	s
Control	15	529	66	23	237	32
Low dose	15	551	71	20	249	54

a. Is there sufficient evidence to conclude that a low dose of toxaphene increases mean weight gain for male rats? Use a significance level of .05.

b. Do you reach the same conclusion as in part (a) for female rats?

10.26 In an effort to assess the risk of brain damage in infants, the authors of the paper "Neuropathologic Documentation of Prenatal Brain Damage" (*Amer. J. of Diseases of Children* (1988): 858–66) studied both infants born with brain damage and infants born without brain damage. They also looked at premature babies as well as those carried to term.

	Sample Size	Mean Birth Weight (g)	Sample Standard Deviation
Term			
Brain damage	12	2998	707
No brain damage	13	2704	627
Premature			
Brain damage	10	1541	448
No brain damage	54	1204	656

a. Does the data indicate that the true mean birth weight for premature infants with brain damage differs from that of premature infants without brain damage? Test the relevant hypotheses using a significance level of .05.
b. Repeat part (a) for infants carried to term.
c. Use a 90% confidence interval to estimate the difference in mean weight between term and premature infants without brain damage.
d. Repeat part (c) for infants with brain damage.

10.27 The effect of plant diversity on beetle density was examined in a series of experiments described in the paper "Effects of Plant Diversity, Host Density, and Host Size on Population Ecology of the Colorado Potato Beetle" (*Environ. Entomology* (1987): 1019–26). Potatoes grown in fallow plots and potatoes grown in plots that also included bean plants and weeds (called a *triculture*) were compared on the basis of the number of beetle eggs found on the plant leaves. Data is given in the following table.

Plots	Number of Plants Inspected	Mean Number of Beetle Eggs Found	Sample Standard Deviation
Fallow	4	7.1	5.40
Triculture	4	14.6	8.30

Is there sufficient evidence to conclude that potatoes grown in fallow plots have a smaller mean number of eggs per plant than those grown in a triculture? Use a .05 level of significance.

10.28 Unionization of university faculty is a fairly recent phenomenon. There has been much speculation about the effect of collective bargaining on faculty and student satisfaction. Eighteen unionized and 23 nonunionized campuses participated in a study described in the paper "The Relationship between Faculty Unionism and Organizational Effectiveness" (*Academy of Management J.* (1982): 6–24). The participating schools were scored on a number of dimensions. Summary statistics on faculty satisfaction and student academic development are given.

Campus	Student Academic Development		Faculty Satisfaction	
Unionized ($n = 18$)	$\bar{x} = 3.71$	$s = .49$	$\bar{x} = 4.49$	$s = .56$
Nonunionized ($n = 23$)	$\bar{x} = 4.36$	$s = .88$	$\bar{x} = 4.85$	$s = .39$

a. Let μ_1 be the mean score on student academic development for all unionized schools and μ_2 the corresponding mean for nonunionized schools. Use a level .05 test to determine if there is a significant difference in mean student academic development score between unionized and nonunionized colleges.
b. Does the data indicate that unionized and nonunionized schools differ significantly with respect to mean faculty satisfaction score? Use a significance level of .05.
c. What P-value is associated with the value of the test statistic computed in part (b)?

10.29 Measurements on a number of physiological variables for samples of 8 male and 8 female adolescent tennis players were reported in "Physiological and Anthropometric Profiles of Elite Prepubescent Tennis Players" (*Sportsmed.* (1984): 111–16). Results are summarized in the accompanying table.

	Boys		Girls	
	$\bar{x}$	s	$\bar{x}$	s
Shoulder flexibility	214.4	12.9	216.3	20.0
Ankle flexibility	71.4	4.6	72.5	8.9
Grip strength	23.9	2.5	22.2	4.1

a. Estimate the difference in true mean grip strength for boys and girls using a 95% confidence interval. Does the confidence interval indicate that precise information about this difference is available?

b. Construct 95% confidence intervals for the difference between boys and girls with respect to both true mean shoulder flexibility and ankle flexibility. Interpret each of the intervals.

10.30 A study concerning the sublethal effects of insecticides ("Effects of Sublethal Doses of DDT and Three Other Insecticides on Tribolium Confusum" *J. of Stored Products Research* (1983): 43–50) reported the accompanying data on oxygen consumption for flour beetles ten days after DDT treatment and for a control sample of untreated beetles.

Group	Sample Size	Sample Mean	Sample Standard Deviation
Untreated	25	5.02	.94
Treated	25	4.37	.98

a. Does the use of the pooled t test seem reasonable in determining whether DDT treatment results in a decrease in average oxygen consumption? Explain.

b. Test the hypotheses suggested in part (a) using a significance level of .05.

c. Obtain as much information as you can about the P-value for the test in part (b), and use the result to reach a conclusion at significance level .01.

10.31 Do certain behaviors result in a severe drain on energy resources because a great deal of energy is expended in comparison to energy intake? The paper "The Energetic Cost of Courtship and Aggression in a Plethodontid Salamander" (*Ecology* (1983): 979–83) reported on one of the few studies concerned with behavior and energy expenditure. The accompanying data is on oxygen consumption (ml/g/hr) for male–female salamander pairs. (The determination of consumption values was rather complicated, and it is partly for this reason that so few studies of this type have been carried out.) Compute a 95% confidence interval for the difference between true average consumption for noncourting pairs and true average consumption for courting pairs. What assumptions about the two consumption distributions are necessary for the validity of the interval?

Behavior	Sample Size	Sample Mean	Sample Standard Deviation
Noncourting	11	.072	.0066
Courting	15	.099	.0071

10.32 The establishment and maintenance of vegetation is an important component in most metalliferous mine waste–reclamation schemes. The ecological consequences for animal populations exposed to pollutants at contaminated reclamation sites is of concern. The paper "Cadmium in Small Mammals from Grassland Established on Metalliferous Mine Waste" (*Environ. Pollution* (1984): 153–62) reported on cadmium concentrations in various body organs for several species of small mammals. The accompanying data is for skull concentrations in samples of one such species both at a reclamation site and at a control site.

Site	Sample Size	Sample Mean	Sample Standard Deviation
Control	20	.59	.13
Reclamation	21	.72	.18

a. Would you recommend using a z test from the previous section to analyze this data? Why or why not?
b. Use the test procedure discussed in *this* section to decide whether true mean cadmium concentrations in skulls differ at the two sites.

10.33   The paper "Pine Needles as Sensors of Atmospheric Pollution" (*Environ. Monitoring* (1982): 273–86) reported on the use of neutron activity analysis to determine pollutant concentrations in pine needles. According to the paper's authors, "These observations strongly indicated that for those elements which are determined well by the analytical procedures, the distribution of concentrations is lognormal. Accordingly, in tests of significance the logarithms of concentrations will be used." The given data refers to bromine concentration in needles taken from a site near an oil-fired steam plant and from a relatively clean site. The summary values are means and standard deviations of the log-transformed observations. Let μ_1 be the true average log concentration at the first site and define μ_2 analogously for the second site. Test for equality of the two concentration distribution means against the alternative that they are different, by using the pooled t test at level .05 with the log-transformed data.

Site	Sample Size	Mean Log Concentration	Standard Deviation of Log Concentration
Steam plant	8	18	4.9
Clean	9	11	4.6

10.34   Information concerning the extent to which various agricultural products are affected by drought can have a great bearing on planting decisions. The paper "Varietal Differences in the Response of Potatoes to Repeated Periods of Water Stress in Hot Climates" (*Potato Research* (1983): 315–21) reported the accompanying data on tuber yield (g/plant) for one particular potato variety.

Treatment	Sample Size	Mean	Standard Deviation
Unstressed	6	376	78.4
Stressed	12	234	65.8

a. Assuming that the yield distributions for the two treatments have the same variance σ^2, compute the pooled estimate of this variance.

b. Does the data suggest that true average yield under water stress is less than for the unstressed condition? Test at level of significance .01.

c. Based on the computed value of the test statistic in part (b), what can you say about the *P*-value?

10.35 Referring to Exercise 10.34, suppose that the investigator wished to know whether water stress lowered true average yield by more than 50 g/plant. State and test the appropriate hypotheses at level .01.

10.3 PAIRED DATA

Two samples are said to be *independent* if the selection of the individuals or objects that make up one of the samples has no bearing on the selection of those in the other sample. In some situations an experiment with independent samples is not the best way to obtain information concerning any possible difference between the populations. For example, suppose that an investigator wants to determine if regular aerobic exercise affects blood pressure. A random sample of people who jog regularly and a second random sample of people who do not exercise regularly are selected independently of one another. The researcher then uses the pooled *t* test to conclude that a significant difference exists between the average blood pressures for joggers and nonjoggers. Is it reasonable to think that jogging influences blood pressure? It is known that blood pressure is related to both diet and body weight. Might it not be the case that joggers in the sample tend to be leaner and adhere to a healthier diet than the nonjoggers and that *this* might account for the observed difference? On the basis of this study, the researcher wouldn't be able to rule out the possibility that the observed difference in blood pressure is explained by weight differences between the people in the two samples and that aerobic exercise itself has no effect.

One way to avoid this difficulty would be to match subjects by weight. The researcher would find pairs of subjects so that the jogger and nonjogger in each pair were similar in weight (although weights for different pairs might vary widely). The factor *weight* could then be ruled out as a possible explanation for an observed difference in average blood pressure between the two groups. Matching the subjects by weight results in two samples for which each observation in the first sample is coupled in a meaningful way with a particular observation in the second sample. Such samples are said to be **paired**.

Experiments can be designed to yield paired data in a number of different ways. Some studies involve using the same group of individuals with measurements recorded both before and after some intervening treatment. Others use naturally occurring pairs such as twins or husbands and wives, and some construct pairs by matching on factors with an effect that might otherwise obscure differences (or the lack of them) between the two populations of interest (as might weight in the jogging example). Paired samples often provide more information than would independent samples because extraneous effects are screened out.

EXAMPLE 7

It has been hypothesized that strenuous physical activity affects hormone levels. The paper "Growth Hormone Increase during Sleep after Daytime Exercise" (*J. of Endocrinology* (1974): 473–8) reported the results of an experiment involving six healthy male subjects. For each participant, blood samples were taken during sleep on two different nights using an indwelling venous catheter. The first blood sample (the control) was drawn after a day that included no strenuous activities, and the second was drawn after a day when the subject engaged in strenuous exercise. The resulting data on growth hormone level follows. The samples are paired rather than independent because both samples are comprised of measurements on the same men.

Growth Hormone Level (mg/ml)

Subject	1	2	3	4	5	6
Postexercise	13.6	14.7	42.8	20.0	19.2	17.3
Control	8.5	12.6	21.6	19.4	14.7	13.6

Let μ_1 denote the mean nocturnal growth hormone level for the population of all healthy males who participated in strenuous activity on the previous day. Similarly, let μ_2 denote the mean nocturnal hormone level for the population consisting of all healthy males whose activities on the previous day did not include any strenuous physical exercise. The hypotheses of interest are then

$$H_0: \mu_1 - \mu_2 = 0 \quad \text{versus} \quad H_a: \mu_1 - \mu_2 \neq 0$$

Notice that in each of the six data pairs, the postexercise hormone level is higher than the corresponding control level. Intuitively this suggests that there may be a difference between the population means.

However, if we were to use the pooled t test for two independent samples (a mistake), the resulting t test statistic value is 1.28. This value does not allow for rejection of the hypothesis $\mu_1 - \mu_2 = 0$, even at level of significance .10. This result might surprise you at first, but remember that this test procedure ignores the fact that the samples are paired. Two plots of the data are given in Figure 2. The first one ignores the pairing, and the two samples look quite similar. The plot in which pairs are identified does suggest a difference, since for each pair the exercise observation exceeds the no-exercise

FIGURE 2

Two plots of the paired data from Example 7.
(a) Pairing ignored
(b) Pairs identified

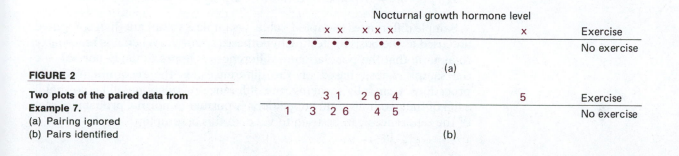

observation. Disregarding the paired nature of the samples results in a loss of information. Nocturnal growth hormone levels vary substantially from one individual to another. It is this variability that obscures the difference in hormone level associated with strenuous exercise when the pooled t test is used.

Example 7 suggests that the methods of inference developed for independent samples are not adequate for dealing with paired samples. When sample observations from the first population are paired in some meaningful way with sample observations from the second population, inferences can be based on the differences between the two observations within each sampled pair. The n sample differences can then be regarded as having been selected from a large population of differences. Thus in Example 7, we can think of the six postexercise–control differences as having been selected from an entire population of differences. Let

μ_d = mean value of the difference population

and

σ_d = standard deviation of the difference population

The relationship between μ_d and the two individual population means is

$\mu_d = \mu_1 - \mu_2$

Therefore, when the samples are paired, inferences about $\mu_1 - \mu_2$ are equivalent to inferences about μ_d. Since inferences about μ_d can be based on the n observed sample differences, the original two-sample problem becomes a familiar one-sample problem.

THE PAIRED t TEST

To compare two population means when the samples are paired, we first translate the hypothesis of interest from one about the value of $\mu_1 - \mu_2$ to an equivalent one involving μ_d.

HYPOTHESIS	EQUIVALENT HYPOTHESIS WHEN SAMPLES ARE PAIRED
$H_0: \mu_1 - \mu_2$ = hypothesized value	$H_0: \mu_d$ = hypothesized value
$H_a: \mu_1 - \mu_2$ > hypothesized value	$H_a: \mu_d$ > hypothesized value
$H_a: \mu_1 - \mu_2$ < hypothesized value	$H_a: \mu_d$ < hypothesized value
$H_a: \mu_1 - \mu_2 \neq$ hypothesized value	$H_a: \mu_d \neq$ hypothesized value

Sample differences (sample 1 value − sample 2 value) are then computed and used as the basis for testing hypotheses about μ_d. When it is reasonable to assume that the population of differences is approximately normal, the one-sample t test, based on the differences, is the recommended test procedure. Generally speaking, the difference population will be normal if each of the two individual populations is normal. A normal probability plot of the differences can be used to validate this assumption.

THE PAIRED *t* TEST

Null hypothesis: H_0: μ_d = hypothesized value

Test statistic:

$$t = \frac{\bar{x}_d - \text{hypothesized value}}{s_d/\sqrt{n}}$$

where n is the number of sample differences and $\bar{x}_d$ and s_d are the sample mean and standard deviation of these differences.

ALTERNATIVE HYPOTHESIS	REJECTION REGION
H_a: μ_d > hypothesized value	t > t critical value (upper-tailed test)
H_a: μ_d < hypothesized value	t < $-t$ critical value (lower-tailed test)
H_a: μ_d ≠ hypothesized value	Either t > t critical value or t < $-t$ critical value (two-tailed test)

Once the level of significance α has been specified, the appropriate t critical value can be obtained from the corresponding column and $n - 1$ df row of Appendix Table III.

EXAMPLE 8

Trace metals in drinking water affect the flavor of the water, and unusually high concentrations can pose a health hazard. The paper "Trace Metals of South Indian River" (*Environ. Studies* (1982): 62–6) reported trace-metal concentrations for both surface water and bottom water at six different river locations. Data on zinc concentration is given here.

	Concentration (mg/l)	
Location	Bottom Water	Top Water
1	.430	.415
2	.266	.238
3	.567	.390
4	.531	.410
5	.707	.605
6	.716	.609

Although zinc concentration varies widely from one location to another, in every case the zinc concentration in bottom water is higher than that in top water. This suggests that bottom water differs from top water in mean zinc concentration. Since this data is paired (by location), we will employ a method of analysis appropriate for paired data.

We form the differences by subtracting top-water zinc concentration from the bottom-water concentration for each location.

Location	Difference, x_d
1	$.430 - .415 = .015$
2	$.266 - .238 = .028$
3	$.567 - .390 = .177$
4	$.531 - .410 = .121$
5	$.707 - .605 = .102$
6	$.716 - .609 = .107$

For the differences, $\sum x_d = .550$ and $\sum x_d^2 = .068832$. Then

$$\bar{x}_d = \frac{\sum x_d}{n} = \frac{.550}{6} = .0917$$

$$s_d^2 = \frac{\sum x_d^2 - n\bar{x}_d^2}{n - 1} = \frac{.068832 - 6(.0917)^2}{5} = .00368$$

$$s_d = \sqrt{s_d^2} = \sqrt{.00368} = .061$$

We now use the paired t test with significance level .05 to test for any difference between mean zinc concentration in top and bottom water.

1. $\mu_d = \mu_1 - \mu_2 =$ mean difference between bottom- and top-water zinc concentrations.
2. $H_0: \mu_d = 0$
3. $H_a: \mu_d \neq 0$
4. Test statistic:

$$t = \frac{\bar{x}_d - 0}{s_d/\sqrt{n}} = \frac{\bar{x}_d}{s_d/\sqrt{n}}$$

5. The nature of H_a implies that a two-tailed rejection region should be used. With level of significance .05 and df $= n - 1 = 6 - 1 = 5$, Table III gives 2.57 as the appropriate critical value. The null hypothesis will therefore be rejected if either $t > 2.57$ or $t < -2.57$.
6. The values of $\bar{x}_d$ and s_d were previously computed to be $\bar{x}_d = .0917$ and $s_d = .061$. Substituting these values into the test-statistic formula yields

$$t = \frac{.0917}{.061/\sqrt{6}} = \frac{.0917}{.0249} = 3.68$$

7. The value $t = 3.68$ exceeds 2.57, implying that H_0 should be rejected in favor of H_a. The data suggests that mean zinc concentration is not the same for bottom water as it is for top water.

Use of the pooled t test on the data in Example 8 would have been incorrect because the top- and bottom-water samples are not independent. Inappropriate use of the pooled t test in this setting would result in a computed t value of

$$t = \frac{(\bar{x}_1 - \bar{x}_2) - 0}{\sqrt{s_p^2 \left(\dfrac{1}{n_1} + \dfrac{1}{n_2} \right)}} = \frac{.5362 - .4445}{\sqrt{.025 \left(\dfrac{1}{6} + \dfrac{1}{6} \right)}} = \frac{.0917}{.091} = 1.01$$

The hypothesis of equal mean concentrations for top and bottom water would not be rejected. When the pairing is ignored, the difference between top- and bottom-water concentrations is obscured by the variability in zinc concentration from one location to another.

The numerators $\bar{x}_d$ and $\bar{x}_1 - \bar{x}_2$ of the two test statistics are always equal. The difference between paired t and pooled t lies in the denominator. The variability in differences is usually much smaller than the variability in each sample separately (because measurements in a pair tend to be similar). As a result, the value of the paired t statistic is usually larger in magnitude than the value of the pooled t statistic—3.68 versus 1.01 in the example we just considered. Pairing typically reduces variability that might otherwise obscure small but nevertheless significant differences.

EXAMPLE 9 Researchers have long been interested in the effects of alcohol on the human body. The authors of the paper "Effects of Alcohol on Hypoxia" (*J. of Amer. Med. Assoc.* (Dec. 13, 1965): 135) examined the relationship between alcohol intake and the time of useful consciousness during high-altitude flight. Ten male subjects were taken to a simulated altitude of 25,000 ft and given several tasks to perform. Each was carefully observed for deterioration in performance due to lack of oxygen, and the time at which useful consciousness ended was recorded. Three days later, the experiment was repeated one hour after the subjects had ingested .5 cc of 100-proof whiskey per pound of body weight. The time (in seconds) of useful consciousness was again recorded. The resulting data appears in the accompanying table.

Time of Useful Consciousness (sec)

Subject	No Alcohol	Alcohol	Difference
1	261	185	76
2	565	375	190
3	900	310	590
4	630	240	390
5	280	215	65
6	365	420	−55
7	400	405	−5
8	735	205	530
9	430	255	175
10	900	900	0

Since the samples are paired rather than independent, we use the paired t test to determine if the data supports the hypothesis that ingestion of the stated amount of alcohol reduces the mean time of useful consciousness at high altitudes. A normal probability plot based on the sample of differences looks approximately linear, suggesting that the assumption of normality is reasonable. We have selected a .05 level of significance for this test.

1. μ_d = difference between the true mean time of useful consciousness when no alcohol is consumed and that when .5 cc of alcohol per pound of body weight is ingested.
2. $H_0: \mu_d = 0$
3. $H_a: \mu_d > 0$
4. Test statistic:

$$t = \frac{\bar{x}_d}{s_d/\sqrt{n}}$$

5. An upper-tailed rejection region is appropriate. With level of significance .05 and df $= n - 1 = 9$, Table III gives the critical value 1.83. The hypothesis H_0 will be rejected if $t > 1.83$.
6. From the ten observed differences, $\bar{x}_d = 195.6$ and $s_d = 230.53$. Then

$$t = \frac{195.6}{230.53/\sqrt{10}} = \frac{195.6}{72.9} = 2.68$$

7. The value $t = 2.68$ exceeds the critical value 1.83, so the null hypothesis should be rejected at level .05. There is sufficient evidence to indicate that ingestion of .5 cc of whiskey per pound of body weight reduces the average time of useful consciousness.

A CONFIDENCE INTERVAL FOR μ_d

The t confidence interval for μ given in Chapter 8 is easily adapted to obtain an interval estimate for μ_d.

When it is reasonable to assume that the difference population is (approximately) normal, the **paired-t confidence interval for μ_d** is

$$\bar{x}_d \pm (t \text{ critical value}) \cdot s_d/\sqrt{n}$$

For a specified confidence level, the $(n - 1)$-df row of Table III gives the appropriate t critical value.

EXAMPLE 10

Cushing's disease is characterized by muscular weakness due to adrenal or pituitary dysfunction. In order to provide effective treatment, it is important to detect childhood Cushing's disease as early as possible. Age at onset of symptoms and age at diagnosis for 15 children suffering from the disease were given in the paper "Treatment of Cushing's Disease in Childhood and Adolescence by Transphenoidal Microadenomectomy" (*New Eng. J. of Med.* (April 5, 1984): 889). Since early diagnosis is crucial for successful treatment, the length of time between onset of symptoms and diagnosis is of interest. Let μ_d be the mean difference between age at onset and age at diagnosis (so that μ_d is a negative number). We use the following data to estimate μ_d with a 90% confidence interval.

	Age (months)							
Patient	**1**	**2**	**3**	**4**	**5**	**6**	**7**	**8**
Onset	84	90	96	108	126	144	156	63
Diagnosis	108	102	151	123	156	204	170	84
Difference	−24	−12	−55	−15	−30	−60	−14	−21
Patient	**9**	**10**	**11**	**12**	**13**	**14**	**15**	
Onset	119	120	132	144	144	144	144	
Diagnosis	167	132	157	197	205	213	224	
Difference	−48	−12	−25	−53	−61	−69	−80	

The summary values are $\bar{x}_d = -38.6$ and $s_d = 23.18$. With $n - 1 = 14$ df, the t critical value for a 90% confidence level is 1.76 (from Table III). The interval is therefore

$$-38.6 \pm (1.76)(23.18/\sqrt{15}) = -38.6 \pm 10.53$$
$$= (-49.13, -28.07)$$

Based on the sample data, we can be 90% confident that the mean elapsed time between onset of symptoms and diagnosis of childhood Cushing's disease is between 28.07 and 49.13 months. It appears that a great deal of time generally passes before Cushing's disease is diagnosed. This is probably due to the fact that childhood Cushing's disease is very rare and so may be overlooked as a possibility until other more common illnesses are ruled out as causes of the symptoms.

When there are two populations to compare in order to draw a conclusion on the basis of sample data, a researcher might choose to use independent samples or paired samples. In many situations, paired data provides a more effective comparison by screening out the effects of extraneous variables that might obscure differences between the two populations or that might suggest a difference when none exists.

EXERCISES 10.36–10.47   **SECTION 10.3**

10.36 Suppose you were interested in investigating the effect of a drug that is to be used in the treatment of patients who have glaucoma in both eyes. A comparison between the mean reduction in eye pressure for this drug and for a standard treatment is desired. Both treatments are applied directly to the eye.
a. Describe how you would go about collecting data for your investigation.
b. Does your method result in paired data?
c. Can you think of a reasonable method of collecting data that would result in independent samples? Would such an experiment be as informative as a paired experiment? Comment.

10.37 Two different underground pipe coatings for preventing corrosion are to be compared. Effect of a coating (as measured by maximum depth of corrosion penetration on a piece of pipe) may vary with depth, orientation, soil type, pipe composition, etc. Describe how an experiment that filters out the effects of these extraneous factors could be carried out.

10.38 Twelve infants paired according to birth weight were used to compare an enriched formula with a standard formula. Weight gains (g) are given.

Pair	Enriched Formula	Standard Formula
1	3604	3140
2	2950	3100
3	3344	2810
4	4022	3761
5	4316	3774
6	3077	2630

a. Let μ_d denote the true average difference in weight gains between the two formulas (enriched − standard). What alternative hypothesis suggests that the enriched formula is more effective than the standard formula in increasing weight? Test $H_0: \mu_d = 0$ against this alternative at significance level .05.

b. Why do you think a paired experiment was chosen for this study?

10.39 The paper "Relative Controllability of Dissimilar Cars" (*Human Factors* (1962): 375–80) reported results of an experiment to compare handling ability for two cars of different design. Time (in seconds) required to parallel park each car was recorded for 14 drivers. Does the data suggest that the mean time to parallel park differs significantly for the two designs? Test the relevant hypotheses using $\alpha = .10$.

Driver	Design 1	Design 2
1	37.0	17.8
2	25.8	20.2
3	16.2	16.8
4	24.2	41.4
5	22.0	21.4
6	33.4	38.4
7	23.8	16.8
8	58.2	32.2
9	33.6	27.8
10	24.4	23.2
11	23.4	29.6
12	21.2	20.6
13	36.2	32.2
14	29.8	53.8

10.40 Recent evidence suggests that nonsmokers who live with smokers and are therefore exposed to what researchers call *sidestream smoke* (as opposed to *mainstream smoke*, which is inhaled by the smoker directly from the cigarette) may have an increased risk of lung disease. Mainstream (M) and sidestream (S) yields (in mg) of tar, nicotine, and carbon monoxide for eight brands of nonfilter cigarettes appeared in the paper "Yields of Tar, Nicotine, and Carbon Monoxide in the Sidestream Smoke from Fifteen Brands of Canadian Cigarettes" (*Amer. J. of Public*

Health (1984): 228–31). Use this data and a significance level of .05 to answer the following questions.

a. Does the data strongly suggest that the mean tar yield of sidestream smoke is higher than that of mainstream smoke?

b. Is there sufficient evidence to suggest that the mean nicotine yield is higher for sidestream smoke than for mainstream smoke?

c. Estimate the mean difference in carbon monoxide yield between sidestream and mainstream smoke using a 95% confidence interval.

	Tar		Nicotine		Carbon Monoxide	
Brand	S	M	S	M	S	M
A	15.8	18.5	2.8	1.2	40.5	18.6
B	16.9	17.0	2.7	1.1	59.8	20.5
C	21.6	17.2	3.7	1.2	42.9	16.8
D	18.8	19.4	2.8	1.0	42.0	17.8
E	29.3	15.6	4.3	1.1	60.8	19.8
F	20.7	16.4	3.9	1.2	45.1	16.4
G	18.9	13.3	3.3	1.0	43.9	13.1
H	25.0	10.2	4.6	1.0	67.3	12.4

10.41 The article "Action of Drugs on Movements of the Rat during Swimming" (*J. of Human Movement Studies* (1984): 225–30) described the effects of the drug ephedrine. Rats were placed in a swimming apparatus where swimming movement triggered rotation of an exercise wheel. The number of revolutions during a fixed time interval was recorded both before and after administration of a dose of 5 mg of ephedrine per kilogram of body weight. The resulting data is as follows.

Before	15	30	3	16	11
After	6	5	3	6	2

Does the data suggest that ephedrine reduces the true mean number of revolutions? Test using a .05 significance level.

10.42 A large amount of alcohol is known to reduce reaction time. To investigate the effects of small amounts of alcohol, reaction time was recorded for seven individuals before and after 2 oz of 90-proof alcohol were consumed by each. Does the following data suggest that 2 oz of alcohol reduces mean reaction time? Use a significance level of .05.

Reaction Time (sec)

Subject	1	2	3	4	5	6	7
Before	.6	.8	.4	.7	.8	.9	.7
After	.7	.8	.6	.8	.8	.8	.9

10.43 Dentists make many people nervous (even more so than statisticians!). To see if such nervousness elevates blood pressure, the blood pressure and pulse rates of 60 subjects were measured in a dental setting and in a medical setting ("The Effect of the Dental Setting on Blood Pressure Measurement" *Am. J. of Public Health* (1983): 1210–14). For each subject, the difference (dental-setting blood pressure minus medical-setting blood pressure) was formed. The analogous differences were also formed for pulse rates. Summary data follows.

	Mean Difference	Standard Deviation of Differences
Systolic blood pressure	4.47	8.77
Pulse (beats/min)	−1.33	8.84

a. Does the data strongly suggest that true mean blood pressure is higher in a dental setting than in a medical setting? Use a level .01 test.

b. Is there sufficient evidence to indicate that true mean pulse rate in a dental setting differs from the true mean pulse rate in a medical setting? Use a significance level of .05.

10.44 Samples of both surface soil and subsoil were taken from eight randomly selected agricultural locations in a particular county. The soil samples were analyzed to determine both pH and subsoil pH, with the following results.

Location	1	2	3	4	5	6	7	8
Surface pH	6.55	5.98	5.59	6.17	5.92	6.18	6.43	5.68
Subsoil pH	6.78	6.14	5.80	5.91	6.10	6.01	6.18	5.88

a. Compute a 90% confidence interval for the true average difference between surface and subsoil pH for agricultural land in this county.

b. What assumptions are necessary to validate the interval in part (a)?

10.45 The paper "Cardiac Output in Preadolescent Competitive Swimmers and in Untrained Normal Children" (*J. of Sports Med.* (1983): 291–9) reported the results of an experiment designed to assess the effect of athletic training on cardiac output. Sixteen children participated in the study. Eight of the subjects were trained competitive swimmers. The other eight children were normal healthy untrained children selected from a large group of volunteers. An untrained subject was chosen to match each trained subject with respect to age, height, weight, and body surface area, giving eight matched pairs of subjects. Resting heart rate (beats/min) and cardiac output (l/min) were measured for each child, resulting in the accompanying data.

	Heart Rate		Cardiac Output	
Pair	Trained	Untrained	Trained	Untrained
1	90	95	3.2	2.9
2	85	75	5.9	5.4
3	75	80	4.2	3.4
4	120	65	7.4	2.8
5	95	82	5.5	4.3
6	105	80	4.5	4.8
7	85	100	4.3	4.3
8	75	85	5.3	4.9

a. Is there sufficient evidence to indicate a difference between trained and untrained children with respect to mean resting heart rate? Use a .05 significance level.

b. Does the data suggest that mean resting cardiac output differs for trained and untrained children? Test the appropriate hypotheses using a .01 significance level.

c. Explain why the researchers used paired samples rather than independent samples.

10.46 Many people who quit smoking complain of weight gain. The results of an investigation into the relationship between smoking cessation and weight gain are given in the paper "Does Smoking Cessation Lead to Weight Gain?" (*Amer. J. of Public Health* (1983): 1303–5). Three hundred twenty-two subjects who successfully participated in a program to quit smoking were weighed at the beginning of the program and again 1 year later. The mean change in weight was 5.15 lb and the standard deviation of the weight changes was 11.45 lb. Is there sufficient evidence to conclude that the true mean change in weight is positive? Use $\alpha = .05$.

10.47 A famous paper on the effects of marijuana smoking ("Clinical and Psychological Effects of Marijuana in Man" *Science* (1968): 1234–41) described the results of an experiment in which the change in heartrate was measured for nine subjects who had never used marijuana before. Measurements were taken both 15 min after smoking at a low-dose level and 15 min after smoking a placebo (untreated) cigarette.

Subject	1	2	3	4	5	6	7	8	9
Placebo	16	12	8	20	8	10	4	−8	8
Low dose	20	24	8	8	4	20	28	20	20

a. Does the data suggest that marijuana smoking leads to a greater increase in heartrate than does smoking a placebo cigarette? Test using $\alpha = .01$.

b. Give as much information as you can about the *P*-value associated with the test in part (a).

10.4 LARGE-SAMPLE INFERENCES CONCERNING A DIFFERENCE BETWEEN TWO POPULATION PROPORTIONS

Large-sample methods for estimating and testing hypotheses about a single population proportion, π, were presented in Chapters 8 and 9. The symbol π was used to represent the true proportion of individuals in the population who possess some characteristic (the "successes"). Inferences about the value of π were based on p, the corresponding sample proportion of successes.

Many investigations are carried out to compare the proportion of successes in one population (or resulting from one treatment) to the proportion of successes in a second population (or from a second treatment). As was the case for means, the subscripts 1 and 2 are used to distinguish between the two population proportions, samples sizes, and sample proportions.

NOTATION

Population 1: Proportion of "successes" = π_1

Population 2: Proportion of "successes" = π_2

	SAMPLE SIZE	PROPORTION OF SUCCESSES
Sample from population 1	n_1	p_1
Sample from population 2	n_2	p_2

When comparing two populations on the basis of "success" proportions, it is natural to focus on the quantity $\pi_1 - \pi_2$, the difference between the two population proportions. Since p_1 provides an estimate of π_1 and p_2 provides an estimate of π_2, the obvious choice for an estimate of $\pi_1 - \pi_2$ is $p_1 - p_2$.

Because p_1 and p_2 each vary in value from sample to sample, so will the difference $p_1 - p_2$. For example, a first sample from each of two populations might yield $p_1 = .74$, $p_2 = .63$, and $p_1 - p_2 = .74 - .63 = .11$. A second sample from each might result in $p_1 = .79$, $p_2 = .67$, and $p_1 - p_2 = .79 - .67 = .12$; and so on. Since the statistic $p_1 - p_2$ will be the basis for drawing inferences about $\pi_1 - \pi_2$, we need to know something about its behavior.

PROPERTIES OF THE SAMPLING DISTRIBUTION OF $p_1 - p_2$

If the two samples are selected independently of one another, the following properties hold:

1. $\mu_{p_1 - p_2} = \pi_1 - \pi_2$

   This says that the sampling distribution of $p_1 - p_2$ is centered at $\pi_1 - \pi_2$, so $p_1 - p_2$ is an unbiased statistic for estimating $\pi_1 - \pi_2$.

2. $\sigma^2_{p_1 - p_2} = \sigma^2_{p_1} + \sigma^2_{p_2} = \dfrac{\pi_1(1 - \pi_1)}{n_1} + \dfrac{\pi_2(1 - \pi_2)}{n_2}$

   and

   $\sigma_{p_1 - p_2} = \sqrt{\dfrac{\pi_1(1 - \pi_1)}{n_1} + \dfrac{\pi_2(1 - \pi_2)}{n_2}}$

3. If both n_1 and n_2 are large [$n_1\pi_1 \geq 5$, $n_1(1 - \pi_1) \geq 5$, $n_2\pi_2 \geq 5$, and $n_2(1 - \pi_2) \geq 5$], then p_1 and p_2 each have approximately normal sampling distributions, so their difference, $p_1 - p_2$, also has approximately a normal sampling distribution.

The properties in the box imply that, when the samples are independently selected and both sample sizes are large, the distribution of the standardized variable

$$z = \frac{p_1 - p_2 - (\pi_1 - \pi_2)}{\sqrt{\dfrac{\pi_1(1 - \pi_1)}{n_1} + \dfrac{\pi_2(1 - \pi_2)}{n_2}}}$$

is described approximately by the standard normal (z) curve.

A LARGE-SAMPLE TEST PROCEDURE

Comparisons of π_1 and π_2 are often based on large, independently selected samples, and we restrict ourselves to this case. The most general null hypothesis of interest has the form

$$H_0: \pi_1 - \pi_2 = \text{hypothesized value}$$

However, when the hypothesized value is something other than zero, the appropriate test statistic differs somewhat from the test statistic used for $H_0: \pi_1 - \pi_2 = 0$. Since this latter H_0 is almost always the relevant one in applied problems, we will focus exclusively on it.

Our basic testing principle has been to use a procedure that controls the probability of a type I error at the desired level α. This requires using a test statistic with a sampling distribution that is known when H_0 is true. That is, the test statistic should be developed under the assumption that $\pi_1 = \pi_2$ (as specified by the null hypothesis $\pi_1 - \pi_2 = 0$). In this case, π can be used to denote the common value of the two population proportions. The z variable obtained by standardizing $p_1 - p_2$ then simplifies to

$$z = \frac{p_1 - p_2}{\sqrt{\dfrac{\pi(1 - \pi)}{n_1} + \dfrac{\pi(1 - \pi)}{n_2}}}$$

Unfortunately, this cannot serve as a test statistic because the denominator cannot be computed: H_0 says that there is a common value π but it does not specify what that value is. A test statistic can be obtained, though, by first estimating π from the sample data and then using this estimate in the denominator of z.

When $\pi_1 = \pi_2$, either p_1 or p_2 separately gives an estimate of the common proportion π. A better estimate than either of these is a weighted average of the two, in which more weight is given to the sample proportion based on the larger sample.

▌▌▌ DEFINITION

The **combined estimate of the common population proportion** is

$$p_c = \left(\frac{n_1}{n_1 + n_2}\right)p_1 + \left(\frac{n_2}{n_1 + n_2}\right)p_2$$

The test statistic for testing $H_0: \pi_1 - \pi_2 = 0$ results from using p_c in place of π in the standardized variable z just given. This z statistic has approximately a standard normal distribution when H_0 is true, so a test that has the desired level of significance α uses the appropriate z critical value to specify the rejection region.

SUMMARY OF LARGE-SAMPLE z TESTS FOR $\pi_1 - \pi_2$

Null hypothesis: $H_0: \pi_1 - \pi_2 = 0$

Test statistic:
$$z = \frac{p_1 - p_2}{\sqrt{\dfrac{p_c(1 - p_c)}{n_1} + \dfrac{p_c(1 - p_c)}{n_2}}}$$

ALTERNATIVE HYPOTHESIS	REJECTION REGION
$H_a: \pi_1 - \pi_2 > 0$	$z > z$ critical value
$H_a: \pi_1 - \pi_2 < 0$	$z \ll z$ critical value
$H_a: \pi_1 - \pi_2 \neq 0$	Either $z > z$ critical value or $z < -z$ critical value

The z critical values corresponding to the most frequently used significance levels appear in the bottom row of Table III.

This test requires large samples and should be used only when all the quantities n_1p_1, $n_1(1 - p_1)$, n_2p_2, and $n_2(1 - p_2)$ are at least 5.

EXAMPLE 11

Many investigators have studied the effect of the wording of questions on survey responses. Consider the following two versions of a question concerning gun control:

1. Would you favor or oppose a law that would require a person to obtain a police permit before purchasing a gun?
2. Would you favor or oppose a law that would require a person to obtain a police permit before purchasing a gun, or do you think that such a law would interfere too much with the right of citizens to own guns?

The extra phrase in Question 2 reminding individuals of the right to bear arms might tend to elicit a smaller proportion of favorable responses than would the first question without the phrase. Does the data suggest that this is the case? We will test the relevant hypotheses using $\alpha = .01$.

Let π_1 denote the proportion of all adults who would respond *favor* when asked Question 1, and define π_2 similarly for Question 2. The paper "Attitude Measurement and the Gun Control Paradox" (*Public Opinion Quarterly* (1977–78): 427–38) reported the accompanying sample data.

Sample size	$n_1 = 615$	$n_2 = 585$
Number who favor	463	403
Sample proportion	$p_1 = \frac{463}{615} = .753$	$p_2 = \frac{403}{585} = .689$

Suppose that $\pi_1 = \pi_2$; let π denote the common value. Then the combined estimate of π (since $n_1 + n_2 = 1200$) is

$$p_c = \frac{615}{1200}(.753) + \frac{585}{1200}(.689) = .722$$

The seven-step procedure can now be used to perform the hypothesis test.
1. $\pi_1 - \pi_2$ is the difference between the true proportions of favorable responses to Questions 1 and 2.
2. $H_0: \pi_1 - \pi_2 = 0 \quad (\pi_1 = \pi_2)$
3. $H_a: \pi_1 - \pi_2 > 0 \quad (\pi_1 > \pi_2)$, in which case the extra phrase *does* result in proportionately fewer favorable responses.
4. Test statistic:

$$z = \frac{p_1 - p_2}{\sqrt{\dfrac{p_c(1 - p_c)}{n_1} + \dfrac{p_c(1 - p_c)}{n_2}}}$$

5. The appropriate test is upper-tailed, and for $\alpha = .01$, Table III gives the z critical value 2.33. In other words, H_0 will be rejected in favor of H_a if $z > 2.33$.
6. $n_1 = 615$, $n_2 = 585$, $p_1 = .753$, $p_2 = .689$, and $p_c = .722$, so

$$z = \frac{.753 - .689}{\sqrt{\dfrac{(.722)(.278)}{615} + \dfrac{(.722)(.278)}{585}}} = \frac{.064}{.0259} = 2.47$$

7. Since $2.47 > 2.33$, the hypothesis H_0 is rejected at level .01. Inclusion of the extra phrase about the right to bear arms *does* seem to result in fewer favorable responses than would be elicited without the phrase.

Because the large-sample test procedure here is a z test, we can compute a P-value in exactly the same way as we did for other z tests. For an upper-tailed test, the P-value is the area under the z curve to the right of the computed z. In the last example, $z = 2.47$ and the 2.47 entry of Table II is .9932, so

$$P\text{-value} = 1 - .9932 = .0068.$$

Since $.0068 < .01$, the hypothesis H_0 would be rejected at significance level .01 but would not be rejected at level .005 or .001.

EXAMPLE 12

Some defendants in criminal proceedings plead guilty and are sentenced without a trial, whereas others plead innocent, are subsequently found guilty, and then are sentenced. In recent years legal scholars have speculated as to whether the sentences of those who plead guilty differ in severity from the sentences of those who plead innocent and are subsequently judged guilty. Consider the accompanying data on a group of defendants from San Francisco County, all of whom were accused of robbery and had previous prison records ("Does It Pay to Plead Guilty? Differential

Sentencing and the Functioning of Criminal Courts" *Law and Society Rev.* (1981–2): 45–69).

	Plea	
	Guilty	Not Guilty
Number judged guilty	$n_1 = 191$	$n_2 = 64$
Number sentenced to prison	101	56
Sample proportion	$p_1 = .529$	$p_2 = .875$

Does this data suggest that the proportion of all defendants in these circumstances who plead guilty and are sent to prison differs from the proportion who are sent to prison after pleading innocent and being found guilty?

Let π_1 and π_2 denote the two population proportions. The hypotheses of interest are

$$H_0: \pi_1 - \pi_2 = 0 \text{ versus } H_a: \pi_1 - \pi_2 \neq 0.$$

A significance level of .01 will be used.

1. $\pi_1 - \pi_2$ is the difference between the true proportions sent to prison for those pleading guilty and for those pleading not guilty.
2. $H_0: \pi_1 - \pi_2 = 0$
3. $H_a: \pi_1 - \pi_2 \neq 0$
4. Test statistic:

$$z = \frac{p_1 - p_2}{\sqrt{\dfrac{p_c(1 - p_c)}{n_1} + \dfrac{p_c(1 - p_c)}{n_2}}}$$

5. Since this is a two-tailed test, with $\alpha = .01$, we will reject H_0 if $z > 2.58$ or $z < -2.58$.
6. The combined estimate of the common population proportion is

$$p_c = \frac{191}{255}(.529) + \frac{64}{255}(.875) = .616$$

The computed value of the test statistic is then

$$z = \frac{.529 - .875}{\sqrt{\dfrac{(.616)(.384)}{191} + \dfrac{(.616)(.384)}{64}}} = \frac{-.346}{.070} = -4.94$$

7. Since $-4.94 < -2.58$, the hypothesis H_0 is rejected at significance level .01. The data suggests that $\pi_1 \neq \pi_2$ and, in particular, that initially pleading guilty may be a good strategy as far as avoiding prison is concerned.

The cited article also reported data on defendants in several other counties. The authors broke down the data by type of crime (burglary or robbery) and by nature of prior record (none, some record but no prison, and prison). In every case the conclusion was the same: among defendants judged guilty, those who pleaded guilty were less likely to receive prison sentences.

A CONFIDENCE INTERVAL

A large-sample confidence interval for $\pi_1 - \pi_2$ is a special case of the general z interval formula

point estimate $\pm$ (z critical value)(estimated standard deviation)

The statistic $p_1 - p_2$ gives a point estimate of $\pi_1 - \pi_2$, and the standard deviation of this statistic is

$$\sigma_{p_1-p_2} = \sqrt{\frac{\pi_1(1-\pi_1)}{n_1} + \frac{\pi_2(1-\pi_2)}{n_2}}$$

An estimated standard deviation is obtained by using the sample proportions p_1 and p_2 in place of π_1 and π_2, respectively, under the square root symbol. Notice that this estimated standard deviation differs from the one used earlier in the test statistic. Here there isn't a null hypothesis that claims $\pi_1 = \pi_2$, so there is no common value of π to estimate.

> A large-sample confidence interval for $\pi_1 - \pi_2$ is given by
>
> $$(p_1 - p_2) \pm (z \text{ critical value}) \sqrt{\frac{p_1(1-p_1)}{n_1} + \frac{p_2(1-p_2)}{n_2}}$$
>
> The interval is valid whenever $n_1 p_1$, $n_1(1-p_1)$, $n_2 p_2$, and $n_2(1-p_2)$ are at least 5.

EXAMPLE 13

A person released from prison before completing the original sentence is placed under the supervision of a parole board. If that person violates specified conditions of good behavior during the parole period, the board can order a return to prison. To what extent is the frequency of parole violations related to the type of crime and various other factors? The paper "Impulsive and Premeditated Homicide: An Analysis of the Subsequent Parole Risk of the Murderer" (*J. of Criminal Law and Criminology* (1978): 108–14) reported the accompanying data on parole behavior. One sample of individuals had served time in prison for impulsive murder and the other sample had served time for premeditated murder.

	Crime	
	Impulsive	Premeditated
Sample size	$n_1 = 42$	$n_2 = 40$
Number with no violation	13	22
Sample proportion	$p_1 = .310$	$p_2 = .550$

Let π_1 denote the proportion of all impulsive murderers who successfully complete parole, and define π_2 analogously for premeditated murderers. The sample sizes are large enough for the large-sample interval to be valid ($n_1 p_1 = 42(.310) \approx 13 \geq 5$, $n_1(1 - p_1) = 42(.690) \approx 29 \geq 5$, etc.). A 98% confidence interval for $\pi_1 - \pi_2$ uses the z critical value 2.33. The resulting interval is

$$(.310 - .550) \pm (2.33)\sqrt{\frac{(.310)(.690)}{42} + \frac{(.550)(.450)}{40}}$$

$$= -.240 \pm (2.33)(.106)$$

$$= -.240 \pm .247$$

$$= (-.487, .007)$$

This interval includes zero, so (with a 98% level of confidence) one of the plausible values of $\pi_1 - \pi_2$ is 0. This implies that there may be no difference in the proportion who successfully complete parole for impulsive and premeditated murderers. The interval is quite wide because of the relatively small sample sizes.

EXERCISES   10.48–10.59   **SECTION 10.4**

10.48   Let π_1 and π_2 denote the proportions of all male and all female shoppers, respectively, who buy only name-brand grocery products (as opposed to generic or store brands).
   a. Test $H_0: \pi_1 - \pi_2 = 0$ versus $H_a: \pi_1 - \pi_2 \neq 0$ at level .05, using the following data: $n_1 = 200$, number of successes (only name-brand purchases) = 87, $n_2 = 300$, number of successes = 96.
   b. Use the data of part (a) to compute a 95% confidence interval for $\pi_1 - \pi_2$.

10.49   Using an electronic process called time compression, a 30-sec television commercial can be broadcast in its entirety in only 24 sec. There is no shift in voice pitch and subjects are not aware that commercials have been altered. The article "Reducing the Costs of TV Commercials by Use of Time Compressions" (*J. of Marketing Research* (1980): 52–7) reported on a study involving recall ability for subjects watching compressed as compared to noncompressed commercials. For one commercial, 15 of the 57 subjects viewing the normal version could subsequently recall the commercial, while 32 of the 74 subjects viewing the compressed version could subsequently recall it. Does this data suggest any difference between true recall proportions for the two versions?

   a. Verify that the sample sizes are large enough to justify using the large-sample procedures.
   b. Carry out a test at level .05 to answer the question posed.
   c. Compute the *P*-value. Based on this value, what would you conclude at level .10?

10.50   The paper "Softball Sliding Injuries" (*Amer. J. of Diseases of Children* (1988): 715–16) provided a comparison of break-away bases (designed to reduce injuries) and stationary bases. Consider the following data (which agrees with summary values given in the paper).

	Number of Games Played	Number of Games Where a Player Suffered a Sliding Injury
Stationary bases	1250	90
Break-away bases	1250	20

Does the use of break-away bases reduce the proportion of games with a player suffering a sliding injury? Answer by performing a level .01 test.

10.51 An experiment to determine the effects of temperature on the survival of insect eggs was described in the paper "Development Rates and a Temperature-Dependent Model of Pales Weevil" (*Environ. Entomology* (1987): 956–62). At 11°C, 73 of 91 eggs survived to the next stage of development. At 30°C, 102 of 110 eggs survived. Do the results of this experiment suggest that the survival rate (proportion surviving) differs for the two temperatures? Test the relevant hypotheses using $\alpha = .05$.

10.52 Ionizing radiation is being given increasing attention as a method for preserving horticultural products. The paper "The influence of Gamma-Irradiation on the Storage Life of Red Variety Garlic" (*J. of Food Processing and Preservation* (1983): 179–83) reported that 153 of 180 irradiated garlic bulbs were marketable (no external sprouting, rotting, or softening) 240 days after treatment, while only 119 of 180 untreated bulbs were marketable after this length of time. Does this data suggest that the true proportion of marketable irradiated bulbs exceeds that for untreated bulbs? Test the relevant hypotheses at level .01.

10.53 The positive effect of water fluoridation on dental health is well documented. One study that validates this is described in the paper "Impact of Water Fluoridation on Children's Dental Health: A Controlled Study of Two Pennsylvania Communities" (*Amer. Stat. Assoc. Proc. of the Social Statistics Section* (1981): 262–5). Two communitites were compared. One had adopted fluoridation in 1966, while the other had no fluoridation program. Of 143 children from the town without fluoridated water, 106 had decayed teeth, while 67 of 119 children from the town with fluoridated water had decayed teeth. Let π_1 denote the true proportion of children drinking fluoridated water who have decayed teeth and let π_2 denote the analogous proportion for children drinking unfluoridated water. Estimate $\pi_1 - \pi_2$ using a 90% confidence interval. Does the interval contain 0? Interpret the interval.

10.54 What psychological factors contribute to the success of competitive athletes? Numerous possibilities are examined in the paper "Elite Divers and Wrestlers: A Comparison between Open- and Closed-Skill Athletes" (*J. of Sport Psych.* (1983): 390–409). Competitive divers participating in qualifying trials were asked whether they exercised within 1 hr of a competition. Suppose that of 20 qualifying divers, 7 exercised within 1 hr of a competition, while 12 of 25 nonqualifying divers exercised within 1 hr of a meet. Is there sufficient evidence to indicate that the true proportion of qualifying divers who exercise within 1 hr of competition differs from the corresponding proportion for nonqualifying divers? Use a level .10 test.

10.55 As part of a class project, two college students from Florida found that people are willing to help strangers in quite surprising circumstances (Associated Press, May 10, 1984). The two students splashed their faces and hands and rinsed their mouths with gin. They told passersby that they were too drunk to unlock their car doors and asked for help. One student was dressed in a business suit, whereas the other wore a dirty T-shirt. Of 50 people approached by the student in the suit, 21 helped unlock the car door and aided him in getting into the car! The student in the T-shirt also approached 50 people, and was assisted by 23 of them! Does the data suggest that the true proportion of people that would assist a well-dressed drunk into a car differs significantly from the true proportion who would assist a drunk in dirty clothes? Use a level .05 test.

10.56 The Associated Press (*Corvallis Gazette Times*, May 12, 1986) reported that "infants whose mothers use cocaine during pregnancy are far more likely than other babies to fall victim to sudden infant death syndrome." This statement was based on the data in the accompanying table. Does the data support the statement made? Test the relevant hypotheses using a .01 significance level.

	n	Number Whose Infants Died
Mothers using cocaine	60	10
Mothers with no cocaine use	1600	5

10.57 How common is driving under the influence of alcohol among teenagers? This question was examined in the paper "Predictors of Driving While Intoxicated Among Teenagers" (*J. of Drug Issues* (1988): 367–84). The accompanying data is compatible with summary statistics in the paper. Is there sufficient evidence to conclude that the proportion of girls who have driven while intoxicated in the past year is smaller than the corresponding proportion of boys? Use a .01 significance level.

	Number Surveyed	Number Who Have Driven While Intoxicated
Boys	100	28
Girls	100	17

10.58 The *San Francisco Chronicle* (May 27, 1983) reported the results of a poll designed to assess public opinion on legalized gambling. Of 750 Californians interviewed, 578 favored a state lottery. A similar survey in 1971 found that of 750 people contacted, 518 favored a state lottery.
  a. Does the data strongly suggest that the true 1983 proportion of Californians who favored a state lottery exceeds the corresponding 1971 proportion? Use a level .01 test.
  b. What *P*-value is associated with the value of the test statistic in part (a)?

10.59 The article "New Stance Taken on Blood Cholesterol" (*Los Angeles Times*, Feb. 12, 1984) states: "Ten thousand patients have been treated surgically and 10,000 comparable patients have been treated medically. Five years later, 91% of the surgical patients are alive and 90% of the medical patients are alive. No difference."
  a. Use the preceding information to test $H_0: \pi_1 - \pi_2 = 0$, where π_1 is the true proportion of patients who survive 5 years when treated surgically and π_2 is defined analogously for those treated medically.
  b. Does the result of your hypothesis test in part (a) agree with the statement of "no difference" that appeared in the article? Do you think this is a case of statistical rather than practical significance? Explain.

SUMMARY OF KEY CONCEPTS AND FORMULAS

TERM OR FORMULA	PAGE	COMMENT
Independent samples	305	Two samples where the individuals or objects in the first sample are selected independently from those in the second sample
Paired samples	306	Two samples for which each observation in the first sample is paired in a meaningful way with a particular observation in the second sample
$z = \dfrac{(\bar{x}_1 - \bar{x}_2) - \text{hyp. value}}{\sqrt{\dfrac{s_1^2}{n_1} + \dfrac{s_2^2}{n_2}}}$	308	The test statistic for testing $H_0: \mu_1 - \mu_2 = $ hypothesized value when both sample sizes are large
Pooled estimate of σ^2: $s_p^2 = \left(\dfrac{n_1 - 1}{n_1 + n_2 - 2}\right)s_1^2 + \left(\dfrac{n_2 - 1}{n_1 + n_2 - 2}\right)s_2^2$	320	s_p^2 is the statistic for estimating the common variance σ^2 when $\sigma_1^2 = \sigma_2^2 = \sigma^2$, a basic assumption required for use of the pooled t test and confidence interval.
$t = \dfrac{(\bar{x}_1 - \bar{x}_2) - \text{hyp. value}}{\sqrt{s_p^2\left(\dfrac{1}{n_1} + \dfrac{1}{n_2}\right)}}$	321	The test statistic for testing $H_0: \mu_1 - \mu_2 = $ hypothesized value when it is reasonable to assume that both population distributions are normal and $\sigma_1^2 = \sigma_2^2$. It is used when at least one sample size is small.
$t = \dfrac{\bar{x}_d - \text{hyp. value}}{s_d/\sqrt{n}}$	333	The test statistic for testing $H_0: \mu_d = $ hypothesized value when samples are paired and it is reasonable to assume that the difference distribution is normal.
Combined estimate of the common population proportion: $p_c = \left(\dfrac{n_1}{n_1 + n_2}\right)p_1 + \left(\dfrac{n_2}{n_1 + n_2}\right)p_2$	343	p_c is the statistic for estimating the common population proportion when $\pi_1 = \pi_2$.
$z = \dfrac{p_1 - p_2}{\sqrt{\dfrac{p_c(1 - p_c)}{n_1} + \dfrac{p_c(1 - p_c)}{n_2}}}$	344	The test statistic for testing $H_0: \pi_1 - \pi_2 = 0$ when both sample sizes are large.
$(\bar{x}_1 - \bar{x}_2) \pm (z \text{ crit. value})\sqrt{\dfrac{s_1^2}{n_1} + \dfrac{s_2^2}{n_2}}$	312	A formula for constructing a confidence interval for $\mu_1 - \mu_2$ when both sample sizes are large.
$(\bar{x}_1 - \bar{x}_2) \pm (t \text{ crit. value})\sqrt{s_p^2\left(\dfrac{1}{n_1} + \dfrac{1}{n_2}\right)}$	324	A formula for constructing a confidence interval for $\mu_1 - \mu_2$ when it is reasonable to assume that the population distributions are normal and $\sigma_1^2 = \sigma_2^2$. It is used when at least one sample size is small.
$\bar{x}_d \pm (t \text{ crit. value}) \cdot s_d/\sqrt{n}$	336	A formula for constructing a confidence interval for μ_d when samples are paired and it is reasonable to assume that the difference distribution is normal.
$(p_1 - p_2) \pm (z \text{ crit. value})\sqrt{\dfrac{p_1(1 - p_1)}{n_1} + \dfrac{p_2(1 - p_2)}{n_2}}$	347	A formula for constructing a confidence interval for $\pi_1 - \pi_2$ when both sample sizes are large.

10.60 Meteorologists classify storms as either single-peak or multiple-peak. The total number of lightning flashes was recorded for seven single-peak and four multiple-peak storms, resulting in the given data ("Lightning Phenomenology in the Tampa Bay Area" *J. of Geophys. Research* (1984): 11,789–805).

Single-peak	117	56	19	40	82	69	80
Multiple-peak	229	197	242	430			

a. Does the data suggest that the true mean number of lightning flashes differs for the two types of storms? Use a .05 significance level.
b. What assumptions about the distribution of number of flashes for each of the two types of storms are necessary in order that your test in part (a) be valid?

10.61 Nine observations of surface-soil pH were made at each of two different locations at the Central Soil Salinity Research Institute experimental farm, and the resulting data appeared in the article "Sodium–Calcium Exchange Equilibria in Soils as Affected by Calcium Carbonate and Organic Matter" (*Soil Sci.* (1984): 109). Does the data suggest that the true mean soil pH values differ for the two locations? Test the appropriate hypotheses using a .05 significance level. Be sure to state any assumptions necessary for the validity of your test.

Site	pH								
Location A	8.53	8.52	8.01	7.99	7.93	7.89	7.85	7.82	7.80
Location B	7.85	7.73	7.58	7.40	7.35	7.30	7.27	7.27	7.23

10.62 The accompanying 1982 and 1983 net earnings (in millions of dollars) of ten food and beverage firms appeared in the article "Capital Expenditures Report" (*Food Engr.* (1984): 93–101). Is there sufficient evidence to indicate that mean net earnings increased from 1982 to 1983? Assume that the ten companies represent a random sample of all food and beverage firms. Perform the appropriate hypothesis test using a .01 significance level.

Firm	1983	1982	Firm	1983	1982
Coors	89.0	33.0	Nestle	113.4	93.7
ConAgra	28.7	20.6	Beatrice	292.1	320.8
ADM	110.2	155.0	Carnation	155.0	137.6
Heinz	237.5	214.3	Hershey	100.2	94.2
General Mills	130.7	134.1	Procter & Gamble	58.5	66.5

10.63 The results of a study on job satisfaction among tenure-track faculty members and librarians employed by the California State University System were described in the paper "Job Satisfaction among Faculty and Librarians: A Study of Gender, Autonomy, and Decision-Making Opportunities" (*J. of Library Admin.* (1984): 43–56). Random samples of 115 male and 105 female academic employees were selected. Each participant completed the Minnesota Satisfaction Questionnaire (MSQ) and was assigned a satisfaction score. The resulting mean and standard deviation were 75.43 and 10.53 for the males and 72.54 and 13.08 for the females. Does the data strongly suggest that male and female academic employees differ with respect to mean score on the MSQ? Use a .01 significance level.

10.64 The paper "An Evaluation of Football Helmets under Impact Conditions" (*Amer. J. of Sports Med.* (1984): 233–7) reported that when 44 padded football helmets and 37 suspension-type helmets were subjected to an impact test (a drop of 1.5 m onto a hard surface), 5 of the padded and 24 of the suspension-type helmets showed damage. Using a .01 significance level, test appropriate hypotheses to determine if there is a difference between the two helmet types with respect to the true proportion of each type that would be damaged by a 1.5-m drop onto a hard surface.

10.65 The paper "Post-Mortem Analysis of Neuropeptides in Brains from Sudden Infant Death Victims" (*Brain Research* (1984): 279–85) reported age (in days) at death for infants who died of sudden infant death syndrome (SIDS). Assuming that age at death for SIDS victims is normally distributed, use the given data to construct a 95% confidence interval for the difference in the true mean age at death for female and male SIDS victims. Interpret the resulting interval. How does the interpretation depend on whether zero is included in the interval?

Age at Death (days)						
Females	55	120	135	154	54	
Males	56	60	60	60	105	140	147

10.66 The discharge of industrial wastewater into rivers affects water quality. To assess the effect of a particular power plant on water quality, 24 water specimens were taken 16 km upstream and 4 km downstream of the plant. Alkalinity (mg/l) was determined for each specimen, resulting in the given summary quantities. Does the data suggest that the true mean alkalinity is higher downstream than upstream? Use a .05 significance level.

Location	n	Mean	Standard Deviation
Upstream	24	75.9	1.83
Downstream	24	183.6	1.70

10.67 The paper "Chronic 60-Hz Electric Field Exposure Induced Subtle Bioeffects on Serum Chemistry" (*J. of Environ. Sci. and Health* (1984): 865–85) described an experiment to assess the effects of exposure to a high-intensity electric field. A group of 45 rats exposed to an electric field from birth to 120 days of age was compared to a control group of 45 rats with no exposure. Summary quantities for various blood characteristics are given. Conduct the hypothesis tests necessary to determine whether the experimental treatment differs from no treatment with respect to true mean glucose, potassium, protein, or cholesterol levels. Use a .05 significance level for each test.

	Control		Experimental	
	Mean	s	Mean	s
Glucose (mg/dl)	136.30	12.70	139.20	16.10
Potassium (mg/dl)	7.44	.53	7.62	.54
Total protein (gm/dl)	6.63	.27	6.61	.34
Cholesterol (mg/dl)	69.00	11.40	67.80	9.38

10.68 The paper "Dyslexic and Normal Readers' Eye Movements" (*J. of Exper. Psych.* (1983): 816–25) reported data on the number of eye movements while reading a particular passage for 34 dyslexic and 36 normal readers. The sample mean number of total movements and corresponding sample standard deviation were 8.6 and .30 for dyslexics and 9.2 and .16 for normal readers.

  a. Does the data indicate a significant difference between dyslexic and normal readers with respect to true average number of eye movements? Use a level .10 test.

  b. Give the *P*-value associated with the test statistic in part (a). Would your conclusions have been any different at significance levels .05 or .01?

10.69 The paper "The Effects of Education on Self-Esteem of Male Prison Inmates" (*J. Correlational Educ.* (1982): 12–18) described the result of an experiment designed to ascertain whether mathematics education increases the self-esteem of prison inmates. Two random samples each of size 40 were selected from the population of prison inmates at Angola, Louisiana. One sample was designated as a control group and the other as an experimental group. Inmates in the experimental group received 18 weeks of mathematics tutoring, whereas those in the control group were not tutored. Both groups were given the Self-Esteem Inventory (SEI) at the beginning and end of the 18-week period. The mean and standard deviation of the change in SEI score were 2.9 and 5.4 for the experimental group and −1.3 and 5.6 for the control group.

  a. Does the data provide sufficient evidence to conclude that mathematics tutoring results in a higher mean change in SEI score? Test the relevant hypotheses using a .01 significance level.

  b. What is the *P*-value associated with the test in part (a)?

10.70 Many researchers have investigated the relationship between stress and reproductive efficiency. One such study is described in the paper "Stress or Acute Adrenocorticotrophin Treatment Suppresses LHRH-Induced LH Release in the Ram" (*J. of Reprod. and Fertility* (1984): 385–93). Seven rams were used in the study, and LH (luteinizing hormone) release (ng/min) was recorded before and after treatment with ACTH (adrenocorticotrophin, a drug that results in stimulation of the adrenal gland). Use the accompanying data and $\alpha = .01$ to determine if there is a significant reduction in mean LH release following treatment with ACTH.

			LH Release (ng/min)				
Ram	1	2	3	4	5	6	7
Before	2400	1400	1375	1325	1200	1150	850
After	2250	1425	1100	800	850	925	700

10.71 Two different methods (ampul and hot plate) for recovering metals from sewage were compared in the paper "Simple Sample Digestion of Sewage Sludge for Multi-Element Analysis" (*J. of Environ. Sci. and Health* (1984): 959–72). Both methods were used to treat oyster tissue and the metal and mineral recovery were recorded. Answer the following questions, assuming that each method was used on 10 tissue specimens. Be sure to state any assumptions that must be true in order for the inferential procedure applied to be valid.

  a. For iron, the mean recovery and standard deviation were 16 ppm and 2.5 ppm for the ampul method and 17.7 ppm and 1.2 ppm for the hot-plate method. Does the evidence suggest that the two methods differ with respect to mean iron recovery? Use a .05 significance level.

b. The sample mean and standard deviation for copper recovery were 62.6 ppm and 3.7 ppm for the ampul method and 65.0 and 3.8 for the hot-plate method. Estimate the true difference in mean copper recovery rate for the two methods using a 90% confidence interval. Does the interval include zero? Interpret the interval.

10.72 An electronic implant that stimulates the auditory nerve has been used to restore partial hearing to a number of deaf people. In a study of implant acceptability (*Los Angeles Times*, Jan. 29, 1985), 250 adults born deaf and 250 adults who went deaf after learning to speak were followed for a period of time after receiving an implant. Of those deaf from birth, 75 had removed the implant, while only 25 of those who went deaf after learning to speak had done so. Does this suggest that the true proportion who remove the implants differs for those that were born deaf and those that went deaf after learning to speak? Test the relevant hypotheses using a .01 significance level.

10.73 Is reading computer output on a terminal screen more tiring than reading output on paper? This question is addressed in the paper "Doing the Same Work with Hard Copy and with Cathode-Ray Tube [CRT] Computer Terminals" (*Human Factors* (1984): 323–37). One measure of eye fatigue was number of blinks during a 1-min interval. Twenty-four clerk-typists were hired from a temporary employment agency to do proof reading. Twelve were asked to proofread material from paper copy, and the other 12 proofread the same text on a CRT screen. After one hour, the number of blinks during a 1-min period was recorded. The average number of blinks was 6.70 for the paper group and 9.32 for the CRT group. Suppose that the corresponding sample standard deviations were 1.2 and 1.4 for the paper and CRT groups, respectively.
a. Does the data suggest that the mean number of blinks per minute differs for those working with paper and those working with the CRT? Use a .05 level of significance.
b. The number of blinks was also recorded after six hours of work. The resulting averages were 9.62 and 12.21 for the paper and CRT groups, respectively. Suppose that the corresponding sample standard deviations were 1.5 and 1.6. Does the data suggest that the mean number of blinks after 6 hr of work differs for the two groups? Use a .05 level of significance. Do you reach the same conclusion as in the test of part (a)?

REFERENCES The book by Devore cited in earlier chapters contains a slightly more general exposition of the two-sample material.

REGRESSION AND CORRELATION: DESCRIPTIVE METHODS

INTRODUCTION It often happens that two numerical variables *x* and *y* are strongly related, but the value of *y* is not completely determined by the value of *x*. As an example, let *x* denote the age of a child and *y* the child's vocabulary size. There is a strong tendency for older children to have a larger vocabulary—that is, for large *y* values to be associated with large *x* values and small *y* values with small *x* values (though there is some question as to whether this tendency persists into the teenage years!). Yet age is certainly not the sole determinant of vocabulary size. Two children who have identical *x* values (ages) may have rather different *y* values (vocabulary sizes).

A picture of sample data, called a *scatter plot*, is suggested in Section 1 as a basis for characterizing the relationship between *x* and *y*. The pattern in a plot will often indicate a relationship that can be described as approximately linear. The principle of *least squares* is used in Section 2 to obtain a line that summarizes such a relationship. The extent to which this line provides an effective summary is the subject of Section 3.

A main objective of regression analysis is to be able to predict the value of *y* associated with a specified *x* value; e.g., to predict vocabulary size for a six-year-old child. Frequently an investigator will have bivariate data and wish simply to assess how strongly *x* and *y* are related. The most widely used measure of assocation is *Pearson's correlation coefficient*, introduced in the last section of the chapter.

11.1 BIVARIATE RELATIONSHIPS AND SCATTER PLOTS

Two variables x and y are said to be **deterministically** related if the value of y is completely determined with no uncertainty by the value of x. Suppose, for example, that a rental car agency charges a flat fee of $20 plus $.25 per mile driven when it rents a certain type of car. If we define two variables by

x = distance driven (in miles)

y = rental charge (in dollars)

then there is a simple equation that relates y to x, namely,

$y = 20 + .25x$

According to this equation, when $x = 100$, then

$y = 20 + .25(100) = 20 + 25 = \$45,$

whereas the x value 250 results in

$y = 20 + .25(250) = \$82.50.$

The value of y is determined solely by x. Two different renters who have identical x values will have identical y values. Once the value of x is specified, there is no uncertainty concerning the associated y value.

Information about the relationship between two variables frequently comes from examining data. Perhaps a sample of individuals or objects is selected and both the x and y values are determined for each one. Alternatively, an investigator may specify some x values at which observations are desired (e.g., values of temperature at which she would like observations on yield from a chemical reaction) and then perform an experiment to obtain a value of y at each such x value.

NOTATION

A bivariate sample consists of n pairs (x, y). These pairs are denoted by $(x_1, y_1), (x_2, y_2), \ldots, (x_n, y_n)$, where

x_1 = the value of x for the first observation

y_1 = the value of y for the first observation

$\vdots$

x_n = the value of x for the nth (last) observation

y_n = the value of y for the nth (last) observation

Once sample data is available, the general nature of any relationship can be revealed by constructing a **scatter plot**. This is a picture in which each

(x, y) observation in the sample is represented as a point on a two-dimensional coordinate system. First a horizontal x axis and vertical y axis are drawn. These axes are then scaled so that all x (y) values in the sample are included in the range on the horizontal (vertical) axis. To locate the point corresponding to (x_1, y_1), move over to the value x_1 on the horizontal axis and then move up or down until opposite the value y_1 on the vertical axis. Points corresponding to other observations are identified in a similar manner.

EXAMPLE 1

Consider again the rental car situation introduced earlier, in which $x =$ distance driven and $y =$ rental charge. A sample of $n = 8$ renters yielded the accompanying data.

				Observation				
	1	2	3	4	5	6	7	8
x	100	40	64	250	116	120	172	66
y	45	30	36	82.5	49	50	63	36.5

The first observation is $(x_1, y_1) = (100, 45)$. The corresponding point lies above 100 on the horizontal axis and to the right of 45 on the vertical axis. This is illustrated in Figure 1(a), which also includes a point for the second observation, (40, 30). The complete scatter plot appears in Figure 1(b). All points in the plot fall exactly on a straight line. This is because x and y are deterministically related via the equation $y = 20 + .25x$, and (x, y) pairs satisfying this relationship plot as a straight line. The sampled x values are all different. However, if there had been a ninth observation with $x_9 = 100$, then of necessity $y_9 = 45$ and the resulting point would be identical to that for the first observation.

FIGURE 1

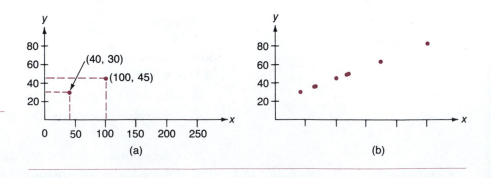

Plotting the data from Example 1.
(a) The first two points in a scatter plot
(b) The complete scatter plot

EXAMPLE 2

Landslides are common events in tree-growing regions of the Pacific Northwest, so their effect on timber growth is of special concern to foresters. The paper "Effects of Landslide Erosion on Subsequent Douglas Fir Growth and Stocking Levels in the Western Cascades, Oregon" (*Soil Science Soc.*

of Amer. J. (1984): 667–71) reported on the results of a study in which growth in a landslide area was compared with growth in a previously clear-cut area. Here we present data on clear-cut growth, with x = tree age (yr) and y = 5-year height growth (cm).

	Observation							
	1	2	3	4	5	6	7	8
x	5	9	9	10	10	11	11	12
y	70	150	260	230	255	165	225	340

	Observation							
	9	10	11	12	13	14	15	16
x	13	13	14	14	15	15	18	18
y	305	335	290	340	225	300	380	400

A scatter plot of this data appears in Figure 2. Notice that for each of the x values 9, 10, 11, 13, 14, 15, and 18, there are two y observations. In each case the two y values are different. For example, $x_2 = x_3 = 9$ whereas $y_2 = 150$ and $y_3 = 260$. We immediately conclude that x and y are not deterministically related, for if they were, repeated observations at the same x value would have to yield identical y values. Yet the plot shows evidence of a rather strong relationship—a tendency for y to increase as x does. Furthermore, the general pattern in the plot can reasonably be described as linear: the points in the plot appear to be distributed about some straight line that slopes upward.

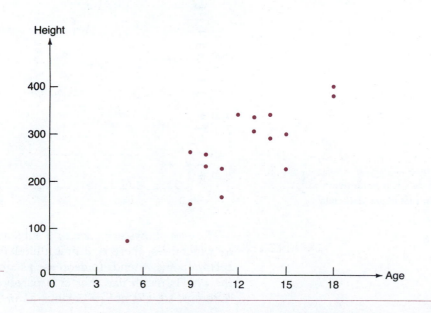

FIGURE 2

Scatter plot for the data of Example 2.

EXAMPLE 3

The focus of many agricultural experiments is to study how the yield of a crop varies with the time at which it is harvested. The accompanying data appears in the paper "Determination of Biological Maturity and Effect of Harvesting and Drying Conditions on Milling Quality of Paddy" (*J. of Agric. Engr. Research* (1975): 353–61). The variables are

x = time between flowering and harvesting (days)

y = yield of paddy, a type of grain farmed in India (kg/hectare)

Obs.	x	y	Obs.	x	y
1	16	2508	9	32	3823
2	18	2518	10	34	3646
3	20	3304	11	36	3708
4	22	3423	12	38	3333
5	24	3057	13	40	3517
6	26	3190	14	42	3214
7	28	3500	15	44	3103
8	30	3883	16	46	2776

Figure 3 displays a scatter plot obtained from the statistical computer package MINITAB. Notice that the axes do not intersect at (0, 0). This is because zero lies well outside both the range of x values (16 to 46) and the range of y values (roughly 2500 to 3900). Had the scaling been chosen so that the axes did intersect at this point, all points in the plot would have been squeezed into the upper right-hand corner, obscuring somewhat the general pattern.

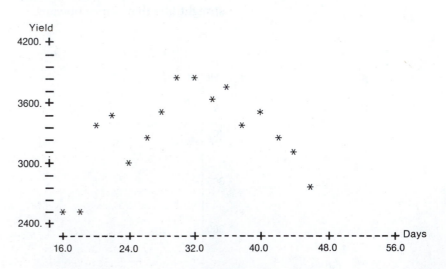

FIGURE 3

A scatter plot of the yield data.

The plot shows evidence of a substantial relationship between x and y: as x increases, there is at first a tendency for y to do likewise, but after a certain point, y tends to decrease as x continues to increase. The shape of the plot is much like that of a parabola. All x values in the sample are different, so it is not as obvious as in Example 2 that the relationship is not deterministic. However, a "simple" deterministic relationship between x and y—one described by an elementary mathematical function such as $y = 20 + 4x$ or $y = \sqrt{x}$ or $y = \sin(2x)$—would imply that all (x, y) points fall exactly on a line or a "nice" smooth curve. This is clearly not the case here.

In the remainder of this chapter we will develop methods for analyzing bivariate data when the scatter plot reveals a linear pattern. A curved pattern like that of Figure 3 necessitates a more complicated analysis.

EXERCISES 11.1–11.8   **SECTION 11.1**

11.1   Manganese (Mn) is thought to be critical to the health of newborn infants. The paper "Manganese Intake and Serum Manganese Concentration of Human Milk-Fed and Formula-Fed Infants" (*Amer. J. of Clinical Nutrition* (1984): 872–8) gave the accompanying data on Mn intake and serum Mn level for eight human milk-fed infants. Use this data to construct a scatter plot. Does the plot suggest that there is a relationship between Mn intake and serum Mn level?

Intake (μg/kg/day)	.34	.35	.39	.39	.41	.41	.49	.68
Serum Mn (μg/l)	2.8	1.9	3.3	5.6	4.2	5.6	4.2	7.9

11.2   Bicyclists are well aware that, even when a street has a bike lane, riding may pose considerable risks if the street and lane are poorly designed. The paper "Effects of Bike Lanes on Driver and Bicyclist Behavior" (*ASCE Trans. Eng. J.* (1977): 243–56) reported the accompanying data on x = available travel space (the distance between a cyclist and the roadway center line) and y = separation distance between a bike and a passing car (determined by photography).

x	12.8	12.9	12.9	13.6	14.5	14.6	15.1	17.5	19.5	20.8
y	5.5	6.2	6.3	7.0	7.8	8.3	7.1	10.0	10.8	11.0

a. Construct a scatter plot in which the horizontal and vertical axes intersect at the point $(0, 0)$.

b. Construct a scatter plot in which the axes intersect at $(12, 5)$. Does this plot seem preferable to the one in part (a) for investigating the relationship? Why or why not?

11.3   The decline of water supplies in certain areas of the U.S. has created the need for increased understanding of relationships between economic factors such as crop yield and hydrologic and soil factors. The paper "Variability of Soil Water Properties and Crop Yield in a Sloped Watershed" (*Water Resources Bull.* (1988): 281–8) gave data on grain sorghum yield (y, in g/m-row) and distance upslope (x, in m) on a sloping watershed. Selected observations appear in the accompanying table.

x	0	10	20	30	45	50	70	80	100	120	140	160	170	190
y	500	590	410	470	450	480	510	450	360	400	300	410	280	350

a. Construct a scatter plot.

b. Does there appear to be a deterministic relationship between x and y? Explain.

c. As distance upslope increases, how does crop yield tend to behave?

11.4 Many of us are all too familiar with the effects of stress on humans. In recent years researchers have focused increasing attention on how stress affects nonhuman behavior. The accompanying data on x = plasma cortisol concentration (ng cortisol/ml plasma) and y = oxygen consumption rate (mg/kg/hr) for juvenile steelhead after three 2-min disturbances was read from a graph in the paper "Metabolic Cost of Acute Physical Stress in Juvenile Steelhead" (*Trans. of Amer. Fish. Soc.* (1987): 257–63); the paper also included data for unstressed fish.

x	25	36	48	59	62	72	80	100	100	137
y	155	184	180	220	280	163	230	222	241	350

a. Is there a deterministic relationship between x and y? How can you tell from the given data?

b. Construct a scatter plot of the data.

c. Does an increase in plasma cortisol concentration appear to have an effect on oxygen consumption rate? Comment.

11.5 The metabolic effect of cross-country skiing was the subject of the research study described in the paper "Metabolic Modifications Caused by Sport Activity: Effect in Leisure-Time Cross-Country Skiers" (*J. of Sports Med.* (1983): 385–92). Subjects were participants in a 24-hr cross-country relay. Age and blood CPK concentration were recorded 12 hr into the relay. Use the given data to construct a scatter plot. Would you describe this plot as approximately linear?

Skier	1	2	3	4	5	6	7	8	9
Age (x)	33	21	19	24	25	32	36	35	36
CPK (y)	180	300	520	480	580	440	380	480	520

Skier	10	11	12	13	14	15	16	17	18
Age (x)	24	25	44	51	50	52	55	62	57
CPK (y)	1040	1360	640	260	360	400	280	300	400

11.6 The paper "Effects of Enhanced UV-B Radiation on Ribulose-1, 5-Biphosphate, Carboxylase in Pea and Soybean" (*Environ. and Exper. Botany* (1984): 131–43) included the accompanying data on distance from an ultraviolet light source and an index of sunburn in pea plants. Use the data to construct a scatter plot. How would you describe the relationship between these two variables?

Distance (cm)	18	21	25	26	30	32	36	40
Sunburn units	4.0	3.7	3.0	2.9	2.6	2.5	2.2	2.0

Distance (cm)	40	50	51	54	61	62	63
Sunburn units	2.1	1.5	1.5	1.5	1.3	1.2	1.1

11.7 A number of research studies have looked at the relationship between water stress and plant productivity. The paper "Water Stress Affecting Nitrate Reduction and

Leaf Diffusive Resistance in Coffea Arabica L. Cultivars'' (*J. of Horticultural Sci.* (1983): 147–52) examined water availability and nitrate activity in five different types of coffee plants. Data on water potential (*x*) and nitrate activity (*y*) for the Angustifolia and Nacional varieties are given.

Angustifolia

x	−10	−11	−11	−14	−15	−15	−16	−16	−16	−17	−18
y	3.2	3.0	5.4	3.5	3.0	3.6	4.8	3.2	2.4	4.6	1.6

x	−19	−20	−21	−22	−23	−23	−23	−24
y	3.4	4.0	1.4	.4	.8	1.8	2.0	.2

Nacional

x	−10	−12	−12	−13	−14	−14	−14	−14	−14	−15	−15	−15
y	9.8	8.0	13.0	6.0	5.0	6.0	7.2	10.4	10.8	7.8	9.0	11.0

x	−15	−16	−16	−17	−18	−18	−18	−18	−18	−18	−19	−20
y	14.0	6.6	14.2	3.4	8.0	8.2	8.6	9.2	11.6	12.0	5.2	9.4

a. Draw a scatter plot for the Angustifolia data. Does the plot look linear?

b. Construct a scatter plot for the Nacional data. Would you describe the relationship between nitrate activity and water potential exhibited by this data set as linear?

c. Discuss the similarities and differences between the scatter plots of nitrate activity versus water potential for Angustifolia and Nacional coffee plants.

11.8 The problem of soil erosion is faced by farmers all over the world. The paper "Soil Erosion by Wind from Bare Sandy Plains in Western Rajasthan, India" (*J. of Arid Environ.* (1981): 15–20) reported on a study of the relationship between *x* = wind velocity (km/hr) and *y* = soil erosion (kg/day) in a very dry environment, where erosion control is especially important. We present selected data extracted from the paper, along with the value of log(*y*) corresponding to each *y* in the data set.

Observation	x	y	$y' = \log(y)$	Observation	x	y	$y' = \log(y)$
1	13.5	5	.6990	8	21	140	2.1461
2	13.5	15	1.1761	9	22	75	1.8751
3	14	35	1.5441	10	23	125	2.0969
4	15	25	1.3979	11	25	190	2.2788
5	17.5	25	1.3979	12	25	300	2.4771
6	19	70	1.8451	13	26	240	2.3802
7	20	80	1.9031	14	27	315	2.4983

a. Construct a scatter plot of the (*x, y*) pairs. Does there appear to be a strong relationship between *x* and *y*? A strong linear relationship?

b. Construct a scatter plot of the (*x, y'*) pairs. How does log(*y*) appear to be related to *x*? Note: When *x* and *y* appear to be related in a curvilinear fashion, it is often possible to transform one or both variables so that the resulting pattern is linear. A logarithmic transformation is frequently used for this purpose.

11.2 FITTING A LINE BY LEAST SQUARES

Given two variables x and y, the general objective of *regression analysis* is to use information about x to draw some type of conclusion concerning y. Often an investigator will want a prediction of the y value that would result from making a single observation at a specified x value—for example, predict product sales y during a given period when shelf space for displaying the product is $x = 6$ sq. ft. The different roles played by the two variables are reflected in standard terminology: y is called the **dependent** or **response variable** and x is referred to as the **independent** or **predictor variable**.

A scatter plot of y versus x (that is, of the (x, y) pairs in a sample) will frequently exhibit a linear pattern. It is natural in such cases to summarize the relationship between the variables by finding a line that is as close as possible to the points in the plot. Recall that the equation of a straight line has the form

$$y = a + bx$$

where a is the *vertical* (or y) *intercept* (the height of the line above the value $x = 0$) and b is the *slope* (the amount by which y changes when x increases by one unit). Thus the line $y = 500 - 10x$ has vertical intercept 500, and a one-unit increase in x is associated with a 10-unit decrease in y.

Figure 4 shows a scatter plot with two lines superimposed on the plot. Line II clearly gives a better fit to the data than does Line I. The line that gives the most effective summary of the approximate linear relation is the one that in some sense is the best-fitting line, the one closest to the sample

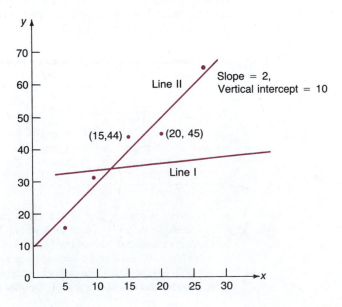

FIGURE 4

Lines I and II give poor and good fits, respectively, to the data.

data. To measure the extent to which a particular line provides a good fit, let's focus on the vertical deviations from the line. For example, Line II in Figure 4 has equation $y = 10 + 2x$ and the third and fourth points from the left are (15, 44) and (20, 45), respectively. The vertical deviations are then

$$3\text{rd deviation} = y_3 - \text{height of the line above } x_3$$
$$= 44 - [10 + 2(15)]$$
$$= 4$$

and

$$4\text{th deviation} = 45 - [10 + 2(20)]$$
$$= -5$$

A positive deviation results from a point that lies above the chosen line, and a negative deviation from a point that lies below this line. A particular line gives a good fit if the deviations from the line are small in magnitude. Line I in Figure 4 fits poorly, because all deviations from that line are larger in magnitude (some are much larger) than the corresponding deviations from Line II.

We now need a way to combine the n deviations into a single measure of fit. The standard approach is to square the deviations (to obtain non-negative numbers) and sum these squared deviations.

 DEFINITION

The criterion for measuring the goodness-of-fit of a line $y = a + bx$ to bivariate data $(x_1, y_1), \ldots, (x_n, y_n)$ is the sum of the squared deviations about the line:

$$\sum [y - (a + bx)]^2 = [y_1 - (a + bx_1)]^2$$
$$+ [y_2 - (a + bx_2)]^2 + \cdots + [y_n - (a + bx_n)]^2$$

The line that gives the best fit to the data is the one that minimizes this sum; it is called the **least-squares line**.*

Fortunately the equation of the least-squares line can be obtained without having to calculate deviations from any particular line. This is because mathematical techniques can be applied to obtain relatively simple formulas for the slope and vertical intercept of the least-squares line.

---

* The least-squares line is frequently referred to as the *sample regression line*.

The slope of the least-squares line is

$$b = \frac{\sum xy - n\bar{x}\bar{y}}{\sum x^2 - n\bar{x}^2}$$

and the y intercept is

$$a = \bar{y} - b\bar{x}$$

We shall write the equation of the least-squares line as

$$\hat{y} = a + bx$$

where the ^ above y emphasizes that $\hat{y}$ is a prediction of y resulting from the substitution of a particular x value into the equation.

Obtaining the least-squares line requires first calculating $\sum x$ (for $\bar{x}$), $\sum y$, $\sum x^2$, and $\sum xy$. In addition, we will shortly need $\sum y^2$ and, of course, n. These six quantities are called *summary statistics*. As illustrated in the next example, computations are expedited by using a tabular format with columns for x, y, x^2, xy, and y^2. After placing the appropriate entries in each column, the column sums are the summary statistics.

EXAMPLE 4

In Example 2 we presented data on

$$x = \text{tree age (yr)}, \qquad y = \text{5-yr height growth (cm)}$$

The scatter plot (Figure 2) shows a substantial linear pattern, suggesting the appropriateness of a straight-line summary. The values of the summary statistics come from the accompanying tabular format.

Obs.	x	y	x^2	xy	y^2
1	5	70	25	350	4900
2	9	150	81	1350	22500
3	9	260	81	2340	67600
4	10	230	100	2300	59900
5	10	255	100	2550	65025
6	11	165	121	1815	27225
7	11	225	121	2475	50625
8	12	340	144	4080	115600
9	13	305	169	3965	93025
10	13	335	169	4355	112225
11	14	290	196	4060	84100
12	14	340	196	4760	115600
13	15	225	225	3375	50625
14	15	300	225	4500	90000
15	18	380	324	6840	144400
16	18	400	324	7200	160000
Sum:	197	4270	2601	56315	1263350
	$\sum x$	$\sum y$	$\sum x^2$	$\sum xy$	$\sum y^2$

Thus,

$$\bar{x} = \frac{\sum x}{n} = \frac{197}{16} = 12.3125$$

$$\bar{y} = \frac{\sum y}{n} = \frac{4270}{16} = 266.8750$$

The slope of the least-squares line is

$$b = \frac{\sum xy - n\bar{x}\bar{y}}{\sum x^2 - n\bar{x}^2}$$

$$= \frac{56315 - 16(12.3125)(266.8750)}{2601 - 16(12.3125)^2}$$

$$= \frac{56315 - 52574.3750}{2601 - 2425.5625}$$

$$= \frac{3740.6250}{175.4375}$$

$$= 21.321696$$

and the vertical intercept is

$$a = \bar{y} - b\bar{x} = 266.8750 - (21.321696)(12.3125)$$
$$= 4.351618$$

This gives us

$$\hat{y} = 4.351618 + 21.321696x$$
$$\approx 4.352 + 21.322x$$

as the equation of the least-squares line. The reason for giving so many digits of decimal accuracy in a and b will become apparent in the next section. The interpretation of b is that for trees that differ in age by one year, the increase in growth associated with the older trees is roughly 21.322 cm. To predict a 5-year height growth when the tree age is 12.5 yr, simply substitute 12.5 into the equation:

$$\hat{y} = 4.352 + (21.322)(12.5) \approx 270.9 \text{ cm}$$

Similarly, the predicted height growth when $x = 10$ yr is

$$\hat{y} = 4.352 + (21.322)(10) \approx 217.6 \text{ cm}$$

Notice that there are two sample observations for which $x = 10$ yr, and the y values in both cases are larger than 217.6. So in the scatter plot, both of these points lie above the least-squares line (see Figure 5).

FIGURE 5

Scatter plot and least squares line for the data of Example 4.

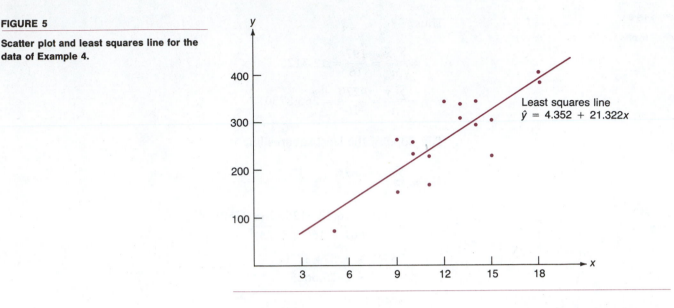

Least squares line
$\hat{y} = 4.352 + 21.322x$

EXAMPLE 5

Millions of Muslims all over the world participate in controlled or partial fasting during the month of Ramadan. The paper "Body Weight Loss and Changes in Blood Lipid Levels in Normal Men on Hypo-caloric Diets during Ramadan Fasting" (*Amer. J. of Clinical Nutr.* (1988): 1197–1210) reported on a study carried out to assess the physiological effects of such fasting. The accompanying table gives selected data (read from a graph) on $x =$ percent change in body weight during phase two of the fasting period (in this phase experimental subjects received a high-fat, low-carbohydrate diet) and $y =$ high-density lipo-protein cholesterol level (mg/dl).

x	−5	−3.2	−2.2	−1.7	−1.6	−1.5	−.9	0	0	1.2	1.6	1.7	2.8
y	43	50	61	63	47	57	51	60	67	76	70	51	74

A scatter plot is displayed in Figure 6. The predominant pattern is linear, though the points in the plot spread out rather substantially about *any* line (even the least-squares line). Nevertheless, we shall follow the paper's authors in using the principle of least squares to obtain a summarizing line. The summary statistics are as follows.

$$n = 13 \qquad \sum x = -8.8 \qquad \sum x^2 = 63.32$$
$$\sum y = 770 \qquad \sum y^2 = 46940 \qquad \sum xy = -325.80$$
$$\bar{x} = -.676923 \qquad \bar{y} = 59.230769$$

The slope of the least-squares line is

$$b = \frac{-325.80 - 13(-.676923)(59.230769)}{63.32 - 13(-.676923)^2}$$

$$= \frac{195.430708}{57.363078}$$

$$= 3.406908$$

FIGURE 6

Scatter plot for the data of Example 5.

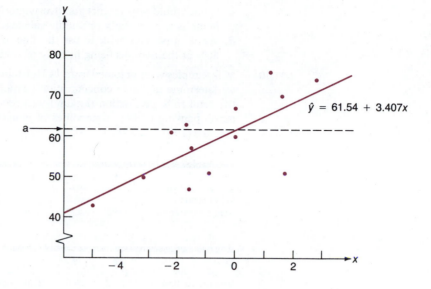

and the vertical intercept is

$$a = 59.230769 - (3.406908)(-.676923)$$
$$= 61.536983$$

This gives us (after rounding)

$$\hat{y} = 61.54 + 3.407x$$

as the equation of the least-squares line. For an experimental subject whose weight decreases by one percent during phase two, the predicted HDL–cholesterol level is

$$\hat{y} = 61.54 + 3.407(-1) \approx 58.1$$

EXERCISES 11.9–11.16   **SECTION 11.2**

11.9   The accompanying data on the percent of red pine scale nymphs (a forest insect) in early and middle stages (x) and overwintering mortality rate (y) appeared in the paper "Population Dynamics of a Pernicious Parasite: Density-Dependent Vitality of Red Pine Scale" (*Ecology* (1983): 710–18).

x	20	32	36	42	43	46	48	51	60
y	81	81.5	83	87	84	84.5	86.5	89	86.5

a. Construct a scatter plot for this data set.
b. The least-squares line relating x and y given in the paper was $\hat{y} = 76.2 + .2x$. Draw this line on your scatter diagram.

c. What would you predict for overwintering mortality rate if 50% of the scale nymphs were in early or middle substages?

d. Would it be reasonable to use this line to predict mortality rate associated with 80% of the nymphs being in early or middle substages? Explain.

11.10 Milk samples were obtained from 14 Holstein–Friesian cows, and each was analyzed to determine uric acid concentration (μ mol/l). In addition to acid concentration, the total milk production (kg/day) was recorded for each cow, as shown in the accompanying table ("Metabolites of Nucleic Acids in Bovine Milk," *J. of Dairy Science* (1983): 723–8).

Cow	1	2	3	4	5	6	7
Milk production	42.7	40.2	38.2	37.6	32.2	32.2	28.0
Acid concentration	92	120	128	110	153	162	202

Cow	8	9	10	11	12	13	14
Milk production	27.2	26.6	23.0	22.7	21.8	21.3	20.2
Acid concentration	140	218	195	180	193	238	213

Let x denote milk production and y uric acid concentration.

a. Draw a scatter plot for this data.

b. Using the following summary quantities, compute the least-squares line.

$$n = 14 \qquad \sum x^2 - n\bar{x}^2 = 762.012 \qquad \sum xy - n\bar{x}\bar{y} = -3964.486$$

c. Draw the least-squares line on your scatter plot. Hint: Substitute $x = 25$ into the equation of the least-squares line to find one point (x, y) on the line. Repeat with $x = 40$. Then draw a line through these two points.

d. What uric acid concentration would you predict for a cow whose total milk production was 30 kg/day?

e. Would you feel comfortable using the least-squares line to make a prediction for a cow whose total milk production was 10 kg/day? Explain your answer.

11.11 Infestation of crops by insects has long been of great concern to farmers and agricultural scientists. The paper "Cotton Square Damage by the Plant Bug, *Lygus hesperus*, and Abscission Rates" (*J. of Econ. Entom.* (1988): 1328–37) reported data on x = age of a cotton plant (days) and y = percent damaged squares. Consider the accompanying $n = 12$ observations (read from a scatter plot in the paper).

x	9	12	12	15	18	18	21	21	27	30	30	33
y	11	12	23	30	29	52	41	65	60	72	84	93

a. Why is the relationship between x and y not deterministic?

b. Does a scatter plot suggest that the least-squares line will effectively summarize the relationship between the two variables?

c. The summary statistics are

$$\sum x = 246 \qquad \sum x^2 = 5742 \qquad \sum y = 572 \qquad \sum y^2 = 35634 \qquad \sum xy = 14022.$$

Determine the equation of the least-squares line.

d. Predict the percent of damaged squares when the age is 20 days.

11.12 Exercise 11.4 gave the accompanying data on x = plasma cortisol concentration and y = oxygen consumption rate for juvenile steelhead.

x	25	36	48	59	62	72	80	100	100	137
y	155	184	180	220	280	163	230	222	241	350

a. Establish a tabular format with columns headed x, y, x^2, xy, and y^2. Place the given data in the first two columns, fill in the entries for the remaining three columns, and add the numbers in each column to obtain summary statistics. Then determine the slope and vertical intercept of the least-squares line, and give the equation of the line.
b. Interpret b in the context of this problem situation.
c. Predict the value of the oxygen consumption rate that would be observed for a single fish having plasma cortisol concentration 50 ng/ml.

11.13 The paper "Ion Beam–Assisted Etching of Aluminum with Chlorine" (*J. of Electrochem. Soc.* (1985): 2010–12) gave the accompanying data (read from a graph) on chlorine flow (x, in SCCM) through a nozzle used in a plasma etching process and etch rate (y, in 100 A/min).

x	1.5	1.5	2.0	2.5	2.5	3.0	3.5	3.5	4.0
y	23.0	24.5	25.0	30.0	33.5	40.0	40.5	47.0	49.0

a. Why is the relationship between x and y clearly not deterministic?
b. Verify that the values of the summary statistics are

$$\sum x = 24.0 \quad \sum y = 312.5 \quad \sum x^2 = 70.50 \quad \sum xy = 902.25 \quad \sum y^2 = 11626.75.$$

c. Obtain the equation of the least-squares line.
d. Predict etch rate for an observation made when chlorine flow is 2.5 A/min.

11.14 Athletes competing in a triathalon participated in a study described in the paper "Myoglobinemia and Endurance Exercise" (*Amer. J. of Sports Med.* (1984): 113–18). The following data on finishing time x (hr) and myoglobin level y (ng/ml) was read from a scatter plot in the paper.

x	4.90	4.70	5.35	5.22	5.20	5.40	5.70	6.00
y	1590	1550	1360	895	865	905	895	910

x	6.20	6.10	5.60	5.35	5.75	5.35	6.00
y	700	675	540	540	440	380	300

a. Obtain the equation of the least-squares line.
b. Interpret the value of b.
c. What happens if the line in part (a) is used to predict the myoglobin level for a finishing time of 8 hr? Is this reasonable? Explain.

11.15 The paper "Increased Oxygen Consumption during the Uptake of Water by the Eversible Vesicles of Petrobius Brevistylis" (*J. of Insect Phys.* (1977): 1285–94) presented the results of a regression involving variables x = weight increase (mg) when a dehydrated insect was allowed access to water and y = increased oxygen uptake (μl) above the mean resting rate. Data read from a graph gave the following summary statistics:

$$n = 20 \qquad \sum x = 63.5 \qquad \sum y = 17.26$$
$$\sum x^2 = 311.74 \qquad \sum xy = 71.51 \qquad \sum y^2 = 19.9625$$

a. Obtain the equation of the least-squares line.

b. The only observation with an x value larger than 7 was $(x_{20}, y_{20}) = (9.8, 1.9)$. It is of interest to know whether this point has greatly influenced the equation of the least-squares line. Compute the least-squares line based on just the remaining 19 observations, and comment on the difference between this line and the line of part (a). Hint: Simply adjust the summary statistics, for example, new $\sum x$ = old $\sum x - 9.8$, etc.

11.16 The paper "Aspects of Food Finding by Wintering Bald Eagles" (*The Auk* (1983): 477–84) examined the relationship between the time that eagles spend aerially searching for food (indicated by the percentage of eagles soaring) and relative food availability. The data is taken from a scatter plot that appeared in this paper. Let x denote salmon availability and y the percent of eagles in the air.

x	0.0	0.0	0.2	0.5	0.5	1.0
y	28.2	69.0	27.0	38.5	48.4	31.1

x	1.2	1.9	2.6	3.3	4.7	6.5
y	26.9	8.2	4.6	7.4	7.0	6.8

a. Construct a scatter plot. Would you feel comfortable in summarizing the relationship between x and y by reporting the equation of the least-squares line?

b. Calculate $\sqrt{x}$ and $\sqrt{y}$ for each observation, and then construct a scatter plot of the $(\sqrt{x}, \sqrt{y})$ pairs. Does there appear to be an approximate linear relationship between these transformed variables?

11.3 ASSESSING THE FIT

Once the best-fit (least-squares) line has been obtained, it is natural to ask how effectively the line summarizes the relationship between x and y. That is, we would like a quantitative indicator of the extent to which y variation can be attributed to the approximate linear relationship between the two variables. Such an assessment is based on the vertical deviations from the least-squares line.

PREDICTED VALUES AND RESIDUALS

If the x value for the first observation is substituted in the equation for the least-squares line, the result is $a + bx_1$, the height of the line above x_1. The point (x_1, y_1) in the scatter plot also lies above x_1, so the difference

$$y_1 - (a + bx_1)$$

is the vertical deviation from this point to the line (see Figure 7). A point lying above the line gives a positive deviation, and a point below the line results in a negative deviation. The remaining vertical deviations come from repeating this process for $x = x_2$, then $x = x_3$, and so on.

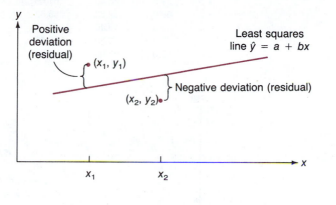

FIGURE 7

Positive and negative deviations (residuals) from the least-squares line.

▌▌▌ **DEFINITION**

The **predicted** or **fitted values** result from substituting each sample x value in turn into the equation for the least-squares line. This gives

$$\hat{y}_1 = \text{1st predicted value } = a + bx_1$$
$$\hat{y}_2 = \text{2nd predicted value} = a + bx_2$$
$$\vdots$$
$$\hat{y}_n = n\text{th predicted value} = a + bx_n$$

The **residuals** from the least-squares line are the n quantities

$$y_1 - \hat{y}_1, \quad y_2 - \hat{y}_2, \ldots, \quad y_n - \hat{y}_n$$

Each residual is a difference between an observed y value and the corresponding predicted y value.

EXAMPLE 6

The ability of proteins to bind fat is important in improving the flavor and texture of meat. The paper "A Simple Turbidimetric Method for Determining the Fat-Binding Capacity of Proteins" (*J. of Agric. and Food Chem.* (1983): 58–63) proposed the use of regression methods for predicting fat-binding capacity (FBC) y from other protein characteristics. The first predictor used in the study was $x =$ surface hydrophobicity (HBCTY), a measure of the extent to which water is not attracted to the protein's surface. The $n = 8$ observations appear in the accompanying tabular format. A scatter plot is displayed in Figure 8.

Obs.	x HBCTY	y FBC	Predicted y Value	Residual
1	6.0	37.70	28.77	8.93
2	28.0	10.10	50.71	−40.61
3	95.0	105.90	117.54	−11.64
4	39.0	85.30	61.68	23.62
5	66.0	92.30	88.62	3.68
6	5.0	19.10	27.77	−8.67
7	47.0	105.80	69.66	36.14
8	55.0	66.20	77.64	−11.44

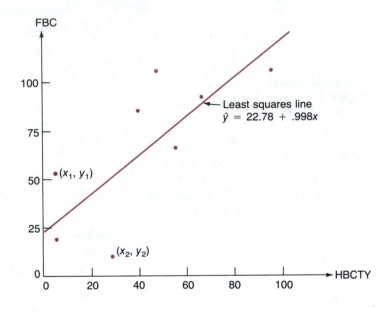

FIGURE 8

Scatter plot of the data from Example 6.

The summary statistics are

$$n = 8 \qquad \sum x = 341 \qquad \sum y = 522.4$$
$$\sum x^2 = 20981 \qquad \sum xy = 28697.1 \qquad \sum y^2 = 44474.38$$

from which we get

$$b = .99750616 \approx .9975,$$

and

$$a = 22.78129993 \approx 22.78$$

The first predicted value is

$$\hat{y}_1 = a + bx_1 = 22.78 + (.9975)(6.0) = 28.77$$

and the corresponding residual is

$$y_1 - \hat{y}_1 = 37.70 - 28.77 = 8.93$$

The second predicted value and residual are

$$\hat{y}_2 = a + bx_2 = 22.78 + (.9975)(28.0) = 50.71$$

and

$$y_2 - \hat{y}_2 = 10.10 - 50.71 = -40.61$$

As Figure 8 shows, this extreme negative residual is a consequence of (x_2, y_2) being far below the estimated regression line. The remaining predicted values and residuals are displayed in the table with the data.

THE COEFFICIENT OF DETERMINATION

Variation in y can effectively be explained by an approximate straight-line relationship when the points in the scatter plot fall close to the least-squares line, i.e., when the residuals are small in magnitude. A natural measure of variation about the least-squares line is the sum of the squared residuals. (Squaring before combining prevents negative and positive residuals from counteracting one another.) A second sum of squares is needed to assess the total amount of variation in observed y values without reference to x.

▌▌ DEFINITION

The **residual sum of squares**,* denoted by **SSResid**, is given by

$$SSResid = \sum (y - \hat{y})^2$$
$$= (y_1 - \hat{y}_1)^2 + (y_2 - \hat{y}_2)^2 + \cdots + (y_n - \hat{y}_n)^2$$

The **total sum of squares**, denoted by **SSTo**, is defined as

$$SSTo = \sum (y - \bar{y})^2$$
$$= (y_1 - \bar{y})^2 + (y_2 - \bar{y})^2 + \cdots + (y_n - \bar{y})^2$$

These sums of squares can be obtained using just the summary statistics via the following computational formulas:

$$SSResid = \sum y^2 - a \sum y - b \sum xy$$
$$SSTo = \sum y^2 - n\bar{y}^2$$

As with any sum of squares in statistics, SSResid and SSTo cannot be negative. The computing formula for SSResid is quite sensitive to rounding. Using as many digits of decimal accuracy as possible in the values of a and b will ensure an accurate result.

---

* Some sources refer to this as the **error sum of squares** and denote it by **SSE**.

EXAMPLE 7

(*Example 6 continued*) The summary statistics and values of a and b for the hydrophobicity–fat-binding capacity data were given earlier. Substitution yields

$$\text{SSResid} = \sum y^2 - a \sum y - b \sum xy$$
$$= 44474.38 - (22.78129993)(522.4) - (.99750616)(28697.1)$$
$$= 44474.38 - 11900.95 - 28625.53$$
$$= 3947.90$$

$$\text{SSTo} = \sum y^2 - n\bar{y}^2$$
$$= 44474.38 - 8(65.3000)^2$$
$$= 44474.38 - 34112.72$$
$$= 10361.66$$

Although SSResid is rather sizable (because several residuals are large in magnitude), it is relatively small when compared to SSTo.

The residual sum of squares is the sum of squared deviations from the least-squares line. As Figure 9 illustrates, SSTo is also a sum of squared deviations from a line—the horizontal line at height $\bar{y}$. Since the least-squares line is by definition the one having the smallest sum of squared deviations, it follows that $\text{SSResid} \le \text{SSTo}$. The two sums of squares will be equal only when the least-squares line *is* the horizontal line.

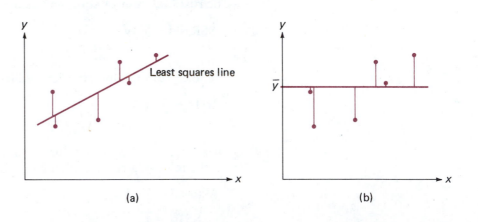

FIGURE 9

Geometric interpretations of SSResid and SSTo.
(a) SSResid = sum of squared deviations from least-squares line
(b) SSTo = Sum of squared deviations from horizontal line at height $\bar{y}$

The residual sum of squares is often referred to as a measure of "unexplained" variation. It is the amount of variation in y that cannot be attributed to the linear relationship between x and y. When $\text{SSResid} = 0$, all points in the scatter plot fall exactly on the least-squares line; all y variation can be attributed to the linear relationship; and no y variation is left unexplained. The larger the value of SSResid, the greater the amount of y variation that cannot be explained by the approximate linear relationship. Similarly, SSTo is interpreted as a measure of total variation. The larger the value of SSTo, the greater the amount of variability in $y_1, y_2, \ldots, y_n$. The ratio SSResid/SSTo is the fraction or proportion of total variation that is unexplained by a straight-line relation. Subtracting this ratio from 1 gives the proportion of total variation that is explained.

█ DEFINITION

The **coefficient of determination**, denoted by r^2, is given by

$$r^2 = \frac{\text{SSTo} - \text{SSResid}}{\text{SSTo}} = 1 - \frac{\text{SSResid}}{\text{SSTo}}$$

It is the proportion of variation in y that can be attributed to a linear relationship between x and y in the sample.

Multiplying r^2 by 100 gives the percentage of y variation attributable to the approximate linear relationship. The closer this percentage is to 100%, the more successful is the relationship in explaining variation in y.

EXAMPLE 8 We found that for the hydrophobicity–fat-binding capacity data of Examples 6 and 7,

SSResid = 3947.90   SSTo = 10361.66

Thus,

$$r^2 = 1 - \frac{\text{SSResid}}{\text{SSTo}}$$

$$= 1 - \frac{3947.90}{10361.66}$$

$$= 1 - .381$$

$$= .619$$

so 61.9% of the sample variation in FBC is explained by the approximate linear relationship with hydrophobicity. In many situations, to have 61.9% of the variation explained would be quite respectable, but in this particular instance the investigators were not satisfied, so they proceeded to an analysis involving more than a single predictor variable.

EXAMPLE 9 The Arabian Sea suffers from oxygen depletion to a greater extent than almost any other part of an open ocean. Dentrification, the process by which nitrates and nitrites are reduced to other forms of nitrogen, is one step in the oxidation of organic matter to carbon dioxide and the consequent reduction of oxygen. The paper "Evidence for and Rate of Dentrification in the Arabian Sea" (*Deep Sea Research* (1978): 431–5) reported on a study in which water samples were selected and x = salinity (%) and y = nitrate level (μM/l) were determined. A regression analysis was then carried out. The accompanying data is a subset of that contained in the article. A scatter plot is shown in Figure 10.

x	35.43	36.10	35.74	35.30	35.40	35.91	35.48	36.28
y	30.0	24.2	25.4	29.8	30.7	24.0	28.5	22.7

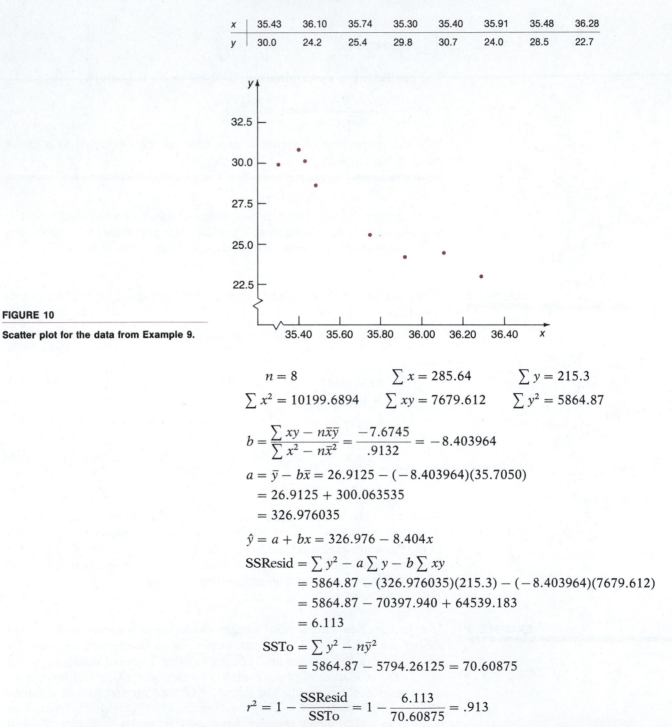

FIGURE 10

Scatter plot for the data from Example 9.

$$n = 8 \qquad \sum x = 285.64 \qquad \sum y = 215.3$$

$$\sum x^2 = 10199.6894 \qquad \sum xy = 7679.612 \qquad \sum y^2 = 5864.87$$

$$b = \frac{\sum xy - n\bar{x}\bar{y}}{\sum x^2 - n\bar{x}^2} = \frac{-7.6745}{.9132} = -8.403964$$

$$a = \bar{y} - b\bar{x} = 26.9125 - (-8.403964)(35.7050)$$

$$= 26.9125 + 300.063535$$

$$= 326.976035$$

$$\hat{y} = a + bx = 326.976 - 8.404x$$

$$\text{SSResid} = \sum y^2 - a\sum y - b\sum xy$$

$$= 5864.87 - (326.976035)(215.3) - (-8.403964)(7679.612)$$

$$= 5864.87 - 70397.940 + 64539.183$$

$$= 6.113$$

$$\text{SSTo} = \sum y^2 - n\bar{y}^2$$

$$= 5864.87 - 5794.26125 = 70.60875$$

$$r^2 = 1 - \frac{\text{SSResid}}{\text{SSTo}} = 1 - \frac{6.113}{70.60875} = .913$$

Thus 91.3% of the observed variation in nitrate level can be attributed to an approximate linear relationship between the nitrate level and salinity.

Use of the least-squares line appears to provide an effective method for predicting nitrate level at any specified value of salinity (within the range of the given data).

EXERCISES 11.17–11.24   **SECTION 11.3**

11.17 The article "Effects of Gamma Radiation on Juvenile and Mature Cuttings of Quaking Aspen" (*Forest Science* (1967): 240–45) reported the accompanying data on x = exposure time to radiation (kR/16 hr) and y = dry weight of roots (mg $\times$ 10^{-1}). The least-squares line $\hat{y} = 126.6 - 6.65x$ was used to obtain the predicted values ($\hat{y}$) given. Compute the residuals for this data set.

x	0	2	4	6	8
y	110	123	119	86	62
Predicted y	127.0	113.7	100.4	87.1	73.8

11.18 The given data on fish survival and ammonia concentration was taken from the paper "Effects of Ammonia on Growth and Survival of Rainbow Trout in Intensive Static-Water Culture" (*Trans. of the Amer. Fisheries Soc.* (1983): 448–54).

Ammonia exposure (mg/L)	10	10	20	20	25	27	27	31	50
Percent survival	85	92	85	96	87	80	90	59	62

With x = ammonia exposure and y = percent survival, the least-squares line is $\hat{y} = 100.79 - .78x$.
a. Compute the predicted values and the residuals. Are there any unusually large residuals? What is the sum of the residuals?
b. Use the residuals in part (a) to compute SSResid.
c. Would the line $y = 100 - 1x$ have a larger or smaller SSResid than the line $y = 100.79 - .78x$? Explain your answer.

11.19 There have been numerous studies on the effects of radiation. Data on the relationship between degree of exposure to ^{242}Cm alpha particles (x) and the percentage of exposed cells without aberrations (y) appeared in the paper "Chromosome Aberrations Induced in Human Lymphocytes by D-T Neutrons" (*Radiation Research* (1984): 561–73).

x	.106	.193	.511	.527	1.08	1.62	1.73	2.36	2.72	3.12	3.88	4.18
y	98	95	87	85	75	72	64	55	44	41	37	40

Summary statistics are

$$n = 12 \qquad \sum x = 22.027 \qquad \sum y = 793$$
$$\sum x^2 = 62.600235 \qquad \sum xy = 1114.5 \qquad \sum y^2 = 57939$$

a. Obtain the equation of the least-squares line.
b. Calculate SSResid.

11.20 The summary quantities for data on x = plasma concentration and y = oxygen consumption rate, given in Exercise 11.12, are

$$n = 10 \qquad \sum x = 719 \qquad \sum x^2 = 61903$$
$$\sum y = 2225 \qquad \sum y^2 = 526415 \qquad \sum xy = 173865$$

from which $a = 124.672917$ and $b = 1.360599$. Calculate SSResid and r^2.

11.21 A study was carried out to investigate the relationship between the hardness of molded plastic (y, in Brinell units) and the amount of time elapsed since termination of the molding process (x, in hr). Summary quantities include

$$n = 15, \qquad \text{SSResid} = 1235.470, \qquad \text{SSTo} = 25321.368$$

Calculate and interpret the coefficient of determination.

11.22 Data on $x =$ age of a cotton plant and $y =$ percent damaged squares was presented in Exercise 11.11. In addition to the summary statistics given there, $a = -19.669528$ and $b = 3.284692$.
  a. Compute the value of residual sum of squares.
  b. What percentage of observed variation in y can be attributed to an approximate linear relationship between the two variables?
  c. Repeat part (a) using the rounded values $a = -19.67$ and $b = 3.28$. Does rounding make a difference?

11.23 Exercise 11.3 gave data on $x =$ distance upslope and $y =$ grain sorghum yield. Summary quantities include

$$n = 14 \qquad \sum x = 1185 \qquad \sum x^2 = 151825 \qquad \sum xy = 449850 \qquad \sum y = 5960$$
$$\sum y^2 = 2631200 \qquad b = -1.060132.$$

  a. Calculate SSResid and SSTo.
  b. Calculate r^2, the proportion of observed variation in yield explained by an approximate linear relationship between distance and yield. Based on this value, do you feel comfortable in describing the relationship as approximately linear? Explain.

11.24 Refer to Example 5 in Section 11.2, in which we carried out a regression of $y =$ cholesterol level on $x =$ percent weight change.
  a. Use the equation of the least-squares line given there to calculate the residuals.
  b. Use the residuals from part (a) to calculate SSResid.
  c. Compute SSResid using the computational formula given in this section along with the values of a and b calculated in Example 5.
  d. Repeat part (c) using the rounded values $a = 61.5$ and $b = 3.4$. Does rounding make a substantial difference?
  e. Is "approximately linear" an accurate description of the relationship between the variables? Explain.

11.4 CORRELATION

Regression methods are often used to infer something about y from knowledge of x (e.g., predicting y or estimating the mean y value). In contrast, many investigations focus only on assessing how strongly the values of two variables are related to one another. For example, x and y might be material strengths determined in two different ways (such as compressive and flexural strengths). Is there a tendency for large values of y to be associated with (paired with) large x values—a *positive* relationship? Can a quantitative assessment of this tendency be obtained? Alternatively, a psychologist might have developed a numerical measure (x) of the extent to which an individual is susceptible to hypnosis. Perhaps it is the case that large values of x are paired with small values of $y =$ IQ—a *negative* relationship—so that less intelligent people tend to be more susceptible to hypnosis. If so, how strong is this tendency? A correlation analysis attempts to answer questions of the sort just posed.

PEARSON'S SAMPLE CORRELATION COEFFICIENT

Let $(x_1, y_1), (x_2, y_2), \ldots, (x_n, y_n)$ denote a sample of (x, y) pairs. Figure 11(a) shows a scatter plot of such data indicating a substantial positive relationship: points on the right part of the plot, which have large x values, tend to be higher (have larger y values) than those on the left. A vertical line through $\bar{x}$ and a horizontal line through $\bar{y}$ divide the plot into four regions. In region I, both x and y are larger than their mean values, so the deviations $x - \bar{x}$ and $y - \bar{y}$ are both positive and so is the product $(x - \bar{x})(y - \bar{y})$. The product of deviations is also positive for any point in region III, because there both deviations are negative and multiplying two negative numbers gives a positive number. Any point in the other two regions gives rise to a negative value for $(x - \bar{x})(y - \bar{y})$, since one term in the product is negative and the other is positive. But because almost all points lie in regions I or III, almost all products are positive, so the sum of products $\sum (x - \bar{x})(y - \bar{y})$ will be a large positive number.

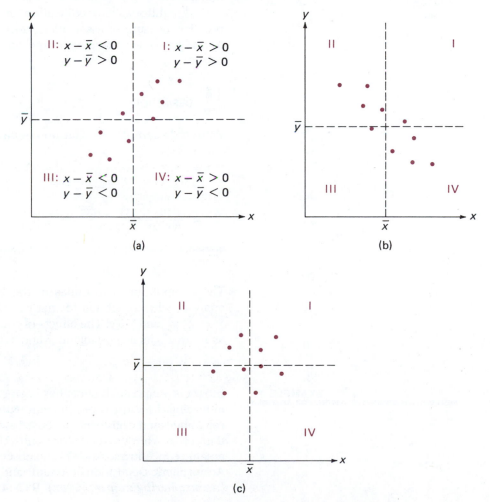

FIGURE 11

Breaking up a scatter plot according to the signs of $x - \bar{x}$ and $y - \bar{y}$.
(a) A positive relation
(b) A negative relation
(c) No strong relation

Similar reasoning for the data displayed in Figure 11(b), which exhibits a strong negative relationship, implies that $\sum (x - \bar{x})(y - \bar{y})$ will be a large negative number. When there is no strong relationship, as in Figure 11(c), positive and negative products tend to counteract one another, producing a value of $\sum (x - \bar{x})(y - \bar{y})$ that is close to zero. In summary, $\sum (x - \bar{x})(y - \bar{y})$ seems to be a reasonable measure of the degree of association between x and y; it will be a large positive number, a large negative number, or a number close to zero according to whether there is a strong positive, a strong negative, or no strong relationship.

Unfortunately our proposed measure has a serious deficiency: its value depends on the choice of unit for measuring either x or y. Suppose, for example, that x is height. Each height expressed in inches is twelve times the corresponding height expressed in feet. It follows that when the unit for x is inches, the value of $\sum (x - \bar{x})(y - \bar{y})$ will be twelve times what it is for x given in feet. A measure of the inherent strength of the relationship between x and y should give the same value whatever the units for the variables; otherwise our impressions can be distorted by the choice of units.

A straightforward modification of $\sum (x - \bar{x})(y - \bar{y})$ leads to the most popular measure of association, one that is free of the defect just alluded to and has other attractive properties.

DEFINITION

Pearson's sample correlation coefficient, denoted by r, is given by

$$r = \frac{\sum (x - \bar{x})(y - \bar{y})}{\sqrt{\sum (x - \bar{x})^2}\sqrt{\sum (y - \bar{y})^2}}$$

$$= \frac{\sum xy - n\bar{x}\bar{y}}{\sqrt{\sum x^2 - n\bar{x}^2}\sqrt{\sum y^2 - n\bar{y}^2}}$$

The computational formulas in the box were used earlier in regression analysis, where a tabular format was recommended for obtaining $\sum x$, $\sum y$, $\sum x^2$, $\sum y^2$, and $\sum xy$. The effects of rounding will be minimized by carrying as many digits as possible in $\bar{x}$ and $\bar{y}$.

EXAMPLE 10

In recent years much effort has been expended by environmental scientists in tracing the sources of acid rain. Nitrates are a major constituent of acid rain, and arsenic has been proposed as a tracer element. The accompanying data on $x =$ nitrate concentration (μM) of a precipitation sample and $y =$ arsenic concentration (nM) is a subset of that contained in the paper "The Atmospheric Deposition of Arsenic and Association with Acid Precipitation" (*Atmospheric Environ.* (1988): 937–43).

Obs.	1	2	3	4	5	6	7	8	9	10	11	12
x	11	13	18	30	36	40	50	58	67	82	91	102
y	1.1	.5	2.4	1.2	2.1	1.2	4.0	2.3	1.7	3.7	3.0	3.9

The scatter plot displayed in Figure 12 shows a clear positive relationship.

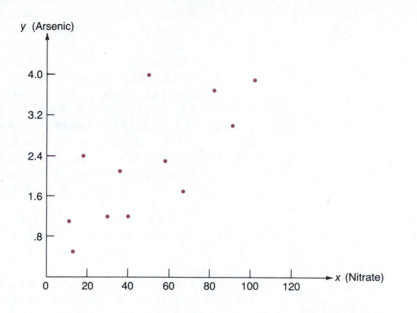

FIGURE 12

Scatter plot of the data from Example 10.

The summary quantities needed for the calculation of r are

$$\sum x = 598 \qquad \sum y = 27.1 \qquad \sum x^2 = 40172$$

$$\sum y^2 = 76.59 \qquad \sum xy = 1642.9$$

from which we get $\bar{x} = 49.833333$, $\bar{y} = 2.258333$, and

$$\sum xy - n\bar{x}\bar{y} = 1642.9 - 12(49.833333)(2.258333)$$
$$= 292.416875$$

$$\sum x^2 - n\bar{x}^2 = 40172 - 12(49.833333)^2$$
$$= 10371.66706$$

$$\sum y^2 - n\bar{y}^2 = 76.59 - 12(2.258333)^2$$
$$= 15.389185$$

Thus

$$r = \frac{292.416875}{\sqrt{10371.66706}\sqrt{15.389185}} = .732$$

To interpret this value and appreciate the strengths and weaknesses of the correlation coefficient, we need to discuss the most important properties of r.

PROPERTIES OF *r*

1. *The value of r does not depend on the unit of measurement for either variable.* If, for example, x is height, the factor of 12 that appears in the numerator when changing from feet to inches will also appear in the denominator, so the two will cancel and leave r unchanged. The same value of r results from height expressed in inches, meters, or miles. If y is temperature, expressing values in °F, °C, or °K will give the same value of r. The correlation coefficient measures the inherent strength of relationship between two numerical variables.

2. *The value of r does not depend on which of the two variables is labeled x.* Thus if we had let $x =$ arsenic concentration and $y =$ nitrate concentration in Example 10, the same value $r = .732$ would have resulted. This is in marked contrast to regression analysis, where most calculated quantities (such as a, b, SSResid) depend on which variable plays the role of the independent variable.

3. *The value of r is between -1 and $+1$.* A value near the upper limit, $+1$, is indicative of a substantial positive relationship, whereas an r close to the lower limit, -1, suggests a prominent negative relationship. A useful informal rule for characterizing the nature of the relationship in everyday language is as follows: We say the relationship is

   strong if either $r \geq .8$ or $r \leq -.8$;

   moderate if either $.5 < r < .8$ or $-.8 < r < -.5$;

   weak if $-.5 \leq r \leq .5$.

   It may seem surprising that a value of r as extreme as $-.5$ or $.5$ should be in the "weak" category; an explanation for this is given in Property 6.

4. *$r = 1$ only when all the points in a scatter plot of the data lie exactly on a straight line that slopes upward. Similarly, $r = -1$ only when all the points lie exactly on a downward-sloping line.* Only when there is a perfect linear relationship between x and y in the sample will r take on one of its two possible extreme values. If there is any deviation from a straight line, then r will be strictly between -1 and $+1$ (that is, $-1 < r < 1$).

5. *The value of r is a measure of the extent to which x and y are linearly related*—i.e., the extent to which the points in the scatter plot fall close to a straight line. A value of r close to zero does not rule out any strong relationship between x and y; there could still be a strong relationship but one that is not linear. (Figure 3 in Section 1 shows data for which this is the case.) So don't stop with r—look at a scatter plot!

6. *The correlation coefficient r and the coefficient of determination r^2 from a regression analysis are related in the obvious way, namely:*

   (correlation coefficient)^2 = coefficient of determination

   Thus, if $r = .8$ or $r = -.8$ then $r^2 = .64$, so that 64% of the observed variation in the dependent variable can be attributed to the linear relationship. Notice that because the value of r does not depend on which variable is labeled x, the same is true of r^2. The coefficient of determination is one the very few quantities from regression whose value remains the same when the role of dependent and independent variables are interchanged. When $r = .5$, we get $r^2 = .25$, so only 25% of the observed variation is explained by a linear relation. This is why values of r between $-.5$ and $.5$ can fairly be described as evidence of a weak relationship.

CORRELATION AND CAUSATION

A value of r close to 1 indicates that relatively large values of one variable tend to be associated with relatively large values of the other variable. This is far from saying that a large value of one variable *causes* the value of the other variable to be large. Correlation (Pearson's or any other) measures the extent of association, but **association does not imply causation**. It frequently happens that two variables are highly correlated not because one is causally related to the other but because they are both strongly related to a third variable. Among all elementary-school children, there is a strong positive relationship between the number of cavities in a child's teeth and the size of his or her vocabulary. Yet no one advocates eating foods that result in more cavities in order to increase vocabulary size (or working to decrease vocabulary size in order to protect against cavities). Number of cavities and vocabulary size are both strongly related to age, so older children tend to have higher values of both variables than do younger ones. Among children of any fixed age, there would undoubtedly be little relationship between number of cavities and vocabulary size.

Scientific experiments can frequently make a strong case for causality by carefully controlling the values of all variables that might be related to the ones under study. Then, if y is observed to change in a "smooth" way as the experimenter changes the value of x, the most plausible explanation would be a causal relationship between x and y. In the absence of such control and ability to manipulate values of one variable, we must admit the possibility that an unidentified underlying third variable is influencing both the variables under investigation. A high correlation in many uncontrolled studies carried out in different settings can marshal support for causality—as in the case of cigarette smoking and cancer—but proving causality is often a very elusive task.

EXERCISES 11.25–11.39   **SECTION 11.4**

11.25 For each of the following pairs of variables, indicate whether you would expect a positive correlation, a negative correlation, or no correlation. Explain your choice.
a. Maximum daily temperature and cooling costs
b. Interest rate and number of loan applications
c. Incomes of husbands and wives when both have full-time jobs
d. Height and IQ
e. Height and shoe size
f. Score on the math section of the SAT exam and score on the verbal section of the same test
g. Time spent on homework and time spent watching television during the same day by elementary-school children

h. Amount of fertilizer used per acre and crop yield (*Hint:* As the amount of fertilizer is increased, yield tends to increase for awhile but then tends to decrease.)

11.26 Is the following statement correct? Explain why or why not.

A correlation coefficient of zero implies that no relationship exists between the two variables under study.

11.27 Draw two scatter plots, one for which $r = 1$ and a second for which $r = -1$.

11.28 A number of different methods for measuring growth rate in lobsters are discussed in the paper "A Comparison of Techniques for the Measurement of Growth in Adult Lobsters" (*Aquaculture* (1984): 195–9). Twenty-three adult female lobsters were included in the study and both dry weight and volume were recorded for each one. Letting x denote dry weight (in g) and y volume (in ml), summary quantities for the data given in the paper are

$$n = 23 \qquad \sum (x - \bar{x})^2 = 557582.1$$
$$\sum (y - \bar{y})^2 = 6347190.0 \qquad \sum (x - \bar{x})(y - \bar{y}) = 1410933.1$$

Use this information to calculate the correlation coefficient. Would you describe the correlation between dry weight and volume as positive or negative? As weak, moderate, or strong?

11.29 Sixteen different air samples were obtained at Herald Square in New York City, and both the carbon monoxide concentration x (ppm) and benzo(a) pyrene concentration y ($\mu g/10^3 m^3$) were measured for each sample ("Carcinogenic Air Pollutants in Relation to Automobile Traffic in New York City" *Environ. Science and Tech.* (1971): 145–50).

x	2.8	15.5	19.0	6.8	5.5	5.6	9.6	13.3
y	.5	.1	.8	.9	1.0	1.1	3.9	4.0

x	5.5	12.0	5.6	19.5	11.0	12.8	5.5	10.5
y	1.3	5.7	1.5	6.0	7.3	8.1	2.2	9.5

Compute the sample correlation coefficient for this data. What does the value of r suggest about the nature of the relationship between x and y?

11.30 The accompanying data on diesel oil consumption rate measured by the drain-weight method (x) and measured by the Cl-trace method (y), both in g/hr, was read from a graph in the paper "A New Measurement Method of Diesel Engine Oil Consumption Rate" (*J. of Soc. Auto Engr.* (1985): 28–33).

x	4	5	8	11	12	16	17	20	22	28	30	31	39
y	5	7	10	10	14	15	13	25	20	24	31	28	39

$$\sum x = 243 \qquad \sum y = 241 \qquad \sum x^2 = 5965 \qquad \sum y^2 = 5731 \qquad \sum xy = 5805$$

Determine the value of the sample correlation coefficient. Does there appear to be good agreement between the two methods?

11.31 Toughness and fibrousness of asparagus is a major determinant of quality. This was the focus of a study reported in "Post-Harvest Glyphosate Application Reduces Toughening, Fiber Content, and Lignification of Stored Asparagus Spears" (*J. of*

Amer. Soc. Horticultural Science (1988): 569–72). The paper reported the following data (read from a graph) on x = shear force (kg) and y = percent fiber dry weight.

x	46	48	55	57	60	72	81	85	94
y	2.18	2.10	2.13	2.28	2.34	2.53	2.28	2.62	2.63

x	109	121	132	137	148	149	184	185	187
y	2.50	2.66	2.79	2.80	3.01	2.98	3.34	3.49	3.26

$$n = 18 \qquad \sum x = 1950 \qquad \sum x^2 = 251970$$
$$\sum y = 47.92 \qquad \sum y^2 = 130.6074 \qquad \sum xy = 5530.92$$

a. Calculate the value of the sample correlation coefficient.
b. If you were to use regression analysis to relate percent dry weight to shear force, what percentage of the observed variation in percent dry weight could be explained by the approximate linear relationship? Answer without doing any regression calculations.

11.32 The article "Reduction in Soluble Protein and Chlorophyll Contents in a Few Plants as Indicators of Automobile Exhaust Pollution" (*Int. J. of Environ. Studies* (1983): 239–44) reported the accompanying data on x = distance from a highway (m) and y = lead content of soil at that distance (ppm).

x	.3	1	5	10	20	25
y	62.75	37.51	29.70	20.71	17.65	15.41

x	25	30	40	50	75	100
y	14.15	13.50	12.11	11.40	10.85	10.85

a. Calculate the value of the sample correlation coefficient r.
b. Construct a scatter plot. Is the value of r an effective summary quantity for describing the relationship between the two variables?

11.33 An employee of an auction house has a list of 25 recently sold paintings. Eight artists were represented in these sales. The sale price of each painting appears on the list. Would the correlation coefficient be an appropriate way to summarize the relationship between artist (x) and sale price (y)? Why or why not?

11.34 A sample of automobiles traversing a certain stretch of highway is selected. Each one travels at roughly a constant rate of speed, though speed does vary from auto to auto. Let x = speed and y = time needed to traverse this segment of highway. Would the sample correlation coefficient be closest to .9, .3, −.3, or −.9? Explain.

11.35 **a.** Show that $(n-1)s_x s_y$ is an equivalent expression for the denominator of r, where s_x and s_y are the sample standard deviations of the x values and y values, respectively.
b. An investigation of the relationship between water temperature x and calling rate y for a particular type of hybrid toad ("The Mating Call of Hybrids of the Fire-Bellied Toad and Yellow-Bellied Toad" *Oecologia* (1974): 61--71) yielded the following summary quantities:

$$n = 17 \qquad s_x = 3.86 \qquad s_y = 8.77 \qquad \sum(x - \bar{x})(y - \bar{y}) = 499.08.$$

Does this data indicate that there is a strong positive correlation between water temperature and calling rate? If water temperature decreases, what can be said about the calling rate?

11.36 Suppose that sample x and y values are first expressed in standard units by means of

$$x' = \frac{(x - \bar{x})}{s_x} \quad \text{and} \quad y' = \frac{(y - \bar{y})}{s_y}$$

How does the value of $\sum x'y'$, the sum of the products of these standardized values, relate to r? Hint: The denominator of r can be written as $(n - 1) s_x s_y$.

11.37 Suppose that x and y are positive variables and that a sample of n pairs results in $r \approx 1$, If the sample correlation coefficient is computed for the n pairs (x_1, y_1^2), $(x_2, y_2^2), \ldots, (x_n, y_n^2)$—i.e., for the (x, y^2) pairs—will the resulting value also be approximately 1? Explain.

11.38 A sample of $n = 5$ (x, y) pairs gives $(1, 1)$, $(2, 2)$, $(3, 3)$, $(4, 4)$, and $(5, y_5)$. It must be the case that both $x \geq 0$ and $y \geq 0$. Could r be negative?

11.39 Nine students currently taking introductory statistics are randomly selected, and both the first midterm exam score (x) and the second midterm exam score (y) are determined. Three of the students have the class at 8 A.M., another three have it at noon, and the remaining three have a night class. The resulting (x, y) pairs are as follows:

8 A.M.:	(70, 60)	(72, 83)	(94, 85)
Noon:	(80, 72)	(60, 74)	(55, 58)
Night:	(45, 63)	(50, 40)	(35, 54)

a. Calculate the sample correlation coefficient for the nine (x, y) pairs.
b. Let $\bar{x}_1$ = the average score on the first midterm for the 8 A.M. students and $\bar{y}_1$ = the average score on the second midterm for these students. Let $\bar{x}_2$ and $\bar{y}_2$ be these averages for the noon students, and $\bar{x}_3$ and $\bar{y}_3$ these averages for the evening students. Calculate r for these three $(\bar{x}, \bar{y})$ pairs.
c. Construct a scatter plot of the nine (x, y) pairs and construct another one of the three $(\bar{x}, \bar{y})$ pairs. Can you see why r in part (a) is smaller than r in part (b)? Does this suggest that a correlation coefficient based on averages (an "ecological" correlation) might be misleading? Explain.

CHAPTER 11 SUMMARY OF KEY CONCEPTS AND FORMULAS

TERM OR FORMULA	PAGE	COMMENT
Scatter plot	357	A picture of bivariate numerical data in which each observation (x, y) is represented as a point located with respect to a horizontal x-axis and a vertical y-axis.
Principle of least squares	365	The method used to select a line that summarizes an approximate linear relationship between x and y. The least-squares line is the line that minimizes $\sum [y - (a + bx)]^2$, the sum of the squared vertical deviations from the points in the scatter plot.
$b = \dfrac{\sum xy - n\bar{x}\bar{y}}{\sum x^2 - n\bar{x}^2}$ $a = \bar{y} - b\bar{x}$	366	The slope and y intercept of the least-squares line.
Predicted (fitted) values $\hat{y}_1, \ldots, \hat{y}_n$	373	Obtained by substituting the x value for each observation into the least-squares line: $\hat{y}_1 = a + bx_1, \ldots, \hat{y}_n = a + bx_n$
Residuals	373	Obtained by subtracting each predicted value from the corresponding observed y value: $y_1 - \hat{y}_1, \ldots, y_n - \hat{y}_n$. These are the vertical deviations from the least-squares line.

(continued on next page)

CHAPTER 11   SUMMARY OF KEY CONCEPTS AND FORMULAS

TERM OR FORMULA	PAGE	COMMENT
Residual (error) sum of squares, $\text{SSResid} = \sum (y - \hat{y})^2$	375	The sum of the squared residuals is a measure of y variation that cannot be attributed to an approximate linear relationship (unexplained variation).
Total sum of squares, $\text{SSTo} = \sum (y - \bar{y})^2$	375	The sum of squared deviations from the sample mean $\bar{y}$ is a measure of total variation in the observed y values.
Coefficient of determination, $r^2 = 1 - \dfrac{\text{SSResid}}{\text{SSTo}}$	377	The proportion of variation in observed y's that can be attributed to an approximate linear relationship
Pearson's sample correlation coefficient, $r = \dfrac{\sum xy - n\bar{x}\bar{y}}{\sqrt{\sum x^2 - n\bar{x}^2}\,\sqrt{\sum y^2 - n\bar{y}^2}}$	382	A measure of the extent to which sample x and y values are linearly related

SUPPLEMENTARY EXERCISES   11.40–11.50

11.40 Silane coupling agents have been used in the rubber industry to improve the performance of fillers in rubber compounds. The accompanying data on y = tensile modulus (in MPa, a measure of silane coupling effectiveness) and x = bound rubber content (%) appeared in the paper "The Effect of the Structure of Sulfur-Containing Silane Coupling Agents on Their Activity in Silica-Filled SBR" (*Rubber Chem. and Tech.* (1984): 675–85).

x	16.1	31.5	21.5	22.4	20.5	28.4	30.3	25.6	32.7	29.2	34.7
y	4.41	6.81	5.26	5.99	5.92	6.14	6.84	5.87	7.03	6.89	7.87

$$n = 11 \qquad \sum x = 292.9 \qquad \sum x^2 = 8141.75$$
$$\sum y = 69.03 \qquad \sum y^2 = 442.1903 \qquad \sum xy = 1890.200$$

a. Construct a scatter plot for this data. Does the plot look linear?
b. Find the equation of the least-squares line.
c. Compute and interpret the value of r^2.

11.41 The relationship between depth of flooding and the amount of flood damage was examined in the paper "Significance of Location in Computing Flood Damage" (*J. of Water Resources Planning and Mgmt.* (1985): 65–81). The data on x = depth of flooding (feet above first-floor level) and y = flood damage (as a percent of structure value) was obtained using a sample of flood insurance claims.

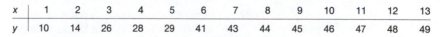

x	1	2	3	4	5	6	7	8	9	10	11	12	13
y	10	14	26	28	29	41	43	44	45	46	47	48	49

a. Obtain the equation of the least-squares line.
b. Construct a scatter plot and draw the least-squares line on the plot. Does it look as though a straight line provides an adequate description of the relationship between y and x? Explain.

c. Predict flood damage for a structure subjected to 6.5 ft of flooding.

d. Would you use the least-squares line to predict flood damage when depth of flooding is 18 ft? Explain.

11.42 An investigation was carried out to study the relationship between speed (ft/sec) and stride rate (number of steps taken/sec) among female marathon runners. Resulting summary quantities included

$$n = 11 \qquad \sum (\text{speed}) = 205.4 \qquad \sum (\text{speed})^2 = 3880.08$$
$$\sum (\text{rate}) = 35.16 \qquad \sum (\text{rate})^2 = 112.681 \qquad \sum (\text{speed})(\text{rate}) = 660.130$$

a. Calculate the equation of the least-squares line that you would use to predict stride rate from speed.

b. Calculate the equation of the least-squares line that you would use to predict speed from stride rate.

c. Calculate the coefficient of determination for the "stride rate on speed" regression of part (a) and for the "speed on stride rate" regression of part (b). How are these related?

11.43 The article "Sex Ratio Variation in Odocoileus: A Critical Review" (*J. of Wildlife Mgmt.* (1983): 573–82) reported the accompanying data on y = percent male fawns and x = fawns/doe for 29 groups of white-tailed, mule, and black-tailed deer.

x	1.03	1.43	1.79	1.74	0.95	1.47	0.54	1.76	1.76	1.09
y	57.6	60.1	45.7	35.6	68.4	41.5	75.0	51.3	50.0	60.5

x	1.97	1.65	1.69	1.41	1.68	1.26	1.70	1.92	1.25	1.71
y	57.0	44.0	52.3	40.7	58.4	43.8	48.8	52.9	48.6	53.1

x	1.22	1.24	1.44	1.62	1.47	1.64	1.85	1.19	1.32
y	51.9	56.0	51.9	57.6	50.0	44.9	44.1	65.7	59.1

$$\sum x = 42.79 \qquad \sum x^2 = 66.1519 \qquad \sum y = 1526.5$$
$$\sum y^2 = 82481.67 \qquad \sum xy = 2205.501$$

a. Determine the equation of the least-squares line.

b. Compute SSResid.

c. Calculate r^2 for this data set and interpret the resulting value.

11.44 Refer to Exercise 11.43 and calculate the residuals. Do any of the residuals appear to be unusually large? Which ones?

11.45 An alternative to Pearson's sample correlation coefficient r for assessing the strength of any relationship (linear or nonlinear) between x and y is **Spearman's rank correlation coefficient r_s**. To obtain r_s, each x value is replaced by its *rank* among the x's—1 for the smallest x, 2 for the next smallest, and so on. Similarly, each y is replaced by its rank among the y's. Then r_s is r applied to the (x rank, y rank) pairs. Calculate the value of r_s for the depth of flooding–flood damage data of Exercise 11.41. Does the result surprise you? Explain.

11.46 Spearman's rank correlation coefficient r_s was defined in Exercise 11.45. Calculate r_s for the rubber content–tensile modulus data of Exercise 11.40.

11.47 Consider the four (x, y) pairs $(0, 0)$, $(1, 1)$, $(1, -1)$, and $(2, 0)$.

a. What is the value of the sample correlation coefficient r?

b. If a fifth observation is made at the value $x = 6$, find a value of y for which $r > .5$.

c. If a fifth observation is made at the value $x = 6$, find a value of y for which $r < -.5$.

11.48   With s_x and s_y denoting the sample standard deviation of the x's and y's, respectively, an alternative form of the least-squares line is

$$\hat{y} = \bar{y} + r\left(\frac{s_y}{s_x}\right)(x - \bar{x})$$

Rearranging gives

$$\frac{\hat{y} - \bar{y}}{s_y} = r \cdot \frac{x - \bar{x}}{s_x}$$

The interpretation of this expression is that however many standard deviations x is from its mean value, we predict that y will be r times that many standard deviations from its mean. For example, if $r = .5$ and x is two standard deviations ($2s_x$) above $\bar{x}$, then $\hat{y}$ will be only one standard deviation (s_y) above its mean.

a. If $r = -.75$, what can be said about the prediction of y for an x value two standard deviations above its mean? For an x value one standard deviation below its mean?

b. A study by the Berkeley Institute of Human Development (see the book *Statistics* by Freedman, Pisani, and Purves listed in the Chapter 1 References) reported the following summary data for a sample of $n = 66$ California boys.

$r \approx .80$

At age 6, average height ≈ 46 in., standard deviation ≈ 1.7 in.

At age 18, average height ≈ 70 in., standard deviation ≈ 2.5 in.

Predict age-18 height when age six height is 50 in.

11.49   The least-squares intercept a and slope b come from solving a system of two linear equations called the **normal equations**:

$$na + \left(\sum x\right)b = \sum y$$

$$\left(\sum x\right)a + \left(\sum x^2\right)b = \sum xy$$

a. Verify by direct substitution that

$$a = \bar{y} - b\bar{x} = \frac{\sum y}{n} - b\frac{\sum x}{n}, \qquad b = \frac{\sum xy - n\bar{x}\bar{y}}{\sum x^2 - n\bar{x}^2}$$

is the solution to this system of equations.

b. Show that the first normal equation implies that the sum of the residuals from the least-squares line is zero, i.e., that $\sum (y - \hat{y}) = 0$.   Hint: Recall that

$$\hat{y} = a + bx.$$

11.50   **a.** Show that SSResid = SSTo $- b \cdot$ (numerator of b).
Hint: Use the computational formula for SSResid and the fact that $a = \bar{y} - b\bar{x}$.

b. Argue that the expression for SSResid in part (a) implies that SSResid $\leq$ SSTo.
Hint: Both b and the numerator of b have the same sign.

REFERENCES   Neter, John, William Wasserman, and Michael Kutner. *Applied Linear Statistical Models*. Homewood, Ill. Richard D. Irwin, Inc., 1985. (The first half of this book gives a comprehensive, up-to-date treatment of regression analysis without overindulging in mathematical development; a highly recommended reference.)

Younger, Mary Sue. *A Handbook for Linear Regression*. Boston, Mass.: Duxbury Press, 1985. (A good, thorough introduction to many aspects of regression; particularly recommended for its discussion of various statistical computer packages.)

C H A P T E R   12

Regression
and
Correlation:
Inferential
Methods

INTRODUCTION Regression and correlation were introduced in Chapter 11 as tools for description and summarization. In this chapter we develop inferential methods to use for drawing conclusions from a regression or correlation analysis. As a first step, the simple linear regression model is introduced in Section 1. The model postulates that there is a *population regression line* such that when (x, y) observations are made, the resulting points in the scatter plot will be spread about the population line in a random fashion. The least-squares line of Chapter 11 then becomes the estimate of the population line. Sections 2 and 3 present various methods for making inferences about model characteristics. Inferences about the population correlation coefficient are discussed in Section 4. The focus of Section 5 is on diagnostic techniques for deciding whether the simple linear regression model (or any other initially specified model) needs to be modified. Finally, multiple regression analysis, in which the model equation includes more than one independent (predictor) variable, is introduced in the last section.

12.1   THE SIMPLE LINEAR REGRESSION MODEL

A deterministic relationship is one in which the value of y is completely determined by the value of an independent variable x. A deterministic relationship can be described using traditional mathematical notation such as $y = f(x)$, where $f(x)$ is a specified function of x. For example, we might write $y = 10 + 2x$ or $y = 4 - (10)^{2x}$. However, the variables of interest are often not deterministically related, and a precise mathematical description of the relationship can be given by specifying a **probabilistic model**. The general form of an **additive probabilistic model** allows y to deviate from $f(x)$ by a random amount. The *model equation* is

$$y = \text{deterministic function of } x + \text{random deviation}$$
$$= f(x) + e$$

Let x^* denote some particular value of x, and suppose that an observation on y is made when $x = x^*$. If the value of the random deviation is positive then y will exceed $f(x^*)$, whereas if e is negative then y will be smaller than $f(x^*)$. Thinking geometrically, if $e > 0$, the observed point (x^*, y) will lie above the graph of $y = f(x)$, and $e < 0$ implies that this point will fall below the graph. This is illustrated in Figure 1. When $f(x)$ is a function used in a probabilistic model relating y to x and observations on y are made for various values of x, the resulting (x, y) points will be distributed about the graph of $f(x)$, some falling above it and others below it.

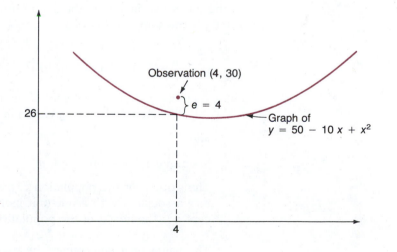

FIGURE 1

A deviation from the deterministic part of a probabilistic model.

SIMPLE LINEAR REGRESSION

The simple linear regression model is a specialization of the general probabilistic model in which the deterministic function $f(x)$ is linear (so its graph is a straight line).

||| **DEFINITION**

The **simple linear regression model** assumes that there is a line with slope β and y intercept α, called the **true** or **population regression line**. When x is fixed and an observation on y is made,

$$y = \alpha + \beta x + e$$

Because there is uncertainty about the value of the random deviation e (it is a random variable), y itself is a random variable.

The assumptions about e are as follows:

1. It has mean value 0 ($\mu_e = 0$).
2. It has standard deviation σ (which does not depend on x).
3. It has a normal distribution.
4. The random deviations $e_1, e_2, \ldots, e_n$ associated with different observations are independent of one another.

Consider properties of y when x has some fixed value x^*. Then

$$y = \alpha + \beta x^* + e$$

where $\alpha + \beta x^*$ is a fixed number. The sum of a fixed number and a normally distributed variable is again a normally distributed variable (the bell-shaped curve is simply relocated), so y itself has a normal distribution. Furthermore, $\mu_e = 0$ implies that the mean value of y is just $\alpha + \beta x^*$, the height of the population regression line above the value x^*. Lastly, because there is no variability in the fixed number $\alpha + \beta x^*$, the standard deviation of y is the same as that of e.

For any fixed x value, y itself has a normal distribution, with

$$\begin{pmatrix} \text{mean } y \text{ value} \\ \text{for fixed } x \end{pmatrix} = \begin{pmatrix} \text{height of the population} \\ \text{regression line above } x \end{pmatrix} = \alpha + \beta x$$

and

standard deviation of $y = \sigma$

The slope β of the population regression line is the *average* change in y associated with a one-unit increase in x, and the vertical intercept α is the height of the population line when $x = 0$.

The value of σ determines the extent to which (x, y) observations deviate from the population line: when σ is small, most observations will be quite close to the line, but with large σ, there are likely to be some substantial deviations.

The key features of the model are illustrated in Figure 2. Notice that the three normal curves in the figure have identical spreads. This is a consequence of $\sigma_e = \sigma$ independent of x.

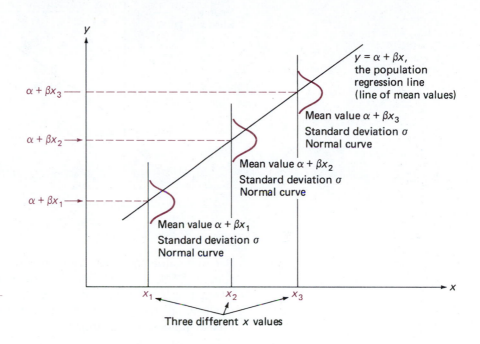

FIGURE 2

Illustration of the simple linear regression model.

EXAMPLE 1

It is often important to have a preliminary assessment of material strength. The article "Some Field Experience in the Use of an Accelerated Method in Estimating 28-Day Strength of Concrete" (*J. of Amer. Concrete Inst.* (1969): 895) suggests the simple linear regression model as a reasonable way to relate the variable

y = 28-day cured strength of concrete   (lb/in.²)

to the variable

x = accelerated strength   (lb/in.²)

Suppose that the actual model has

$\beta = 1.25,$ $\alpha = 1800,$ $\sigma = 350$ lb/in.²

The population regression line is shown in Figure 3. Thus for any fixed x, the variable y is normally distributed, with

$\mu_y = 1800 + 1.25x$

$\sigma_y = 350$

For example, when $x = 2000$, then 28-day strength has mean value

$\mu_y = 1800 + (1.25)(2000) = 4300$ lb/in.²

FIGURE 3

The population regression line for
Example 1.

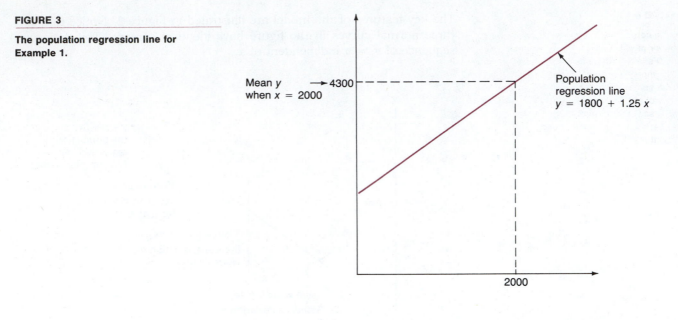

Because values between 3600 and 5000 are within 2σ of this mean value, roughly 95% of all 28-day strength observations made when accelerated strength is 2000 lb/in? will be between these limits. The slope $\beta = 1.25$ is the average increase in 28-day strength associated with a one-lb/in? increase in accelerated strength.

An intuitive way to appreciate the implications of the simple linear regression model is to think of the population of all (x, y) pairs as consisting of many smaller populations. Each one of these smaller populations contains pairs for which x has a fixed value. Suppose, for example, that the variables

x = grade point average in major courses

and

y = starting salary after graduation

are related according to the simple linear regression model. Then there is the population of all pairs with $x = 3.20$, the population of all pairs having $x = 2.75$, and so on. The model assumes that for each such population, y is normally distributed with the same standard deviation, and the *mean y* value (rather than y itself) is linearly related to x.

In practice, the judgment as to whether or not the simple linear regression model is appropriate must be based on sample data and a scatter plot. The plot should show a linear rather than a curved pattern, and the vertical spread of points should be relatively homogeneous throughout the range of x values. Figure 4 shows plots with three different patterns, only one of which is consistent with the model.

FIGURE 4

Some commonly encountered patterns in scatter plots.
(a) A scatter plot consistent with the simple linear regression model
(b) A scatter plot that suggests a nonlinear probabilistic model
(c) A scatter plot that suggests that variability in y is not the same for all x values

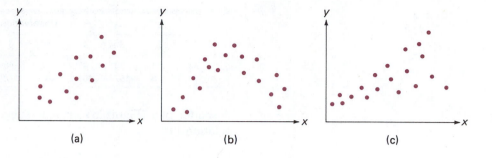

(a)   (b)   (c)

ESTIMATING THE POPULATION REGRESSION LINE

The values of α and β (vertical intercept and slope of the population regression line) will almost never be known to an investigator. Instead, these values must first be estimated from the sample data $(x_1, y_1), \ldots, (x_n, y_n)$. We now assume that these n (x, y) pairs were obtained independently of one another, where each observed y is related to the corresponding x via the model equation for simple linear regression.

Let a and b denote point estimates of α and β, respectively. These estimates come from applying the method of least squares.

The point estimates of β, the slope, and α, the y intercept of the population regression line, are the slope and y intercept, respectively, of the least-squares line. That is,

$$b = \text{point estimate of } \beta = \frac{\sum xy - n\bar{x}\bar{y}}{\sum x^2 - n\bar{x}^2}$$

$$a = \text{point estimate of } \alpha = \bar{y} - b\bar{x}$$

The estimated regression line is then just the least-squares line

$$\hat{y} = a + bx$$

Let x^* denote a specified value of the predictor variable x. Then $a + bx^*$ plays two different roles:

1. It is a (point) estimate of the mean y value when $x = x^*$.
2. It is a (point) prediction of an individual y value to be observed when $x = x^*$.

EXAMPLE 2

When anthropologists analyze human skeletal remains, an important piece of information is living stature (the height of the person when alive). Since skeletons are usually quite incomplete, inferences about stature are commonly based on statistical methods that utilize measurements on small bones. The paper "The Estimation of Adult Stature from Metacarpal Bone Length"(*Amer. J. of Phys. Anthro.* (1978): 113–20) presented data to validate one such method. Consider the accompanying representative data, where $x =$ metacarpal bone I length (cm) and $y =$ stature (cm).

	Observation									
	1	2	3	4	5	6	7	8	9	10
x	45	51	39	41	52	48	49	46	43	47
y	171	178	157	163	183	172	183	172	175	173

A scatter plot (Figure 5) strongly suggests the appropriateness of a linear relationship.

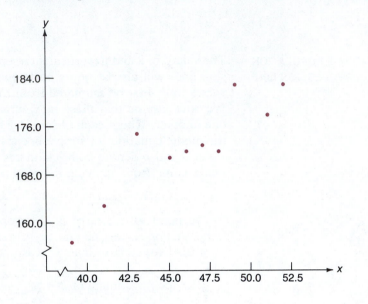

FIGURE 5

Scatter plot of the data from Example 2.

The summary statistics are

$$n = 10 \qquad \sum x = 461 \qquad \sum y = 1727$$

$$\sum x^2 = 21411 \qquad \sum xy = 79886 \qquad \sum y^2 = 298843$$

The estimated slope is

$$b = \frac{\sum xy - n\bar{x}\bar{y}}{\sum x^2 - n\bar{x}^2} = \frac{79886 - 10(46.1)(172.7)}{21411 - 10(46.1)^2}$$

$$= \frac{271.30}{158.90}$$

$$= 1.707363$$

$$\approx 1.707$$

and the estimated y intercept is

$$a = \bar{y} - b\bar{x} = 172.7 - (1.707363)(46.1)$$

$$= 93.990566$$

$$\approx 93.99$$

This gives us

$$\hat{y} = 93.99 + 1.707x$$

as the equation of the estimated regression line. A point estimate of the average stature for all population members having a metacarpal bone length of 45 cm results from substituting $x = 45$ into the estimated equation:

$$\begin{pmatrix} \text{estimate of average } y \\ \text{when } x = 45 \end{pmatrix} = a + b(45)$$

$$= 93.99 + (1.707)(45)$$

$$= 170.81 \text{ cm}$$

If an anthropologist obtains a 45-cm bone from an incomplete skeleton (of a population member), that individual's predicted stature would be

$$\begin{pmatrix} \text{predicted } y \text{ value} \\ \text{when } x = 45 \end{pmatrix} = a + b(45)$$

$$= 93.99 + (1.707)(45)$$

$$= 170.81 \text{ cm}$$

The point estimate and point prediction $a + bx^*$ are not as informative as we might like, since no information about the precision of estimation or prediction is conveyed. In Section 3 we will present both a confidence interval and a prediction interval to remedy this defect.

In Example 2, the x values in the sample ranged from 39 to 52. An estimate or prediction should not be calculated for any x value much outside this range. Without sample data for such values, there is no hard evidence that the estimated linear relationship can be extrapolated very far. Statisticians refer to this potential pitfall as the **danger of extrapolation**.

ESTIMATING σ^2 AND σ The value of σ determines the extent to which observed points (x, y) tend to fall close to or far away from the population regression line. A point estimate of σ is based on SSResid, which measures the extent to which the sample data spreads out about the estimated regression line.

▌▌▌ **DEFINITION**

The statistic for estimating the variance σ^2 is

$$s_e^2 = \frac{\text{SSResid}}{n - 2}$$

where $\text{SSResid} = \sum(y - \hat{y})^2$.

The estimate of σ is the **estimated standard deviation**

$$s_e = \sqrt{s_e^2}$$

It is customary to call $n - 2$ the number of degrees of freedom associated with estimating σ^2 or σ in simple linear regression.

The estimates and number of degrees of freedom here have analogues in our earlier work involving a single sample $x_1, x_2, \ldots, x_n$. The sample variance s^2 had numerator $\sum (x - \bar{x})^2$, a sum of squared deviations (residuals), and the denominator $n - 1$ was the number of degrees of freedom associated with s^2 and s. The use of $\bar{x}$ as an estimate of μ in the formula for s^2 reduced the number of degrees of freedom by one, from n to $n - 1$. In simple linear regression, estimation of α and β results in a loss of 2 degrees of freedom, leaving $n - 2$ as the d.f. for SSResid, s_e^2, and s_e.

The coefficient of determination, r^2, gives one assessment of how well the simple linear regression model is performing. It can now be interpreted as the proportion of observed y variation that can be explained by (or attributed to) the model relationship. The estimate s_e also gives another assessment of model performance. Roughly speaking, the value of σ represents the magnitude of a typical deviation of a point (x, y) in the population from the population regression line. Similarly, in a rough sense, s_e is the magnitude of a typical sample deviation (residual) from the least-squares line. The smaller the value of s_e, the closer the points in the sample fall to the line, and the better the line does in predicting y from x.

EXAMPLE 3

Forest managers are increasingly concerned about the damage done to natural (animal) populations when forests are clearcut. Woodpeckers are a valuable forest asset both because they provide nest and roost holes for other animals and birds and because they prey on many forest insect pests. The paper "Artificial Trees as a Cavity Substrate for Woodpeckers" (*J. of Wildlife Mgmt.* (1983): 790–98) reported on a study of how woodpeckers behaved when provided with polystyrene cylinders as an alternative roost and nest cavity substrate. We give selected values of $x =$ ambient temperature (°C) and $y =$ cavity depth (cm); these values were read from a scatter plot that appeared in the paper. Our plot (Figure 6), as well as the original plot, gives evidence of a strong negative linear relationship between x and y.

Row	Temp	Depth	Pred. y Value	Residual
1	−6	21.1	22.195	−1.095
2	−3	26.0	21.160	4.840
3	−2	18.0	20.815	−2.815
4	1	19.2	19.780	−0.580
5	6	16.9	18.055	−1.155
6	10	18.1	16.675	1.425
7	11	16.8	16.330	0.470
8	19	11.8	13.569	−1.769
9	21	11.0	12.879	−1.879
10	23	12.1	12.189	−0.089
11	25	14.8	11.499	3.301
12	26	10.5	11.154	−0.654

FIGURE 6

Scatter plot from MINITAB for Example 3.

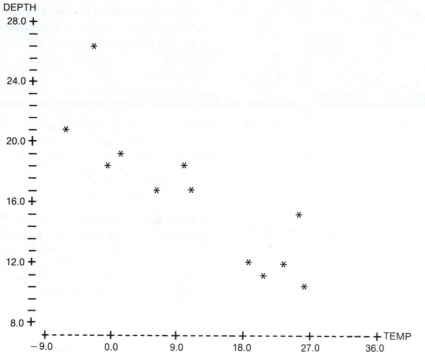

The summary statistics are

$$n = 12 \qquad \sum x = 131 \qquad \sum y = 196.3$$

$$\sum x^2 = 2939 \qquad \sum xy = 1622.3 \qquad \sum y^2 = 3445.25$$

from which we calculate

$$b = -.345043 \qquad a = 20.125053$$

$$\text{SSResid} = 54.4655 \qquad \text{SSTo} = 234.1093$$

Thus

$$r^2 = 1 - \frac{\text{SSResid}}{\text{SSTo}} = 1 - \frac{54.4655}{234.1093} = 1 - .233 = .767$$

$$s_e^2 = \frac{\text{SSResid}}{n - 2} = \frac{54.4655}{10} = 5.447$$

and

$$s_e = \sqrt{5.447} = 2.33$$

Approximately 76.7% of the observed variation in cavity depth y can be attributed to the probabilistic linear relationship with ambient temperature. The magnitude of a typical sample deviation from the least-squares line is about 2.3, which is reasonably small in comparison to the y values themselves. The model appears to be useful for estimation and prediction.

A key assumption of the simple linear regression model is that the random deviation e in the model equation is normally distributed. In Section 5 we will indicate how the residuals can be used to check whether this is plausible.

EXERCISES 12.1–12.12   **SECTION 12.1**

12.1 Let x be the size of a house (sq. ft) and y the amount of natural gas used (therms) during a specified period. Suppose that for a particular community, x and y are related according to the simple linear regression model with

$$\beta = \left(\begin{array}{c}\text{slope of population}\\\text{regression line}\end{array}\right) = .017, \qquad \alpha = \left(\begin{array}{c}y \text{ intercept of}\\\text{population regression line}\end{array}\right) = -5.0$$

a. What is the equation of the population regression line?

b. Graph the population regression line by first finding the point on the line corresponding to $x = 1000$, then the point corresponding to $x = 2000$, and drawing a line through these points.

c. What is the mean value of gas usage for houses with 2100 sq. ft of space?

d. What is the average change in usage associated with a one-square-foot increase in size?

e. What is the average change in usage associated with a 100-square-foot increase in size?

f. Would you use the model to predict usage for a 500-square-foot house? Why or why not? Note: There are no small houses in the community in which this model is valid.

12.2 The flow rate y (m^3/min) in a device used for air-quality measurement depends on the pressure drop x (in. of water) across the device's filter. Suppose that for x values between 5 and 20, these two variables are related according to the simple linear regression model with true regression line $y = -.12 + .095x$.

a. What is the true average (i.e., expected) flow rate for a pressure drop of 10 inches? A drop of 15 inches?

b. What is the true average change in flow rate associated with a 1-inch increase in pressure drop? Explain.

c. What change in flow rate can be expected when pressure drop decreases by 5 inches?

12.3 Data presented in the paper "Manganese Intake and Serum Manganese Concentration of Human Milk-Fed and Formula-Fed Infants" (*Amer. J. of Clinical Nutr.* (1984): 872–8) suggests that a simple linear regression model is reasonable for describing the relationship between $y = $ serum manganese (Mn) and $x = $ Mn intake (μg/kg/day). Suppose that the true regression line is $y = -2 + 1.4x$ and that $\sigma = 1.2$. Then for a fixed x value, y has a normal distribution with mean $-2 + 1.4x$ and standard deviation 1.2.

a. What is the mean value of serum Mn when Mn intake is 4.0? When Mn intake is 4.5?

b. What is the probability that an infant whose Mn intake is 4.0 will have serum Mn greater than 5?

c. Approximately what proportion of infants whose Mn intake is 5 will have a serum Mn greater than 5? Less than 3.8?

12.4 Suppose that a simple linear regression model is appropriate for describing the relationship between $y = $ house price and $x = $ house size (ft^2) for houses in a large city. The true regression line is $y = 23,000 + 47x$ and $\sigma = 5000$.

a. What is the average (i.e., expected) change in price associated with one extra square foot of space? With 100 extra square feet of space?

b. What proportion of 1800-ft^2 homes would be priced over $110,000? Under $100,000?

12.5   **a.** Explain the difference between the line $y = \alpha + \beta x$ and the line $\hat{y} = a + bx$.
     b. Explain the difference between β and b.
     c. Let x^* denote a particular value of the independent variable. Explain the difference between $\alpha + \beta x^*$ and $a + bx^*$.

12.6   Explain the difference between σ and s_e.

12.7   The given summary quantities were obtained from a study that used regression analysis to investigate the relationship between pavement deflection and surface temperature of the pavement at various locations on a state highway. Let $x = $ temperature (°F) and $y = $ deflection adjustment factor. (A scatter plot in the paper "Flexible Pavement Evaluation and Rehabilitation" *Trans. Eng. J.* (1977): 75–85, displayed many more than 15 observations.)

$$n = 15 \qquad \sum x = 1425 \qquad \sum y = 10.68$$
$$\sum xy = 987.645 \qquad \sum x^2 = 139037.25 \qquad \sum y^2 = .78518$$

a. Assuming that the simple linear regression model is appropriate, calculate point estimates of α and β, and then give the equation of the estimated regression line.

b. What is the estimated average change in adjustment factor associated with a one-degree increase in temperature? With a ten-degree increase in temperature?

12.8   The authors of the paper "Age, Spacing and Growth Rate of Tamarix as an Indication of Lake Boundary Fluctuations at Sebkhet Kelbia, Tunisia" (*J. of Arid Environ.* (1982): 43–51) used a simple linear regression model to describe the relationship between $y = $ vigor (average width in centimeters of the last two annual rings) and $x = $ stem density (stems/m^2). Data on which the estimated model was based is as follows.

x	4	5	6	9	14	15	15	19	21	22
y	.75	1.20	.55	.60	.65	.55	0	.35	.45	.40

a. Construct a scatter plot for the data.

b. Summary quantities are

$$\sum x = 130 \qquad \sum x^2 = 2090 \qquad \sum y = 5.5 \qquad \sum y^2 = 3.875 \qquad \sum xy = 59.95.$$

Find the estimated regression line and draw it on your scatter plot.

c. What is your estimate of the average change in vigor associated with a 1-unit increase in stem density?

d. What would you predict vigor to be for a plant whose density was 17 stems/m^2?

12.9   The accompanying summary quantities resulted from a study in which x was the number of photocopy machines serviced during a routine service call and y was the total service time (min):

$$n = 16 \qquad \sum (y - \bar{y})^2 = 22398.05 \qquad \sum (y - \hat{y})^2 = 2620.57$$

a. What proportion of observed variation in total service time can be explained by a linear probabilistic relationship between total service time and the number of machines serviced?

b. Calculate the value of the estimated standard deviation s_e. What is the number of degrees of freedom associated with this estimate?

12.10 Exercise 11.21 described a regression situation in which y = hardness of molded plastic and x = amount of time elapsed since termination of the molding process. Summary quantities included n = 15, SSResid = 1235.470, and SSTo = 25321.368.
  a. Calculate a point estimate of σ. On how many degrees of freedom is the estimate based?
  b. What percentage of observed variation in hardness can be explained by the simple linear regression model relationship between hardness and elapsed time?

12.11 The data below on x = advertising share and y = market share for a particular brand of cigarettes during ten randomly selected years appeared in the paper "Testing Alternative Econometric Models on the Existence of Advertising Threshold Effect" (*J. of Marketing Research* (1984): 298–308).

x	.103	.072	.071	.077	.086	.047	.060	.050	.070	.052
y	.135	.125	.120	.086	.079	.076	.065	.059	.051	.039

  a. Construct a scatter plot for this data. Do you think the simple linear regression model would be appropriate for describing the relationship between x and y?
  b. Calculate the equation of the estimated regression line and use it to obtain the predicted market share when the advertising share is .09.
  c. Compute r^2. How would you interpret this value?
  d. Calculate a point estimate of σ. On how many degrees of freedom is your estimate based?

12.12 Periodic measurements of salinity and water flow were taken in North Carolina's Pamlico Sound, resulting in the given data (*J. Amer. Stat. Assoc.* (1980): 828–38).

Water flow (x)	23	24	26	25	30	24	23	22
Salinity (y)	7.6	7.7	4.3	5.9	5.0	6.5	8.3	8.2

Water flow (x)	22	24	25	22	22	22	24
Salinity (y)	13.2	12.6	10.4	10.8	13.1	12.3	10.4

$$n = 15 \qquad \sum x = 358 \qquad \sum x^2 = 8608$$
$$\sum y = 136.3 \qquad \sum y^2 = 1362.59 \qquad \sum xy = 3195.0$$

  a. Find the equation of the estimated regression line.
  b. What would you predict salinity to be when water flow is 25?
  c. Estimate the mean salinity for times when the water flow is 29.
  d. What percentage of observed variation in salinity can be explained by the simple linear regression model?
  e. Roughly speaking, what is the magnitude of a typical deviation from the estimated regression line?

12.2 INFERENCES CONCERNING THE SLOPE OF THE POPULATION REGRESSION LINE

The slope β in the simple linear regression model is the average or expected change in the dependent variable y associated with a one-unit increase in the value of the independent variable x. Examples include the average change in vocabulary size associated with an age increase of one year, the expected change in yield associated with the use of an additional gram of catalyst, and the average change in annual maintenance expense associated with using a word processing system for one additional hour per week (all presuming that the simple linear regression model is appropriate).

For any specified population, the value of β will be a fixed number, but this value will almost never be known to an investigator. Instead, a sample of n independently selected observations $(x_1, y_1), \ldots, (x_n, y_n)$ will be available, and inferences concerning β are based on this data. In particular, substitution into the formula for b given earlier yields a point estimate of β. As with any point estimate, though, it is desirable to have some indication of how accurately b estimates β. In some situations, the value of the statistic b may vary greatly from sample to sample, so b computed from a single sample may well be rather different from β. In other situations, almost all possible samples may yield b values quite close to β, so the error of estimation is almost sure to be small. To proceed further, we need some facts about the sampling distribution of the statistic b—information concerning the shape of the sampling distribution curve, where the curve is centered relative to β, and how much the curve spreads out about its center.

PROPERTIES OF THE SAMPLING DISTRIBUTION OF b

1. The mean value of b is β. That is, $\mu_b = \beta$, so the sampling distribution of b is always centered at the value of β. Thus b is an unbiased statistic for estimating β.

2. The standard deviation of the statistic b is

$$\sigma_b = \frac{\sigma}{\sqrt{\sum (x - \bar{x})^2}} = \frac{\sigma}{\sqrt{\sum x^2 - n\bar{x}^2}}$$

3. The statistic b has a normal distribution (a consequence of assuming that the random deviation e is normally distributed).

Unbiasedness of b by itself is not a guarantee that the resulting estimate will be close to β. If σ_b is large, the normal sampling distribution curve will be quite spread out around β, and an estimate far from β may well result. For σ_b to be small, the numerator σ should be small (little variability about the population line) and/or the denominator should be large. This latter condition is equivalent to $\sum (x - \bar{x})^2$ being large. Because $\sum (x - \bar{x})^2$ is a measure of how much the observed x values spread out, we conclude that β will tend to be more precisely estimated when the x values in our sample are spread out than when they are close together.

Normality of b implies that the standardized variable $(b - \beta)/\sigma_b$ has a standard normal distribution. However, inferential methods cannot be based on this variable because the value of σ_b cannot be calculated from sample data (since the unknown σ appears in the numerator of σ_b). The obvious way out of this dilemma is to replace σ by s_e, yielding an estimated standard deviation.

The **estimated standard deviation of the statistic b** is

$$s_b = \frac{s_e}{\sqrt{\sum x^2 - n\bar{x}^2}}$$

The probability distribution of the standardized variable

$$t = \frac{b - \beta}{s_b}$$

is the t distribution with $n - 2$ degrees of freedom.

In the same way that $t = (\bar{x} - \mu)/(s/\sqrt{n})$ was used in Chapter 8 to develop a confidence interval for μ, the t variable in the preceding box can be employed to give a confidence interval (interval estimate) for β.

A **confidence interval for β**, the slope of the population regression line, has the form

$$b \pm (t \text{ critical value}) \cdot s_b$$

where the t critical value is based on $n - 2$ degrees of freedom. Appendix Table III gives critical values corresponding to the most frequently used confidence levels.

The interval estimate of β is centered at b and extends out from the center by an amount that depends on the sampling variability of b. When s_b is small, the interval will be narrow, implying that the investigator has precise knowledge of β.

EXAMPLE 4 Durable-press cotton fabric is produced by a chemical reaction involving formaldehyde. For economic reasons, finished fabric usually receives its first wash at home rather than at the manufacturing plant. Because the pH of in-home wash water varies greatly from location to location, textile researchers are interested in how pH affects different fabric properties. The paper "Influence of pH in Washing on the Formaldehyde-Release Properties of Durable-Press Cotton" (*Textile Research J.* (1981): 263–70) reported the accompanying data, read from a scatter plot, on $x =$ wash-water pH and $y =$ formaldehyde release (in ppm). The scatter plot (Figure 7) suggests the appropriateness of the simple linear regression model. The slope β in this context is the average change in formaldehyde release associated with a one-unit pH increase.

				Observation					
	1	**2**	**3**	**4**	**5**	**6**	**7**	**8**	**9**
x	5.3	6.8	7.1	7.1	7.2	7.6	7.6	7.7	7.7
y	545	770	780	790	680	760	790	795	935

				Observation					
	10	**11**	**12**	**13**	**14**	**15**	**16**	**17**	**18**
x	7.8	7.9	8.1	8.6	9.1	9.2	9.4	9.4	9.5
y	780	935	830	1015	1190	1030	1045	1250	1075

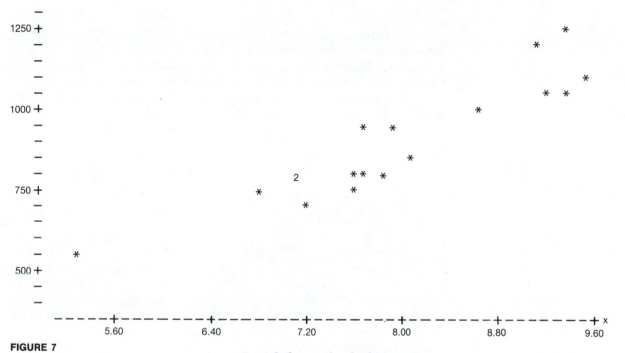

FIGURE 7

MINITAB scatter plot of the data from Example 4.

Straightforward calculation gives us

$$\sum x = 143.1 \qquad \bar{x} = 7.95 \qquad \sum y = 15995 \qquad \bar{y} = 888.611111$$

$$\sum x^2 = 1158.33 \qquad \sum xy = 130281.5 \qquad \sum y^2 = 14781675$$

$$b = 150.894368 \approx 150.894$$

$$a = -310.999115 \approx -311.00$$

$$\text{SSResid} = 97361.24 \qquad \text{SSTo} = 568340.2801$$

$$r^2 = .829$$

$$s_e^2 = 6085.0775 \qquad s_e = 78.01$$

$$s_b = \frac{s_e}{\sqrt{\sum x^2 - n\bar{x}^2}} = \frac{78.01}{\sqrt{20.685}} = 17.152$$

Relative to the magnitude of b itself, this estimated standard deviation is not particularly large. The 95% confidence interval based on 16 degrees of freedom requires t critical value = 2.12. The resulting interval is then

$$b \pm (t \text{ critical value}) \cdot s_b \approx 150.89 \pm (2.12)(17.15)$$
$$\approx 150.89 \pm 36.36$$
$$= (114.53, 187.25)$$

The investigator can be quite confident that, on the average, formaldehyde release will increase by from 114.53 ppm to 187.25 ppm when the pH is increased by one unit.

Output from any of the standard statistical computer packages routinely includes the computed values of a, b, SSResid, s_e, SSTo, r^2, and s_b. Figure 8 displays MINITAB output for the data of Example 4. The format from other packages is very similar. Rounding will occasionally lead to small discrepancies between hand-calculated and computer-calculated values, but there are no such discrepancies for this example.

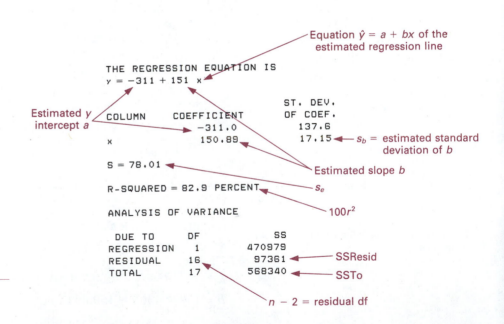

FIGURE 8

Partial MINITAB output for the data of Example 4.

HYPOTHESIS TESTS CONCERNING β

Hypotheses about β can be tested using a t test very similar to the t tests discussed in earlier chapters. The null hypothesis states that β has a specific hypothesized value. The t statistic results from standardizing b, the point estimate of β, under the assumption that H_0 is true. When H_0 is true, the sampling distribution of this statistic is the t distribution with $(n-2)$ df. The level of significance (type I error probability) is then controlled through the use of an appropriate t critical value from Table III.

SUMMARY OF HYPOTHESIS TESTS CONCERNING β

Null hypothesis: $H_0: \beta =$ hypothesized value

Test statistic: $t = \dfrac{b - \text{hypothesized value}}{s_b}$

ALTERNATIVE HYPOTHESIS	REJECTION REGION
$H_a: \beta >$ hypothesized value	$t > t$ critical value
$H_a: \beta <$ hypothesized value	$t < -t$ critical value
$H_a: \beta \neq$ hypothesized value	Either $t > t$ critical value or $t < -t$ critical value

---

Very frequently, the null hypothesis of interest is that $\beta = 0$. When this is the case, the population regression line is a horizontal line, and knowledge of x is of no use in predicting y. On the other hand, when $\beta \neq 0$ there *is* a useful linear relationship between x and y. The test of $H_0: \beta = 0$ versus $H_a: \beta \neq 0$ is often called the **model utility test** in simple linear regression. Since the hypothesized value is zero, the test statistic is the *t ratio* $t = b/s_b$. If a scatter plot and the value of r^2 do not provide convincing evidence for a useful linear relationship, it is recommended that the model utility test be carried out before the estimated model is used to make other inferences.

EXAMPLE 5

Data on surface hydrophobicity (x) for various proteins and FBC, fat-binding capacity (y), was given in Examples 6–8 of the previous chapter. Summary quantities include

$$n = 8 \qquad \sum x = 341 \qquad \sum y = 522.4 \qquad \sum x^2 = 20981$$

$$\sum xy = 28697.1 \qquad \sum y^2 = 44474.38$$

$$b = .9975 \qquad a = 22.78 \qquad \text{SSResid} = 3947.90$$

The fact that $r^2 = .619$ indicates that a substantial amount of variation in FBC cannot be explained by a linear relationship. Does the data suggest that there is indeed a useful linear relationship between x and y? Let's carry out a test at significance level .01 using the sequence of steps employed in earlier chapters.

1. $\beta =$ the average change in fat binding capacity associated with a one-unit increase in hydrophobicity
2. $H_0: \beta = 0$ (There is not a useful linear relationship.)
3. $H_a: \beta \neq 0$ (There is a useful linear relationship.)
4. Test statistic: $t = b/s_b$
5. Rejection region: The inequality in H_a calls for a two-tailed test based on $n - 2 = 8 - 2 = 6$ df. From Table III, the required critical value is 3.71. Thus H_0 will be rejected if either $t > 3.71$ or $t < -3.71$.

6. Computations: From the given information, $\sum x^2 - n\bar{x}^2 = 6445.875$, so

$$s_e = \sqrt{3947.90/(8-2)} = 25.651$$

$$s_b = \frac{s_e}{\sqrt{\sum x^2 - n\bar{x}^2}} = \frac{25.651}{\sqrt{6445.875}} = .3195$$

$$t = b/s_b = .9975/.3195 = 3.12$$

7. Conclusion: Because 3.12 neither exceeds 3.71 nor is less than -3.71, the hypothesis H_0 cannot be rejected at significance level .01. At this level, the data does not provide compelling evidence for a useful linear relationship. From Table III, the t critical value for a test with significance level .02 is 3.14. Since $3.12 \approx 3.14$, the P-value for the test is approximately .02, and H_0 would be rejected at level .05. (H_0 can be rejected whenever the P-value is smaller than the significance level.) Thus the use of a 5% significance level confirms the utility of the simple linear regression model. The test statistic value falls in that grey area where different investigators might reach opposite conclusions.

When $H_0: \beta = 0$ cannot be rejected by the model utility test at a reasonably small significance level, the search for a useful model must continue. One possibility is to relate x to y via a nonlinear model, an appropriate strategy if the scatter plot shows curvature. Details can be found in one of the chapter references. Alternatively, a multiple regression model using more than one predictor variable can be developed. Section 6 gives a very brief introduction to such models.

EXERCISES 12.13–12.28 **SECTION 12.2**

12.13 a. What is the difference between σ and σ_b?
b. What is the difference between σ_b and s_b?

12.14 a. Suppose that a single y observation is made at each of the x values 5, 10, 15, 20, and 25. If $\sigma = 4$, what is σ_b, the standard deviation of the statistic b?
b. Now suppose that a second y observation is made at every x value listed in part (a) (making a total of 10 observations). Is the resulting value of σ_b half of what it was in part (a)?
c. How many observations at each x value in part (a) are required to yield a σ_b value that is half the value calculated in part (a)? Verify your conjecture.

12.15 Exercise 11.13 described an experiment carried out to study the relationship between $x =$ chlorine flow and $y =$ etch rate. Summary quantities included

$$n = 9 \qquad \sum x = 24.0 \qquad \sum x^2 = 70.50$$

If a primary objective had been to estimate β as accurately as possible, would it have been better to make $n = 8$ observations at x values $x_1 = 1.5$, $x_2 = 1.5$, $x_3 = 2.0$, $x_4 = 2.0$, $x_5 = 3.5$, $x_6 = 3.5$, $x_7 = 4.0$, $x_8 = 4.0$?

12.16 Exercise 12.10 of Section 12.1 presented information from a study in which y was the hardness of molded plastic and x was the time elapsed since termination of the molding process. Summary quantities included

$$n = 15 \qquad b = 2.50 \qquad \text{SSResid} = 1235.470 \qquad \sum (x - \bar{x})^2 = 4024.20$$

a. Calculate s_b, the estimated standard deviation of the statistic b.

b. Obtain a 95% confidence interval for β, the slope of the true regression line.

c. Does the interval in part (b) suggest that β has been precisely estimated? Explain.

12.17 A study was carried out to relate sales revenue y (in thousands of dollars) to advertising expenditure x (also in thousands of dollars) for fast food outlets during a three-month period. A sample of 15 outlets yielded the accompanying summary quantities.

$$\sum x = 14.10 \qquad \sum y = 1438.50 \qquad \sum x^2 = 13.92 \qquad \sum y^2 = 140{,}354$$

$$\sum xy = 1387.20 \qquad \sum (y - \bar{y})^2 = 2401.85 \qquad \sum (y - \hat{y})^2 = 561.46$$

a. What proportion of observed variation in sales revenue can be attributed to the linear relationship between revenue and advertising expenditure?

b. Calculate s_e and s_b.

c. Obtain a 90% confidence interval for β, the average change in revenue associated with a thousand-dollar (i.e., one-unit) increase in advertising expenditure.

12.18 An experiment to study the relationship between $x =$ time spent exercising (min) and $y =$ amount of oxygen consumed during the exercise period resulted in the following summary statistics:

$$n = 20 \qquad \sum x = 50 \qquad \sum y = 16705 \qquad \sum x^2 = 150$$
$$\sum xy = 44194 \qquad \sum y^2 = 14{,}194{,}231$$

a. Estimate the slope and y intercept of the population regression line.

b. One sample observation on oxygen usage was 757 for a two-minute exercise period. What amount of oxygen would you predict for this exercise period, and what is the corresponding residual?

c. Compute a 99% confidence interval for the true average change in oxygen usage associated with a one-minute increase in exercise time.

12.19 The paper "Bumblebee Response to Variation in Nectar Availability" (*Ecology* (1981): 1648–61) reported a positive linear relationship between y, a measure of bumblebee abundance, and x, a measure of nectar availability. Representative data is given here.

x	3	8	11	10	23	23	30	35
y	4	6	12	18	11	24	22	37

a. Assuming that the simple linear regression model is valid, estimate β, the true average change in abundance associated with a one unit increase in availability, using a 90% confidence interval.

b. Does the interval in part (a) support the paper's claim of the existence of a positive linear relationship between x and y? Explain.

12.20 The paper "Effects of Enhanced UV-B Radiation on Ribulose-1,5-Biphosphate, Carboxylase in Pea and Soybean" (*Environ. and Exper. Botany* (1984): 131–43) included the accompanying pea plant data, with $y =$ sunburn index and $x =$ distance (cm) from an ultraviolet light source.

x	18	21	25	26	30	32	36	40	40	50	51	54	61	62	63
y	4.0	3.7	3.0	2.9	2.6	2.5	2.2	2.0	2.1	1.5	1.5	1.5	1.3	1.2	1.1

$$\sum x = 609 \qquad \sum y = 33.1 \qquad \sum x^2 = 28037$$
$$\sum xy = 1156.8 \qquad \sum y^2 = 84.45$$

Estimate the mean change in the sunburn index associated with an increase of 1 cm in distance in a way that includes information about the precision of estimation.

12.21 Exercise 12.17 described a regression analysis in which y = sales revenue and x = advertising expenditure. Summary quantities given there yield

$$n = 15 \quad b = 52.57 \quad s_b = 8.05$$

a. Test the hypotheses $H_0: \beta = 0$ vs. $H_a: \beta \neq 0$ using a significance level of .05. What does your conclusion say about the nature of the relationship between x and y?

b. Consider the hypotheses $H_0: \beta = 40$ vs. $H_a: \beta > 40$. The null hypothesis states that the average change in sales revenue associated with a one-unit increase in advertising expenditure is (at most) \$40,000. Carry out a test using significance level .01.

12.22 The accompanying data on fish survival and ammonia concentration is taken from the paper "Effects of Ammonia on Growth and Survival of Rainbow Trout in Intensive Static-Water Culture" (*Trans. of Amer. Fisheries Soc.* (1983): 448–54). Let x = ammonia exposure (mg/l) and y = percent survival.

x	10	10	20	20	25	27	27	31	50
y	85	92	85	96	87	80	90	59	62

a. Estimate the slope and intercept of the true regression line.

b. Is the simple linear regression model useful for predicting survival from knowledge of ammonia exposure?

c. Predict percent survival when ammonia exposure is 30 mg/l.

d. Estimate β using a 90% confidence interval.

12.23 A paper in the 1943 issue of the *Journal of Experimental Psychology* entitled "A Study of the Relationship between Hypnotic Susceptibility and Intelligence" reported on a study in which both an intelligence score x and a hypnotic susceptibility score y were obtained for each of 32 subjects. The computed values for summary quantities were:

$$\sum (x - \bar{x})^2 = 4929.22 \quad \sum (y - \bar{y})^2 = 1531.88$$
$$\sum (x - \bar{x})(y - \bar{y}) = 1192.69$$

Assuming that the simple linear regression model is appropriate, test $H_0: \beta = 0$ vs. $H_a: \beta > 0$ to determine if there is a *positive* relationship between intelligence and hypnotic susceptibility. Use level of significance .01.

12.24 The paper "Technology, Productivity and Industry Structure" (*Tech. Forecasting and Social Change* (1983): 1–13) included the accompanying data on x = research and development expenditure and y = growth rate for eight different industries.

x	2024	5038	905	3572	1157	327	378	191
y	1.90	3.96	2.44	.88	.37	−.90	.49	1.01

a. Would a simple linear regression model provide useful information for predicting growth rate from research and development expenditure? Use a .05 level of significance.

b. What can be said about the P-value associated with the test statistic value in part (a)?

c. Use a 90% confidence interval to estimate the average change in growth rate associated with a one-unit increase in expenditure. Interpret the resulting interval.

12.25 The paper "Effect of Temperature on the pH of Skim Milk" (*J. of Dairy Research* (1988): 277–80) reported on a study involving x = temperature (°C) under specified experimental conditions and y = milk pH. The accompanying data (read from a graph) is a representative subset of that which appeared in the paper.

x	4	4	24	24	25	38	38	40
y	6.85	6.79	6.63	6.65	6.72	6.62	6.57	6.52

x	45	50	55	56	60	67	70	78
y	6.50	6.48	6.42	6.41	6.38	6.34	6.32	6.34

$$\sum x = 678 \qquad \sum y = 104.54 \qquad \sum x^2 = 36056$$
$$\sum xy = 4376.36 \qquad \sum y^2 = 683.4470$$

Does this data strongly suggest that there is a negative (inverse) linear relationship between temperature and pH? State and test the relevant hypotheses using a significance level of .01.

12.26 The accompanying data on x = soil pH and y = Cl^- ion retention (ml/100 g) is from the paper "Single Equilibration Method for Determination of Cation and Anion Retention by Variable Charge Soils" (*Soil Science Plant Nutr.* (1984): 71–6).

x	6.15	6.11	5.88	6.45	5.80	6.06	5.83	6.33	7.35
y	0.14	0.37	1.47	1.12	2.08	1.79	3.18	2.15	0.51

x	8.18	7.69	7.29	6.53	5.01	5.34	6.19	5.81
y	0.32	0.76	2.13	2.75	6.69	5.59	2.87	4.22

$$n = 17 \qquad \sum x = 108.00 \qquad \sum x^2 = 697.2968$$
$$\sum y = 38.14 \qquad \sum y^2 = 140.9262 \qquad \sum xy = 225.3963$$

a. Obtain the equation of the estimated regression line.
b. Does the data indicate that the simple linear regression model is useful for predicting ion retention from pH? Use a level .01 test.
c. What can you say about the *P*-value associated with the computed value of the test statistic in part (b)?

12.27 In anthropological studies, an important characteristic of fossils is cranial capacity. Frequently skulls are at least partially decomposed, so it is necessary to use other characteristics to obtain information about capacity. One such measure that has been used is the length of the lambda–opisthion chord. A paper that appeared in the 1971 *Amer. J. of Phys. Anthro.* entitled "Vertesszollos and the Presapiens Theory" reported the following data for $n = 7$ *Homo erectus* fossils.

x (chord length in mm)	78	75	78	81	84	86	87
y (capacity in cm³)	850	775	750	975	915	1015	1030

Suppose that from previous evidence, anthropologists had believed that for each 1-mm increase in chord length, cranial capacity would be expected to increase by 20 cm³. Does this new experimental data strongly contradict prior belief? That is, should $H_0: \beta = 20$ be rejected in favor of $H_a: \beta \neq 20$? Use a .05 level of significance.

12.28 The article "Hydrogen, Oxygen, and Nitrogen in Cobalt Metal" (*Metallurgia* (1969): 121–7) contains a plot of the following data pairs, where x = pressure of extracted gas (in microns) and y = extraction time (min).

x	40	130	155	160	260	275	325	370	420	480
y	2.5	3.0	3.1	3.3	3.7	4.1	4.3	4.8	5.0	5.4

a. Suppose that the investigators had believed prior to the experiment that $\beta = .006$. Does the data contradict this prior belief? Use a significance level of .10.
b. Give an upper and/or lower bound on the *P*-value associated with the test statistic value in part (a).
c. Compute and interpret a 95% confidence interval for the slope of the true regression line.

12.3 INFERENCES BASED ON THE ESTIMATED REGRESSION LINE

We have seen how, for any specified x value, the estimated regression line $\hat{y} = a + bx$ gives either an estimate of the corresponding average y value or a prediction of a single y value. How precise is the resulting estimate or prediction? That is, how close might $a + bx$ be to the actual mean value $\alpha + \beta x$ or to a particular y observation? Because both a and b vary in value from sample to sample (each one is a statistic), for a fixed x the statistic $a + bx$ also has different values for different samples. The way in which this statistic varies in value with different samples is summarized by its sampling distribution. Properties of the sampling distribution are used to obtain both a confidence interval formula for $\alpha + \beta x$ and a prediction interval formula for a particular y observation. The narrowness of the corresponding interval conveys information about the precision of the estimate or prediction.

PROPERTIES OF THE SAMPLING DISTRIBUTION OF $a + bx$ FOR A FIXED x VALUE

Let x^* denote a particular value of the independent variable x. Then the sampling distribution of the statistic $a + bx^*$ has the following properties:

1. The mean value of $a + bx^*$ is $\alpha + \beta x^*$, so that $a + bx^*$ is an unbiased statistic for estimating the average y value when $x = x^*$.
2. The standard deviation of the statistic $a + bx^*$, denoted by σ_{a+bx^*}, is given by

$$\sigma_{a+bx^*} = \sigma \sqrt{\frac{1}{n} + \frac{(x^* - \bar{x})^2}{\sum x^2 - n\bar{x}^2}}$$

3. The assumption that the random deviation e in the model has a normal distribution implies that $a + bx^*$ is normally distributed.

The standard deviation of $a + bx^*$ is larger when $(x^* - \bar{x})^2$ is large than when it is small. That is, $a + bx^*$ tends to be a more precise estimate of $\alpha + \beta x^*$ when x^* is close to the center of the x values at which observations were made than when x^* is far from the center.

The standard deviation σ_{a+bx^*} cannot be calculated from the sample data because the value of σ is unknown. It can, however, be estimated by using s_e in place of σ. Using this estimated standard deviation to standardize $a + bx^*$ gives a variable with a t distribution.

> The **estimated standard deviation of the statistic $a + bx^*$**, denoted by s_{a+bx^*}, is given by
>
> $$s_{a+bx^*} = s_e \sqrt{\frac{1}{n} + \frac{(x^* - \bar{x})^2}{\sum x^2 - n\bar{x}^2}}$$
>
> The probability distribution of the standardized variable
>
> $$t = \frac{a + bx^* - (\alpha + \beta x^*)}{s_{a+bx^*}}$$
>
> is the t distribution with $n - 2$ degrees of freedom.

INFERENCES ABOUT THE MEAN VALUE $\alpha + \beta x^*$

Previous z and t standardized variables were manipulated to give confidence intervals of the form

(point estimate) $\pm$ (critical value) $\cdot$ (estimated standard deviation)

A parallel argument leads immediately to the following interval.

> A **confidence interval for $\alpha + \beta x^*$**, the average y value when x has value x^*, is
>
> $$a + bx^* \pm (t \text{ critical value}) \cdot s_{a+bx^*}$$
>
> where the t critical value is based on $(n - 2)$ df. Table III gives critical values corresponding to the most frequently used confidence levels.

Because of the dependence of s_{a+bx^*} on $(x^* - \bar{x})^2$, as discussed earlier, the confidence interval for $\alpha + \beta x^*$ gets wider as x^* moves farther from the center of the data.

EXAMPLE 6

Dairy scientists have recently carried out several studies on protein biosynthesis in milk and the accompanying decomposition of nucleic acids into various constituents. The paper "Metabolites of Nucleic Acids in Bovine Milk" (*J. of Dairy Science* (1984): 723–8) reported the accompanying data on x = milk production (kg/day) and y = milk protein (kg/day) for Holstein–Friesan cows.

	Observation						
	1	2	3	4	5	6	7
x	42.7	40.2	38.2	37.6	32.2	32.2	28.0
y	1.20	1.16	1.07	1.13	.96	1.07	.85

	Observation						
	8	9	10	11	12	13	14
x	27.2	26.6	23.0	22.7	21.8	21.3	20.2
y	.87	.77	.74	.76	.69	.72	.64

A scatter plot (Figure 9) gives strong support for using the simple linear regression model. In addition, it is easily verified that

$$b = .024576 \qquad a = .175571 \qquad SSResid = .021140$$
$$SSTo = .481436 \qquad s_e = .0420 \qquad \sum x^2 - n\bar{x}^2 = 762.012$$

from which we get $r^2 = .956$ and the t ratio for testing model utility:

$$t = \frac{b}{s_b} = \frac{.0246}{.042/\sqrt{762.012}} = \frac{.0246}{.00152} = 16.2$$

This calculated t considerably exceeds the upper-tailed critical value 3.06 for a two-tailed test with significance level .01. This confirms the utility of the model, so further inferences can now be carried out.

FIGURE 9

MINITAB scatter plot of the data from Example 6.

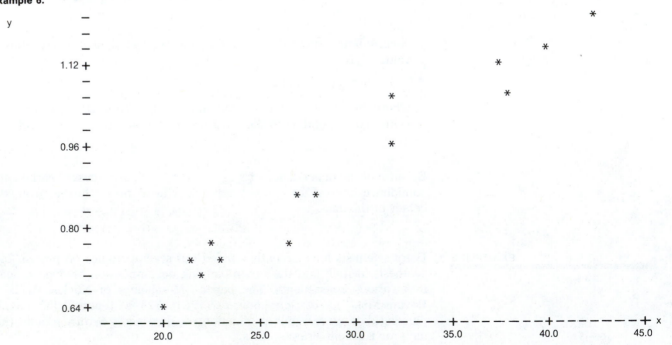

Let's use the data to compute a 99% confidence interval for average milk protein when milk production is 30 kg/day, that is, for $\alpha + \beta(30)$. The point estimate of $\alpha + \beta(30)$ is

$$a + b(30) = .176 + (.0246)(30) = .914$$

With $\bar{x} = 29.564$, the estimated standard deviation of $a + b(30)$ is

$$s_{a+b(30)} = s_e \sqrt{\frac{1}{n} + \frac{(30 - \bar{x})^2}{\sum x^2 - n\bar{x}^2}}$$

$$= .0420 \sqrt{\frac{1}{14} + \frac{(.436)^2}{762.012}}$$

$$= .0113$$

The t critical value for 99% confidence based on 12 df is 3.06. The confidence interval is

$$a + b(30) \pm (t \text{ crit. value}) \cdot s_{a+b(30)} = .914 \pm (3.06)(.0113)$$

$$= .914 \pm .035$$

$$= (.879, .949)$$

Even with the very high confidence level, this interval is relatively narrow, partly because $x = 30$ is very close to $\bar{x}$ and partly because the model fits the data so well ($s_e = .0420$ and $r^2 = .956$).

A PREDICTION INTERVAL FOR A SINGLE y

Suppose that an investigator is contemplating making a single observation on y when x has the value x^* at some future time. Let y^* denote the resulting future observation. Recall that the point prediction for y^* is $a + bx^*$, and this is also the point estimate for $\alpha + \beta x^*$, the mean y value when $x = x^*$. Consider now the errors of estimation and prediction:

estimation error = estimate − true value

$$= a + bx^* - (\alpha + \beta x^*)$$

prediction error = prediction − true value

$$= a + bx^* - y^*$$

In the estimation error, only $a + bx^*$ is subject to sampling variability, since $\alpha + \beta x^*$ is a fixed (albeit unknown) number. However, both $a + bx^*$ and the observation y^* in the prediction error are subject to sampling variability. This implies that there is more uncertainty associated with predicting a single value y^* than with estimating a mean value $\alpha + \beta x^*$.

We can obtain an assessment of how precise the prediction $a + bx^*$ is by computing a prediction interval for y^*. If the resulting interval is narrow, there is little uncertainty in y^*, and the prediction $a + bx^*$ is quite precise.

The interpretation of a prediction interval is very similar to the interpretation of a confidence interval. A 95% prediction interval for y^* is constructed using a method for which 95% of all possible samples would yield interval limits capturing y^*; only 5% of all samples would give an interval that did not include y^*.

Manipulation of a standardized variable similar to the one from which a confidence interval was obtained gives the following prediction interval.

A **prediction interval for y^***, a single y observation made when $x = x^*$, has the form

$$a + bx^* \pm (t \text{ critical value}) \cdot \sqrt{s_e^2 + s_{a+bx^*}^2}$$

The prediction interval and confidence interval are centered at exactly the same place, $a + bx^*$. The inclusion of s_e^2 under the square root symbol makes the prediction interval wider—often substantially so—than the confidence interval.

EXAMPLE 7

In Example 6 we computed a 99% confidence interval for average milk protein when milk production is 30 kg/day. Suppose that a single cow is randomly selected and its milk production on that day is found to be 30 kg. Let's compute a 99% prediction interval for y^*, the amount of protein in this milk. The necessary quantities are

$$a + b(30) = .176 + (.0246)(30) = .914$$

$$s_e^2 = (.0420)^2 = .001764$$

$$s_{a+b(30)}^2 = (.0113)^2 = .000128$$

The t critical value for 12 df and a 99% prediction level is 3.06. Substitution then gives the interval

$$.914 \pm (3.06)\sqrt{.001764 + .000128} = .914 \pm (3.06)(.0435)$$
$$= .914 \pm .133$$
$$= (.781, 1.047)$$

We can be quite confident that an individual cow with a daily milk production of 30 kg will have a daily milk protein yield of between .781 and 1.047 kg.

The confidence interval for the mean value $\alpha + \beta(30)$ is $.914 \pm .035$. The prediction interval is almost four times as wide as the confidence interval. Even with a rather precise estimate of $\alpha + \beta(30)$, there is a relatively wide range of plausible values for y^*.

Many statistical computer packages will give both the confidence interval and the prediction interval for a specified x^* upon request.

EXERCISES 12.29–12.44   **SECTION 12.3**

12.29 Explain the difference between a confidence interval and a prediction interval. How can a prediction level of 95% be interpreted?

12.30 Suppose that a regression data set is given and your are asked to obtain a confidence interval. How would you tell from the phrasing of the request whether the interval is for β or for $\alpha + \beta x^*$?

12.31 In Exercises 12.18, we considered a regression of $y =$ oxygen consumption on $x =$ time spent exercising. Summary quantities given there yield

$$n = 20 \qquad \bar{x} = 2.50 \qquad \sum x^2 - n\bar{x}^2 = 25$$
$$b = 97.26 \qquad a = 592.10 \qquad s_e = 16.486$$

a. Calculate $s_{a+b(2.0)}$, the estimated standard deviation of the statistic $a + b(2.0)$.
b. Without any further calculation, what is $s_{a+b(3.0)}$ and what reasoning did you use to obtain it?
c. Calculate the estimated standard deviation of the statistic $a + b(2.8)$.
d. For what value x^* is the estimated standard deviation of $a + bx^*$ smallest, and why?

12.32 The data of Exercise 12.25, in which $x =$ milk temperature and $y =$ milk pH, yields

$$n = 16 \qquad \bar{x} = 42.375 \qquad \sum x^2 - n\bar{x}^2 = 7325.75$$
$$b = -.00730608 \qquad a = 6.843345 \qquad s_e = .0356$$

a. Obtain a 95% confidence interval for $\alpha + \beta(40)$, the true average milk pH when the milk temperature is 40°C.
b. Calculate a 99% confidence interval for the true average milk pH when the milk temperature is 35°C.
c. Would you recommend using the data to calculate a 95% confidence interval for the true average pH when the temperature is 90°C? Why or why not?

12.33 Return to the regression of $y =$ milk pH on $x =$ milk temperature described in Exercise 12.32.
a. Obtain a 95% prediction interval for a single pH observation to be made with milk temperature = 40°C.
b. Calculate a 99% prediction interval for a single pH observation when milk temperature = 35°C.
c. When the milk temperature is 60°C, would a 99% prediction interval be wider than the intervals of parts (a) and (b)? Answer without calculating the interval.

12.34 The sugar content of certain types of fruit is a critical factor in determining when harvesting should begin. One method for assessing sugar content involves taking a measurement using a refractometer. The paper "Use of Refractometer to Determine Soluble Solids of Astringent Fruits of Japanese Persimmons" (*J. of Horticultural Science* (1983): 241–6) examined the relationship between $y =$ total sugar content (%) and $x =$ refractometer reading for persimmons. The estimated regression equation for predicting total sugar content from refractometer readings was given in the paper as $\hat{y} = -7.52 + 1.15x$. Suppose that $n = 50$, $\bar{x} = 17$, $s_e^2 = 1.1$, and $\sum (x - \bar{x})^2 = 112.5$.
a. Use a 95% confidence interval to estimate the mean percent of sugar for all persimmons with a refractometer reading of 18.
b. Construct a 90% prediction interval for the percent of sugar of an individual persimmon with a refractometer reading of 20.

c. Would a 90% prediction interval for the percent of sugar when the refractometer reading is 15 be narrower or wider than the interval of part (b)? Answer without computing the interval.

12.35 High blood-lead levels are associated with a number of different health problems. The paper "A Study of the Relationship between Blood Lead Levels and Occupational Lead Levels" (*Am. Stat.* (1983): 471) gave data on x = air-lead level ($\mu g/m^3$) and y = blood-lead level ($\mu g/dl$). Summary quantities (based on a subset of the data given in a plot appearing in the paper) are

$$n = 15 \qquad \sum x = 1350 \qquad \sum y = 600$$
$$\sum x^2 = 155{,}400 \qquad \sum y^2 = 24869.33 \qquad \sum xy = 57760$$

a. Find the equation of the estimated regression line.
b. Estimate the mean blood-lead level for people who work where the air-lead level is 100 $\mu g/m^3$ using a 90% interval.
c. Construct a 90% prediction interval for the blood-lead level of a particular person who works where the air-lead level is 100 $\mu g/m^3$.
d. Explain the difference in interpretation of the intervals computed in parts (b) and (c).

12.36 The paper "Digestive Capabilities in Elk Compared to White-Tailed Deer" (*J. of Wildlife Mgmt.* (1982): 22–29) examined the relationship between y = digestible amount of detergent-solubles (g) and x = amount of detergent-solubles in feed (%). Data for white-tailed deer is given.

x	30	40	40	48	56	60
y	15	28	27	29	33	38

a. Assuming that the simple linear regression model is appropriate, find a 95% confidence interval for the mean digestible amount of detergent-solubles when feed is composed of 36% detergent-solubles.
b. When x is 46%, would a 95% confidence interval for the mean y value be wider or narrower than the interval in part (a)? Explain.

12.37 The shelf life of packaged food depends on many factors. Dry cereal is considered to be a moisture-sensitive product (no one likes soggy cereal!) with the shelf life determined primarily by moisture content. In a study of the shelf life of one particular brand of cereal, x = time on shelf (stored at 73°F and 50% relative humidity) and y = moisture content were recorded. The resulting data is from "Computer Simulation Speeds Shelf Life Assessments" (*Package Engr.* (1983): 72–73).

x	0	3	6	8	10	13	16
y	2.8	3.0	3.1	3.2	3.4	3.4	3.5

x	20	24	27	30	34	37	41
y	3.1	3.8	4.0	4.1	4.3	4.4	4.9

a. Summary quantities are

$$\sum x = 269 \qquad \sum y = 51 \qquad \sum xy = 1081.5$$
$$\sum x^2 = 7445 \qquad \sum y^2 = 190.78$$

Find the equation of the estimated regression line for predicting moisture content from time on the shelf.
b. Does the simple linear regression model provide useful information for predicting moisture content from knowledge of shelf time?

c. Find a 95% interval for the moisture content of an individual box of cereal that has been on the shelf 30 days.

d. According to the paper, taste tests indicate that this brand of cereal is unacceptably soggy when the moisture content exceeds 4.1. Based on your interval in part (c), do you think that a box of cereal that has been on the shelf 30 days will be acceptable? Explain.

12.38 For the cereal data of Exercise 12.37 the average x value is 19.21. Would a 95% confidence interval with $x^* = 20$ or $x^* = 17$ be wider? Explain. Answer the same question for a prediction interval.

12.39 Rhizobia is a small soil bacteria that forms nodules on the roots of legumes and aids in fixing nitrogen. The number of viable rhizobia per clover seed at the time of sowing is of interest to crop scientists. The paper "Survival of Rhizobia on Commercially Lime-Pelleted White Clover and Lucerne Seed" (*N. Zeal. J. of Exp. Agric.* (1983): 275–8) gave the following data on $x =$ time stored (in weeks) and $y =$ number of viable rhizobia per seed for clover.

x	1	8	12	16	20	24	32	44
y	41	40	35	32	28	28	25	24

a. Find the equation of the estimated regression line.

b. Use a 95% confidence interval to estimate the mean number of rhizobia per seed for seeds stored 18 weeks.

c. Estimate the mean number of rhizobia for seed stored 22 weeks using a 95% confidence interval.

12.40 A regression of $y =$ sunburn index for a pea plant on $x =$ distance from an ultraviolet light source was considered in Exercise 12.20. The data and summary statistics presented there give

$$n = 15 \qquad \bar{x} = 40.60 \qquad \sum (x - \bar{x})^2 = 3311.60$$
$$b = -.0565 \qquad a = 4.500 \qquad \text{SSResid} = .8430$$

a. Calculate a 95% confidence interval for the true average sunburn index when the distance from the light source is 35 cm.

b. When two 95% confidence intervals are computed, it can be shown that the *simultaneous confidence level* is at least $[100 - 2(5)]\% = 90\%$. That is, if both intervals are computed for a first sample, then for a second sample, yet again for a third, and so on, in the long run at least 90% of the samples will result in intervals *both* of which capture the values of the corresponding population characteristics. Calculate confidence intervals for the true mean sunburn index when the distance is 35 cm and when the distance is 45 cm in such a way that the simultaneous confidence level is at least 90%.

c. If two 99% intervals were computed, what do you think could be said about the simultaneous confidence level?

12.41 The $n = 10$ observations from Exercise 12.28 on $x =$ pressure of extracted gas and $y =$ extraction time yield

$$b = .0068 \qquad a = 2.142 \qquad s_e = .106$$
$$s_{a+b(200)} = .0384 \qquad s_{a+b(250)} = .0349$$

a. Obtain a 95% prediction interval for a single observation to be made on extraction time when the pressure is 200 microns.

b. Obtain a 99% prediction interval for an observation on time when the pressure is 250 microns.

12.42 By analogy with the discussion in Exercise 12.40, when two different prediction intervals are computed, each at the 95% prediction level, the *simultaneous prediction level* is at least $[100 - 2(5)]\% = 90\%$.

a. Return to Exercise 12.41 and obtain prediction intervals for extraction time both when pressure is 200 microns and when pressure is 250 microns; so that the simultaneous prediction level is at least 90%.

b. If three different 99% prediction intervals were calculated for pressures of 200, 250, and 300 microns, what can be said about the simultaneous prediction level?

12.43 The article "Performance Test Conducted for a Gas Air-Conditioning System" (*Am. Soc. of Heating, Refrigerating, and Air Cond. Engr.* (1969): 54) reported the following data on maximum outdoor temperature (x) and hours of chiller operation per day (y) for a 3-ton residential gas air-conditioning system.

x	72	78	80	86	88	92
y	4.8	7.2	9.5	14.5	15.7	17.9

Suppose that the system is actually a prototype model, and the manufacturer does not wish to produce this model unless the data strongly indicates that when maximum outdoor temperature is 82°F, the true average number of hours of chiller operation is less than 12. The appropriate hypotheses are then H_0: $\alpha + \beta(82) = 12$ vs. H_a: $\alpha + \beta(82) < 12$. Use the statistic $t = [a + b(82) - 12]/s_{a+b(82)}$, which has a t distribution based on $(n - 2)$ df when H_0 is true, to test the hypotheses at significance level .01.

12.44 The paper "The Incorporation of Uranium and Silver by Hydrothermally Synthesized Galena" (*Econ. Geol.* (1964): 1003–24) reported on the determination of silver content of galena crystals grown in a closed hydrothermal system over a range of temperatures. With x = crystallization temperature (°C) and y = silver content (%), the data is

x	398	292	352	575	568	450	550
y	.15	.05	.23	.43	.23	.40	.44

x	408	484	350	503	600	600
y	.44	.45	.09	.59	.63	.60

Summary quantities are

$$\sum x = 6130 \qquad \sum y = 4.73 \qquad \sum xy = 2418.74$$
$$\sum x^2 = 3{,}022{,}050 \qquad \sum y^2 = 2.1785$$

When temperature equals 400°C, does the true average silver content appear to differ significantly from .25? Test the appropriate hypotheses at the .01 level of significance. Hint: Examine the hypotheses and test procedure given in Problem 12.43.

12.4 INFERENCES ABOUT THE POPULATION CORRELATION COEFFICIENT

The sample correlation coefficient r measures how strongly the x and y values in a *sample* of pairs are linearly related to one another. There is an analogous measure of how strongly x and y are related in the entire *population* of pairs from which the sample $(x_1, y_1), \ldots, (x_n, y_n)$ was obtained. It is called the **population correlation coefficient** and is denoted by ρ (notice again the use of a Greek letter for a population characteristic and Roman letter for a sample characteristic). Properties 1–5 of Chapter 11, Section 4, remain valid if ρ is substituted for r in every statement. In particular, ρ measures the extent of any *linear* association in the population. To have $\rho = 1$ or -1, all (x, y) pairs in the population must lie exactly on a straight line.

The relationship between r and ρ is similar to the relationship between $\bar{x}$ and μ, between p and π, and between b and β. The first-listed quantity in each pair is a statistic with a value that varies from sample to sample. A first sample of pairs might yield $r = .57$, a second might give $r = .65$, a third $r = .48$, and so on. On the other hand, the second quantity in each case has a fixed value characteristic of the population being studied—e.g., $\mu = 60.0$ or $\beta = 12.5$ or $\rho = .55$—but its value is typically unknown.

A TEST FOR INDEPENDENCE ($\rho = 0$)

Investigators are often interested in detecting not just linear association but association of *any* kind. When there is no association of any type between x and y values, statisticians say that the two variables are *independent*. In general, $\rho = 0$ is not equivalent to the independence of x and y. However, there is one special, yet frequently occurring, situation in which the two conditions ($\rho = 0$ and independence) are identical. This is when the pairs in the population have what is called a **bivariate normal distribution**. The essential feature of such a distribution is that for *any* fixed x value, the distribution of associated y values is normal, *and* for any fixed y value, the distribution of x values is normal. As an example, suppose that height x and weight y have a bivariate normal distribution in the American adult male population. (There is good empirical evidence for this.) Then when $x = 68$ in., weight y has a normal distribution; when $x = 72$ in., weight is normally distributed; when $y = 160$ lb, height x has a normal distribution; when $y = 175$ lb, height has a normal distribution; and so on. In this example, of course, x and y are not independent, since a large height value tends to be associated with large weight values and a small height value with small weight values.

There is no easy way to check the assumption of bivariate normality, especially when the sample size n is small. A partial check can be based on the following property: If (x, y) has a bivariate normal distribution, then x alone has a normal distribution and so does y. This suggests doing a normal probability plot of $x_1, x_2, \ldots, x_n$ and a separate normal probability plot of $y_1, \ldots, y_n$. If either plot shows a substantial departure from a straight line, bivariate normality is a questionable assumption. If both plots are reasonably straight, bivariate normality is plausible, although no guarantee can be given.

The test of independence (zero correlation) is a t test. The formula for the test statistic essentially involves standardizing the estimate r under the assumption that $\rho = 0$.

A TEST FOR INDEPENDENCE IN A BIVARIATE NORMAL POPULATION

Null hypothesis: $H_0: \rho = 0$ (x and y are independent)

Test statistic: $t = \dfrac{r}{\sqrt{(1 - r^2)/(n - 2)}}$

ALTERNATIVE HYPOTHESIS	REJECTION REGION
$H_a: \rho > 0$ (positive dependence)	$t > t$ critical value
$H_a: \rho < 0$ (negative dependence)	$t < -t$ critical value
$H_a: \rho \neq 0$ (dependence)	Either $t > t$ critical value or $t < -t$ critical value

The t critical value is based on $n - 2$ degrees of freedom.

EXAMPLE 8

An accurate assessment of soil productivity is an essential input to rational land-use planning. Unfortunately, as the author argues in the article "Productivity Ratings Based on Soil Series" (*Prof. Geographer* (1980): 158–63), an acceptable soil productivity index is not easy to come by. One difficulty is that productivity is determined partly by the crop that is planted, and the relationship between the yields of two different crops planted in the same soil may not be very strong. To illustrate, the paper presents the accompanying data on corn yield, x, and peanut yield, y (mT/Ha).

	Observation							
	1	2	3	4	5	6	7	8
x	2.4	3.4	4.6	3.7	2.2	3.3	4.0	2.1
y	1.33	2.12	1.80	1.65	2.00	1.76	2.11	1.63

The scatter plot in Figure 10 certainly casts doubt on the possibility of any relationship between the two yields. Assuming that the (x, y) pairs were

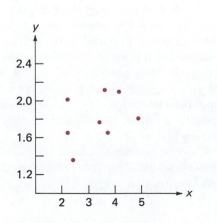

FIGURE 10

Scatter plot of y = peanut yield versus x = corn yield.

drawn from a bivariate normal population, let's carry out a test of independence using significance level .10.

1. ρ = the correlation between corn yield and peanut yield in the population from which the given eight observations were selected.
2. H_0: $\rho = 0$
3. H_a: $\rho \neq 0$
4. Test statistic:

$$t = \frac{r}{\sqrt{(1 - r^2)/(n - 2)}}$$

5. The t critical value for a two-tailed test based on $n - 2 = 6$ df is 1.94, so H_0 will be rejected in favor of H_a if either $t > 1.94$ or $t < -1.94$.

6. Summary quantities are

$$\sum x = 25.7 \qquad \sum y = 14.40 \qquad \sum x^2 = 88.31$$
$$\sum y^2 = 26.4324 \qquad \sum xy = 46.856$$

from which, after some further calculation, we get

$$r = \frac{.5960}{\sqrt{5.74875}\sqrt{.51240}} = .347$$

The test statistic value is

$$t = \frac{.347}{\sqrt{[1 - (.347)^2]/6}} = \frac{.347}{\sqrt{.880/6}} = .91$$

7. Because .91 neither exceeds 1.94 nor is less than -1.94, it does not fall in the rejection region. The data is quite consistent with the assertion that $\rho = 0$, so there does not appear to be any dependence between the two yields.

In the context of regression analysis, the hypothesis of no linear relationship (H_0: $\beta = 0$) was tested using the t ratio b/s_b. Some algebraic manipulation shows that $r/\sqrt{(1 - r^2)/(n - 2)} = b/s_b$, so the two test procedures are completely equivalent. The reason for using the formula for t that involves r is that when interest lies only in correlation, the extra effort involved in computing the regression quantities b, a, SSResid, s_e, and s_b need not be expended.

Other inferential procedures for drawing conclusions about ρ—a confidence interval or a test of hypotheses with nonzero hypothesized value—are somewhat complicated. One of the chapter references can be consulted for details.

EXERCISES 12.45–12.55 **SECTION 12.4**

12.45 Discuss the difference between r and ρ.

12.46 **a.** If the sample correlation coefficient is equal to 1, is it necessarily true that $\rho = 1$?

b. If $\rho = 1$, is it necessarily true that $r = 1$?

12.47 The paper "A Dual-Buffer Titration Method for Lime Requirement of Acid Mine-soils" (*J. of Environ. Qual.* (1988): 452–6) reported on the results of a study relating to revegetation of soil at mine reclamation sites. With $x =$ KCl extractable aluminum and $y =$ amount of lime required to bring soil pH to 7.0, data in the paper resulted in the accompanying summary statistics.

$$n = 24 \qquad \sum x = 48.15 \qquad \sum x^2 = 155.4685$$
$$\sum y = 263.5 \qquad \sum y^2 = 3750.53 \qquad \sum xy = 658.455$$

Carry out a test at significance level .01 to see whether the population correlation coefficient is something other than zero.

12.48 The accompanying summary quantities for $x =$ particulate pollution ($\mu g/m^3$) and $y =$ luminance (.01 cd/m^2) were calculated from a representative sample of data that appeared in the paper "Luminance and Polarization of the Sky Light at Seville (Spain) Measured in White Light" (*Atmos. Environ.* (1988): 595–9):

$$n = 15 \qquad \sum x = 860 \qquad \sum y = 348$$
$$\sum x^2 = 56700 \qquad \sum y^2 = 8954 \qquad \sum xy = 22265$$

a. Test to see whether there is a positive correlation between particulate pollution and luminance in the population from which the data was selected.
b. What proportion of observed variation in luminance can be attributed to the approximate linear relationship between luminance and particulate pollution?

12.49 In a study of bacterial concentration in surface and subsurface water ("Pb and Bacteria in a Surface Microlayer" *J. of Marine Research* (1982): 1200–6), the following data was obtained.

Concentration ($\times 10^6$/ml)								
Surface	48.6	24.3	15.9	8.29	5.75	10.8	4.71	8.26	9.41
Subsurface	5.46	6.89	3.38	3.72	3.12	3.39	4.17	4.06	5.16

Summary quantities are

$$\sum x = 136.02 \qquad \sum y = 39.35 \qquad \sum xy = 673.65$$
$$\sum x^2 = 3602.65 \qquad \sum y^2 = 184.27$$

a. Using a significance level of .05, determine whether the data supports the hypothesis of a linear relationship between surface and subsurface concentration.
b. Give an upper and/or lower bound on the P-value associated with the computed test statistic value in part (a).

12.50 Physical properties of six flame-retardant fabric samples were investigated in the paper "Sensory and Physical Properties of Inherently Flame-Retardant Fabrics" (*Textile Research* (1984): 61–8). Use the accompanying data and a .05 significance level to determine if a linear relationship exists between stiffness and thickness.

Stiffness (mg-cm)	7.98	24.52	12.47	6.92	24.11	35.71
Thickness (mm)	.28	.65	.32	.27	.81	.57

12.51 The April 11, 1983, issue of *Advertising Age* gave the accompanying data on $x =$ memory size (K) and $y =$ retail price (in dollars) for 13 of the many home computer systems on the market. Does this data suggest a linear relationship between x and y? Use a .10 significance level.

x	2	1	4	5	16	16	16
y	80	80	100	200	300	300	400

x	16	64	128	32	48	64
y	450	595	795	995	679	899

12.52 The paper "Chronological Trend in Blood Lead Levels" (*New. Eng. J. of Med.* (1983):1373–7) gave the following data on y = average blood-lead level of white children ages 6 months to 5 years and x = amount of lead used in gasoline production (in 1000 tons) for ten 6-month periods.

x	48	59	79	80	95	95	97	102	102	107
y	9.3	11.0	12.8	14.1	13.6	13.8	14.6	14.6	16.0	18.2

a. Construct separate normal probability plots for x and y. Do you think that it is reasonable to assume that the (x, y) pairs are from a bivariate normal population?

b. Does the data provide sufficient evidence to conclude that there is a linear relationship between blood-lead level and the amount of lead used in gasoline production? Use $\alpha = .01$

c. If a simple linear regression analysis were carried out, what proportion of observed variation in blood-lead level could be explained by the model relationship?

12.53 The paper "The Mechanics of Swimming Muskrats" (*J. of Exp. Biology* (1984): 183–201) contained a scatter plot of y, the arc (in degrees) through which the hind feet were swept during the power phase, versus x, the swimming velocity (m/sec). Selected data is given.

x	.25	.30	.35	.40	.45	.50	.50
y	98	92	87	97	101	116	96

x	.55	.55	.60	.65	.70	.75
y	115	114	110	115	123	133

$$\sum x = 6.55 \qquad \sum y = 1397 \qquad \sum x^2 = 3.5775$$
$$\sum y^2 = 152283 \qquad \sum xy = 725.35$$

a. Compute the value of the sample correlation coefficient.

b. Does the data suggest that muskrats increase swimming speed in a linear fashion by increasing the sweep arc of their hind feet? State and test the appropriate hypotheses at level of significance .05.

c. What can you say about the P-value corresponding to the test statistic value computed in part (b)?

d. How would your conclusion change if x were expressed in ft/sec? Explain.

12.54 A sample of n = 500 (x, y) pairs was collected and a test of $H_0: \rho = 0$ vs. $H_a: \rho \neq 0$ was carried out. The resulting P-value was computed to be .00032.

a. What conclusion would be appropriate at level of significance .001?

b. Does this small P-value indicate that there is a very strong linear relationship between x and y (a value of ρ that differs considerably from zero)? Explain.

12.55 A sample of n = 10,000 (x, y) pairs resulted in r = .022. Test $H_0: \rho = 0$ vs. $H_a: \rho \neq 0$ at level. 05. Is the result statistically significant? Comment on the practical significance of your analysis.

12.5   CHECKING ADEQUACY OF A LINEAR REGRESSION MODEL

The simple linear regression model equation, introduced in Section 12.1, is given by

$$y = \alpha + \beta x + e$$

where e represents the random deviation of an observed y value from the population regression line $y = \alpha + \beta x$. The inferential methods presented in the previous sections require some assumptions about e. Key assumptions are

1. e has a normal distribution;
2. The standard deviation of e is σ, which does not depend on x.

Inferences based on the simple linear regression model continue to be reliable when model assumptions are slightly violated (e.g., mild non-normality of the random deviation distribution). However, use of an estimated model in the face of grossly violated assumptions can result in very misleading conclusions being drawn. Therefore, it is desirable to have easily applied methods available for identifying such serious violations and for suggesting how a satisfactory model can be obtained.

RESIDUAL ANALYSIS   If the deviations $e_1, e_2, \ldots, e_n$ from the population line were available, they could be examined for any inconsistencies with model assumptions. For example, a normal probability plot would suggest whether the normality assumption was tenable. But, since

$$e_1 = y_1 - (\alpha + \beta x_1)$$
$$\vdots$$
$$e_n = y_n - (\alpha + \beta x_n)$$

these deviations can be calculated only if the equation of the population line is known. In practice, this will never be the case. Instead, diagnostic checks must be based on the residuals

$$y_1 - \hat{y}_1 = y_1 - (a + bx_1)$$
$$\vdots$$
$$y_n - \hat{y}_n = y_n - (a + bx_n)$$

which are the deviations from the estimated line.

Before a sample has been selected, any particular residual $y_i - \hat{y}_i$ is a random variable because its value varies from sample to sample. When all model assumptions are met, the mean value of any residual is zero. Any observation that gives a very large positive or negative value should be examined carefully for any anomalous circumstances, such as a recording error or exceptional experimental conditions. Identifying residuals with unusually large magnitudes is made easier by inspecting **standardized residuals**.

Recall that a quantity is standardized by subtracting its mean value (zero in this case) and dividing by its standard deviation. Thus

$$\text{standardized residual} = \frac{\text{residual}}{\text{standard deviation of residual}}$$

The value of a standardized residual tells how many standard deviations the corresponding residual lies from its expected value, zero.

Since each residual $y_i - \hat{y}_i$ has a different standard deviation (depending on the value of x_i for that observation), computing the standardized residuals can be tiresome. Fortunately, many computer regression programs provide standardized residuals as part of the output.

In Section 6.3, the normal probability plot was introduced as a technique for deciding if the n observations in a random sample could plausibly have come from a normal population distribution. To check the assumption that $e_1, e_2, \ldots, e_n$ all come from the same normal distribution, we recommend a normal probability plot of the standardized residuals.

EXAMPLE 9

Example 2 of Chapter 11 introduced data on $x =$ tree age and $y = $ 5-year growth. A simple linear regression model analysis was carried out. The residuals, their standard deviations, and the standardized residuals are given here in Table 1. Except for $x = 5$ and $x = 18$, the two most extreme values, the residuals have roughly equal standard deviations. The residual with the largest magnitude, -99.2, initially seems quite extreme, but the corresponding standardized residual is only -2.04. That is, the residual is approximately two standard deviations below its expected value zero, which is not terribly unusual in a sample this size. On the standardized scale, no residual here is surprisingly large. Before standardization, there are some large residuals simply because there appears to be a substantial amount of variability about the true regression line ($s_e = 51.43$, $r^2 = .683$).

TABLE 1

Residuals and Standardized Residuals for Example 9

Obs.	x	y	$\hat{y}$	Residual	Standard Deviation of Residual	Standardized Residual
1	5	70	111.0	−41.0	40.9	−1.00
2	9	150	196.2	−46.2	48.1	−.96
3	9	260	196.2	63.8	48.1	1.33
4	10	230	217.6	12.4	49.0	.25
5	10	255	217.6	37.4	49.0	.76
6	11	165	238.9	−73.9	49.5	−1.49
7	11	225	238.9	−13.9	49.5	−.28
8	12	340	260.2	79.8	49.8	1.60
9	13	305	281.5	23.5	49.7	.47
10	13	335	281.5	53.5	49.7	1.08
11	14	290	302.9	−12.9	49.4	−.26
12	14	340	302.9	37.1	49.4	.75
13	15	225	324.2	−99.2	48.7	−2.04
14	15	300	324.2	−24.2	48.7	−.50
15	18	380	388.1	−8.1	44.6	−.18
16	18	400	388.1	11.9	44.6	.27

Figure 11 displays a normal probability plot of the standardized residuals. Few plots are straighter than this one! The plot casts no doubt on the normality assumption.

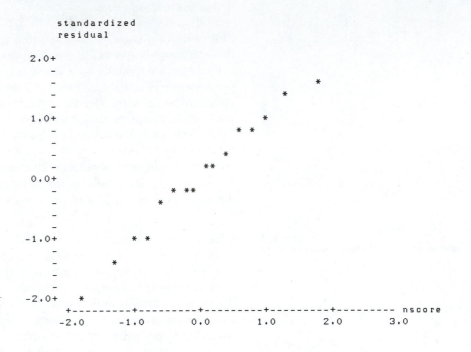

FIGURE 11

Normal probability plot of the standardized residuals from Example 9 (from MINITAB).

PLOTTING THE RESIDUALS

A plot of the (x, standardized residual) pairs, called a **standardized residual plot**, is often helpful in identifying unusual or highly influential observations and in checking for violations of model assumptions. A desirable plot is one that exhibits no particular pattern (such as curvature or a much greater spread in one part of the plot than in another), and one that has no point that is far removed from all the others. A point falling far above or below the horizontal line at height zero corresponds to a large standardized residual, which may indicate some kind of unusual behavior, such as a recording error, a nonstandard experimental condition, or an atypical experimental subject. A point that has an x value that differs greatly from others in the data set could have exerted excessive influence in determining the fitted line.

A plot such as the one pictured in Figure 12(a) is desirable, since no point lies much outside the horizontal band between -2 and 2 (so there is no unusually large residual corresponding to an outlying observation); there is no point far to the left or right of the others (thus no observation that might greatly influence the fit), and there is no pattern to indicate that the model should somehow be modified. When the plot has the appearance of Figure 12(b), the fitted model should be changed to incorporate curvature (a nonlinear model). The details of fitting such models are beyond the scope of this text.

The increasing spread from left to right in Figure 12(c) suggests that the variance of y is not the same at each x value but rather increases with x.

A straight-line model may still be appropriate, but the best-fit line should be selected by using weighted least squares rather than ordinary least squares, as described earlier. This involves giving more weight to observations in the region exhibiting low variability and less weight to observations in the region exhibiting high variability. A specialized regression analysis text or a knowledgeable statistician should be consulted for details.

The standardized residual plots of Figures 12(d) and (e) show an extreme outlier and a potentially influential observation, respectively. Consider deleting the observation corresponding to such a point from the data set and refitting the same model. Substantial changes in estimates and various other quantities warn of instability in the data. The investigator should certainly carry out a more careful analysis and perhaps collect more data before drawing any firm conclusions. Improved computing power has allowed statisticians to develop and implement a variety of diagnostics for identifying unusual observations in a regression data set.

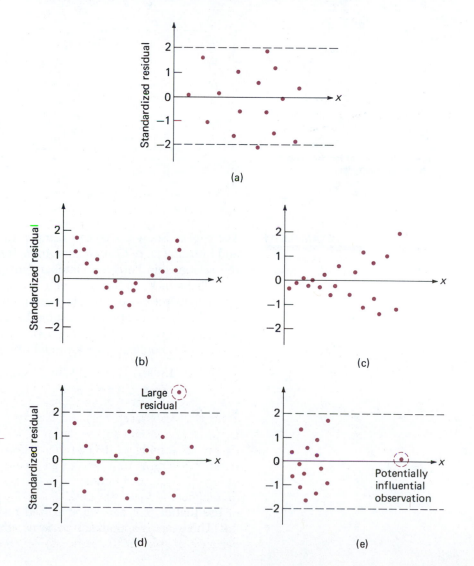

FIGURE 12

Examples of residual plots.
(a) A satisfactory plot
(b) A plot suggesting that a curvilinear regression model is needed
(c) A plot indicating nonconstant variance
(d) A plot showing a large residual
(e) A plot showing a potentially influential observation

EXAMPLE 10

Figure 13 displays a standardized residual plot for the tree age–five-year growth data of Example 2 from Chapter 11. The first observation was at $x_1 = 5$ and the corresponding standardized residual was -1.00, so the first plotted point is $(5, -1.00)$. Other points are similarly obtained and plotted. The plot shows no unusual behavior that might call for model modification or further analysis.

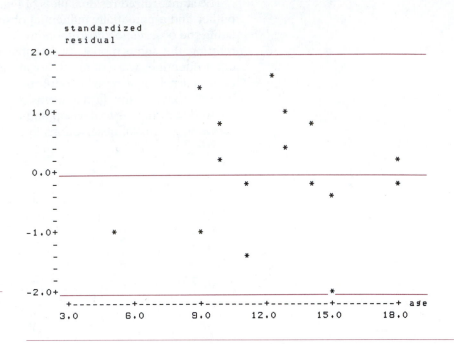

FIGURE 13

Standardized residual plot for the data of Example 10 (from MINITAB).

EXAMPLE 11

The paper "Snow Cover and Temperature Relationships in North America and Eurasia" (*J. of Climate and Appl. Meteorology* (1983): 460–9) explored the relationship between October–November continental snow cover (x, in millions of km^2) and December–February temperature (y, in °C). The given data refers to Eurasia during the $n = 13$ time periods 1969–70, 1970–71, . . . , 1981–82. A simple linear regression analysis done by the authors yielded $r^2 = .52$ ($r = -.72$), suggesting a substantial linear relationship. This is confirmed by a model utility test.

x	13.00	12.75	16.70	18.85	16.60	15.35	13.90
y	−13.5	−15.7	−15.5	−14.7	−16.1	−14.6	−13.4
St. resid.	−.11	−2.19	−.36	1.23	−.91	−.12	.34

x	22.40	16.20	16.70	13.65	13.90	14.75
y	−18.9	−14.8	−13.6	−14.0	−12.0	−13.5
St. resid.	−1.54	.04	1.25	−.28	1.54	.58

The scatter plot and standardized residual plot are displayed in Figure 14. There are no unusual patterns, though one standardized residual,

FIGURE 14

Plots for the data of Example 11 (from MINITAB).
(a) Scatter plot
(b) Standardized residual plot

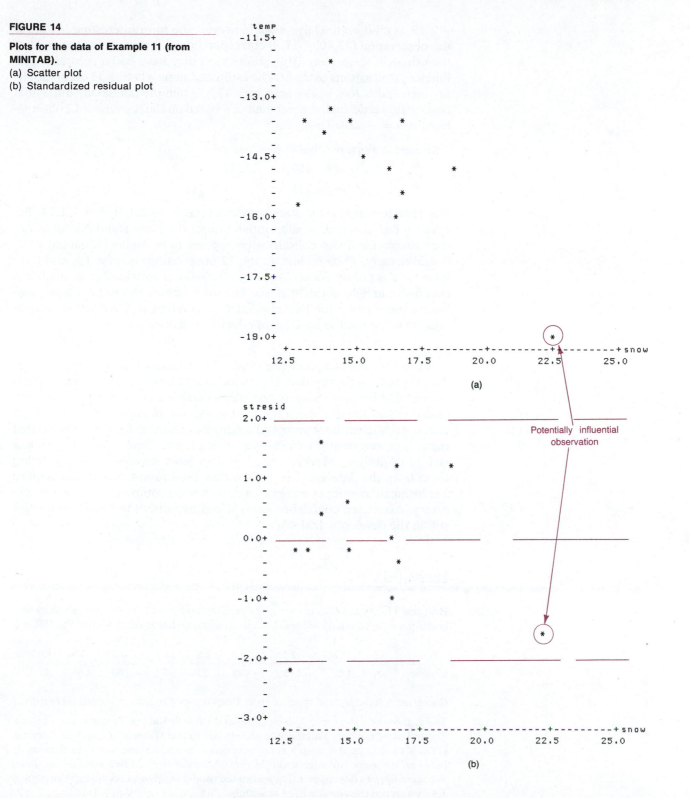

−2.19, is a bit on the large side. However, the most interesting feature is the observation (22.40, −18.9) corresponding to a point far to the right of the others in these plots. This observation may have had a substantial influence on all aspects of the fit. The estimated slope when all 13 observations are included is $b = -.459$, and $s_b = .133$. When the potentially influential observation is deleted, the estimate of β based on the remaining 12 observations is $b = -.228$. Thus,

$$
\begin{aligned}
\text{change in slope} &= \text{original } b - \text{new } b \\
&= -.459 - (-.228) \\
&= -.231
\end{aligned}
$$

The change expressed in standard deviations is $-.231/.133 = -1.74$. Because b has changed by substantially more than one standard deviation, the observation under consideration appears to be highly influential.

Additionally, r^2 based just on the 12 observations is only .13, and the t ratio for β is not significant. Evidence for a linear relationship is much less conclusive in light of this analysis. The investigators should seek a climatological explanation for the influential observation and collect more data which can be used in seeking an effective relationship.

When the distribution of the random deviation e has heavier tails than does the normal distribution, observations with large standardized residuals are not that unusual. Such observations can have great effects on the estimated regression line when the least-squares approach is used. In recent years, statisticians have proposed a number of alternative methods—called **robust**, or **resistant**, methods—for fitting a line. Such methods give less weight to outlying observations than does least squares without deleting them from the data set. The most widely used robust procedures require a substantial amount of computation, so a good computer program is necessary. Associated confidence-interval and hypothesis-testing formulas are still in the developmental stage.

EXERCISES 12.56–12.61

SECTION 12.5

12.56 Exercise 11.13 gave data on $x =$ chlorine flow and $y =$ etch rate. The x values and corresponding standardized residuals from a simple linear regression are as follows:

x	1.5	1.5	2.0	2.5	2.5	3.0	3.5	3.5	4.0
st. resid.	.31	1.02	−1.15	−1.23	.23	.73	−1.36	1.53	.07

Construct a standardized residual plot. Does the plot exhibit any unusual features?

12.57 The authors of the paper "Age, Spacing and Growth Rate of Tamarix as an Indication of Lake Boundary Fluctuations at Sebkhet Kelbia, Tunisia" (*J. of Arid Environ.* (1982): 43–51) used a simple linear regression model to describe the relationship between $y =$ vigor (average width in centimeters of the last two annual rings) and $x =$ stem density (stems/m²). The estimated model was based on the following data. Also given are the standardized residuals.

x	4	5	6	9	14	15	15	19	21	22
y	.75	1.20	.55	.60	.65	.55	0	.35	.45	.40
St. resid.	−.28	1.92	−.90	−.28	.54	.24	−2.05	−.12	.60	.52

a. What assumptions are required in order that the simple linear regression model be appropriate?

b. Construct a normal probability plot of the standardized residuals. Does the assumption that the random deviation distribution is normal appear to be reasonable? Explain.

c. Construct a standardized residual plot. Are there any unusually large residuals?

d. Is there anything about the standardized residual plot that would cause you to question the use of a simple linear regression model to describe the relationship between x and y?

12.58 The article "Effects of Gamma Radiation on Juvenile and Mature Cuttings of Quaking Aspen" (*Forest Science* (1967): 240–5) reported the following data on x = exposure time to radiation (kR/16 hr) and y = dry weight of roots (mg × 10^{-1}).

x	0	2	4	6	8
y	110	123	119	86	62

a. Construct a scatter plot for this data. Does the plot suggest that a simple linear regression model might be appropriate?

b. The estimated regression line for this data data is $\hat{y} = 127 - 6.65x$ and the standardized residuals are as given.

x	0	2	4	6	8
St. resid.	−1.55	.68	1.25	−.05	−1.06

Construct a standardized residual plot. What does the plot suggest about the adequacy of the simple linear regression model?

12.59 Carbon aerosols have been identified as a contributing factor in a number of air quality problems. In a chemical analysis of diesel engine exhaust, x = mass ($\mu g/cm^2$) and y = elemental carbon ($\mu g/cm^2$) were recorded ("Comparison of Solvent Extraction and Thermal Optical Carbon Analysis Methods: Application to Diesel Vehicle Exhaust Aerosol" *Environ. Science Tech.* (1984): 231–4). The estimated regression line for this data set is $\hat{y} = 31 + .737x$. Given are the observed x and y values and the corresponding standardized residuals.

x	164.2	156.9	109.8	111.4	87.0	82.9	78.9
y	181	156	115	132	96	90	86
St. resid.	2.52	0.82	0.27	1.64	0.08	−0.18	−0.27
x	161.8	230.9	106.5	97.6	79.7	100.8	387.8
y	170	193	110	94	77	88	310
St. resid.	1.72	−0.73	0.05	−0.77	−1.11	−1.49	−0.89
x	118.7	248.8	102.4	64.2	89.4	117.9	135.0
y	106	204	98	76	89	130	141
St. resid.	−1.07	−0.95	−0.73	−0.20	−0.68	1.05	0.91
x	108.1	89.4	76.4	131.7			
y	102	91	97	128			
St. resid.	−0.75	−0.51	0.85	0.00			

a. Construct a standardized residual plot. Are there any unusually large residuals? Do you think that there are any influential observations?

b. Is there any pattern in the standardized residual plot that would indicate that the simple linear regression model is not appropriate?

c. Based on your plot in part (a), do you think that it is reasonable to assume that the variance of y is the same at each x value? Explain.

12.60 An investigation of the relationship between traffic flow x (thousands of cars per 24 hr) and lead content y of bark on trees near the highway ($\mu g/g$ dry weight) yielded the accompanying data. A simple linear regression model was fit, and the resulting estimated regression line was $\hat{y} = 28.7 + 33.3x$. Both residuals and standardized residuals are also given.

x	8.3	8.3	12.1	12.1	17.0
y	227	312	362	521	640
Residual	−78.1	6.9	−69.6	89.4	45.3
St. resid.	−0.99	0.09	−0.81	1.04	0.51
x	17.0	17.0	24.3	24.3	24.3
y	539	728	945	738	759
Residual	−55.7	133.3	107.2	−99.8	−78.8
St. resid.	−0.63	1.51	1.35	−1.25	−0.99

a. Plot the (x, residual) pairs. Does the resulting plot suggest that a simple linear regression model is an appropriate choice? Explain your reasoning.

b. Construct a standardized residual plot. Does the plot differ significantly in general appearance from the plot in part (a)?

12.61 The accompanying data on x = U.S. population (millions) and y = crime index (millions) appeared in the paper "The Normal Distribution of Crime" (*J. of Police Science and Admin.* (1975): 312–8).

Year	1963	1964	1965	1966	1967	1968	1969	1970	1971	1972	1973
x	188.5	191.3	193.8	195.9	197.9	199.9	201.9	203.2	206.3	208.2	209.9
y	2.26	2.60	2.78	3.24	3.80	4.47	4.99	5.57	6.00	5.89	8.64

The author comments that "The simple linear regression analysis remains one of the most useful tools for crime prediction." When observations are made sequentially in time, the residuals or standardized residuals should be plotted in time order (that is, first the one for time $t = 1$ (1963 here), then the one for time $t = 2$, etc.). Notice that here x increases with time, so an equivalent plot is of residuals or standardized residuals versus x. Using $\hat{y} = -47.26 + .260x$, calculate the residuals and plot the (x, residual) pairs. Does the plot exhibit a pattern that casts doubt on the appropriateness of the simple linear regression model? Explain.

12.6 MULTIPLE REGRESSION

In many situations there will not be a strong relationship between a dependent variable y and any single predictor variable x, yet knowing the values of *several* independent variables (predictors) may considerably reduce uncertainty about y. For example, some variation in house prices in a large city can certainly be attributed to house size, but knowledge of size by itself would not enable a bank appraiser to accurately predict a home's value.

Price is also determined to some extent by values of other variables, such as age, lot size, number of bedrooms and bathrooms, and so on. A multiple regression model generalizes simple linear regression by allowing the deterministic part of the model equation to involve more than one independent variable. We shall use the letter k to represent the number of independent variables, and the variables themselves will be denoted by $x_1, x_2, \ldots, x_k$ (so x_1 and x_2 are now two different variables, such as age and lot size, not two different values of the same variable, as was previously the case).

▌▌▌ DEFINITION

A **general additive multiple regression model** relating a dependent variable y to k predictor variables $x_1, x_2, \ldots, x_k$ is specified by the model equation

$$y = \alpha + \beta_1 x_1 + \beta_2 x_2 + \cdots + \beta_k x_k + e$$

The random deviation e is assumed to be normally distributed with mean value 0 and standard deviation σ, whatever the values of $x_1, x_2, \ldots, x_k$. This implies that for fixed $x_1, x_2, \ldots, x_k$ values, y itself has a normal distribution with standard deviation σ and

mean y value for fixed $x_1, x_2, \ldots, x_k = \alpha + \beta_1 x_1 + \beta_2 x_2 + \cdots + \beta_k x_k$

The individual β_i's are called **population regression coefficients**, and the expression for the mean y value is often referred to as the **population regression function**.

The value of β_i gives the average change in y when x_i increases by one unit while the values of all other predictors are held fixed.*

EXAMPLE 12 The paper "The Value of Information for Selected Appliances" (*J. of Marketing Research* (1980): 14–25) suggests the plausibility of the general multiple regression model for relating the dependent variable

y = the price of an air conditioner

to $k = 3$ independent variables

x_1 = Btu/hr rating
x_2 = energy efficiency ratio
x_3 = number of settings

Suppose that the model equation is

$$y = -70 + .025x_1 + 20x_2 + 7.5x_3 + e$$

---
* There are some multiple regression models for which a change in the value of x_i will cause a change in the value of another predictor. In such situations the above interpretation of β_i is not legitimate.

and that $\sigma = 20$. Then the population regression function is

$$\begin{pmatrix} \text{mean } y \text{ value for} \\ \text{fixed values of } x_1,\, x_2,\, x_3 \end{pmatrix} = -70 + .025x_1 + 20x_2 + 7.5x_3$$

Thus

$$\begin{pmatrix} \text{mean } y \text{ value when} \\ x_1 = 6000,\, x_2 = 8.0,\, x_3 = 5 \end{pmatrix} = -70 + (.025)(6000) + (20)(8.0) + (7.5)(5)$$

$$= 277.50$$

Because y is normally distributed with $2\sigma = 40$, a y observation for these values of x_1, x_2, and x_3 is quite likely to be between $277.50 - 40 = 237.50$ and $277.50 + 40 = 317.50$. Since $\beta_2 = 20$, the average price increase associated with a one-unit increase in energy efficiency ratio x_2 while both Btu rating x_1 and number of settings x_3 remain fixed is \$20. Similar interpretations apply to the regression coefficients $\beta_1 = .025$ and $\beta_3 = 7.5$.

ESTIMATING THE POPULATION REGRESSION FUNCTION

The values of the coefficients α, β_1, β_2, . . . , β_k are virtually never known to an investigator. Instead, these coefficients must be estimated from sample data, after which the population regression function $\alpha + \beta_1 x_1 + \cdots + \beta_k x_k$ can be estimated. As before, n will denote the number of observations in the sample. In a simple linear regression (a multiple regression with $k = 1$), each observation is an (x, y) pair. For $k \geq 2$, each observation will consist of $k + 1$ numbers: a value of x_1, a value of x_2, . . . , a value of x_k, and the corresponding y value. Thus when $k = 3$ (as in Example 12), each observation will consist of four numbers, (x_1, x_2, x_3, y).

Let a, b_1, . . . , b_k represent estimates of α and the β's. Then the corresponding candidate for the estimated regression function is $a + b_1 x_1 + \cdots + b_k x_k$. The x values from any particular observation can be substituted into this expression to obtain a prediction for y. Now we can form the sum of the squared deviations:

$$\sum (\text{observed } y - \text{predicted } y \text{ using estimated regression function})^2$$

$$= \sum [y - (a + b_1 x_1 + b_2 x_2 + \cdots + b_k x_k)]^2$$

where the sum is over all n observations. The **least-squares estimates** of α, β_1, . . . , β_k are those values of a, b_1, . . . , b_k that make this sum of squared deviations as small as possible. For $k \geq 2$, formulas for the estimates are complicated. Fortunately these formulas have been programmed into many statistical computer packages. After the data has been properly entered, a regression command instructs the computer to perform all the required calculations and then display the estimates along with much other useful information.

EXAMPLE 13

Soil and sediment adsorption, the extent to which chemicals collect in a condensed form on the surface, is an important characteristic because it influences the effectiveness of pesticides and various argricultural chemicals. The paper "Adsorption of Phosphates, Arsenate, Methanearsenate, and Cacodylate by Lake and Stream Sediments: Comparisons with Soils (*J. of Environ. Qual.* (1984): 499–504) presented the accompanying data and proposed the model

$$y = \alpha + \beta_1 x_1 + \beta_2 x_2 + e$$

for relating

y = phosphate adsorption index

x_1 = amount of extractable iron

x_2 = amount of extractable aluminum

Obs.	x_1 = extractable iron	x_2 = extractable aluminum	y = adsorption index
1	61	13	4
2	175	21	18
3	111	24	14
4	124	23	18
5	130	64	26
6	173	38	26
7	169	33	21
8	169	61	30
9	160	39	28
10	244	71	36
11	257	112	65
12	333	88	62
13	199	54	40

Figure 15 displays MINITAB output from a regression command requesting that the model be fit to this data. (We named the dependent variable HPO, an abbreviation for H_2PO_4 (dihydrogen phosphate), and the two predictor variables x_1 and x_2 were named Fe and Al, the standard abbreviations for iron and aluminum, respectively.) Focus on the column labeled COEFFICIENT in the table near the top of the figure. The three numbers in this column are the estimated model coefficients:

$a = -7.351$ (the estimate of the constant term α),

$b_1 = .11273$ (the estimate of the coefficient β_1),

$b_2 = .34900$ (the estimate of the coefficient β_2).

Thus the estimated regression function is

$$\left(\begin{array}{c}\text{estimated mean } y \text{ value for}\\ \text{specified } x_1 \text{ and } x_2 \text{ values}\end{array}\right) = -7.351 + .11273x_1 + .349x_2$$

This equation is given (with coefficients rounded slightly) using the named variables at the very top of the MINITAB output. A point estimate for true average adsorption index when extractable iron = 150 and extractable

FIGURE 15

MINITAB output for the regression analysis of Example 13.

```
THE REGRESSION EQUATION IS
HPO = - 7.35 + 0.113 FE + 0.349 AL

                                   ST. DEV.      T-RATIO =
COLUMN        COEFFICIENT         OF COEF.       COEF/S.D.
                   -7.351            3.485          -2.11
FE              0.11273            0.02969           3.80
AL              0.34900            0.07131           4.89


S = 4.379


R-SQUARED = 94.8 PERCENT
R-SQUARED = 93.8 PERCENT, ADJUSTED FOR D.F.

ANALYSIS OF VARIANCE

  DUE TO      DF            SS         MS=SS/DF
REGRESSION     2         3529.9         1765.0
RESIDUAL      10          191.8           19.2
TOTAL         12         3721.7

FURTHER ANALYSIS OF VARIANCE
SS EXPLAINED BY EACH VARIABLE WHEN ENTERED IN THE ORDER GIVEN
  DUE TO      DF            SS
REGRESSION     2         3529.9
FE             1         3070.5
AL             1          459.4

                      Y       PRED. Y    ST.DEV.
ROW       FE         HPO       VALUE     PRED. Y    RESIDUAL    ST.RES.
 1        61        4.00        4.06       2.43       -0.06      -0.02
 2       175       18.00       19.71       2.31       -1.71      -0.46
 3       111       14.00       13.54       1.72        0.46       0.11
 4       124       18.00       14.66       1.67        3.34       0.83
 5       130       26.00       29.64       2.62       -3.64      -1.04
 6       173       26.00       25.41       1.41        0.59       0.14
 7       169       21.00       23.22       1.56       -2.22      -0.54
 8       169       30.00       32.99       1.60       -2.99      -0.73
 9       160       28.00       24.30       1.30        3.70       0.89
10       244       36.00       44.94       1.71       -8.94      -2.22R
11       257       65.00       60.71       3.20        4.29       1.44
12       333       62.00       60.90       3.19        1.10       0.37
13       199       40.00       33.93       1.29        6.07       1.45

R DENOTES AN OBS. WITH A LARGE ST. RES.
```

aluminum = 25 is

$$\begin{pmatrix} \text{estimated mean } y \text{ value} \\ \text{when } x_1 = 150, x_2 = 25 \end{pmatrix} = -7.351 + (.11273)(150) + (.349)(25)$$

$$\approx 18.3$$

This is also the prediction of a single adsorption index value to be observed when Fe = 150 and Al = 25.

IS THE MODEL USEFUL?

In simple linear regression, a scatter plot will convey a preliminary impression of the model's usefulness. Unfortunately, multiple regression offers no simple picture that can give this type of information. Instead we must proceed as before, by first calculating residuals, sums of squares, and a coefficient of determination, and then performing a model utility test.

DEFINITION

The ith predicted value $\hat{y}_i$ is obtained by taking the values of the predictor variables $x_1, x_2, \ldots, x_k$ for the ith sample observation and substituting these values into the estimated regression function. Doing this successively for $i = 1, 2, 3, \ldots, n$ yields the **predicted values** $\hat{y}_1, \hat{y}_2, \ldots, \hat{y}_n$. The **residuals** are then the differences $y_1 - \hat{y}_1, y_2 - \hat{y}_2, \ldots, y_n - \hat{y}_n$ between the observed and predicted y values. The **residual sum of squares**, **SSResid**, and **the total sum of squares, SSTo,** are given by

$$\text{SSResid} = \sum (y - \hat{y})^2, \qquad \text{SSTo} = \sum (y - \bar{y})^2$$

where $\bar{y}$ is the mean of the y observations in the sample. The number of degrees of freedom associated with SSResid is $n - (k + 1)$, because $k + 1$ df are lost in estimating the $k + 1$ coefficients $\alpha, \beta_1, \ldots, \beta_k$. An estimate of the random deviation variance σ^2 is given by

$$s_e^2 = \frac{\text{SSResid}}{n - (k + 1)}$$

and s_e is the estimate of σ. The **coefficient of multiple determination**, R^2, interpreted as the proportion of variation in observed y values that is explained by the fitted model, is

$$R^2 = 1 - \frac{\text{SSResid}}{\text{SSTo}}$$

Generally speaking, a desirable model is one that results in both a large R^2 value and a small s_e value. However, there is a catch: these two conditions can be achieved by fitting a model that contains a large number of predictors. Such a model may be successful in explaining y variation, but it almost always specifies a relationship that is unrealistic and difficult to interpret. What we really want is a simple model, one with relatively few predictors whose roles are easily interpreted, and one that also does a good job of explaining variation in y.

Hand calculation of the defined quantities is quite tedious, but all statistical computer packages with multiple regression capabilities will perform the computations and display the results.

EXAMPLE 14

(*Example 13 continued*) Looking again at Figure 15, which contains MINITAB output for the adsorption data fit by a two-predictor model, the residual sum of squares is

$$\text{SSResid} = (-.06)^2 + (-1.71)^2 + (.46)^2 + \cdots + (6.07)^2$$
$$= 191.8337$$

This value also appears in rounded form in the RESIDUAL row and SS column of the table in Figure 15 headed ANALYSIS OF VARIANCE. The associated number of degrees of freedom is

$$n - (k + 1) = 13 - (2 + 1) = 10$$

which appears in the DF column just to the left of SSResid. The sample average y value is $\bar{y} = 29.85$, so

$$
\begin{aligned}
\text{SSTo} &= \sum (y - \bar{y})^2 \\
&= (4 - 29.85)^2 + \cdots + (40 - 29.85)^2 \\
&= 3721.69
\end{aligned}
$$

This value appears in the TOTAL row and SS column of the ANALYSIS OF VARIANCE table just under the value of SSResid. The values of s_e^2, s_e, and R^2 are then

$$
s_e^2 = \frac{\text{SSResid}}{n - (k + 1)} = \frac{191.83}{10} = 19.18 \approx 19.2
$$

(in the MS column of the MINITAB output)

$$
s_e = \sqrt{19.18} = 4.379
$$

($S = 4.379$ appears near the top of the output)

$$
\begin{aligned}
R^2 = 1 - \frac{\text{SSResid}}{\text{SSTo}} &= 1 - \frac{191.83}{3721.69} \\
&= 1 - .052 \\
&= .948
\end{aligned}
$$

Thus the percentage of variation explained is $100R^2 = 94.8\%$, which appears on the output as R-SQUARED = 94.8 PERCENT. The values of R^2 and s_e suggest that the chosen model has been very successful in relating y to the predictors.

THE *F* TEST FOR MODEL UTILITY

In the simple linear model with regression function $\alpha + \beta x$, if $\beta = 0$, there is no useful linear relationship between y and the single predictor variable x. Similarly, if all k coefficients $\beta_1, \beta_2, \ldots, \beta_k$ are zero in the general k-predictor multiple regression model, there is no useful linear relationship between y and any of the predictor variables $x_1, x_2, \ldots, x_k$ included in the model. Before using an estimated model to make further inferences (e.g., predictions and estimates of mean values), it is desirable to confirm the model's utility through a formal test procedure. The procedure description involves a new type of probability distribution called an *F distribution*, so we first digress briefly to describe some general properties of *F* distributions.

An *F* distribution always arises in connection with a ratio. The numerator of the ratio has an associated df and so does the denominator. A particular *F* distribution is determined by fixing values of the numerator df and the denominator df. For example, there is an *F* distribution based on 4 numerator df and 12 denominator df, another *F* distribution based on 3 numerator df and 20 denominator df, and so on. A typical *F* curve for fixed numerator and denominator df appears in Figure 16.

All *F* tests presented in this book are upper-tailed, so associated rejection regions require only upper-tailed *F* critical values. A particular *F* critical value depends not only on the desired tail area but also on both numerator

FIGURE 16

A typical *F* probability distribution.

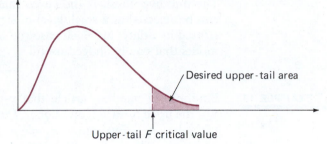

Desired upper-tail area

Upper-tail *F* critical value

and denominator df. Appendix Table IV(a) gives the critical values for tail area .05 and Table IV(b) gives those for tail area .01. Different columns of the tables are identified with different numerator df and different rows are associated with different denominator df. For example, the .05 critical value for the *F* distribution with 4 numerator df and 6 denominator df is found in the column of Table IV(a) labeled 4 and the row labeled 6; that value is 4.53. The .05 critical value for 6 numerator df and 4 denominator df is 6.16 (so don't accidentally interchange numerator and denominator df!).

The model utility test statistic has the form

$$F = \frac{R^2/\text{numerator df}}{(1 - R^2)/\text{denominator df}}$$

When all β's have value zero (so there is no useful linear relationship), this test statistic can be shown to have an *F* distribution. Intuitively, the assertion of no useful relationship should be rejected when R^2 is large relative to $1 - R^2$. This corresponds to a large *F* value (notice that *F* cannot be negative). Hence the test is upper-tailed.

THE *F* TEST FOR UTILITY OF THE MODEL $y = \alpha + \beta_1 x_1 + \cdots + \beta_k x_k + e$

Null hypothesis: $H_0: \beta_1 = \beta_2 = \cdots = \beta_k = 0$
(There is no useful linear relationship between *y* and any of the predictors.)

Alternative hypothesis: H_a: At least one among $\beta_1, \ldots, \beta_k$ is not zero.
(There is a useful linear relationship between *y* and at least one of the predictors.)

Test Statistic: $F = \dfrac{R^2/k}{(1 - R^2)/[n - (k + 1)]}$

Rejection region: $F > F$ critical value. The *F* critical value is based on *k* numerator df and $n - (k + 1)$ denominator df, respectively. Table IV(a) gives critical values corresponding to level of significance .05 and Table IV(b) gives level .01 critical values.

The null hypothesis is the claim that the model is not useful. Unless H_0 can be rejected at a small level of significance, the model has not demonstrated its utility, in which case the investigator must search further for a model that can be judged useful.

EXAMPLE 15

Example 12 cited an article that reported on a regression analysis with $y =$ the price of an air conditioner. Data was collected on three independent variables: $x_1 =$ Btu rating, $x_2 =$ energy efficiency ratio, and $x_3 =$ number of settings. The model with just these three predictors was fit. The resulting estimated regression function, based on $n = 19$ observations, was

$$\begin{pmatrix} \text{estimated mean } y \text{ value} \\ \text{for specified } x_1, x_2, x_3 \end{pmatrix} = -68.326 + .023x_1 + 19.729x_2 + 7.653x_3$$

and R^2 for this model was .84. The high R^2 value certainly suggests a useful model, but let's carry out a formal test at level of significance .01.

1. The fitted model was $y = \alpha + \beta_1 x_1 + \beta_2 x_2 + \beta_3 x_3 + e$.
2. $H_0: \beta_1 = \beta_2 = \beta_3 = 0$
3. H_a: At least one of the three β_i's is not zero
4. Test statistic:

$$F = \frac{R^2/k}{(1 - R^2)/[n - (k + 1)]}$$

5. Rejection region: $F > F$ critical value. For $k = 3$ and $n = 19$, there are 3 numerator df and $19 - 4 = 15$ denominator df. Table IV(b) gives the level .01 critical value as 5.42, so H_0 will be rejected if $F > 5.42$.
6. $R^2 = .84$ and $1 - R^2 = .16$, so

$$F = \frac{.84/3}{.16/15} = \frac{.28}{.0107} = 26.2$$

7. The computed F value does fall in the rejection region ($26.2 > 5.42$), so H_0 is rejected in favor of the conclusion that the postulated model is indeed useful.

Once a fitted model has been judged useful, it can be used to draw further conclusions. These include calculation of a confidence interval for a single β_i, a test of hypotheses about a single β_i, a confidence interval for the mean y value when $x_1 = x_1^*, \ldots, x_k = x_k^*$ (fixed predictor values), and a prediction interval for a future y. For example, denote $a + bx_1^* + \cdots + b_k x_k^*$ by $\hat{y}$ and let $s_{\hat{y}}$ denote the estimated standard deviation of this statistic. Then a confidence interval formula for the mean y value is

$$\hat{y} \pm (t \text{ critical value based on } n - (k + 1) \text{ df}) \cdot s_{\hat{y}}$$

The expression for $s_{\hat{y}}$ is quite complicated, but the better statistical computer packages will provide either $\hat{y}$ and $s_{\hat{y}}$ or the interval itself upon request. A prediction interval has the same general form except that s_y is replaced by the larger quantity $\sqrt{s_e^2 + s_{\hat{y}}^2}$, resulting in a wider interval.

EXAMPLE 16

Returning to the phosphate data and the MINITAB output of Figure 15, we see from row 9 of the large table that when $x_1 = 160$ and $x_2 = 39$ (predictor values for the ninth observation), $\hat{y} = 24.30$ and $s_{\hat{y}} = 1.30$. Additionally, $s_e = 4.379$ and $df = 10$, and a 95% interval requires t critical value 2.23. The confidence interval for mean y when $x_1 = 160$ and $x_2 = 39$ is

$$24.30 \pm (2.23)(1.30) = 24.30 \pm 2.90$$
$$= (21.40, 27.20)$$

The corresponding prediction interval for a single y value is

$$24.30 \pm (2.23)\sqrt{(4.378)^2 + (1.30)^2} = 24.30 \pm 10.18$$
$$= (14.12, 34.48)$$

The width of this prediction interval suggests that there is substantial uncertainty concerning the value of a future observation to be made when $x_1 = 160$ and $x_2 = 39$.

CHECKING THE ADEQUACY OF A MULTIPLE REGRESSION MODEL

The key assumptions of the multiple regression model can be checked by examining residuals and standardized residuals. A normal probability plot of the standardized residuals will again provide a check on the assumption that the random deviation e is normally distributed. When the model contains k predictors, the single standardized residual plot in simple linear regression is replaced by k different plots: one of the $(x_1,$ standardized residual) pairs, a second of the $(x_2,$ standardized residual) pairs, and so on. These plots should be examined for any patterns or discrepant points. A particularly good source for more information on these matters is the book by Neter et al., cited in the References of Chapter 11.

Any of the chapter references can be consulted for more information on inferential procedures and various other aspects of multiple regression. Our presentation in this section has barely scratched the surface of a rich, beautiful, and extremely useful subject.

EXERCISES 12.62–12.75   **SECTION 12.6**

12.62 Explain the difference between a deterministic and a probabilistic model. Give an example of a dependent variable y and two or more independent variables that might be related to y deterministically. Give an example of a dependent variable y and two or more independent variables that might be related to y in a probabilistic fashion.

12.63 The paper "The Influence of Temperature and Sunshine on the Alpha-Acid Contents of Hops" (*Agric. Meteorology* (1974): 375–82) used a multiple regression model to relate

$y =$ yield of hops

$x_1 =$ mean temperature between date of coming into hop and date of picking

$x_2 =$ mean percentage of sunshine during the same period.

The model equation proposed is

$$y = 415.11 - 6.60x_1 - 4.50x_2 + e$$

a. Suppose that this equation does indeed describe the true relationship. What mean yield corresponds to a mean temperature of 20 and a mean sunshine percentage of 40?

b. What is the mean yield when the mean temperature and percentage of sunshine are 18.9 and 43, respectively?

12.64 The multiple regression model $y = \alpha + \beta_1 x_1 + \beta_2 x_2 + e$ can be used to describe the relationship between

y = profit margin of savings and loan companies in a given year

x_1 = net revenues in that year

x_2 = number of branch offices

Based on data given in the paper "Entry and Profitability in a Rate-Free Savings and Loan Market" (*Quarterly Rev. of Econ. and Business* (1978): 87–95), a reasonable model equation is

$$y = 1.565 + .237x_1 - .0002x_2 + e$$

a. How would you interpret the values of β_1 and β_2?

b. If the number of branch offices remains fixed and net revenue increases by 1, what is the average change in profit margin?

c. What would the mean profit margin be for a year in which net revenue is 4.0 and number of branch offices is 6500?

12.65 The multiple regression model

$$y = .69 + 4.70x_1 + .00041x_2 - .72x_3 + .023x_4 + e$$

where

y = stock purchase tender premium (as a percentage of closing market price one week prior to offer date)

x_1 = fee per share (as a percentage of price)

x_2 = percentage of shares sought

x_3 = relative change in Dow Jones industrial average

x_4 = volume of shares traded (as a percentage of those outstanding)

is based on data appearing in the paper "Factors Influencing the Pricing of Stock Repurchase Tenders" (*Quarterly Rev. of Econ. and Business* (1978): 31–9).

a. What is the mean stock tender premium when $x_1 = 2$, $x_2 = .1$, $x_3 = 1.2$, and $x_4 = 6$?

b. If the variables x_1, x_3, and x_4 are held fixed, would the mean tender premium increase or decrease as the percentage of shares sought increases? Explain.

12.66 The paper "Readability of Liquid Crystal Displays: A Response Surface" (*Human Factors* (1983): 185–90) used a multiple regression model with four independent variables, where

y = error percentage for subjects reading a four-digit liquid crystal display

x_1 = level of backlight (ranging from 0 to 122 cd/m^2)

x_2 = character subtense (ranging from .025° to 1.34°)

x_3 = viewing angle (ranging from 0° to 60°)

x_4 = level of ambient light (ranging from 20 to 1500 lx)

The estimated regression function is

$$\hat{y} = 1.52 + .02x_1 - 1.40x_2 + .02x_3 - .0006x_4$$

a. Calculate a point estimate of the mean value of y when $x_1 = 10$, $x_2 = .5$, $x_3 = 50$, and $x_4 = 100$.
b. Predict error percentage for a backlight level of 20, character subtense of .5, viewing angle of 10, and ambient light level of 30.
c. From a table in the paper, $n = 30$, SSResid $= 20.0$, and SSTo $= 39.2$. Calculate R^2 and s_e for this model. Interpret these values.
d. Does the estimated regression equation specify a useful relationship between y and the independent variables? Use the model utility test with a .05 significance level.

12.67 Factors affecting breeding success of puffins were examined in the paper "Breeding Success of the Common Puffin on Different Habitats at Great Island, Newfoundland" (*Ecol. Monographs* (1972): 239–66). Data given in the paper was used to estimate the regression model

$$y = \alpha + \beta_1 x_1 + \beta_2 x_2 + \beta_3 x_3 + \beta_4 x_4 + e$$

where

$y = $ puffin nest density (number per 9 m^2),

$x_1 = $ grass cover (%)

$x_2 = $ mean soil depth (cm)

$x_3 = $ angle of slope (degrees)

$x_4 = $ distance from cliff edge (m)

The estimated regression equation (using least squares) was

$$\hat{y} = 12.3 - .0186x_1 - .0430x_2 + .224x_3 - .182x_4$$

and SSTo $= 1914.21$, SSResid $= 264.19$, and $n = 38$.
a. Calculate R^2 and s_e and interpret these values.
b. For this same data set, would the equation $\hat{y} = 12 - .02x_1 - .05x_2 + .2x_3 - .2x_4$ result in a larger or smaller sum of squared residuals than the estimated regression equation given? Explain.
c. Do the independent variables provide information that is useful for predicting y? Use the model utility test with a significance level of .01.

12.68 The paper "The Caseload Controversy and the Study of Criminal Courts" (*J. of Criminal Law and Criminology* (1979): 89–101) used a multiple regression analysis to help assess the impact of judicial caseload on the processing of criminal court cases. Data was collected in the Chicago criminal courts on the following variables:

$y = $ number of indictments

$x_1 = $ number of cases on the docket

$x_2 = $ number of cases pending in the criminal court trial system

The estimated regression equation (based on $n = 367$ observations) was

$$\hat{y} = 28 - .05x_1 - .003x_2 + .00002x_3$$

where $x_3 = x_1 x_2$.
a. The reported value of R^2 was .16. Conduct the model utility test. Use a .05 significance level.
b. Given the results of the test in part (a), does it surprise you that the R^2 value is so low? Can you think of a possible explanation for this?

12.69 Suppose that a multiple regression data set consists of $n = 20$ observations. For what values of k, the number of model predictors, would the corresponding model with $R^2 = .90$ be judged useful at significance level .05? Does such a large R^2 value necessarily imply a useful model? Explain.

12.70 When a scatter plot of bivariate data (x_1, y_1), (x_2, y_2), . . . , (x_n, y_n) shows a parabolic rather than a linear pattern, a **quadratic regression model** can be fit to the data. The model equation is

$$y = \alpha + \beta_1 x + \beta_2 x^2 + e$$

from which the mean value of y is $\alpha + \beta_1 x + \beta_2 x^2$, a quadratic (rather than linear) function of x. This can be regarded as a special case of multiple regression with k = number of predictor variables = 2, $x_1 = x$, and $x_2 = x^2$. The paper "The Influence of Mount St. Helens Ash on Wheat Growth and Phosphorous, Sulfur, Calcium, and Magnesium Uptake" (*J. of Environ. Quality* (1984): 91–6) used a quadratic regression model to describe the relationship between y = biomass production of wheat (g/pot) and x = percent volcanic ash in the soil. Data from a greenhouse experiment in which x ranged between 0 and 75 was used to estimate α, β_1, and β_2. The resulting least-squares estimates a, b_1, and b_2 were .067, .054, and $-.00052$, respectively.

a. Write out the estimated regression equation.
b. What would you predict biomass to be when the percent of volcanic ash is 20? When it is 40?
c. Graph the parabola $\hat{y} = .067 + .054x - .00052x^2$. Use values of x ranging between 0 and 75. Based on this graph, at approximately what point does an increasing percent of volcanic ash begin to have a detrimental effect on average biomass?

12.71 The paper "Effect of Manual Defoliation on Pole Bean Yield" (*J. of Econ. Entomology* (1984): 1019–23) used a quadratic regression model (see Exercise 12.70) to describe the relationship between y = yield (kg/plot) and x = defoliation level (a proportion between zero and one). The estimated regression equation based on $n = 24$ was

$$\hat{y} = 12.39 + 6.67x_1 - 15.25x_2,$$

where $x_1 = x$ and $x_2 = x^2$. The paper also reported that R^2 for this model was .902. Does the quadratic model specify a useful relationship between y and x? Carry out the appropriate test using a .01 level of significance.

12.72 In the paper "An Ultracentrifuge Flour Absorption Method" (*Cereal Chem.* (1978): 96–101), the authors discussed the relationship between water absorption for wheat flour and various characteristics of the flour. A multiple regression model was used to relate

y = absorption (%)

x_1 = flour protein (%)

x_2 = starch damage (Farrand units)

a. The model $y = \alpha + \beta_1 x_1 + \beta_2 x_2 + e$ was fit to the data, resulting in the equation

$$\hat{y} = 19.44 + 1.4423x_1 + .33563x_2$$

With $n = 28$, SSResid = 29.93, and SSTo = 842.31, calculate the proportion of observed variation in absorption that can be explained by the model relationship.
b. Carry out a test of model utility.
c. An estimate of the mean water absorption for wheat with 10.2% protein and a starch damage of 20 is desired. The estimated standard deviation of the statistic

$\hat{y} = a + b_1(10.2) + b_2(20)$ is $s_{\hat{y}} = .318$. Use this to compute a 95% confidence interval for $\alpha + \beta_1(10.2) + \beta_2(20)$.

d. Compute a 90% confidence interval for $\alpha + \beta_1(11.7) + \beta_2(57)$ if the estimated standard deviation of $a + b_1(11.7) + b_2(57)$ is .522. Interpret the resulting interval.

e. A single shipment of wheat is received. For this particular shipment, $x_1 = 11.7$ and $x_2 = 57$. Predict the water absorption for this shipment (a single y value) using a 90% interval.

12.73 The paper "Predicting Marathon Time for Anaerobic Threshold Measurements" (*The Physician and Sportsmed.* (1984): 95–8) gave data on $y =$ maximum heart rate (beats/min), $x_1 =$ age, and $x_2 =$ weight (kg) for $n = 18$ marathon runners. The estimated regression equation for the model

$$y = \alpha + \beta_1 x_1 + \beta_2 x_2 + e$$

was

$$\hat{y} = 179 - .8x_1 + .5x_2$$

and SSTo $= 1187.78$, SSResid $= 538.03$.

a. Is the model useful for predicting maximum heart rate? Use a significance level of .10.

b. Predict the maximum heart rate of a particular runner who is 43 years old and weighs 65 kg using a 99% interval. The estimated standard deviation of $\hat{y} = a + b_1(43) + b_2(65)$ is $s_{\hat{y}} = 3.52$.

c. Use a 90% interval to estimate the average maximum heart rate for all marathon runners who are 30 years old and weigh 77.2 kg. The estimated standard deviation of $a + b_1(30) + b_2(77.2)$ is 2.97.

d. Would a 90% prediction interval for a single 30-year-old runner weighing 77.2 kg be wider or narrower than the interval computed in part (c)? Explain. (You need not compute the interval.)

12.74 A confidence interval for β_1, the population regression coefficient of x_1, is

$$b_1 \pm (t \text{ critical value}) \cdot s_{b_1}$$

where the t critical value is based on $n - (k + 1)$ degrees of freedom. The formula for the estimated standard deviation s_{b_1} is complicated, but almost any computer package with multiple regression capabilities will include the value of s_{b_1} as part of the output. A confidence interval for any other population regression coefficient will have the same general form. Twenty-six observations appearing in the paper "Multiple Regression Analysis for Forecasting Critical Fish Influxes at Power Station Intakes" (*J. of Appl. Ecol.* (1983): 33–42) were used to fit a multiple regression model relating $y =$ number of fish at intake to the independent variables $x_1 =$ water temperature (°C), $x_2 =$ number of pumps running, $x_3 =$ sea state (taking values 0, 1, 2, or 3), and $x_4 =$ speed (knots). Results included

$$b_1 = -2.179, \qquad b_2 = -19.189, \qquad b_3 = -9.378, \qquad b_4 = 2.321$$
$$s_{b_1} = 1.087, \qquad s_{b_2} = 9.215, \qquad s_{b_3} = 4.356, \qquad s_{b_4} = .769$$

a. Construct a 95% confidence interval for β_3, the coefficient of $x_3 =$ sea state. Interpret the resulting interval.

b. Construct a 90% confidence interval for the mean change in y associated with a 1° increase in temperature when the number of pumps, sea state, and speed remain fixed.

12.75 Much interest in management circles has recently focused on how employee compensation is related to various company characteristics. The paper "Determinants of R and D Compensation Strategies'' (*Personnel Psych.* (1984): 635–50) proposed a quantitiative scale for y = base salary for employees of high-tech companies. The following estimated multiple regression equation was then presented:

$$\hat{y} = 2.60 + .125x_1 + .893x_2 + .057x_3 - .014x_4$$

where

x_1 = sales volume (in millions of dollars)

x_2 = stage in product life cycles (1 = growth, 0 = mature)

x_3 = profitability (%),

x_4 = attrition rate (%).

Note: x_2 is called an *indicator* or *dummy variable*; in this example it indicates in which of two categories a firm belongs.

a. There were $n = 33$ firms in the sample and $R^2 = .69$. Is the fitted model useful?

b. Predict base compensation for a growth-stage firm with a sales volume of $50 million, profitability 8%, and attrition rate 12%.

c. The coefficient β_3 is the difference between the average base compensations for growth-stage and mature-stage firms when all other predictors are held fixed. Use the fact that $s_{b_3} = .014$ to calculate a 95% confidence interval for β_3.

CHAPTER 12 SUMMARY OF KEY CONCEPTS AND FORMULAS

TERM OR FORMULA	PAGE	COMMENT
Simple linear regression model, $y = \alpha + \beta x + e$	394	This model assumes that there is a line with slope β and y intercept α, called the population (true) regression line, such that an observation deviates from the line by a random amount e. The random deviation is assumed to have a normal distribution with mean zero and standard deviation σ, and random deviations for different observations are assumed independent of one another.
Estimated regression line, $\hat{y} = a + bx$	397	The least-squares line introduced in Chapter 11
$s_e = \sqrt{\dfrac{SSResid}{n-2}}$	399	The point estimate of the standard deviation σ, with associated degrees of freedom $n - 2$
$s_b = \dfrac{s_e}{\sqrt{\sum x^2 - n\bar{x}^2}}$	406	The estimated standard deviation of the statistic b
$b \pm (t \text{ crit. val.}) \cdot s_b$	406	A confidence interval for the slope β of the population regression line, where the t critical value is based on $(n - 2)$ df
$t = \dfrac{b - \text{hyp. value}}{s_b}$	409	The test statistic for testing hypotheses about β: The calculated value of t is compared to a t critical value.
Model utility test, with test statistic $t = b/s_b$	409	A test of $H_0: \beta = 0$, which asserts that there is no useful linear relationship between x and y, versus $H_a: \beta \neq 0$, the claim that there is a useful linear relationship
$s_{a+bx^*} = s_e\sqrt{\dfrac{1}{n} + \dfrac{(x^* - \bar{x})^2}{\sum x^2 - n\bar{x}^2}}$	415	The estimated standard deviation of the statistic $a + bx^*$, where x^* denotes a particular value of x
$a + bx^*$ $\pm (t \text{ crit. val.}) \cdot s_{a+bx^*}$	415	A confidence interval for $\alpha + \beta x^*$, the average value of y when $x = x^*$

CHAPTER 12 SUMMARY OF KEY CONCEPTS AND FORMULAS (continued)

TERM OR FORMULA	PAGE	COMMENT
$a + bx^*$ $\pm$ (t crit. val.) $\cdot \sqrt{s_e^2 + s_{a+bx^*}^2}$	418	A prediction interval for a single y value to be observed when $x = x^*$
Population correlation coefficient ρ	423	A measure of the extent to which the x and y values in an entire population are linearly related
$t = \dfrac{r}{\sqrt{(1 - r^2)/(n - 2)}}$	424	The test statistic for testing H_0: $\rho = 0$, according to which (assuming a bivariate normal population distribution) x and y are independent of one another
Residual analysis	428	Methods, based on the residuals, for checking the assumptions of the simple linear regression model or any other regression model
Standardized residual	429	A residual divided by its standard deviation
Standardized residual plot	430	A plot of the (x, st. resid.) pairs, i.e., of standarized residuals versus x values, where a pattern (other than pure randomness) in this plot suggests a problem with the model or data
Additive multiple regression model, $y = \alpha + \beta_1 x_1 + \cdots + \beta_k x_k + e$	437	This equation specifies a general probabilistic relationship between y and k predictor variables $x_1, x_2, \ldots, x_k$, where $\beta_1, \ldots, \beta_k$ are population regression coefficients and $\alpha + \beta_1 x_1 + \cdots + \beta_k x_k$ is the population regression function (the mean value of y for fixed values of $x_1, \ldots, x_k$).
Estimated regression function, $\hat{y} = a + b_1 x_1 + \cdots + b_k x_k$	438	The estimates $a, b_1, \ldots, b_k$ of $\alpha, \beta_1, \ldots, \beta_k$ result from applying the principle of least squares.
Coefficient of multiple determination, $R^2 = 1 - \dfrac{\text{SSResid}}{\text{SSTo}}$	441	The proportion of observed y variation that can be explained by the model relation, where SSResid is defined as in simple linear regression but is now based on $n - (k + 1)$ degrees of freedom
F distribution	442	A type of probability distribution used in many different inferential procedures. A particular F distribution results from specifying both numerator df and denominator df.
$F = \dfrac{R^2/k}{(1 - R^2)/[n - (k + 1)]}$	443	The test statistic for testing H_0: $\beta_1 = \beta_2 = \cdots = \beta_k = 0$, which asserts that there is no useful linear relationship between y and any of the model predictors. The statistic F is compared to an F critical value based on k numerator df and $n - (k + 1)$ denominator df.
$\hat{y} \pm$ (t crit. val.) $\cdot s_{\hat{y}}$ $\hat{y} \pm$ (t crit. val.) $\cdot \sqrt{s_e^2 + s_{\hat{y}}^2}$	444	A confidence interval for a mean y value and a prediction interval for a single y value when $x_1 = x_1^*, \ldots, x_k = x_k^*$

SUPPLEMENTARY EXERCISES   12.76–12.93

12.76 Exercise 11.40 in the previous chapter gave data on x = bound rubber content (%) and y = tensile modulus (MPa), a measure of coupling effectiveness. A scatter plot shows a substantial linear pattern. Summary quantities are

$$n = 11 \qquad \sum x = 292.9 \qquad \sum x^2 = 8141.75$$
$$\sum y = 69.03 \qquad \sum y^2 = 442.1903 \qquad \sum xy = 1890.200$$

a. Calculate point estimates for the slope and vertical intercept of the population regression line.

b. Calculate a point estimate for the standard deviation σ.

c. Using a .01 significance level, does the data suggest the existence of a useful linear relationship between x and y?

12.77 Return to the bound rubber content–tensile modulus data of the Exercise 12.76.

a. Use a 95% confidence interval to estimate the true mean tensile modulus when the bound rubber content is 20%.

b. Use a 95% prediction interval to predict the value of the tensile modulus resulting from a single observation made when the rubber content is 20%.

c. What aspect of the problem description in part (a) tells you that the requested confidence interval is not for β_1?

12.78 Data on $x =$ depth of flooding and $y =$ flood damage was given in Exercise 11.41. Summary quantities are

$$n = 13 \qquad \sum x = 91 \qquad \sum x^2 = 819$$
$$\sum y = 470 \qquad \sum y^2 = 19118 \qquad \sum xy = 3867$$

a. Does the data suggest the existence of a *positive* linear relationship (one in which an increase in y tends to be associated with an increase in x)? Test using a .05 significance level.

b. Predict flood damage resulting from a claim made when depth of flooding is 3.5 ft, and do so in a way that conveys information about the precision of the prediction.

12.79 Eye weight (g) and cornea thickness (μm) were recorded for nine randomly selected calves, and the resulting data from the paper "The Collagens of the Developing Bovine Cornea" (*Exper. Eye Research* (1984): 639–52) is given. Use this data and a .05 significance level to test the null hypothesis of no correlation between eye weight and cornea thickness against the alternative hypothesis of a positive correlation.

Eye weight	.2	1.4	2.2	2.7	4.9	5.3	8.0	8.8	9.6
Thickness	416	673	733	801	957	1035	883	736	567

12.80 Eight surface soil samples were analyzed to determine physiochemical properties. The following data on $x =$ calcium–sodium exchange rate and $y =$ percent of sodium in the soil was read from a scatter plot that appeared in the paper "Sodium–Calcium Exchange Equilibria in Soils as Affected by Calcium Carbonate and Organic Matter" (*Soil Science* (1984): 109).

x	.641	.611	.463	.375	.260	.184	.182	.089
y	3.4	3.0	3.0	2.2	2.2	2.0	1.9	1.6

$$\sum x = 2.805 \qquad \sum x^2 = 1.281697 \qquad \sum y = 19.3$$
$$\sum y^2 = 49.41 \qquad \sum xy = 7.6546$$

a. Find the equation of the estimated regression line.

b. Compute the predicted values and the corresponding residuals. Use the residuals to calculate SSResid and s_e.

c. Using a significance level of .05, test to determine if the data suggests the existence of a linear relationship between the variables x and y.

d. Compute and interpret the values of r and r^2.

e. Predict the value of percent sodium for a future observation when exchange rate is .300, and do this in a way that conveys information about precision of the prediction.

12.81 The given observations on $x =$ body weight (kg) and $y =$ water intake (l/day) for $n = 7$ subjects appeared in the paper "Validation of a Metabolic Model for Tritium"

(*Radiation Research* (1984): 503–9). Calculate the correlation coefficient r and use it to test the null hypothesis of no correlation between body weight and water intake. Use a .01 significance level.

			Subject				
	1	2	3	4	5	6	7
x	95	52	73	50	82	68	60
y	3.94	1.03	1.71	1.75	1.76	2.01	.97

12.82 The paper "Photocharge Effects in Dye Sensitized Ag[Br,I] Emulsions at Millisecond Range Exposures" (*Photographic Science and Engr.* (1981): 138–44) gave the accompanying data on x = % light absorption at 5800 A and y = peak photovoltage.

x	4.0	8.7	12.7	19.1	21.4	24.6	28.9	29.8	30.5
y	.12	.28	.55	.68	.85	1.02	1.15	1.34	1.29

$$\sum x = 179.7 \qquad \sum y = 7.28 \qquad \sum x^2 = 4334.41$$
$$\sum y^2 = 7.4028 \qquad \sum xy = 178.683$$

a. Construct a scatter plot of the data. What does it suggest?

b. Assuming that the simple linear regression model is appropriate, obtain the equation of the estimated regression line.

c. How much of the observed variation in peak photovoltage can be explained by the model relationship?

d. Predict peak photovoltage when percent absorption is 19.1, and compute the value of the corresponding residual.

e. The paper's authors claimed that there is a useful linear relationship between the two variables. Do you agree? Carry out a formal test.

f. Give an estimate of the average change in peak photovoltage associated with a 1% increase in light absorption. Your estimate should convey information about the precision of estimation.

g. Give an estimate of true average peak photovoltage when % light absorption is 20, and do so in a way that conveys information about precision.

12.83 Reduced visual performance with increasing age has been a much-studied phenomenon in recent years. This decline is due partly to changes in optical properties of the eye itself and partly to neural degeneration throughout the visual system. As one aspect of this problem, the paper "Morphometry of Nerve Fiber Bundle Pores in the Optic Nerve Head of the Human" (*Exp. Eye Research* (1988): 559–68) presented the accompanying data on x = age and y = percentage of the cribriform area of the lamina scleralis occupied by pores.

x	22	25	27	39	42	43	44	46	46
y	75	62	50	49	54	49	59	47	54

x	48	50	57	58	63	63	74	74
y	52	58	49	52	49	31	42	41

a. Suppose that the researchers had believed *a priori* that the average decrease in percentage area associated with a one-year age increase was .5%. Does the data contradict this prior belief? State and test the appropriate hypotheses using a .10 significance level.

b. Estimate true average percent area covered by pores for all 50-year-olds in the population in a way that conveys information about the precision of estimation.

12.84 Occasionally an investigator may wish to compute a confidence interval for α, the y intercept of the true regression line, or test hypotheses about α. The estimated y intercept is just the height of the estimated line when $x = 0$, since $a + b(0) = a$. This implies that s_a, the estimated standard deviation of the statistic a, results from substituting $x^* = 0$ in the formula for s_{a+bx^*}. The desired confidence interval is then $a \pm (t \text{ crit. value})s_a$, whereas a test statistic is $t = (a - \text{hyp. value})/s_a$.

a. The paper "Comparison of Winter-Nocturnal Geostationary Satellite Infrared-Surface Temperature with Shelter-Height Temperature in Florida" (*Remote Sensing of the Environ.* (1983): 313–27) used the simple linear regression model to relate surface temperature as measured by a satellite (y) to actual air temperature (x) as determined from a thermocouple placed on a traversing vehicle. Selected data is given (read from a scatter plot in the paper).

x	−2	−1	0	1	2	3	4	5	6	7
y	−3.9	−2.1	−2.0	−1.2	0	1.9	.6	2.1	1.2	3.0

Estimate the true regression line.

b. Compute the estimated standard deviation s_a. Carry out a test at level of significance .05 to see whether the y intercept of the true regression line differs from zero.

c. Compute a 95% confidence interval for α. Does the result indicate that $\alpha = 0$ is plausible? Explain.

12.85 An experiment to measure $y = $ magnetic relaxation time in crystals (μ sec) as a function of $x = $ strength of the external magnetic field (KG) resulted in the following data ("An Optical Faraday Rotation Technique for the Determination of Magnetic Relaxation Times" *IEEE Trans. Magnetics* (1968): 175–8).

x	11.0	12.5	15.2	17.2	19.0	20.8
y	187	225	305	318	367	365

x	22.0	24.2	25.3	27.0	29.0
y	400	435	450	506	558

Summary quantities are

$$\sum x = 223.2 \qquad \sum y = 4116 \qquad \sum (x - \bar{x})^2 = 348.569$$
$$\sum xy - n\bar{x}\bar{y} = 6578.718 \qquad \sum (y - \bar{y})^2 = 126{,}649.636$$

a. Use the given information to compute the equation of the estimated regression line.

b. Is the simple linear regression model useful for predicting relaxation time from knowledge of the strength of the magnetic field?

c. Estimate the mean relaxation time when the field strength is 18 kg in a way that conveys information about the precision of estimation.

12.86 In some studies an investigator has n (x, y) pairs sampled from one population and m (x, y) pairs from a second population. Let β and β' denote the slopes of the first and second population lines, respectively, and let b and b' denote the estimated slopes calculated from the first and second samples, respectively. The investigator may then wish to test the null hypothesis $H_0: \beta - \beta' = 0$ (i.e., $\beta = \beta'$) against an appropriate alternative hypothesis. Suppose that σ^2, the variance about the population line, is the same for both populations. Then this common variance can be estimated by

$$s^2 = \frac{\text{SSResid} + \text{SSResid}'}{n + m - 4}$$

where SSResid and SSResid' are the residual sums of squares for the first and second samples, respectively. With SS_x and SS_x' denoting $\sum (x - \bar{x})^2$ for the first and second samples, respectively, the test statistic is

$$t = \frac{b - b'}{\sqrt{\dfrac{s^2}{SS_x} + \dfrac{s^2}{SS_x'}}}$$

When H_0 is true, this statistic has a t distribution based on $(n + m - 4)$ df.

The given data is a subset of the data in the paper "Diet and Foraging Mode of *Bufa marinus* and *Leptodactylus ocellatus*" (*J. of Herpetology* (1984): 138–46). The independent variable x is body length (cm) and the dependent variable y is mouth width (cm), with $n = 9$ observations for one type of nocturnal frog and $m = 8$ observations for a second type. Test at level .05 to see whether or not the slopes of the true regression lines for the two different frog populations are identical.

Leptodactylus ocellatus	x	3.8	4.0	4.9	7.1	8.1	8.5	8.9	9.1	9.8
	y	1.0	1.2	1.7	2.0	2.7	2.5	2.4	2.9	3.2

Bufa marinus	x	3.8	4.3	6.2	6.3	7.8	8.5	9.0	10.0
	y	1.6	1.7	2.3	2.5	3.2	3.0	3.5	3.8

	LEPTODACTYLUS	**BUFA**
Sample size:	9	8
$\sum x$	64.2	55.9
$\sum x^2$	500.78	425.15
$\sum y$	19.6	21.6
$\sum y^2$	47.28	62.92
$\sum xy$	153.36	163.36

12.87 Consider the following four (x, y) data sets: the first three have the same x values, so these values are listed only once (from Frank Anscombe, "Graphs in Statistical Analysis" *Amer. Statistician* (1973): 17–21).

Data set	1–3	1	2	3	4	4
Variable	x	y	y	y	x	y
	10.0	8.04	9.14	7.46	8.0	6.58
	8.0	6.95	8.14	6.77	8.0	5.76
	13.0	7.58	8.74	12.74	8.0	7.71
	9.0	8.81	8.77	7.11	8.0	8.84
	11.0	8.33	9.26	7.81	8.0	8.47
	14.0	9.96	8.10	8.84	8.0	7.04
	6.0	7.24	6.13	6.08	8.0	5.25
	4.0	4.26	3.10	5.39	19.0	12.50
	12.0	10.84	9.13	8.15	8.0	5.56
	7.0	4.82	7.26	6.42	8.0	7.91
	5.0	5.68	4.74	5.73	8.0	6.89

For each of these data sets, the values of the summary quantities $\bar{x}$, $\bar{y}$, $\sum (x - \bar{x})^2$, $\sum (y - \bar{y})^2$, and $\sum (x - \bar{x})(y - \bar{y})$ are identical, so all quantities computed from these will be identical for the four sets—the estimated regression line, SSResid, s_e^2, r^2, and so on. The summary quantities provide no way of distinguishing among the four data sets. Based on a scatter plot for each set, comment on the appropriateness or inappropriateness of fitting a simple linear regression model.

12.88 The given scatter diagram, based on 34 sediment samples with x = sediment depth (cm) and y = oil and grease content (mg/kg), appeared in the paper "Mined Land Reclamation Using Polluted Urban Navigable Waterway Sediments" (*J. of Environ. Quality* (1984): 415–22). Discuss the effect that the observation (20, 33000) will have on the estimated regression line. If this point were omitted, what can you say about the slope of the estimated regression line? What do you think will happen to the slope if this observation is included in the computations?

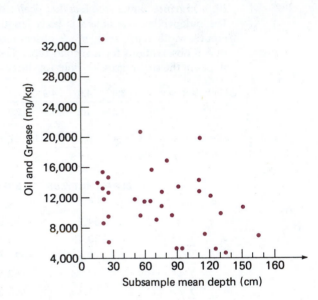

12.89 The paper "Improving Fermentation Productivity with Reverse Osmosis" (*Food Tech.* (1984): 92–6) gave the following data (read from a scatter plot) on y = glucose concentration (g/l) and x = fermentation time (days) for a blend of malt liquor.

x	1	2	3	4	5	6	7	8
y	74	54	52	51	52	53	58	71

a. Use the data to calculate the estimated regression line.
b. Does the data indicate a linear relationship between y and x? Test using a .10 significance level.
c. Using the estimated regression line of part (a), compute the residuals and construct a plot of the residuals versus x (i.e., of the (x, residual) pairs).
d. Based on the plot in part (c), do you think that a linear model is appropriate for describing the relationship between y and x? Explain.

12.90 The employee relations manager of a large company was concerned that raises given to employees during a recent period might not have been based strictly on objective performance criteria. A sample of $n = 20$ employees was selected, and the values of x, a quantitative measure of productivity, and y, the percentage salary increase, were determined for each one. A computer package was used to fit the simple linear regression model, and the resulting output gave the P-value = .0076 for the model utility test. Does the percentage raise appear to be linearly related to productivity? Explain.

12.91 The paper "Statistical Comparison of Heavy Metal Concentrations in Various Louisiana Sediments" (*Environ. Monitoring and Assessment* (1984): 163–70) gave the accompanying data on depth (m), zinc concentration (ppm), and iron concentration (%) for 17 core samples.

Core	Depth	Zinc	Iron
1	.2	86	3.4
2	2.0	77	2.9
3	5.8	91	3.1
4	6.5	86	3.4
5	7.6	81	3.2
6	12.2	87	2.9
7	16.4	94	3.2
8	20.8	92	3.4
9	22.5	90	3.1
10	29.0	108	4.0
11	31.7	112	3.4
12	38.0	101	3.6
13	41.5	88	3.7
14	60.0	99	3.5
15	61.5	90	3.4
16	72.0	98	3.5
17	104.0	70	4.8

a. Using a .05 significance level, test appropriate hypotheses to determine if a correlation exists between depth and zinc concentration.

b. Using a .05 significance level, does the data strongly suggest a correlation between depth and iron concentration?

c. Calculate the slope and intercept of the estimated regression line relating y = iron concentration and x = depth.

d. Use the estimated regression equation to construct a 95% prediction interval for the iron concentration of a single core sample taken at a depth of 50 m.

e. Compute and interpret a 95% interval estimate for the true average iron concentration of core samples taken at 70 m.

12.92 Give a brief answer, comment, or explanation for each of the following:

a. What is the difference between $e_1, e_2, \ldots, e_n$ and the n residuals?

b. The simple linear regression model states that $y = \alpha + \beta x$.

c. Does it make sense to test hypotheses about b?

d. SSResid is always positive.

e. A student reported that a data set consisting of $n = 6$ observations yielded residuals 2, 0, 5, 3, 0, and 1 from the least-squares line.

f. A research report included the following summary quantities obtained from a simple linear regression analysis:

$$\sum (y - \bar{y})^2 = 615, \qquad \sum (y - \hat{y})^2 = 731$$

12.93   Some straightforward but slightly tedious algebra shows that

$$\text{SSResid} = (1 - r^2) \sum (y - \bar{y})^2,$$

from which it follows that

$$s_e = \sqrt{\frac{(n-1)}{(n-2)}} \sqrt{1 - r^2} \; s_y$$

Unless n is quite small, $(n-1)/(n-2) \approx 1$, so $s_e \approx \sqrt{1 - r^2} \; s_y$.

a. For what value of r is s_e as large as s_y? What is the least-squares line in this case?

b. For what values of r will s_e be much smaller than s_y?

REFERENCES   See the References at the end of Chapter 11.

THE ANALYSIS OF VARIANCE

INTRODUCTION Methods for testing $H_0: \mu_1 - \mu_2 = 0$ (that is, $\mu_1 = \mu_2$), where μ_1 and μ_2 are the means of two different populations or the true average responses when two different treatments are applied, were discussed in Chapter 10. Many investigations involve a comparison of more than two population or treatment means. For example, let μ_1, μ_2, μ_3, and μ_4 denote true average burn times under specified conditions for four different fabrics used in children's sleepwear. Data from an appropriate experiment could be used to test the null hypothesis that $\mu_1 = \mu_2 = \mu_3 = \mu_4$ (no difference in true average burn times) against the alternative hypothesis that there are differences among the values of the μ's.

The characteristic that distinguishes the populations or treatments from one another is called the *factor* under investigation. In this example, the factor is fabric type. An experiment might be carried out to compare three different methods for teaching reading (three different treatments), in which case the factor of interest is teaching method. Both fabric type and teaching method are qualitative factors. If growth of fish raised in waters having different salinity levels—0%, 10%, and 30%—is of interest, the factor *salinity level* is quantitative.

In this chapter we focus on single-factor analysis of variance (ANOVA), in which the null hypothesis states that all population means or true average treatment responses are equal. The assumptions and test procedures are presented in Section 1. Section 2 discusses ANOVA computations and their summarization in a tabular format called an ANOVA table. We also briefly indicate how ANOVA methods can be extended to experiments involving more than one factor.

13.1 SINGLE-FACTOR ANOVA AND THE *F* TEST

The decision concerning whether or not the null hypothesis of single-factor ANOVA should be rejected depends on how substantially the samples from the different populations or treatments differ from one another. Figure 1 displays two possible data sets that might arise when observations are selected from each of three populations under study. Each data set consists of five observations from the first population, four from the second, and six observations from the third. For both data sets the three resulting sample means are located by vertical line segments. The means of the two samples from population 1 are identical, and a similar statement holds for the two samples from population 2 and those from population 3.

Almost anyone, after looking at the data set in Figure 1(a), would readily agree that the claim $\mu_1 = \mu_2 = \mu_3$ is false. Not only are the three sample means different, but there is very clear separation between the samples. Put another way, differences between the three sample means are quite large relative to variability within each sample. (If all data sets gave such clear-cut messages, statisticians would not be in such great demand.) The situation pictured in Figure 1(b) is much less clear-cut. While the sample means are as different as they were in the first data set, there is now considerable overlap among the three samples. The separation between sample means here can plausibly be attributed to substantial variability in the populations (and therefore the samples) rather than to differences between μ_1, μ_2, and μ_3. The phrase *analysis of variance* springs from the idea of analyzing variability in the data to see how much can be attributed to differences in the μ's and how much is due to variability in the individual populations. In Figure 1(a) there is little within-sample variability relative to the amount of between-samples variability, whereas in Figure 1(b), a great deal more of the total variability is due to variation within each sample. If differences between the sample means can be explained by within-sample variability, there is no compelling reason for rejecting H_0.

FIGURE 1

Two possible ANOVA data sets when three
populations are under investigation.
● = Observation from Population 1
x = Observation from Population 2
○ = Observation from Population 3

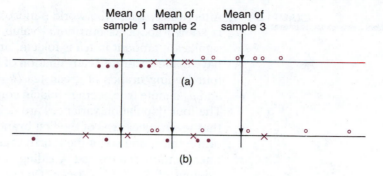

NOTATION AND ASSUMPTIONS   Notation in single-factor ANOVA is a natural extension of the notation used
in Chapter 10 for comparing two population or treatment means.

ANOVA NOTATION AND HYPOTHESES

k = the number of populations or treatments being compared

Population or treatment	1	2	$\cdots$	k
Population or treatment mean	μ_1	μ_2	$\cdots$	μ_k
Population or treatment variance	σ_1^2	σ_2^2	$\cdots$	σ_k^2
Sample size	n_1	n_2	$\cdots$	n_k
Sample mean	$\bar{x}_1$	$\bar{x}_2$	$\cdots$	$\bar{x}_k$
Sample variance	s_1^2	s_2^2	$\cdots$	s_k^2

In addition, let $N = n_1 + n_2 + \cdots + n_k$ be the total number of observations in the data set, and let $\bar{\bar{x}}$ denote the average of all N
observations ($\bar{\bar{x}}$ is the "grand mean").

The hypotheses to be tested are

$H_0: \mu_1 = \mu_2 = \cdots = \mu_k$

versus

H_a: at least two of the μ's are different

A decision between H_0 and H_a will be based on examining the $\bar{x}$'s to see
whether observed discrepancies are small enough to be attributable simply
to sampling variability or whether an alternative explanation for the differences is necessary.

EXAMPLE 1

After water, tea is the world's most widely consumed beverage, yet little is known about its nutritional value. The only B vitamin present in any significant amount in tea is folacin, and recent advances in assay methods have made accurate determination of folacin content feasible. Consider the four leading brands of green tea ($k = 4$ populations), and let μ_1, μ_2, μ_3, and μ_4 denote true average folacin contents for brewed tea of these brands. The four population variances are denoted by σ_1^2, σ_2^2, σ_3^2, and σ_4^2. Suppose that seven specimens of the first brand are obtained, five of the second, six of the third, and six of the fourth brand. The folacin content of each specimen is then determined, yielding the following data (based on "Folacin Content of Tea" *J. of Amer. Dietetic Assoc.* (1983): 627–32; the authors do not give raw data, but their summary values are quite close to those given here).

BRAND	OBSERVATIONS						
1	7.9	6.2	6.6	8.6	8.9	10.1	9.6
2	5.7	7.5	9.8	6.1	8.4		
3	6.4	7.1	7.9	4.5	5.0	4.0	
4	6.8	7.5	5.0	5.3	6.1	7.4	

The summary quantities appear in the accompanying table.

BRAND	SAMPLE SIZE	SAMPLE TOTAL	SAMPLE MEAN	SAMPLE VARIANCE
1	$n_1 = 7$	57.9	$\bar{x}_1 = 8.27$	$s_1^2 = 2.14$
2	$n_2 = 5$	37.5	$\bar{x}_2 = 7.50$	$s_2^2 = 2.83$
3	$n_3 = 6$	34.9	$\bar{x}_3 = 5.82$	$s_3^2 = 2.41$
4	$n_4 = 6$	38.1	$\bar{x}_4 = 6.35$	$s_4^2 = 1.12$

Since the sum of all 24 observations (the *grand total*) is 168.4, the grand mean is

$$\bar{\bar{x}} = \frac{\text{grand total}}{N} = \frac{168.4}{24} = 7.02$$

The hypotheses of interest are

$$H_0: \mu_1 = \mu_2 = \cdots = \mu_4$$

versus

H_a: at least two among the four μ's are different

Notice that there are differences among the four $\bar{x}$'s. Are these differences small enough to be explained solely by sampling variability? Or are they of sufficient magnitude so that a more plausible explanation is that the μ's are not all equal? Our conclusion will depend on how variation among the $\bar{x}$'s (based on their deviations from the grand mean, $\bar{\bar{x}}$) compares to variation within the four samples.

Assumptions about the population or treatment response distributions and the resulting samples are analogous to those on which the two-sample (pooled) t test for testing $H_0: \mu_1 = \mu_2$ were based.

ASSUMPTIONS FOR THE ANOVA *F* TEST

1. Each of the k population or treatment response distributions is normal.
2. $\sigma_1^2 = \sigma_2^2 = \cdots = \sigma_k^2$ (The k normal distributions have identical variances.)
3. The observations in the sample from any particular one of the k populations or treatments are selected independently of one another. (Each sample is a random sample.)
4. The k random samples are selected independently of one another.

In practice, the test based on these assumptions will work well as long the assumptions are not too badly violated. Typically, sample sizes are so small that a separate normal probability plot for each sample is of little value in checking normality. A single combined plot results from first subtracting $\bar{x}_1$ from each observation in the first sample, $\bar{x}_2$ from each value in the second sample, etc., and then constructing a normal probability plot of all N deviations. The plot should be reasonably straight. There are formal procedures for testing the equality of population variances, but for reasons discussed in Chapter 10, we do not favor them. Provided that the largest of the k sample variances is not too many times larger than the smallest, the F test can safely be used. In Example 1, the largest variance (2.83) is less than three times the smallest (1.12). For small sample sizes, this is not at all surprising when all four σ^2's are equal. More on checking assumptions and alternative methods of analysis when these assumptions are judged implausible can be found in the excellent book *Beyond ANOVA: Basics of Applied Statistics*, listed in the chapter references.

MEAN SQUARES AND THE *F* TEST

The standard measure of variation among the $\bar{x}$'s—*between-samples* variation—is based on the deviations $\bar{x}_1 - \bar{\bar{x}}$, $\bar{x}_2 - \bar{\bar{x}}$, . . . , $\bar{x}_k - \bar{\bar{x}}$ from the grand mean. The larger these deviations, the more doubt is cast on H_0. Our benchmark for comparison is a measure of *within-samples* variation obtained by forming a weighted average of the individual sample variances $s_1^2, \ldots, s_k^2$.

 DEFINITION

A measure of disparity among the sample means is the **mean square for treatments**, denoted by **MSTr** and given by

$$\text{MSTr} = \frac{n_1(\bar{x}_1 - \bar{\bar{x}})^2 + n_2(\bar{x}_2 - \bar{\bar{x}})^2 + \cdots + n_k(\bar{x}_k - \bar{\bar{x}})^2}{k - 1}$$

The number of degrees of freedom associated with MSTr is $k - 1$.

A measure of variation within the k samples, called **mean square for error** and denoted by **MSE**, is given by

$$MSE = \frac{(n_1 - 1)s_1^2 + (n_2 - 1)s_2^2 + \cdots + (n_k - 1)s_k^2}{N - k}$$

where $N = n_1 + n_2 + \cdots + n_k$. The quantity MSE has $N - k$ degrees of freedom associated with it.

The terminology "mean square" will be explained in the next section when we discuss ANOVA computations. The fact that MSTr is based on the k deviations $\bar{x}_1 - \bar{\bar{x}}, \ldots, \bar{x}_k - \bar{\bar{x}}$ might initially suggest k degrees of freedom. However, there is a restriction on these deviations, namely, $\sum n(\bar{x} - \bar{\bar{x}}) = 0$. Thus once any $k - 1$ of the deviations are known, the remaining one is completely determined, so MSTr has $(k - 1)$ df. Now consider MSE. The first sample contributes $(n_1 - 1)$ df to MSE (the df for s_1^2), the second sample $(n_2 - 1)$ df, and so on. Because the samples are independent, the df's can be added to yield

$$\begin{aligned}
\text{df for MSE} &= (n_1 - 1) + (n_2 - 1) + \cdots + (n_k - 1) \\
&= n_1 + n_2 + \cdots + n_k - (1 + 1 + \cdots + 1) \\
&= N - k
\end{aligned}$$

Notice also that specializing MSE to the case $k = 2$ gives s_p^2, the pooled estimate of variance used in the two-sample t test, with $df = n_1 + n_2 - 2$.

EXAMPLE 2 Let's return to the folacin-content data of Example 1. With $\bar{x}_1 = 8.27$, $\bar{x}_2 = 7.50$, $\bar{x}_3 = 5.82$, $\bar{x}_4 = 6.35$, $\bar{\bar{x}} = 7.02$, and sample sizes 7, 5, 6, and 6, respectively,

$$\begin{aligned}
MSTr &= \frac{n_1(\bar{x}_1 - \bar{\bar{x}})^2 + \cdots + n_k(\bar{x}_k - \bar{\bar{x}})^2}{k - 1} \\[6pt]
&= \frac{7(8.27 - 7.02)^2 + 5(7.50 - 7.02)^2 + 6(5.82 - 7.02)^2 + 6(6.35 - 7.02)^2}{4 - 1} \\[6pt]
&= \frac{10.9375 + 1.1520 + 8.6400 + 2.6934}{3} \\[6pt]
&\approx 7.81
\end{aligned}$$

Since $s_1^2 = 2.14$, $s_2^2 = 2.83$, $s_3^2 = 2.41$, and $s_4^2 = 1.12$,

$$\begin{aligned}
MSE &= \frac{(n_1 - 1)s_1^2 + \cdots + (n_k - 1)s_k^2}{N - k} \\[6pt]
&= \frac{(7 - 1)(2.14) + (5 - 1)(2.83) + (6 - 1)(2.41) + (6 - 1)(1.12)}{(7 + 5 + 6 + 6) - 4} \\[6pt]
&= \frac{41.81}{20} \\[6pt]
&\approx 2.09
\end{aligned}$$

Both MSTr and MSE are quantities whose values can be calculated once sample data is available, that is, they are statistics. Each of these statistics will vary in value from data set to data set. For example, repeating the experiment of Example 1 with the same sample sizes might yield $\bar{x}_1 = 7.85$, $\bar{x}_2 = 8.03$, $\bar{x}_3 = 6.40$, and $\bar{x}_4 = 6.52$, giving MSTr ≈ 3.14. If these samples also yielded $s_1^2 = 1.83$, $s_2^2 = 3.05$, $s_3^2 = 2.70$, and $s_4^2 = 1.82$, then MSE ≈ 2.29. Both MSTr and MSE have sampling distributions and, in particular, mean (expected) values. The following box describes the key relationship between $E(\text{MSTr})$, the *expected value of the mean square for treatments*, and $E(\text{MSE})$, the *expected value of the mean square for error*.

When H_0 is true ($\mu_1 = \mu_2 = \cdots = \mu_k$),

$E(\text{MSTr}) = E(\text{MSE})$

However, when H_0 is false,

$E(\text{MSTr}) > E(\text{MSE})$,

and the greater the differences among the μ's, the larger $E(\text{MSTr})$ will be relative to $E(\text{MSE})$.

According to this result, when H_0 is true we expect the two mean squares to be close to one another, whereas we expect MSTr to substantially exceed MSE when some μ's differ greatly from others. Thus a calculated MSTr that is much larger than MSE casts doubt on H_0. In Example 2, MSTr $= 7.81$ and MSE $= 2.09$, so MSTr is about 3.7 times as large as MSE. Can this be attributed just to sampling variability, or is the ratio MSTr/MSE of sufficient magnitude to suggest that H_0 is false? Before we can describe a formal test procedure, it is necessary to introduce a new type of probability distribution.

Many ANOVA test procedures are based on sampling distributions called F distributions.* An F distribution is characterized by a number of degrees of freedom associated with the numerator and a number of df associated with the denominator. Thus a particular F distribution may have 3 numerator df and 20 denominator df or 5 numerator df and 18 denominator df.

Figure 2 pictures a typical F curve obtained by specifying both numerator and denominator df's. The numerator and denominator of an F ratio are never negative, so an F curve has positive height only for positive values of an F variable. We need critical values for different F distributions, i.e., values that capture specified tail areas underneath F curves. Notice that an F curve is not symmetric, so knowledge of an upper-tail critical value does not immediately yield the corresponding lower-tail value. However,

---

* F distributions were introduced in Chapter 12 as a basis for the model utility test in multiple regression. Since you may not have covered that material, our discussion here presumes that this is your first confrontation with these distributions. The use of an upper case F is traditional.

ANOVA F tests will always be upper-tailed (we reject H_0 if the F ratio exceeds a specified critical value), so only upper-tailed critical values need be tabulated here.

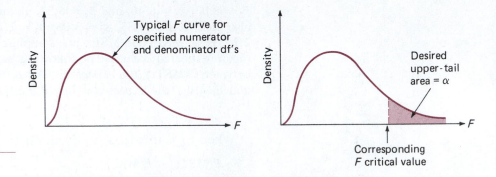

FIGURE 2

An F curve and F critical value.

Table IV(a) in the appendix gives F critical values that capture an upper-tail area of .05, and Table IV(b) gives values for area .01. To obtain a critical value for a particular F distribution, go to Table IV(a) or IV(b), depending on the desired tail area. Then locate the column corresponding to the numerator df, the row corresponding to the denominator df, and finally the critical value at the intersection of this column and row. For example, the critical value for tail area .05, numerator df = 3, and denominator df = 20 is 3.10, the number at the intersection of the column marked 3 and the row marked 20. Similarly the 1% critical value for numerator df = 3 and denominator df = 8 is 7.59, whereas that for numerator df = 8 and denominator df = 3 is 27.49. Do not accidentally reverse numerator and denominator df!

The single-factor ANOVA F test uses the F ratio MSTr/MSE as the test statistic. The theoretical result that underlies the test procedure is this: when H_0 is true, this F ratio has an F sampling distribution with a numerator df of $k - 1$ and a denominator df of $N - k$. The hypothesis H_0 should be rejected when F exceeds an appropriate F critical value, since when this happens the value of MSTr exceeds that of MSE by enough to cast considerable doubt on H_0.

THE SINGLE-FACTOR ANOVA F TEST

Test statistic: $F = \dfrac{\text{MSTr}}{\text{MSE}}$

Rejection region: $F > F$ critical value, where the F critical value is based on $k - 1$ numerator df and $N - k$ denominator df. The critical value is taken from Table IV(a) when a significance level of .05 is desired and from Table IV(b) when a significance level of .01 is appropriate.

EXAMPLE 3

The calculations for the folacin-content data were done in Example 2. The value of F is

$$F = \frac{\text{MSTr}}{\text{MSE}} = \frac{7.81}{2.09} = 3.74$$

The test is based on $k - 1 = 3$ numerator df and $N - k = 20$ a denominator df. For $\alpha = .05$, Table IV(a) gives F critical value 3.10. Since $3.74 > 3.10$, the calculated F falls in the rejection region. We reject H_0 in favor of the conclusion that there are differences among μ_1, μ_2, μ_3, and μ_4.

EXAMPLE 4

An individual's *critical flicker frequency* (cff) is the highest frequency (in cps) at which the flicker in a flickering light source can still be detected. At frequencies above the cff, the light source appears to be continuous even though it is actually flickering. An investigation carried out to see if true average cff depends on iris color yielded the following data (based on the article "The Effect of Iris Color on Critical Flicker Frequency" *J. of Gen. Psych.* (1973): 91–5).

IRIS COLOR	DATA				n	$\bar{x}$	s^2
1. Brown	26.8	27.9	23.7	25.0	8	25.59	1.86
	26.3	24.8	25.7	24.5			
2. Green	26.4	24.2	28.0	26.9	5	26.92	3.40
	29.1						
3. Blue	25.7	27.2	29.9	28.5	6	28.17	2.33
	29.4	28.3					

Also, $N = 19$, grand total $= 508.30$, and $\bar{\bar{x}} = 508.30/19 = 26.75$.

Let's perform a level .05 test to see if true average cff does vary with iris color.

1. Let μ_1, μ_2, and μ_3 denote the true mean cff for individuals having brown, green, and blue iris colors, respectively.
2. H_0: $\mu_1 = \mu_2 = \mu_3$
3. H_a: at least two among μ_1, μ_2, μ_3 are different
4. Test statistic: $F = \dfrac{\text{MSTr}}{\text{MSE}}$
5. Rejection region: Numerator df $= k - 1 = 2$ and denominator df $= N - k = 19 - 3 = 16$. For $\alpha = .05$, Table IV(a) gives 3.63 as the desired critical value. The hypothesis H_0 will now be rejected in favor of H_a if $F > 3.63$.

6. Computations:

$$\text{MSTr} = \frac{8(25.59 - 26.75)^2 + 5(26.92 - 26.75)^2 + 6(28.17 - 26.75)^2}{3 - 1}$$

$$= 11.50$$

$$\text{MSE} = \frac{(8 - 1)(1.86) + (5 - 1)(3.40) + (6 - 1)(2.33)}{19 - 3}$$

$$= 2.39$$

Thus

$$F = \frac{\text{MSTr}}{\text{MSE}} = \frac{11.50}{2.39} = 4.81$$

7. Since $4.81 > 3.63$, we reject H_0 in favor of H_a at level .05. The data does suggest that true average cff varies with iris color, a conclusion agreeing with that given in the paper cited.

The F test was introduced to test the equality of $\mu_1, \mu_2, \ldots, \mu_k$ for $k > 2$. However, the test can also be used when $k = 2$, that is, to test the hypothesis $H_0: \mu_1 = \mu_2$. The formulas for MSTr and MSE remain valid in this case, and Table IV gives F critical values for a numerator df of $k - 1 = 1$ and various denominator df's. The pooled t test can also be used to test $H_0: \mu_1 = \mu_2$. It doesn't matter which test is used when the alternative hypothesis is $H_a: \mu_1 \neq \mu_2$. Irrespective of what samples happen to result, the two test procedures yield exactly the same conclusion—reject H_0 or don't reject H_0—when used at the same level of significance. However, the F test can be used only for $H_a: \mu_1 \neq \mu_2$, while a one-tailed t test can also be used for $H_a: \mu_1 > \mu_2$ or $H_a: \mu_1 < \mu_2$. For this reason, the t test is more flexible when $k = 2$.

EXERCISES   13.1–13.14   **SECTION 13.1**

13.1   What assumptions about the k population or treatment response distributions must you be willing to make in order for the ANOVA F test to be an appropriate method of analysis?

13.2   State the rejection region for the ANOVA F test in each case.
  a. Numerator df $= 5$ and denominator df $= 18$
  b. There are three populations and six observations from each.
  c. There are six treatments and each is applied to three subjects.
  d. Six observations are made on each of the first two treatments and seven observations are made on each of the last three treatments.

13.3   Employees of a certain state university system can choose from among four different health plans. Each plan differs somewhat from the others in terms of hospitalization coverage. Four samples of recently hospitalized individuals were selected, each sample consisting of people covered by a different one of the health plans. The length of the hospital stay (number of days) was determined for each individual selected.

a. What hypotheses would you test to decide whether average length of stay was related to health plan? Note: Be sure to define carefully the population characteristic of interest.

b. If each sample consisted of eight individuals and the value of the ANOVA F statistic was $F = 4.37$, what conclusion would be appropriate for a test with $\alpha = .01$?

c. Answer the question posed in part (b) if the F value given there resulted from sample sizes $n_1 = 9$, $n_2 = 8$, $n_3 = 7$, and $n_4 = 8$.

13.4 A study was carried out to see whether brands of rollerball pens differed with respect to writing lifetime. Five different brands were used in the study. A sample of six pens of each brand was selected, and the pens were inserted in a machine that applied constant pressure on a certain writing surface. The number of hours of use was determined for each pen.

a. What are the hypotheses of interest?

b. If MSTr = 237.5 and MSE = 44.3, carry out a test at significance level .05.

13.5 The paper "Utilizing Feedback and Goal Setting to Increase Performance Skills of Managers" (*Academy of Mgmt. J.* (1979): 516–26) reported the results of an experiment to compare three different interviewing techniques for employee evaluations. One method allowed the employee being evaluated to discuss previous evaluations, the second involved setting goals for the employee, and the third did not allow either feedback or goal setting. After the interviews were concluded, the evaluated employee was asked to indicate how satisfied he or she was with the interview. (A numerical scale was used to quantify level of satisfaction.) The authors used ANOVA to compare the three interview techniques. An F statistic value of 4.12 was reported.

a. Suppose that a total of 33 subjects were used, with each technique applied to 11 of them. Use this information to conduct a level .05 test of the null hypothesis of no difference in mean satisfaction level for the three interview techniques.

b. The actual number of subjects on which each technique was used was 45. After studying the F table, explain why the conclusion in part (a) still holds.

13.6 Mercury is a very hazardous pollutant, and many studies have been carried out in an attempt to assess its toxic effects. One such study is described in the paper "Comparative Responses of the Action of Different Mercury Compounds on Barley" (*Intl. J. of Environ. Studies* (1983): 323–7). Ten different concentrations of mercury (0, 1, 5, 10, 50, 100, 200, 300, 400, and 500 mg/l) were compared with respect to their effects on average dry weight (per 100 seven-day-old seedlings). The basic experiment was replicated four times for a total of 40 dry-weight observations (four for each treatment level). The paper reported an ANOVA F statistic value of 1.895. Using a significance level .05 and the usual seven-step procedure, test the null hypothesis that the true mean dry weight is the same for all ten concentration levels.

13.7 In an experiment to compare tensile strength of five different types of wire, four specimens of each type were tested. MSTr and MSE were computed to be 2573.3 and 1394.2, respectively. Use the ANOVA F test with significance level .01 to test $H_0: \mu_1 = \mu_2 = \mu_3 = \mu_4 = \mu_5$ versus H_a: at least two of these μ's are different.

13.8 Measurement of athletes' body characteristics have been used to establish optimal playing weights for use in training programs. The paper "Prediction of Body Composition in Female Athletes" (*J. of Sports Med.* (1983): 333–41) compared average lean body mass (kg) for female basketball players, cross-country runners, and swimmers. From summary quantities given in the paper, MSTr and MSE were calculated to be 803.9 and 27.5, respectively. These figures were based on three random samples, each of size 10. Does this data suggest that the true average lean body mass differs by sport for female athletes? Use a .01 significance level.

13.9 Three different laboratory methods for determining the concentration of a contaminant in water are to be compared. A gallon of well water is divided into nine equal parts, and each is placed in a container. The containers are then randomly assigned to the three methods. Summary quantities are:

	n	$\bar{x}$	s
Method 1	3	1.914	.212
Method 2	3	1.949	.095
Method 3	3	2.327	.198

Use this data and a significance level of .05 to test the null hypothesis of no difference in mean concentration determination for the three methods.

13.10 The paper "Computer-Assisted Instruction Augmented with Planned Teacher Student Contacts" (*J. of Exp. Educ.* (Winter 1980/81): 120–26) compared five different methods for teaching descriptive statistics. The five methods were traditional lecture and discussion (L/D), programmed textbook instruction (R), programmed text with lectures (R/L), computer instruction (C), and computer instruction with lectures (C/L). Forty-five students were randomly assigned, nine to each method. After completion of the course, a 1-hour exam was given. In addition, a 10-minute retention test was administered six weeks later. Summary quantities are given.

		Exam		Retention Test	
Method	n	$\bar{x}$	s	$\bar{x}$	s
L/D	9	29.3	4.99	30.20	3.82
R	9	28.0	5.33	28.80	5.26
R/L	9	30.2	3.33	26.20	4.66
C	9	32.4	2.94	31.10	4.91
C/L	9	34.2	2.74	30.20	3.53

The grand mean for the exam was 30.82 and the grand mean for the retention test was 29.30.

a. Does the data suggest that there is a difference between the five teaching methods with respect to true mean exam score? Use $\alpha = .05$.

b. Using a .05 significance level, test the null hypothesis of no difference between the true mean retention test scores for the five different teaching methods.

13.11 Growing interest in trout farming has prompted a number of experiments designed to compare various growing conditions. One factor of interest is the salinity of the water. The effect of salinity on the growth of rainbow trout (measured by increase in weight) was examined in the paper "Growth, Training and Swimming Ability of Young Trout Maintained under Different Salinity Conditions" (*J. of Marine Biological Assoc. of U.K.* (1982): 699–708). Full-strength seawater (32% salinity), brackish water (18% salinity), and freshwater (.5% salinity) were used, and the following summary quantities were obtained.

Salinity	Number of Fish	Mean Weight Gain	s
Fresh	12	8.078	1.786
18%	12	7.863	1.756
32%	8	6.468	1.339

Does the data provide sufficient evidence to conclude that the mean weight gain is not the same for the three salinity levels? Use a significance level of .01.

13.12 The *fog index* is a measure of reading difficulty based on the average number of words per sentence and the percentage of words with three or more syllables. High values of the fog index are associated with difficult reading levels. Independent random samples of six advertisements were taken from three different magazines and fog indices were computed to obtain the given data ("Readability Levels of Magazine Advertisements" *J. of Ad. Research* (1981): 45–50).

Scientific American	15.75	11.55	11.16	9.92	9.23	8.20
Fortune	12.63	11.46	10.77	9.93	9.87	9.42
New Yorker	9.27	8.28	8.15	6.37	6.37	5.66

Use a significance level of .01 to test the null hypothesis of no difference between the mean fog index levels for advertisements appearing in the three magazines.

13.13 The paper "Chemical Factors Affecting Soiling and Soil Release from Cotton DP Fabric" (*Amer. Dyestuff Reporter* (1983): 25–30) gave the accompanying data on the degree of soiling of fabric copolymerized with three different mixtures of methacrylic acid (MAA).

	Degree of Soiling				
Mixture 1	0.52	1.12	0.90	1.07	1.04
Mixture 2	0.76	0.82	0.80	0.78	0.81
Mixture 3	0.52	1.08	1.07	1.09	0.93

Is there sufficient evidence to indicate that the true mean degree of soiling is not the same for all three MAA mixtures? Use $\alpha = .05$.

13.14 The given observations are tomato yields (kg/plot) for four different levels of electrical conductivity (EC) of the soil. Chosen EC levels were 1.6, 3.8, 6.0, and 10.2 nmhos/cm.

EC Level			Yield		
1.6	59.5	53.3	56.8	63.1	58.7
3.8	55.2	59.1	52.8	54.5	
6.0	51.7	48.8	53.9	49.0	
10.2	44.6	48.5	41.0	47.3	46.1

Use the ANOVA *F* test with $\alpha = .05$ to test for any differences in true average yield due to the different EC levels.

13.2 ANOVA COMPUTATIONS AND OTHER ISSUES

The formulas for MSTr and MSE given in Section 13.1 are somewhat cumbersome. An alternative method for efficiently computing F involves introducing quantities called *sums of squares*. The accompanying box gives the defining formula, computational formula, and associated df for each of the three sums of squares for single-factor ANOVA.

SUMS OF SQUARES IN SINGLE-FACTOR ANOVA

1. *Total Sum of Squares* (SSTo)

    Defining formula: $\displaystyle\sum_{\substack{\text{all } N \\ \text{obs.}}} (x - \bar{\bar{x}})^2$

    Computing formula: $\displaystyle\sum_{\substack{\text{all } N \\ \text{obs.}}} x^2 - N(\bar{\bar{x}})^2$

    Associated df: $N - 1$

2. *Treatment Sum of Squares* (SSTr)

    Defining formula:

    $$\sum n(\bar{x} - \bar{\bar{x}})^2 = n_1(\bar{x}_1 - \bar{\bar{x}})^2 + \cdots + n_k(\bar{x}_k - \bar{\bar{x}})^2$$

    Computing formula:

    $$\sum n(\bar{x})^2 - N(\bar{\bar{x}})^2 = n_1(\bar{x}_1)^2 + \cdots + n_k(\bar{x}_k)^2 - N(\bar{\bar{x}})^2$$

    Associated df: $k - 1$

3. *Error Sum of Squares* (SSE)

    Defining formula:

    $$\sum_{\substack{\text{1st} \\ \text{sample}}} (x - \bar{x}_1)^2 + \cdots + \sum_{\substack{k\text{th} \\ \text{sample}}} (x - \bar{x}_k)^2$$

    Computing formula: $\text{SSE} = \text{SSTo} - \text{SSTr}$

    Associated df: $N - k$ (total df − treatment df)

Before indicating how F is related to these sums of squares, several preliminary observations are in order. First of all, the defining formula for each SS shows quite clearly that $\text{SSTo} \geq 0$, $\text{SSTr} \geq 0$, and $\text{SSE} \geq 0$. More generally, it is impossible for any sum of squares in statistics to have a negative value. The defining formulas are burdensome for hand calculation because in each case many subtractions are necessary to obtain the deviations that are then squared and summed. This preliminary subtraction is avoided in the computational formulas. In both SSTo and SSTr, the single subtraction of $N(\bar{\bar{x}})^2$ is done as the last step in the computation.

Once SSTo and SSTr have been obtained, the latter is subtracted from the former to give SSE. Since SSE ≥ 0, this implies that SSTo $\geq$ SSTr. As we will see shortly, SSTr and SSE are the two SS's needed for F; SSTo is introduced to give an efficient method for calculating SSE. An additional benefit lies in interpreting the relationship between the three sums of squares. Since SSTo $-$ SSTr $=$ SSE, we can write

$$\text{SSTo} = \text{SSTr} + \text{SSE}$$

This is often called the *fundamental identity for single-factor ANOVA*. The quantity SSTo, the sum of squared deviations about the grand mean, is a measure of total variability in the data set consisting of all k samples. The quantity SSE results from measuring variability separately within each sample and then combining. Such within-sample variability is present regardless of whether or not H_0 is true. The magnitude of SSTr, on the other hand, has much to do with the status of H_0 (whether it's true or false). The more the μ's differ from one another, the larger SSTr will tend to be. Thus SSTr represents variation that can (at least to some extent) be explained by any differences between means. An informal paraphrase of the fundamental identity is

total variation = explained variation + unexplained variation

The F statistic is the ratio of two mean squares. A mean square is an "average square," namely, the sum of squares divided by its associated df. In particular,

$$\text{MSTr} = \frac{\text{SSTr}}{k - 1} \qquad \text{MSE} = \frac{\text{SSE}}{N - k}$$

Thus to carry out the F test, SSTo and SSTr are obtained from the computing formulas, SSE is calculated by subtraction, then the two mean squares are computed, and finally $F = \text{MSTr}/\text{MSE}$.

The ANOVA computations are frequently summarized in a tabular format called an **ANOVA table**. The general form of such a table is displayed in Table 1.

TABLE 1

General Format for a Single-Factor ANOVA Table

Source of Variation	df	Sum of squares	Mean square	F
Treatments	$k - 1$	SSTr	$\text{MSTr} = \dfrac{\text{SSTr}}{k - 1}$	$F = \dfrac{\text{MSTr}}{\text{MSE}}$
Error	$N - k$	SSE	$\text{MSE} = \dfrac{\text{SSE}}{N - k}$	
Total	$N - 1$	SSTo		

EXAMPLE 5

Parents are frequently concerned when their child seems slow to begin walking (although when the child finally walks, the resulting havoc sometimes has the parents wishing they could turn back the clock!). The paper "Walking in the Newborn" (*Science* 176 (1972): 314–5) reported on an experiment in which the effects of several different treatments on the age at which a child first walks were compared. Children in the first group were given special walking exercises for 12 minutes per day beginning at age one week and lasting seven weeks. The second group of children received daily exercises but not the walking exercises administered to the first group. The third and fourth groups were control groups—they received no special treatment and differed only in that the third group's progress was checked weekly, whereas the fourth group's progress was checked just once at the end of the study. Observations on age (in months) when the children first walked are given.

Treatment	Age						n	$\bar{x}$
1	9.00	9.50	9.75	10.00	13.00	9.50	6	10.1250
2	11.00	10.00	10.00	11.75	10.50	15.00	6	11.3750
3	11.50	12.00	9.00	11.50	13.25	13.00	6	11.7083
4	13.25	11.50	12.00	13.50	11.50		5	12.3500

$$N = 23 \qquad \bar{\bar{x}} = \frac{261.00}{23}$$
$$= 11.3478$$

Let's carry out a level .05 test to see whether true average age at which a child first walks depends on which treatment is given.

1. $\mu_1, \mu_2, \mu_3, \mu_4$ are the true average ages of first walking for the four treatments.
2. $H_0: \mu_1 = \mu_2 = \mu_3 = \mu_4$
3. H_a: at least two among the four μ's are different
4. Test statistic: $F = \text{MSTr}/\text{MSE}$
5. Rejection region: $F > F$ critical value; with $\alpha = .05$, treatment df $= k - 1 = 3$, and error df $= N - k = 23 - 4 = 19$, Table IV(a) gives an F critical value of 3.13.
6. Computations: first

$$\sum x^2 = (9.00)^2 + (9.50)^2 + \cdots + (11.50)^2 = 3020.25$$

and

$$N(\bar{\bar{x}})^2 = 23(11.3478)^2 = 2961.77$$

Thus

$$\text{SSTo} = 3020.25 - 2961.77 = 58.48$$

$$\text{SSTr} = n_1 \bar{x}_1^2 + \cdots + n_k \bar{x}_k^2 - N\bar{\bar{x}}^2$$

$$= 6(10.1250)^2 + 6(11.3750)^2 + 6(11.7083)^2 + 5(12.3500)^2$$
$$- 2961.77$$
$$= 2976.56 - 2961.77$$
$$= 14.79$$

$$\text{SSE} = 58.48 - 14.79 = 43.69$$

The remaining computations are summarized in the accompanying ANOVA table.

Source of Variation	df	Sum of Squares	Mean Square	F
Treatments	3	14.79	4.93	$\dfrac{4.93}{2.30} = 2.14$
Error	19	43.69	2.30	
Total	22	58.48		

7. The computed F ratio, 2.14, does not exceed the critical value 3.13, so H_0 is not rejected at level of significance .05. The data does not suggest that there are differences in true average responses among the treatments.

All of the commonly used statistical computer packages will perform a single-factor ANOVA upon request and summarize the results in an ANOVA table. As an example, Table 2 resulted from the use of MINITAB to analyze the data of Example 5. (There are slight differences due to rounding in the hand calculations.)

TABLE 2

An ANOVA Table from MINITAB

```
Analysis of Variance
Due to    df     SS     MS=SS/DF   F-Ratio
Factor     3   14.78     4.93       2.14
Error     19   43.69     2.30
Total     22   58.47
```

P-VALUES As with other test procedures, the *P*-value for an F test is the smallest level of significance at which H_0 can be rejected. Because our F table contains F critical values only for levels .05 and .01, information from these tables about the *P*-value is limited to one of the following three statements:

1. *P*-value > .05 (if F < critical value for $\alpha = .05$);
2. .01 < *P*-value < .05 (if F falls between the two critical values);
3. *P*-value < .01 (if F > critical value for $\alpha = .01$).

However, several of the most widely available statistical computer packages will provide an exact P-value for the test. Table 3 displays output from the package SPSS for the data of Example 5. The P-value appears in the far right column under F PROB as .129. This is consistent with our earlier decision not to reject H_0 at level .05. Even at level of significance .10, we would not reject H_0 because P-value $= .129 > .10$.

TABLE 3

ANOVA Table with P-Value from SPSS

Analysis of Variance

Source	df	Sum of Sq.	Mean Sq.	F Ratio	F Prob
Between Groups	3	14.778	4.926	2.142	.129
Within Groups	19	43.690	2.299		
Total	22	58.467			

MULTIPLE COMPARISONS

When $H_0: \mu_1 = \mu_2 = \cdots = \mu_k$ is rejected by the F test, we believe that there are differences among the k population means. A natural question to ask at this point is "Which means differ?" For example, with $k = 4$, it might be the case that $\mu_1 = \mu_2 = \mu_4$ with μ_3 different from the other three. Another possibility is that $\mu_1 = \mu_4$ and $\mu_2 = \mu_3$. Still another possibility is that all four means are different from one another. A **multiple comparison procedure** is a method for identifying differences among the μ's once the hypothesis of overall equality has been rejected. We will present one such method, called the **Bonferroni multiple comparisons** procedure.

The Bonferroni procedure is based on computing confidence intervals for the difference between each possible pair of μ's. In the case $k = 3$, there are three differences to consider: $\mu_1 - \mu_2$, $\mu_1 - \mu_3$, and $\mu_2 - \mu_3$. (The difference $\mu_2 - \mu_1$ is not considered since the interval for $\mu_1 - \mu_2$ provides the same information. Similarly, intervals for $\mu_3 - \mu_1$ and $\mu_3 - \mu_2$ are not necessary.) Once all confidence intervals have been computed, each is examined to determine whether the interval includes zero. If a particular interval does not include zero, the two means are declared to be "significantly different" from one another. An interval that does include zero supports the conclusion that there is no significant difference between the means involved.

Suppose, for example, that $k = 3$ and the three Bonferroni confidence intervals are:

DIFFERENCE	BONFERRONI INTERVAL
$\mu_1 - \mu_2$	$(-0.9, 3.5)$
$\mu_1 - \mu_3$	$(\ 2.6, 7.0)$
$\mu_2 - \mu_3$	$(\ 1.2, 5.7)$

Since the interval for $\mu_1 - \mu_2$ includes zero, we judge that μ_1 and μ_2 do not differ significantly. The other two intervals do not include zero, so we conclude $\mu_1 \neq \mu_3$ and $\mu_2 \neq \mu_3$.

The following box provides the formulas for constructing the Bonferroni intervals.

THE BONFERRONI MULTIPLE COMPARISON PROCEDURE

When there are k populations to be compared, $k(k-1)/2$ Bonferroni 95% confidence intervals are computed:

For $\mu_1 - \mu_2$: $(\bar{x}_1 - \bar{x}_2) \pm \left(\begin{array}{c}\text{Bonferroni } t \\ \text{critical value}\end{array}\right)\sqrt{\dfrac{\text{MSE}}{n_1} + \dfrac{\text{MSE}}{n_2}}$

$\vdots$

For $\mu_{k-1} - \mu_k$: $(\bar{x}_{k-1} - \bar{x}_k) \pm \left(\begin{array}{c}\text{Bonferroni } t \\ \text{critical value}\end{array}\right)\sqrt{\dfrac{\text{MSE}}{n_{k-1}} + \dfrac{\text{MSE}}{n_k}}$

where MSE is from the analysis of variance, and the Bonferroni t critical value comes from Appendix Table V (using error df).

Two means are judged to differ significantly if the corresponding interval does not include zero.

The Bonferroni confidence intervals are similar in form to the independent-samples confidence interval for $\mu_1 - \mu_2$ introduced in Chapter 10. That interval had the form

$$\bar{x}_1 - \bar{x}_2 \pm (t \text{ critical value})\sqrt{\frac{s_p^2}{n_1} + \frac{s_p^2}{n_2}}$$

where s_p^2 was the pooled estimate of the common variance σ^2, obtained by combining the two sample variances s_1^2 and s_2^2. In the Bonferroni intervals, s_p^2 is replaced by MSE, a combined estimate of the common population variance based on all k samples. A second difference is the use of a Bonferroni t critical value from Table V rather than a t critical value from the t table. This is done in order to achieve a *simultaneous confidence level* of 95% for all intervals computed. This means that the Bonferroni multiple comparison procedure has an overall error rate of 5%. That is, if the procedure is used on many different data sets, about 5% of the time at least one pair of means will incorrectly be declared significantly different. If separate 95% t confidence intervals were used (each constructed using a method that has a 5% error rate), the chance that at least one incorrectly identifies a difference between two sample means increases dramatically with the number of intervals computed.

EXAMPLE 6

Most large companies have established grievance procedures for their employees. One question of interest to employers is why certain groups within a company have higher grievance rates than others. The study described in the paper "Grievance Rates and Technology" (*Academy of Mgmt.* (1979): 810–5) distinguished four types of jobs. These types were labeled apathetic, erratic, strategic, and conservative. Suppose that a total of 52 work groups were selected (13 of each type) and a measure of grievance rate was determined for each one. The resulting sample means and analysis of variance table follow.

Group:	Apathetic	Erratic	Strategic	Conservative
Sample mean:	2.96	5.05	8.74	4.91

ANOVA Table

Source of Variation	df	Sum of Squares	Mean Square	F
Treatments	3	175.9034	58.6344	5.56
Error	48	506.1936	10.5457	
Total	51	682.0970		

The F critical value for $\alpha = .05$, numerator df $= 3$, and denominator df $= 48$ is approximately 2.84, so $H_0: \mu_1 = \mu_2 = \mu_3 = \mu_4$ is rejected at the .05 level of significance.

To determine which means differ, we compute $4(3)/2 = 6$ Bonferroni intervals using MSE $= 10.5457$ and the Bonferroni t critical value of 2.78 (from the column labeled 6 and the row labeled 40—the closest entry to 48 error df—in Table V.) The interval for $\mu_1 - \mu_2$ is computed to be

$$(\bar{x}_1 - \bar{x}_2) \pm \left(\begin{array}{c}\text{Bonferroni } t \\ \text{critical value}\end{array}\right) \sqrt{\frac{\text{MSE}}{n_1} + \frac{\text{MSE}}{n_2}}$$

$$= (2.96 - 5.05) \pm (2.78) \sqrt{\frac{10.5457}{13} + \frac{10.5457}{13}}$$

$$= -2.09 \pm 3.54$$

$$= (-5.63, 1.45)$$

The other five intervals follow.

DIFFERENCE	INTERVAL	CONCLUSION
$\mu_1 - \mu_2$	$(-5.63, \quad 1.45)$	not significantly different
$\mu_1 - \mu_3$	$(-9.32, -2.24)$	$\mu_1 \neq \mu_3$
$\mu_1 - \mu_4$	$(-5.49, \quad 1.59)$	not significantly different
$\mu_2 - \mu_3$	$(-7.23, -0.15)$	$\mu_2 \neq \mu_3$
$\mu_2 - \mu_4$	$(-3.40, \quad 3.68)$	not significantly different
$\mu_3 - \mu_4$	$(\quad 0.29, \quad 7.37)$	$\mu_3 \neq \mu_4$

Based on these intervals, we conclude that the mean grievance rate for the strategic group (μ_3) is significantly different from the other three groups. None of the other means differ significantly from one another.

An effective display for summarizing the results of the Bonferroni multiple comparison procedure involves listing the $\bar{x}$'s and underscoring pairs judged not significantly different. The details of the construction of such a display are described in the accompanying box.

SUMMARIZING THE RESULTS OF THE BONFERRONI PROCEDURE

1. List the sample means in increasing order, identifying the corresponding population just above the value of each $\bar{x}$.

2. Use the Bonferroni intervals to determine the group of means that do not differ significantly from the first in the list. Draw a horizontal line extending from the smallest mean to the last mean in the group identified. For example, if there are five means, arranged in order,

population	3	2	1	4	5
sample mean	$\bar{x}_3$	$\bar{x}_2$	$\bar{x}_1$	$\bar{x}_4$	$\bar{x}_5$

   and μ_3 is judged not significantly different from μ_2 or μ_1 but is judged significantly different from μ_4 and μ_5, draw the following line:

population	3	2	1	4	5
sample mean	$\bar{x}_3$	$\bar{x}_2$	$\bar{x}_1$	$\bar{x}_4$	$\bar{x}_5$

3. Use the Bonferroni intervals to determine the group of means that are not significantly different from the second smallest. (You need only consider means that appear to the right of the mean under consideration.) If there is already a line connecting the second smallest mean with all means in the new group identified, no new line need be drawn. If this entire group of means is not underscored with a single line, draw a line extending from the second smallest to the last mean in the new group. Continuing with our example, if μ_2 is not significantly different from μ_1 but is significantly different from μ_4 and μ_5, no new line need be drawn. However, if μ_2 was not significantly different from either μ_1 or μ_4 but was judged different from μ_5, a second line is drawn, as shown:

population	3	2	1	4	5
sample mean	$\bar{x}_3$	$\bar{x}_2$	$\bar{x}_1$	$\bar{x}_4$	$\bar{x}_5$

4. Continue considering the means in the order listed, adding new lines as needed.

To illustrate this summary procedure, suppose that four samples with $\bar{x}_1 = 19$, $\bar{x}_2 = 27$, $\bar{x}_3 = 24$, and $\bar{x}_4 = 10$ are used to test $H_0: \mu_1 = \mu_2 = \mu_3 = \mu_4$, and that this hypothesis is rejected. Suppose that the Bonferroni confidence intervals indicate that μ_4 is significantly different from μ_1, μ_2, and μ_3, that μ_1 is significantly different from μ_2, and that there are no other significant differences. The resulting summary display would then be

population	4	1	3	2
sample mean	10	19	24	27

EXAMPLE 7

(*Example 6 continued*): For the grievance data of Example 6, the resulting summary display is

	(1)	(4)	(2)	(3)
group	Apathetic	Conservative	Erratic	Strategic
sample mean	2.96	4.91	5.05	8.74

A line is drawn below the means of samples 1, 4, and 2 because these means were judged by the Bonferroni procedure to be not significantly different.

RANDOMIZED BLOCK EXPERIMENTS

In Chapter 10 we saw that when two treatments were to be compared, a paired experiment was often more effective than one involving two independent samples. This is because pairing can considerably reduce the extraneous variation in subjects or experimental units. A similar result can be achieved when more than two treatments are to be compared. Suppose that four different pesticides (the treatments) are being considered for application to a particular crop. There are 20 plots of land available for planting. If five of these plots are randomly selected to receive pesticide #1, five of the remaining 15 randomly selected for #2, and so on, the result is a *completely randomized* experiment and the data should be analyzed using single-factor ANOVA. The disadvantage of this experiment is that if there are any substantial differences in characteristics of the plots that could affect yield, a separate assessment of any differences between treatments won't be possible.

Here is an alternative experiment. Consider separating the 20 plots into five groups, each consisting of four plots. Within each group, the plots are as alike as possible with respect to characteristics affecting yield. Then within each group, one plot is randomly selected for pesticide #1, a second plot is randomly chosen to receive pesticide #2, etc. The homogeneous groups are called **blocks** and the random allocation of treatments within each block as described gives a **randomized block experiment**.

The key to analyzing data from such an experiment is to represent SSTo, which measures total variation, as a sum of three pieces: SSTr, SSE, and a block sum of squares SSBl. This latter SS incorporates any variation due to differences between the blocks, which will be substantial if there was great heterogeneity in experimental units (the plots) prior to creating the blocks. Once the sums of squares have been computed, the test statistic is again an F ratio MSTr/MSE, but error df is no longer $N - k$, as in single-factor ANOVA. Any of the chapter references will provide more information on this important topic.

TWO-FACTOR ANOVA

Suppose an investigator wishes to determine whether yield of a product in a chemical reaction is influenced either by the reaction temperature (factor A) or by the type of catalyst used (factor B). Four different reaction temperatures and three different catalysts are under consideration. These can be combined into 12 different AB pairs: $A_1B_1, A_1B_2, A_1B_3, \ldots, A_4B_3$. The investigator has resources that allow two yield observations to be made at each temperature–catalyst combination. Data from this experiment can be displayed in a rectangular table, as shown in Figure 3.

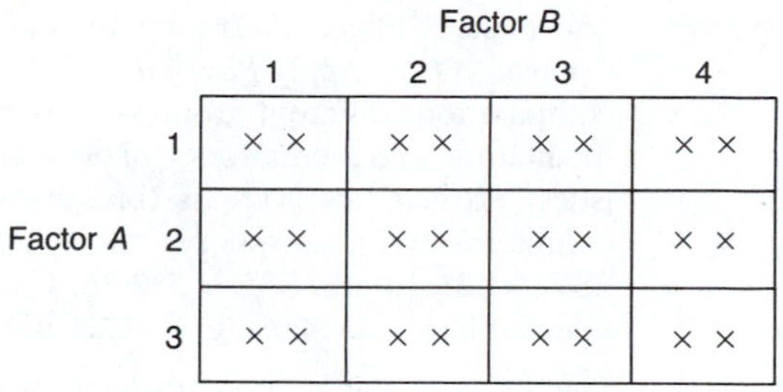

Factor B

FIGURE 3

Representation of data from a two-factor experiment.

Total variation SSTo can be decomposed into several different pieces reflecting random variation (error) and variation due to any differences in levels or categories of the factors. Conclusions are again based on calculating ratios of mean squares and carrying out F tests. Furthermore, these concepts and methods can be extended to situations involving more than two factors. Such ANOVA methods give investigators very powerful tools for analyzing data from complex experiments. One of the chapter references can be consulted for more information.

EXERCISES 13.15–13.25   **SECTION 13.2**

13.15 In an experiment to investigate the performance of four different brands of sparkplugs intended for use on a 125-cc motorcycle, five plugs of each brand were tested and the number of miles (at a constant speed) until failure was observed. A partially completed ANOVA table is given. Fill in the missing entries and test the relevant hypotheses using a .05 level of significance.

Source of Variation	df	Sum of Squares	Mean Square	F
Treatments				
Error		235,419.04		
Total		310,500.76		

13.16 The partially completed ANOVA table given in this problem is taken from the article "Perception of Spatial Incongruity" (*J. of Nervous and Mental Disease* (1961): 222) in which the abilities of three different groups to identify a perceptual incongruity were assessed and compared. All individuals in the experiment had been hospitalized to undergo psychiatric treatment. There were 21 individuals in the depressive group, 32 individuals in the functional "other" group, and 21 individuals in the brain-damaged group. Complete the ANOVA table. Carry out the appropriate test of hypotheses (use $\alpha = .01$) and interpret your results.

Source of Variation	df	Sum of Squares	Mean Square	F
Treatments		152.18		
Error				
Total		1123.14		

13.17 The paper "Effect of Transcendental Meditation on Breathing and Respiratory Control" (*J. of Appl. Physiology* (1984): 607–11) reported on an experiment to compare four different groups—alert nonmeditators, relaxed nonmeditators, alert meditators, and meditators while meditating—with respect to breathing characteristics. Sixteen observations (breaths/min) were made for each condition. Data compatible with summary values given in the paper was used to compute SSTr = 136.14 and SSE = 532.26. Construct an ANOVA table. State and test the relevant hypotheses using a .01 level of significance.

13.18 The accompanying summary quantities are representative of data on professional productivity given in the article "Research Productivity in Academia: A Comparative Study of the Sciences, Social Sciences and Humanities" (*Soc. of Educ.* (1981): 238–53). Randomly selected faculty members in each of the three disciplines were asked to indicate the number of years that they had been teaching. Suppose that 30 faculty members from each subject area were included in the study. Construct an ANOVA table. Is there sufficient evidence to indicate that the true mean number of years of teaching experience is not the same for the three subject areas? Use a .05 significance level.

Natural Sciences	Social Sciences	Humanities	
$\sum (x - \bar{x}_1)^2 = 4468.92$	$\sum (x - \bar{x}_2)^2 = 4138.34$	$\sum (x - \bar{x}_3)^2 = 4629.44$	
$\bar{x}_1 = 12.65$	$\bar{x}_2 = 10.41$	$\bar{x}_3 = 10.93$	$\bar{\bar{x}} = 11.33$

13.19 An article in the British scientific journal *Nature* ("Sucrose Induction of Hepatic Hyperplasis in the Rat" (Aug. 25, 1972: 461) reported on an experiment in which five groups, each consisting of six rats, were put on diets with different carbohydrates. At the conclusion of the experiment, liver DNA content (mg/g) was determined, with the following results:

$$\sum (x - \bar{\bar{x}})^2 = 3.61 \qquad \bar{\bar{x}} = 2.448$$
$$\bar{x}_1 = 2.58 \qquad \bar{x}_2 = 2.63 \qquad \bar{x}_3 = 2.13 \qquad \bar{x}_4 = 2.41 \qquad \bar{x}_5 = 2.49$$

Does the data indicate that the true average DNA content is affected by the type of carbohydrate in the diet? Construct an ANOVA table and use a .05 level of significance.

13.20 College students were assigned to various study methods in an experiment to determine the effect of study technique on learning. The given data was generated to be consistent with summary quantities found in the paper "The Effect of Study Techniques, Study Preferences and Familiarity on Later Recall" (*J. of Exper. Educ.* (1979): 92–5). The study methods compared were reading only, reading and underlining, and reading and taking notes. One week after studying the paper "Love in Infant Monkeys" by Harlow, students were given an exam on the article. Test scores are given in the accompanying table.

Technique	Test Score					
Read only	15	14	16	13	11	14
Read and underline	15	14	25	10	12	14
Read and take notes	18	18	18	16	18	20

a. Compute SSTo, SSTr, and SSE.
b. Construct an ANOVA table.
c. Use a .05 level of significance to test the null hypothesis of no difference between the true mean exam scores for the three study methods.
d. Which of the following statements can be made about the P-value associated with the computed value of the ANOVA F statistic?

  i. P-value $> .05$
  ii. $.01 < P$-value $< .05$
  iii. P-value $< .01$

e. Based on your answer in part (d), would the conclusion of the hypothesis test of part (c) have been any different if a .01 significance level had been used? Explain.

13.21 Some investigators think that the concentration ($\mu g/mL$) of a particular antigen in supernatant fluids could be related to onset of meningitis in infants. The accompanying data is typical of that given in plots appearing in the paper "Type-Specific Capsular Antigen Is Associated with Virulence in Late-Onset Group B Streptococcal Type III Disease" (*Infection and Immunity* (1984): 124–9).

Asymptomatic infants	1.56	1.06	.87	1.39	.71	.87	.95	1.51
Infants with late-onset sepsis	1.51	1.78	1.45	1.13	1.87	1.89	1.07	1.72
Infants with late-onset meningitis	1.21	1.34	1.95	2.00	2.27	.88	1.67	2.57

Construct an ANOVA table and use it to test the null hypothesis of no difference in mean antigen concentrations for the three groups.

13.22 Leaf surface area is an important variable in plant gas-exchange rates. The paper "Fluidized Bed Coating of Conifer Needles with Glass Beads for Determination of Leaf Surface Area" (*Forest Sci.* (1980): 29–32) included an analysis of dry matter per unit surface area (mg/cm^2) for trees raised under three different growing conditions. Let μ_1, μ_2 and μ_3 represent the true mean dry matter per unit surface area for the growing conditions 1, 2, and 3, respectively. The given 95% simultaneous confidence intervals are based on summary quantities that appear in the paper.

Difference	Confidence Interval
$\mu_1 - \mu_2$	$(-3.11, -1.11)$
$\mu_1 - \mu_3$	$(-4.06, -2.06)$
$\mu_2 - \mu_3$	$(-1.95, \ 0.05)$

Which of the following four statements do you think describes the relationship between μ_1, μ_2, and μ_3? Explain your choice.

  i. $\mu_1 = \mu_2$ and μ_3 differs from μ_1 and μ_2.
  ii. $\mu_1 = \mu_3$ and μ_2 differs from μ_1 and μ_3.
  iii. $\mu_2 = \mu_3$ and μ_1 differs from μ_2 and μ_3.
  iv. All three μ's are different from one another.

13.23 Suppose that three different treatments for a particular disease are under consideration. The success of any one of these treatments may depend on such factors as age, weight, blood pressure, and other physiological characteristics. How might the experiment be carried out so as to control for variation due to these extraneous factors?

13.24 Land-treatment wastewater-processing systems work by removing nutrients and thereby discharging water of better quality. The land used is often planted with a crop such as corn, because plant uptake removes nitrogen from the water and sale of the crop helps reduce the costs of wastewater treatment. The concentration of nitrogen in the treated water was observed from 1975 to 1979 under wastewater application rates of none, .05 m/week, and .1 m/week. A randomized block ANOVA was performed with the 5 years serving as blocks. A partially completed ANOVA table is given ("Quality of Percolate Water after Treatment of a Municipal Wastewater Effluent by a Crop Irrigation System" *J. of Environ. Quality* (1984): 256–64).

Source of Variation	df	Sum of Squares	Mean Square	F
Treatments		1835.2	Blocks	
Error				
Total	14	2134.1		

When a randomized block experiment involves k treatments and b blocks, the number of degrees of freedom associated with the four sums of squares SSTo, SSTr, SSBl, and SSE are $bk - 1$, $k - 1$, $b - 1$, and $(k - 1)(b - 1)$, respectively.

a. Complete the ANOVA table.

b. Is there sufficient evidence to reject the null hypothesis of no difference between the true mean nitrogen concentrations for the three application rates? Use $\alpha = .05$.

13.25　Exercise 13.21 presented data that was used to provide support for the theory that the mean antigen concentration was not the same for all three populations considered. Use the Bonferroni multiple comparison procedure to construct simultaneous 95% confidence intervals for $\mu_1 - \mu_2$, $\mu_1 - \mu_3$, and $\mu_2 - \mu_3$. What conclusions can be drawn regarding the three populations?

CHAPTER 13 SUMMARY OF KEY CONCEPTS AND FORMULAS

TERM OR FORMULA	PAGE	COMMENT
Single-factor analysis of variance (ANOVA)	461	A test procedure for determining if there are significant differences among k population or treatment means. The hypotheses tested are $H_0: \mu_1 = \cdots = \mu_k$ versus H_a: at least two μ's differ
Mean square for treatments, $\text{MSTR} = \dfrac{n_1(\bar{x}_1 - \bar{\bar{x}})^2 + \cdots + n_k(\bar{x}_k - \bar{\bar{x}})^2}{k - 1}$	463	A measure of how different the k sample means $\bar{x}_1, \bar{x}_2, \ldots, \bar{x}_k$ are from one another
Mean square for error $\text{MSE} = \dfrac{(n_1 - 1)s_1^2 + \cdots + (n_k - 1)s_k^2}{N - k}$	464	A measure of the amount of variability within the individual samples
$F = \dfrac{\text{MSTr}}{\text{MSE}}$	466	The test statistic for testing $H_0: \mu_1 = \mu_2 = \cdots = \mu_k$ in a single-factor ANOVA: when H_0 is true, F has an F distribution with a numerator df of $k - 1$ and a denominator df of $N - k$.
Sums of squares in single factor ANOVA: SSTo, SSTr, SSE	472	SSTr and SSE are the basis for the computational formulas for the mean squares. The sums of squares satisfy the fundamental identity SSTo = SSTr + SSE.

(*continued next page*)

CHAPTER 13 SUMMARY OF KEY CONCEPTS AND FORMULAS (continued)

TERM OR FORMULA	PAGE	COMMENT
$SSTo = \sum (x - \bar{\bar{x}})^2$ $= \sum x^2 - N(\bar{\bar{x}})^2$	472	The defining and computational formulas for the total sum of squares
$SSTr = \sum n(\bar{x} - \bar{\bar{x}})^2$ $= \sum n\bar{x}^2 - N(\bar{\bar{x}})^2$	472	The defining and computational formulas for the treatment sum of squares
$SSE = SSTo - SSTr$	472	A computational formula for the error sum of squares
$MSTr = \dfrac{SSTr}{k-1}$	473	The formula for the mean square for treatments
$MSE = \dfrac{SSE}{N-k}$	473	The formula for the mean square for error
Bonferroni multiple comparison procedure	477	A procedure for identifying significant differences among the μ's once $H_0: \mu_1 = \mu_2 = \cdots = \mu_k$ has been rejected by the ANOVA F test.

SUPPLEMENTARY EXERCISES   13.26–13.39

13.26 Samples of six different brands of diet or imitation margarine were analyzed to determine the level of physiologically active polyunsaturated fatty acids (PAPUFA, in percent), resulting in the following data:

Imperial	14.1	13.6	14.4	14.3	
Parkay	12.8	12.5	13.4	13.0	12.3
Blue Bonnet	13.5	13.4	14.1	14.3	
Chiffon	13.2	12.7	12.6	13.9	
Mazola	16.8	17.2	16.4	17.3	18.0
Fleischmann's	18.1	17.2	18.7	18.4	

(This data is fictitious, but the sample means agree with data reported in the January 1975 issue of *Consumer Reports*.) Test for differences among the true average PAPUFA percentages for the different brands. Use $\alpha = .05$, and summarize your calculations in an ANOVA table.

13.27 Refer to Exercise 13.26, which introduced data on the level of polyunsaturated fatty acids in margarine. The ANOVA F test showed that $H_o: \mu_1 = \mu_2 = \cdots = \mu_6$ should be rejected. Use the Bonferroni multiple comparison procedure to compute the simultaneous 95% confidence intervals. Summarize the results of the Bonferroni procedure graphically.

13.28 The scores of 24 hard-of-hearing children on a test of basic concepts are given in the accompanying table ("Performance of Young Hearing-Impaired Children on a Test of Basic Concepts" *J. of Speech and Hearing Research* (1974): 342–51).

Age 6	17	20	24	34	34	38					
Age 7	23	25	27	34	38	47					
Age 8	22	23	26	32	34	34	36	38	38	42	48	50

Use an ANOVA F test with significance level .05 to determine if true mean score depends on age.

13.29 Controlling a filling operation with multiple fillers requires adjustment of the individual units. Data resulting from a sample of size five from each pocket of a 12-pocket filler was given in the paper "Evaluating Variability of Filling Operations" (*Food Technology* (1984): 51–55). Data for the first five pockets is given.

Pocket	Fill (oz)				
1	10.2	10.0	9.8	10.4	10.0
2	9.9	10.0	9.9	10.1	10.0
3	10.1	9.9	9.8	9.9	9.7
4	10.0	9.7	9.9	9.7	9.6
5	10.2	9.8	9.9	9.7	9.8

Use the ANOVA F test to determine if the null hypothesis of no difference in the mean fill weights of the five pockets can be rejected.

13.30 Eye inflammation can be induced by the endotoxin lipopolysaccharide (LPS). A random sample of 35 rats was randomly divided into five groups of seven rats each (a completely randomized design). Rats within each group received the same dose of LPS. The accompanying data on vascular permeability was read from plots that appeared in the paper "Endotoxin-Induced Uveitis in the Rat: Observations on Altered Vascular Permeability, Clinical Findings, and Histology" (*Exper. Eye Research* (1984): 665–76). Use analysis of variance with $\alpha = .05$ to test the null hypothesis of no difference between the five treatment means.

Treatment Dose (μg)	Vascular Permeability (ocular to serum fluorescence ratio)						
0	8	3	2	1	0	0	0
1	4.5	4	4	3.5	3	2	0
10	5	5	4	3.5	1	0	0
100	13	12	12	9	8	4	2
500	13	12	9	7.5	7	5	4

13.31 Suppose that each observation in a single-factor ANOVA data set is multiplied by a constant c (a change in units; e.g., $c = 2.54$ changes observations from inches to centimeters). How does this affect MSTr, MSE, and the test statistic F? Is this reasonable? Explain.

13.32 Is it true that the grand mean $\bar{\bar{x}}$ is the ordinary average of $\bar{x}_1, \bar{x}_2, \ldots, \bar{x}_k$? That is, that

$$\bar{\bar{x}} = \frac{\bar{x}_1 + \cdots + \bar{x}_k}{k}$$

Under what conditions on $n_1, n_2, \ldots, n_k$ will such a relationship be true?

13.33 The results of a study on the effectiveness of line drying on the smoothness and stiffness of fabric was summarized in the paper "Line-Dried vs. Machine-Dried Fabrics: Comparison of Appearance, Hand, and Consumer Acceptance" (*Home Ec. Research J.* (1984): 27–35). Smoothness scores were given for nine different types of fabric and five different drying methods: (1) machine dry, (2) line dry, (3) line dry followed by a 15-min machine tumble, (4) line dry with softener, and (5) line dry with air movement. Would you use the single-factor ANOVA F test to see if there is a difference in the true mean smoothness scores for the drying methods? Explain your reasoning.

			Drying Method		
Fabric	1	2	3	4	5
Crepe	3.3	2.5	2.8	2.5	1.9
Double knit	3.6	2.0	3.6	2.4	2.3
Twill	4.2	3.4	3.8	3.1	3.1
Twill mix	3.4	2.4	2.9	1.6	1.7
Terry	3.8	1.3	2.8	2.0	1.6
Broadcloth	2.2	1.5	2.7	1.5	1.9
Sheeting	3.5	2.1	2.8	2.1	2.2
Corduroy	3.6	1.3	2.8	1.7	1.8
Denim	2.6	1.4	2.4	1.3	1.6

13.34 Three different brands of automobile batteries, each one having a 42-month warranty, were included in a study of battery lifetime. A random sample of batteries of each brand was selected and lifetime (in months) was determined, resulting in the following data.

Brand 1: 45 38 52 47 45 42 43
Brand 2: 39 44 50 54 48 46 40
Brand 3: 50 46 43 48 57 44 48

State and test the appropriate hypotheses using a significance level of .05. Be sure to summarize your calculations in an ANOVA table.

13.35 Let $c_1, c_2, \ldots, c_k$ denote k specified numbers, and consider the quantity θ defined by

$$\theta = c_1\mu_1 + c_2\mu_2 + \cdots + c_k\mu_k$$

A confidence interval for θ is then

$$c_1\bar{x}_1 + \cdots + c_k\bar{x}_k \pm (t \text{ critical value}) \cdot \sqrt{\text{MSE}\left(\frac{c_1^2}{n_1} + \cdots + \frac{c_k^2}{n_k}\right)}$$

where the t critical value is based on an error df of $N - k$.

For example, in a study carried out to compare pain relievers with respect to true average time to relief, suppose that brands 1, 2, and 3 are nationally available whereas brands 4 and 5 are sold only by two large chains of drug stores. An investigator might then wish to consider the function

$$\theta = \tfrac{1}{3}\mu_1 + \tfrac{1}{3}\mu_2 + \tfrac{1}{3}\mu_3 - \tfrac{1}{2}\mu_4 - \tfrac{1}{2}\mu_5$$

which, in essence, compares the average effect of the national brands to the average for the house brands.

Refer back to Exercise 13.34, and suppose that brand 1 is a store brand and 2 and 3 are national brands. Obtain a 95% confidence interval for $\mu_1 - \tfrac{1}{2}\mu_2 - \tfrac{1}{2}\mu_3$.

13.36 One of the assumptions that underlies the validity of the ANOVA F test is that the population or treatment response variances $\sigma_1^2, \sigma_2^2, \ldots, \sigma_k^2$ should be identical whether or not H_0 is true—the assumption of constant variance across populations or treatments. In some situations the x values themselves may not satisfy this assumption, yet a transformation using some specified mathematical function (e.g., taking the logarithm or the square root) will give observations that have (approximately) constant variance. The ANOVA F test can then be applied to the transformed data. When observations are made on a counting variable ($x = $ # of something), statisticians have found that taking the square root will frequently "stabilize the variance.'' In an experiment to compare the quality of four different brands of video tape, cassettes of a specified length were selected and the number of flaws in each was determined.

Brand 1:   10   14   5   12   8
Brand 2:   17   14   8   9   12
Brand 3:   13   18   15   18   10
Brand 4:   14   22   12   16   17

Make a square root transformation and analyze the resulting data by using the ANOVA F test at significance level .01.

13.37 The accompanying table appeared in the paper "Effect of SO_2 on Transpiration, Chlorophyll Content, Growth and Injury in Young Seedlings of Woody Angiosperms" (*Canadian J. of Forest Research* (1980): 78–81). Water loss of plants (species: *Acer saccharinum*) exposed to 0, 2, 4, 8, and 16 hours of fumigation was recorded, and a multiple comparison procedure was used to detect differences among the mean water losses for the different fumigation durations. How would you interpret this display?

Duration of fumigation   16   0   8   2   4
Sample mean water loss   27.57   28.23   30.21   31.16   36.21

13.38 A particular county employs three assessors who are responsible for determining the value of residential property in the county. To see whether or not these assessors differ systematically in their appraisals, five houses are selected and each assessor is asked to determine the market value of each house. Explain why a randomized block experiment (with blocks corresponding to the five houses) was used rather than a completely randomized experiment involving a total of 15 houses with each assessor asked to appraise five different houses (a different group of five for each assessor).

13.39 A partially completed ANOVA table for the experiment described in Exercise 13.38 (with houses representing blocks and assessors representing treatments) is given.

Source of Variation	df	Sum of Squares	Mean Square	F
Treatments		11.7		
Block		113.5		
Error				
Total		250.8		

a. Fill in the missing entries in the ANOVA table. Hint: the df for blocks is $b - 1$, where b is the number of blocks.
b. Use the ANOVA F statistic and a .05 level of significance to test the null hypothesis of no difference between assessors.

REFERENCES Hicks, Charles. *Fundamental Concepts in the Design of Experiments*. New York: Holt, Rinehart & Winston, 1982. (Discusses the analysis of data arising from many different types of designed experiments; the focus is more on methods than concepts.)

Miller, Rupert. *Beyond ANOVA: Basics of Applied Statistics*. New York: John Wiley and Sons, 1986. (A wealth of information concerning violations of basic assumptions and alternative methods of analysis.)

Neter, John, William Wasserman, and Michael Kutner. *Applied Linear Statistical Models*. Homewood, Ill.: Richard D. Irwin, Inc., 1985. (The latter half of the book contains a readable survey of ANOVA and experimental design.)

Ott, Lyman. *In Introduction to Statistical Methods and Data Analysis*. Boston: Duxbury Press, 1988. (A good source for learning more about the methods most frequently employed to analyze experimental data, including various multiple comparison techniques.)

C H A P T E R 14

CATEGORICAL DATA AND GOODNESS-OF-FIT TESTS

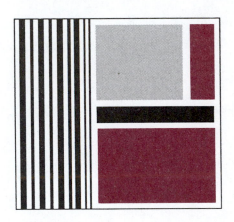

INTRODUCTION   Most of the techniques presented in earlier chapters are designed for numerical data. Different methods must be used when information has been collected on categorical variables. As with numerical data, categorical data sets can be univariate (consisting of observations on a single categorical variable), bivariate (observations on two categorical variables), or even multivariate. In this chapter we first consider inferential methods for analyzing univariate categorical data sets (Section 14.1) and then turn to techniques appropriate for use with bivariate data (Section 14.2).

14.1 CHI-SQUARED TESTS FOR UNIVARIATE CATEGORICAL DATA

Univariate categorical data sets arise in a variety of different settings. If each printed circuit board in a sample of 50 is classified as either defective or nondefective, data on a categorical variable with two categories results. If specifications require that the thickness of the boards be between .055 and .065 inches, boards might be classified into three categories—undersized, meeting specification, and oversized. Each registered voter in a sample of 100 selected from those registered in a particular city might be asked which of five city council members he or she favors for mayor. This would yield observations on a categorical variable with five categories.

Univariate categorical data is most conveniently summarized in a **one-way frequency table**. Suppose, for example, that each item returned to a department store is classified according to disposition—cash refund, credit to charge account, merchandise exchange, or return refused. Records of 100 randomly selected returns are examined and each disposition recorded. The first few observations might be

cash refund

exchange

exchange

return refused

cash refund

credit to account

Counting the number of observations of each type might then result in the accompanying one-way table.

	Disposition			
	Cash	Credit	Exchange	Refused
Frequency	34	18	31	17

For a categorical variable with k possible values (k different levels or categories), sample data is summarized in a one-way frequency table consisting of k cells displayed horizontally or vertically. (This is really nothing more than a frequency distribution, as discussed in Chapter 2.)

In this section, we consider testing hypotheses about the proportion of the population falling into each of the possible categories. For example, the customer relations manager for a department store might be interested in determining whether the four possible dispositions for a return request occur with equal frequency. If this is indeed the case, the long-run proportion of returns falling into each of the four categories is $\frac{1}{4}$ or .25. The test procedure presented shortly would allow the manager to decide whether this hypothesis is plausible.

NOTATION

k = number of categories of a categorical variable

π_1 = true proportion for category 1

π_2 = true proportion for category 2

$\vdots$

π_k = true proportion for category k

(Note: $\pi_1 + \pi_2 + \cdots + \pi_k = 1$)

The hypotheses to be tested are

$H_0: \pi_1$ = hypothesized proportion for category 1

π_2 = hypothesized proportion for category 2

$\vdots$

π_k = hypothesized proportion for category k

$H_a:$ H_0 is not true: at least one of the true category proportions differs from the corresponding hypothesized value

For the example involving department store returns,

π_1 = proportion of all returns that result in a cash refund

π_2 = proportion of all returns that result in an account credit

π_3 = proportion of all returns that result in an exchange

π_4 = proportion of all returns that are refused

The null hypothesis of interest is then

$$H_0: \pi_1 = .25, \quad \pi_2 = .25, \quad \pi_3 = .25, \quad \pi_4 = .25$$

A null hypothesis of the type just described will be tested by first selecting a random sample of size n and then classifying each sample member into one of the k possible categories. In order to decide whether the sample data is compatible with the null hypothesis, the observed cell counts (frequencies) are compared to the cell counts that would have been expected when the null hypothesis is true. In general, the expected cell counts are $n\pi_1$ for category 1, $n\pi_2$ for category 2, etc. The expected cell counts when H_0 is true result from substituting the corresponding hypothesized proportion for each π.

EXAMPLE 1

The paper "Birth Order and Political Success" (*Psych. Reports* (1971): 239–42) reported that in a sample of 31 candidates for political office, 12 were firstborn children, 11 were middleborn, and 8 were lastborn. Since birth position is related to family size, all 31 candidates considered were from families with exactly four children. The author of this paper was interested in ascertaining whether any ordinal position was overrepresented, as this would indicate that those with certain birth orders are more likely to enter political life. In fact, the author thought that first- and lastborns would be overrepresented. Let's suppose that first-, middle-, and lastborns are equally likely to run for political office. In this case, 25% of the candidates from families with four children should be firstborn, 25% lastborn, and 50% middleborn. With π_1, π_2, and π_3 representing the true proportion of candidates for political office from families with four children who are first-, middle-, and lastborn, respectively, the hypotheses of interest are

$$H_0: \pi_1 = .25, \quad \pi_2 = .5, \quad \pi_3 = .25$$

H_a: H_0 is not true

When H_0 is true, the expected counts are

$$\left(\begin{array}{c}\text{expected count} \\ \text{for category 1}\end{array}\right) = n\left(\begin{array}{c}\text{hypothesized proportion} \\ \text{for category 1}\end{array}\right) = 31(.25) = 7.75$$

$$\left(\begin{array}{c}\text{expected count} \\ \text{for category 2}\end{array}\right) = n\left(\begin{array}{c}\text{hypothesized proportion} \\ \text{for category 2}\end{array}\right) = 31(.5) = 15.50$$

$$\left(\begin{array}{c}\text{expected count} \\ \text{for category 3}\end{array}\right) = n\left(\begin{array}{c}\text{hypothesized proportion} \\ \text{for category 3}\end{array}\right) = 31(.25) = 7.75$$

Observed and expected cell counts are given in the accompanying table.

Category	Observed Cell Count	Expected Cell Count
(1) Firstborn	12	7.75
(2) Middleborn	11	15.50
(3) Lastborn	8	7.75

Since the observed cell counts are based on a sample of candidates from families with four children, we would not expect to see exactly 25% falling in the first cell, 50% in the second, and 25% in the third cell, even when H_0 is true. If the differences between the observed and expected cell counts can reasonably be attributed to sampling variation, the data would be considered compatible with H_0. On the other hand, if the discrepancy between the observed and expected cell counts is too large to be attributed solely to chance differences from one sample to another, H_0 should be rejected in favor of H_a. Thus we need an assessment of how different the observed and expected counts are.

The goodness-of-fit statistic, denoted by X^2, is a quantitative measure of the extent to which the observed counts differ from those expected when H_0 is true.*

---

* The Greek letter χ (chi) is often used in place of X. The symbol X^2 is referred to as the *chi-squared* (χ^2) statistic. In using X^2 rather than χ^2, we are adhering to our convention of denoting sample quantities by Roman letters.

The **goodness-of-fit statistic**, X^2, results from first computing the quantity

$$\frac{(\text{observed cell count} - \text{expected cell count})^2}{\text{expected cell count}}$$

for each cell, where, for a sample of size n,

$$\left(\begin{array}{c}\text{expected cell} \\ \text{count}\end{array}\right) = n\left(\begin{array}{c}\text{hypothesized value of corresponding} \\ \text{population proportion}\end{array}\right)$$

Then X^2 is the sum of these quantities for all k cells:

$$X^2 = \sum_{\substack{\text{all} \\ \text{cells}}} \frac{(\text{observed cell count} - \text{expected cell count})^2}{\text{expected cell count}}$$

The value of the X^2 statistic reflects the magnitude of the discrepancies between observed and expected cell counts. When the differences are sizable, the value of X^2 tends to be large. Therefore, large values of X^2 suggest rejection of H_0. A small value of X^2 (it can never be negative) occurs when the observed cell counts are quite similar to those expected when H_0 is true and so would lend support to H_0.

In general, H_0 will be rejected when X^2 exceeds a specified critical value (so the test is upper-tailed). As with previous test procedures, the critical value is chosen to control the probability of a type I error (rejecting H_0 when H_0 is true). This requires information concerning the sampling distribution of X^2 when H_0 is true. A key result is that when the null hypothesis is correct and the sample size is sufficiently large, the behavior of X^2 is described approximately by what statisticians call a **chi-squared distribution**. A chi-squared curve has no area associated with negative values and is asymmetric, with a longer tail on the right. There are actually many chi-squared distributions, each one identified with a different number of degrees of freedom. Curves corresponding to several chi-squared distributions are shown in Figure 1.

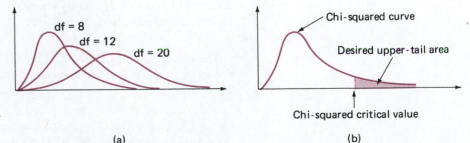

FIGURE 1

Chi-squared curves and an upper-tailed critical value.

Upper-tailed chi-squared critical values, which capture specified upper-tail areas under the corresponding chi-squared curves, are given in Table VI in the appendix. The columns of this table are headed by various areas captured in the upper tail. For example, the critical value 5.99 captures an upper-tail area of .05 under the chi-squared curve with 2 df.

GOODNESS-OF-FIT TEST

As long as none of the expected cell counts are too small, when H_0 is true, the X^2 goodness-of-fit statistic has approximately a chi-squared distribution with $k - 1$ degrees of freedom. It is generally agreed that use of the chi-squared distribution is appropriate when the sample size is large enough so that every expected cell count is at least 5. If any of the expected cell frequencies are less than 5, categories may be combined in a sensible way to create acceptable expected cell counts. Just remember to compute df based on the reduced number of categories.

GOODNESS-OF-FIT TEST PROCEDURE

H_0: π_1 = hypothesized proportion for category 1

$\vdots$

π_k = hypothesized proportion for category k

H_a: H_0 is not true

Test statistic: $X^2 = \sum \dfrac{(\text{observed count} - \text{expected count})^2}{\text{expected count}}$

Rejection Region: Reject H_0 if $X^2 >$ chi-squared critical value. The chi-squared critical value is obtained from Table VI and is based on df $= k - 1$.

EXAMPLE 2

We will use the birth-order data of Example 1 to test the researcher's hypothesis. Let's employ a .05 level of significance and use the seven-step hypothesis-testing procedure illustrated in earlier chapters.

1. Let π_1, π_2, and π_3 denote the true proportions of political candidates who are first-, middle-, and lastborn, respectively (in families with four children).
2. H_0: $\pi_1 = .25$, $\pi_2 = .5$, $\pi_3 = .25$
3. H_a: H_0 is not true
4. Test statistic: X^2
5. Rejection region: Since all expected cell counts are at least 5, the chi-squared critical value that specifies the appropriate rejection region is 5.99 (from the .05 column and the $k - 1 = 3 - 1 = 2$ df row of Table VI). The hypothesis H_0 will be rejected if $X^2 > 5.99$.

6. $$X^2 = \frac{(12 - 7.75)^2}{7.75} + \frac{(11 - 15.50)^2}{15.50} + \frac{(8 - 7.75)^2}{7.75}$$

$$= 2.33 + 1.31 + .01$$

$$= 3.65$$

7. Since 3.65 is less than 5.99, the hypothesis H_0 is not rejected. There is not substantial evidence to suggest that any birth position is over-represented among political candidates. The data does not support the researcher's premise that first- and lastborn children would be more likely to enter politics than those who are middleborn.

EXAMPLE 3

The paper "Environmentalism, Values, and Social Change" (*Brit. J. of Soc.* (1981): 103) investigated characteristics that distinguish environmentalists from the general public. Each member of a sample of 437 environmentalists was classified into one of nine occupational categories. The resulting data appears in the accompanying table. With the nine categories ordered as in the table, the proportions of the general public falling in the nine categories were given as .140, .116, .031, .117, .311, .088, .155, .022, and .020. If the same proportions hold for environmentalists, the corresponding expected cell counts will be

$$\left(\begin{array}{c}\text{first expected}\\\text{count}\end{array}\right) = 437(.140) = 61.18$$

$$\left(\begin{array}{c}\text{second expected}\\\text{count}\end{array}\right) = 437(.116) = 50.69$$

$$\vdots$$

$$\left(\begin{array}{c}\text{ninth expected}\\\text{count}\end{array}\right) = 437(.020) = 8.74.$$

These expected cell counts have been entered in the accompanying table.

Cell	Occupation	Observed Cell Count	Expected Cell Count
1	Professional	67	61.18
2	Clerical	31	50.69
3	Self-employed	42	13.55
4	Service, welfare	190	51.13
5	Manual	23	135.91
6	Retired	11	38.46
7	Housewife	33	67.74
8	Unemployed	6	9.61
9	Student	34	8.74

The X^2 goodness-of-fit test and a .01 level of significance will be used to test the null hypothesis that the true proportion of environmentalists falling into each of the nine categories is the same as that for the general population.

1. Let $\pi_1, \pi_2, \ldots, \pi_9$ denote the true proportions of all environmentalists falling into the nine occupational categories.

2. H_0: $\pi_1 = .140$, $\pi_2 = .116$, $\pi_3 = .031$, $\pi_4 = .117$, $\pi_5 = .311$, $\pi_6 = .088$, $\pi_7 = .155$, $\pi_8 = .022$, $\pi_9 = .020$

3. H_a: H_0 is not true

4. Test statistic:

$$X^2 = \sum \frac{(\text{observed count} - \text{expected count})^2}{\text{expected count}}$$

5. Rejection region: All expected cell counts exceed 5, so the rejection region can be based on a chi-squared distribution with $9 - 1 = 8$ df. With a .01 significance level, the appropriate critical value is obtained from Table VI as 20.09. Therefore, H_0 will be rejected if $X^2 > 20.09$.

6. Computations:

$$X^2 = \frac{(67 - 61.18)^2}{61.18} + \frac{(31 - 50.69)^2}{50.69} + \cdots + \frac{(34 - 8.74)^2}{8.74}$$

$$= .554 + 7.648 + \cdots + 73.005$$

$$= 650.696$$

7. Since $650.696 > 20.09$, the hypothesis H_0 is rejected. There is strong evidence to indicate that at least one of the true cell proportions for environmentalists differs from that of the general public.

EXERCISES 14.1–14.9   **SECTION 14.1**

14.1 What is the rejection region for a chi-squared test based on
a. Five categories carried out at significance level .05?
b. Five categories carried out at significance level .01?
c. Ten categories carried out at significance level .01?

14.2 A particular paperback book is published in a choice of four different covers. A certain bookstore keeps copies of each cover on its racks. To test the hypothesis that sales are equally divided among the four choices, a random sample of 100 purchases is identified.
a. If the resulting X^2 value is 6.4, what conclusion would you reach when using a test with significance level .05?
b. What conclusion would be appropriate at significance level .01 if $X^2 = 15.3$?
c. If there were six different covers rather than just four, what would you conclude if $X^2 = 13.7$ and a test with $\alpha = .05$ were used?

14.3 Packages of mixed nuts made by a certain company contain four types of nuts. The percentages of nuts of types 1, 2, 3, and 4 are supposed to be 40%, 30%, 20%, and 10%, respectively. A random sample of nuts is selected and each one is categorized by type.
a. If the sample size is 200 and the resulting test statistic value is $X^2 = 19.0$, what conclusion would be appropriate for a significance level of .001?
b. If the random sample had consisted of only 40 nuts, would you use the chi-squared test here? Explain your reasoning.

14.4 When public opinion surveys are conducted by mail, a cover letter explaining the purpose of the survey is usually included. To determine if the wording of the cover letter influences the response rate, three different cover letters were used in a survey of students at a Midwestern university ("The Effectiveness of Cover-Letter Appeals" *J. of Soc. Psych.* (1984): 85–91). Suppose that each of the three cover letters accompanied questionnaires sent to an equal number of students. Returned questionnaires were then classified according to the type of cover letter (I, II, or III). Use the given data to test the hypothesis that $\pi_1 = \frac{1}{3}$, $\pi_2 = \frac{1}{3}$, and $\pi_3 = \frac{1}{3}$, where π_1, π_2, and π_3 are the true proportions of all returned questionnaires accompanied by cover letters I, II, and III, respectively. Use a .05 significance level.

	Cover-Letter Type		
	I	II	III
Frequency	48	44	39

14.5 Criminologists have long debated whether there is a relationship between weather and violent crime. The author of the paper "Is There a Season for Homicide?" (*Criminology* (1988): 287–96) classified 1361 homicides according to season, resulting in the following data. Does this data support the theory that the homicide rate is not the same over the four seasons? Test using a significance level of .01.

Season			
Winter	Spring	Summer	Fall
328	334	372	327

14.6 The *Los Angeles Times* (Oct. 17, 1984) reported that the color distribution for plain M&M's was: 40% brown, 20% yellow, 20% orange, 10% green, and 10% tan. Each piece of candy in a random sample of 100 plain M&M's was classified according to color, resulting in the given data. Using a significance level of .05, test to determine if the data suggests that the published color distribution is incorrect.

	Color				
	Brown	Yellow	Orange	Green	Tan
Frequency	45	13	17	7	18

14.7 A certain genetic characteristic of a particular plant can appear in one of three forms (phenotypes). A researcher has developed a theory, according to which the hypothesized proportions are $\pi_1 = .25$, $\pi_2 = .50$, and $\pi_3 = .25$. A sample of 200 plants yields $X^2 = 4.63$.
  a. Carry out a test of the null hypothesis that the theory is correct using level of significance $\alpha = .05$.
  b. Suppose that a sample of 300 plants had resulted in the same value of X^2. How would your analysis and conclusion differ from the analysis and conclusion in part (a)?

14.8 The paper "Linkage Studies of the Tomato" (*Trans. Royal Canad. Inst.* (1931): 1–19) reported the accompanying data on phenotypes resulting from crossing tall cut-leaf tomatoes with dwarf potato-leaf tomatoes. There are four possible phenotypes: (1) tall cut-leaf, (2) tall potato-leaf, (3) dwarf cut-leaf, and (4) dwarf potato-leaf. Mendel's laws of inheritance imply that $\pi_1 = \frac{9}{16}$, $\pi_2 = \frac{3}{16}$, $\pi_3 = \frac{3}{16}$, and $\pi_4 = \frac{1}{16}$. Is the data from this experiment consistent with Mendel's laws? Use a .01 significance level.

	Phenotype			
	1	2	3	4
Frequency	926	288	293	104

14.9 It is hypothesized that when homing pigeons are disoriented in a certain manner, they will exhibit no preference for any direction of flight after takeoff. To test this, 120 pigeons are disoriented, let loose, and the direction of flight of each is recorded. The resulting data is given. Use the goodness-of-fit test with significance level .10 to determine if the data supports the hypothesis.

Direction	0°–45°	45°–90°	90°–135°	135°–180°
Frequency	12	16	17	15

Direction	180°–225°	225°–270°	270°–315°	315°–360°
Frequency	13	20	17	10

14.2 TESTS FOR HOMOGENEITY AND INDEPENDENCE IN A TWO-WAY TABLE

Data resulting from observations made on two different categorical variables can also be summarized using a tabular format. As an example, suppose that residents of a particular city can watch national news on ABC, CBS, NBC, or PBS (the public television network) affiliate stations. A researcher wishes to know whether there is any relationship between political philosophy (liberal, moderate, or conservative) and preferred news program among those residents who regularly watch the national news. Let x denote the variable *political philosophy* and y the variable *preferred network*. A random sample of 300 regular watchers is to be selected, and each one will be asked for his or her x and y values. The data set is bivariate and might initially be displayed as follows:

Observation	x Value	y Value
1	Liberal	CBS
2	Conservative	ABC
3	Conservative	PBS
⋮	⋮	⋮
299	Moderate	NBC
300	Liberal	PBS

Bivariate categorical data of this sort can most easily be summarized by constructing a **two-way frequency table**, or **contingency table**. This is a rectangular table that consists of a row for each possible value of x (each category specified by this variable) and a column for each possible value of y. There is then a cell in the table for each possible (x, y) combination. Once such a table has been constructed, the number of times each particular (x, y) combination occurs in the data set is determined and these numbers (frequencies) are entered in the corresponding cells of the table. The resulting numbers are called **observed cell counts**. The table for the *political philosophy–preferred news program* example discussed earlier contains 3 rows and 4 columns (because x and y have 3 and 4 possible values, respectively). Table 1 is one possible table, with the rows and columns labelled with the possible x and y "values". These are often referred to as the *row* and *column categories*.

TABLE 1

An Example of a 3 × 4 Frequency Table

	ABC	CBS	NBC	PBS	Row Marginal Total
Liberal	20	20	25	15	80
Moderate	45	35	50	20	150
Conservative	15	40	10	5	70
Column Marginal Total	80	95	85	40	300

Marginal totals are obtained by adding the observed cell counts in each row and also in each column of the table. The row and column marginal totals, along with the total of all observed cell counts in the table—the **grand total**—have been included in Table 1. The marginal totals provide information on the distribution of observed values for each variable separately. In this example, the row marginal totals reveal that the sample consisted of 80 liberals, 150 moderates, and 70 conservatives. Similarly, column marginal totals indicated how often each of the preferred program categories occurred: 80 preferred ABC news, 95 preferred CBS, and so on. The grand total, 300, is the number of observations in the bivariate data set, in this case the sample size (although occasionally such a table results from a census of the entire population).

Two-way frequency tables are often characterized by the number of rows and columns in the table (specified in that order: rows first, then columns). Table 1 is called a 3 × 4 table. The smallest two-way frequency table is a 2 × 2 table, which has only two rows and two columns and thus four cells.

Bivariate categorical data arises naturally in two different types of investigations. A researcher may be interested in comparing two or more groups on the basis of a categorical variable and so may obtain a sample separately from each group. For example, data could be collected at a university in order to compare students, faculty, and staff on the basis of primary mode of transportation to campus (car, bicycle, motorcycle, bus, or by foot). One random sample of 200 students, another of 100 faculty, and a third of 150 staff might be chosen, and the selected individuals could be interviewed in order to obtain the needed transportation information. Data from such a study could easily be summarized in a 3 × 5 two-way frequency table with row categories of student, faculty, and staff and column categories corresponding to the five possible modes of transportation. The observed cell counts could then be used to gain insight into differences and similarities between the three groups with respect to the means of transportation. This type of bivariate categorical data is characterized by having one set of marginal totals fixed (the sample sizes from the different groups), whereas each total in the other set is random. In the 3 × 5 situation just discussed, the row totals would be fixed at 200, 100, and 150, respectively.

Bivariate data also arises when the values of two different variables are observed for all individuals or items in a single sample. For example, a sample of 500 registered voters might be selected. Each voter could then be asked both if he or she favored a particular property tax initiative and if he or she was a registered Democrat, Republican, or Independent. This would result in a bivariate data set with x representing *political affiliation* (with categories Democrat, Republican, and Independent) and y representing *response* (favor initiative or oppose initiative). The corresponding 3 × 2 frequency table could then be used to investigate any association between position on the tax initiative and political affiliation. This type of bivariate categorical data is characterized by having both sets of marginal totals random and only the grand total fixed.

COMPARING TWO OR MORE POPULATIONS

When the value of a categorical variable is recorded for members of separate random samples obtained from each population under study, the central issue is whether the category proportions are the same for all of the populations. As in Section 14.1, the test procedure uses a chi-squared statistic that compares the observed counts to those that would be expected if there were no differences between the populations.

EXAMPLE 4

Until recently, a number of professions were prohibited from advertising. In 1977, the U.S. Supreme Court ruled that prohibiting doctors and lawyers from advertising violated their right to free speech. The paper "Should Dentists Advertise?" (*J. of Ad. Research* (June 1983): 33–8) compared the attitudes of consumers and dentists toward the advertising of dental services. Separate samples of 101 consumers and 124 dentists were asked to respond to the following statement: I favor the use of advertising by dentists to attract new patients. Possible responses were: strongly agree, agree, neutral, disagree, and strongly disagree. The data presented in the paper appears in the accompanying 2 × 5 frequency table. The authors were interested in determining whether the two groups—consumers and dentists—differed in their attitudes toward advertising.

Group	Response Strongly Agree	Agree	Neutral	Disagree	Strongly Disagree	Row Marg. Total
Consumers	34	49	9	4	5	101
Dentists	9	18	23	28	46	124
Col. Marg. Total	43	67	32	32	51	225
Col. % of Total	19.11	29.78	14.22	14.22	22.67	

Estimates of expected cell counts can be reasoned out in the following manner: There was a total of 225 responses, of which 43 were "strongly agree." The proportion of the total responding "strongly agree" is then $43/225 = .1911$, or 19.11%. If there were no difference in response for consumers and dentists, we would then expect about 19.11% of the consumers and 19.11% of the dentists to have strongly agreed. Therefore, the *expected cell counts** for the two cells in the "strongly agree" column are

$$\left(\begin{array}{c}\text{expected count for}\\ \text{consumer–strongly agree cell}\end{array}\right) = .1911(101) = 19.30$$

$$\left(\begin{array}{c}\text{expected count for}\\ \text{dentist–strongly agree cell}\end{array}\right) = .1911(124) = 23.70$$

Note that the expected cell counts need not be whole numbers. The expected cell counts for the remaining cells can be computed in a similar manner. For example, $67/225 = .2978$, or 29.78%, of all responses were in the agree category, so

$$\left(\begin{array}{c}\text{expected count for}\\ \text{consumer–agree cell}\end{array}\right) = .2978(101) = 30.08$$

$$\left(\begin{array}{c}\text{expected count for}\\ \text{dentist–agree cell}\end{array}\right) = .2978(124) = 36.93$$

It is common practice to display the observed cell counts and the corresponding estimated expected cell counts in the same table, with the estimated expected cell counts enclosed in parentheses. Expected cell counts for the remaining six cells have been computed and entered into the table that follows. Except for small differences due to rounding, each marginal total for expected cell counts is identical to that for the corresponding observed counts.

---

* In this section, all expected cell counts will be estimated from sample data. All references to "expected counts" should be interpreted as *estimated* expected counts.

Observed and Expected Cell Counts

Group	Strongly Agree	Agree	Neutral	Disagree	Strongly Disagree	Row Marg. Total
Consumers	34 (19.30)	49 (30.08)	9 (14.36)	4 (14.36)	5 (22.89)	101
Dentists	9 (23.70)	18 (36.92)	23 (17.64)	28 (17.64)	46 (28.11)	124
Col. Marg. Total	43	67	32	32	51	225

A quick comparison of the observed and expected cell counts reveals large discrepancies. It appears that most consumers are more favorable toward advertising since the observed counts in the consumer–strongly agree and consumer–agree cells are substantially higher than expected, whereas the observed cell counts for consumers in the disagree and strongly disagree cells are lower than expected (when no difference exists). The opposite relationship between observed and expected counts is exhibited by the dentists.

In Example 4, the expected count for a cell corresponding to a particular group–response combination was computed in two steps. First the response *marginal proportion* was computed (e.g., 43/225 for the strongly agree response). This was then multiplied by a marginal group total (e.g., 101(43/225) for the consumer group). This is equivalent to first multiplying the row and column marginal totals and then dividing by the grand total:

$$\frac{(101)(43)}{225}$$

To compare two or more groups on the basis of a categorical variable, calculate an **expected cell count** for each cell by selecting the corresponding row and column marginal totals and then computing

$$\left(\begin{array}{c}\text{expected} \\ \text{cell count}\end{array}\right) = \frac{(\text{column marginal total})(\text{row marginal total})}{\text{grand total}}$$

These quantities represent what would be expected when there is no difference between the groups under study.

The X^2 statistic, introduced in Section 14.1, can now be used to compare the observed cell counts to the expected cell counts. A large value of X^2 results when there are substantial discrepancies between the observed and expected counts, and suggest that the hypothesis of no differences between the populations should be rejected. A formal test procedure is described in the accompanying box.

COMPARING TWO OR MORE POPULATIONS USING THE X^2 STATISTIC

Null hypothesis: H_0: The true category proportions are the same for all of the populations (homogeneity of populations)

Alternative hypothesis: H_a: The true category proportions are not the same for all of the populations

Test statistic:

$$X^2 = \sum_{\substack{all \\ cells}} \frac{(\text{observed cell count} - \text{expected cell count})^2}{\text{expected cell count}}$$

The expected cell counts are estimated from the sample data (assuming that H_0 is true) using the formula

$$\left(\begin{array}{l}\text{expected} \\ \text{cell count}\end{array}\right) = \frac{(\text{row marginal total})(\text{column marginal total})}{\text{grand total}}$$

Rejection region: When H_0 is true and all expected cell counts are at least 5, X^2 has approximately a chi-squared distribution with

$$df = (\text{number of rows} - 1)(\text{number of columns} - 1)$$

Therefore, H_0 should be rejected if $X^2 >$ chi-squared critical value. (If some expected cell counts are less than 5, row or columns of the table may be combined to achieve a table with satisfactory expected counts.) The critical value comes from the column of Table VI corresponding to the desired level of significance.

EXAMPLE 5

The following table of observed and expected cell counts appeared in Example 4.

Group	Strongly Agree	Agree	Neutral	Disagree	Strongly Disagree
Consumers	34 (19.30)	49 (30.08)	9 (14.36)	4 (14.36)	5 (22.89)
Dentists	9 (23.70)	18 (36.92)	23 (17.64)	28 (17.64)	46 (28.11)

The hypotheses to be tested are:

H_0: proportions in each response category are the same for consumers and dentists;

H_a: H_0 is not true.

A significance level of .05 will be used. All expected cell counts exceed 5, so use of the X^2 test statistic is appropriate.

The critical value comes from the .05 column and the row corresponding to df = $(2 - 1)(5 - 1) = 4$ of Table VI; it is 9.49, so H_0 will be rejected if $X^2 > 9.49$. In fact,

$$X^2 = \frac{(34 - 19.30)^2}{19.30} + \cdots + \frac{(46 - 28.11)^2}{28.11} = 84.47$$

and since $84.47 > 9.49$, the hypothesis H_0 is rejected. There is strong evidence to support the claim that the proportions in each response category are not the same for dentists and consumers.

EXAMPLE 6

The results of an experiment to assess the effects of crude oil on fish parasites were described in the paper "Effects of Crude Oils on the Gastrointestinal Parasites of Two Species of Marine Fish" (*J. of Wildlife Diseases* (1983): 253–8). Three treatments (corresponding to three populations) were compared: (1) no contamination, (2) contamination by 1-year-old weathered oil, and (3) contamination by new oil. For each treatment condition, a sample of fish was taken and then each fish was classified as either parasitized or not parasitized. Data compatible with that in the paper is given; expected cell counts (computed under the hypothesis of no treatment differences) appear in parentheses. Does the data strongly indicate that the three treatments differ with respect to the true proportion of parasitized and nonparasitized fish? A significance level of .01 will be used to test the relevant hypotheses.

Treatment	Parasitized	Nonparasitized	Row Marginal Total
Control	30 (23.0)	3 (10.0)	33
Old oil	16 (16.7)	8 (7.3)	24
New oil	16 (22.3)	16 (9.7)	32
Column marginal Total	62	27	89

The hypotheses are

H_0: proportions of parasitized and of nonparasitized fish are the same for all three treatments

H_a: H_0 is not true

Since all expected cell counts are at least 5, the X^2 statistic can be used. The appropriate critical value, from the .01 column and $(3 - 1)(2 - 1) = 2$ df row of Table VI is 9.21. Therefore, H_0 will be rejected if $X^2 > 9.21$. The computed value of X^2 is

$$X^2 = \frac{(30 - 23.0)^2}{23.0} + \cdots + \frac{(16 - 9.7)^2}{9.7} = 13.0$$

Since $13.0 > 9.21$, the hypothesis H_0 is rejected. The data strongly indicates that the proportions of parasitized and nonparasitized fish are not the same for all three treatments.

TESTING FOR INDEPENDENCE IN A TWO-WAY TABLE

The X^2 test statistic and test procedure can also be used to investigate association between two categorical variables in a single population. Suppose that each population member has a value of a first categorical variable and also of a second such variable. As an example, television viewers in a particular city might be categorized both with respect to preferred network (ABC, CBS, NBC, or PBS) and with respect to favorite type of programming (comedy, drama, or information–news). The question of interest is often whether knowledge of one variable's value provides any information about the value of the other variable–that is, are the two variables independent? Continuing the example, suppose that those who favor ABC prefer the three types of programming in proportions .4, .5, and .1 and that these proportions are also correct for individuals favoring any of the other three networks. Then learning an individual's preferred network provides no (extra) information concerning that individual's favorite type of programming. The categorical variables *preferred network* and *favorite program type* are independent.

To see how the expected counts are obtained in this situation, first recall from our probability discussion in Chapter 4 the condition for independence of two events. The events A and B are said to be independent if

$$P(A \text{ and } B) = P(A) \cdot P(B)$$

so that the proportion of time that they occur together in the long run is the product of the two individual long-run relative frequencies. Similarly, two categorical variables are independent in a population if for *any* particular category of the first variable and *any* particular category of the second variable,

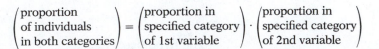

$$\begin{pmatrix} \text{proportion} \\ \text{of individuals} \\ \text{in both categories} \end{pmatrix} = \begin{pmatrix} \text{proportion in} \\ \text{specified category} \\ \text{of 1st variable} \end{pmatrix} \cdot \begin{pmatrix} \text{proportion in} \\ \text{specified category} \\ \text{of 2nd variable} \end{pmatrix}$$

Thus if 30% of all viewers prefer ABC and the proportions of program type preferences are as previously given, assuming the two variables are independent, the proportion of individuals who both favor ABC and prefer comedy is $(.3)(.4) = .12$ (or 12%).

Multiplying the right-hand side of this expression by the sample size gives us the expected number of individuals in the sample who are in both specified categories of the two variables when the variables are independent.

However, these expected counts cannot be calculated because the individual population proportions are not known. The resolution of this dilemma is to estimate each population proportion by the corresponding sample proportion:

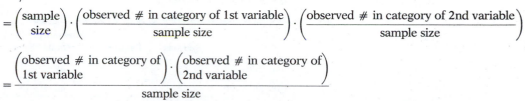

$$\begin{pmatrix} \text{estimated expected number} \\ \text{in specified categories} \\ \text{of the two variables} \end{pmatrix}$$

$$= \begin{pmatrix} \text{sample} \\ \text{size} \end{pmatrix} \cdot \left(\frac{\text{observed \# in category of 1st variable}}{\text{sample size}} \right) \cdot \left(\frac{\text{observed \# in category of 2nd variable}}{\text{sample size}} \right)$$

$$= \frac{\begin{pmatrix} \text{observed \# in category of} \\ \text{1st variable} \end{pmatrix} \cdot \begin{pmatrix} \text{observed \# in category of} \\ \text{2nd variable} \end{pmatrix}}{\text{sample size}}$$

Suppose that the observed counts are displayed in a rectangular table in which rows correspond to the different categories of the first variable and columns to the categories of the second variable. Then the numerator in the preceding expression for estimated expected counts is just the product of the row and column marginal totals. This is exactly how estimated expected counts were computed in the test for homogeneity of several populations.

TESTING FOR INDEPENDENCE OF TWO CATEGORICAL VARIABLES

Null hypothesis: H_0: The two variables are independent

Alternative hypothesis: H_a: The two variables are not independent

Test statistic:

$$X^2 = \sum_{\substack{\text{all} \\ \text{cells}}} \frac{(\text{observed cell count} - \text{expected cell count})^2}{\text{expected cell count}}$$

The expected cell counts are estimated (assuming H_0 is true) by the formula

$$\begin{pmatrix} \text{expected} \\ \text{cell count} \end{pmatrix} = \frac{(\text{row marginal total})(\text{column marginal total})}{\text{grand total}}$$

Rejection region: When H_0 is true and all expected cell counts are at least 5, then X^2 has approximately a chi-squared distribution, with

df = (number of rows $- 1$)(number of columns $- 1$).

The hypothesis H_0 should be rejected if $X^2 >$ chi-squared critical value where chi-squared critical values are given in Table VI.

EXAMPLE 7

The paper "Impulsive and Premeditated Homicide: An Analysis of Subsequent Parole Risk of the Murderer" (*J. of Criminal Law and Criminology* (1978): 108–14) investigated the relationship between the circumstances surrounding a murder and subsequent parole success of the murderer. A sample of 82 convicted murderers was selected, and each murderer was categorized according to type of murder (impulsive, premeditated) and parole outcome (success, failure). The resulting data is given in the accompanying 2 × 2 table.

	Success	Failure	Row Marg. Total
Impulsive	13	29	42
Premeditated	22	18	40
Col. Marg. Total	35	47	82

The authors were interested in determining whether there is an association between type of murder and parole outcome. Using a .05 level of significance, we will test

H_0: type of murder and parole outcome are independent

H_a: the two variables are not independent

Expected cell counts are as follows:

Cell Row	Column	Expected Cell Count
1	1	$\frac{(35)(42)}{82} = 17.93$
1	2	$\frac{(47)(42)}{82} = 24.07$
2	1	$\frac{(35)(40)}{82} = 17.07$
2	2	$\frac{(47)(40)}{82} = 22.93$

The observed and expected counts are given together in the accompanying table.

	Success	Failure	Row Marg. Total
Impulsive	13 (17.93)	29 (24.07)	42
Premeditated	22 (17.07)	18 (22.93)	40
Col. Marg. Total	35	47	82

Since all expected cell counts exceed 5, the X^2 statistic can be used. With level of significance $\alpha = .05$ and df $= (2 - 1)(2 - 1) = 1$, the chi-squared critical value is 3.84. Thus H_0 will be rejected if $X^2 > 3.84$.

The computed value of X^2 is

$$X^2 = \frac{(13 - 17.93)^2}{17.93} + \cdots + \frac{(18 - 22.93)^2}{22.93} = 4.85$$

Since $4.85 > 3.84$, the hypothesis H_0 is rejected. The data supports the existence of an association between type of murder and parole outcomes.

EXAMPLE 8

The accompanying two-way frequency table appeared in the paper "Marijuana Use in College" (*Youth and Society* (1979): 323–34). Four hundred and forty-five college students were classified according to both frequency of marijuana use and parental use of alcohol and psychoactive drugs. Expected cell counts (computed under the assumption of no association between student marijuana use and parental use) appear in parentheses in the given table.

Parental Use of Alc. and Drugs	Student Level of Marijuana Use			Row Marg. Total
	Never	Occasional	Regular	
Neither	141 (119.3)	54 (57.6)	40 (58.1)	235
One	68 (82.8)	44 (39.9)	51 (40.3)	163
Both	17 (23.9)	11 (11.5)	19 (11.6)	47
Col. Marg. Total	226	109	110	445

The X^2 test with a .01 significance level will be used to determine if there is an association between marijuana use and parental use of drugs and alcohol.

H_0: student marijuana use is independent of parental drug and alcohol use

H_a: the two variables are not independent

Since all expected cell counts are at least 5, the X^2 test can be used. With $\alpha = .01$ and df $= (3 - 1)(3 - 1) = 4$, the chi-squared critical value is 13.28 and H_0 will be rejected if $X^2 > 13.28$. The computed value of X^2 is

$$X^2 = \frac{(141 - 119.3)^2}{119.3} + \cdots + \frac{(19 - 11.6)^2}{11.6} = 22.45$$

Since $22.45 > 13.28$, the hypothesis H_0 is rejected. There *does* appear to be an association between student use of marijuana and parental use of alcohol and drugs.

Most statistical computer packages will calculate both estimated expected cell counts and the value of the X^2 test statistic. MINITAB output for this data follows. (The discrepancy between $X^2 = 22.45$ and $X^2 = 22.37$ is due to rounding.)

```
EXPECTED FREQUENCIES ARE PRINTED BELOW OBSERVED FREQUENCIES
        I  C1   I  C2   I  C3    ITOTALS
-------I-------I-------I-------I-------
    1  I  141  I   54  I   40  I    235
       I  119.3I  57.6I  58.1I
-------I-------I-------I-------I-------
    2  I   68  I   44  I   51  I    163
       I  82.8I   39.9I  40.3I
-------I-------I-------I-------I-------
    3  I   17  I   11  I   19  I     47
       I  23.9I   11.5I  11.6I
-------I-------I-------I-------I-------
TOTALS I  226  I  109  I  110  I    445

TOTAL CHI SQUARE =

        3.93 + 0.22 + 5.63 +
        2.64 + 0.42 + 2.85 +
        1.98 + 0.02 + 4.69 +

            = 22.37

DEGREES OF FREEDOM = (3 - 1) × (3 - 1) = 4
```

In some investigations, values of more than two categorical variables are recorded for each individual in a sample. For example, in addition to the variables *student marijuana use* and *parental use of drugs and alcohol*, the researchers in the study referenced in Example 8 might also have recorded political affiliation for each student in the sample. A number of interesting questions could then be explored: Are all three variables independent of one another? Is it possible that student use and parental use are dependent but that the relationship between them does not depend on political affiliation? For a particular political affiliation, are student use and parental use independent?

The X^2 test procedure described in this section for analysis of bivariate categorical data can be extended for use with *multivariate categorical data*. Appropriate hypothesis tests can then be used to provide insight into the relationships between variables. However, the computations required to calculate estimated expected cell counts and to compute the value of X^2 are quite tedious and so are seldom done without the aid of a computer. Several statistical computer packages (including BMDP and SAS) can perform this type of analysis. The chapter references can be consulted for further information on the analysis of categorical data.

14.10   A particular state university system has six campuses. On each campus, a random sample of students will be selected and each student will be categorized with respect to political philosophy as liberal, moderate, or conservative. The null hypothesis of interest is that the proportion of students falling in these three categories is the same at all six campuses.

     a. On how many degrees of freedom will the resulting X^2 test be based, and what is the rejection region for significance level .01?

     b. How do your answers in part (a) change if there are seven campuses rather than six?

     c. How do your answers in part (a) change if there are four rather than three categories for political philosophy?

14.11   A random sample of 1000 registered voters in a certain county is selected and each voter is categorized with respect to both educational level (four categories) and preferred candidate in an upcoming supervisorial election (five possibilities). The hypothesis of interest is that educational level and preferred candidate are independent factors.

     a. If it is decided to use a test with significance level .05, what is the rejection region?

     b. If $X^2 = 7.2$, what would you conclude at significance level .10?

     c. If there were only four candidates vying for election, what would you conclude if $X^2 = 14.5$ and a test with $\alpha = .05$ is appropriate?

14.12   An increasing number of people are spending their working hours in front of a video display terminal (VDT). The paper "VDT Workstation Design: Preferred Settings and Their Effects" (*Human Factors* (1983): 161–75) summarized a study of adjustable VDT screens. Sixty-five workers using nonadjustable screens and 66 workers using adjustable were asked if they experienced annoying reflections from the screens. The resulting data is given in the accompanying table.

Screen Type	Annoying Reflection	
	No	Yes
Nonadjustable	15	50
Adjustable	28	38

     a. The investigators were interested in whether the proportion experiencing annoying reflections was the same for both types of VDT screens. Does this problem situation involve comparing two populations or testing for independence? Explain.

     b. Use a .05 significance level and the X^2 statistic to test the appropriate hypotheses.

     c. Can you think of another test statistic that could be used to answer the researchers' question? (*Hint*: See Chapter 10.) If the researchers were interested in determining whether the proportion experiencing annoying reflection was smaller for the adjustable VDT screens, which test statistic would you recommend? Explain. (*Hint*: Look at the alternative hypothesis for the X^2 test.)

14.13   The paper "The Liking and Viewing of Regular TV Series" (*J. of Consumer Research* (1987): 63–70) raised the issue of the extent to which people actually like the TV programs they watch. Viewers of two comedy series were asked to respond to the question "How interesting do you find this show?" Data in the accompanying table is compatible with percentages given in the paper. Using a significance level of .05, test the null hypothesis

H_0: The true proportion of responses in the five response categories are the same for regular viewers of the two programs.

	Response				
Series	Extremely Interesting	Very Interesting	Fairly Interesting	Not Very Interesting	Not at all Interesting
Three's Company	19	24	30	10	17
Hello Larry	9	22	46	37	36

14.14 The paper "Color Associations of Male and Female Fourth-Grade School Children" (*J. of Psych.* (1988): 383–8) asked children to indicate what emotion they associated with the color red. The response and the sex of the child were noted, and the data is summarized in the accompanying two-way table. Is there strong evidence of an association between response and gender? Use a .01 level of significance.

	Gender	
Emotion	Male	Female
Anger	34	27
Pain	28	17
Happiness	12	19
Love	38	39

14.15 The U.S. Census Bureau contends ". . . the American people are the most educated people in the world." ("Ticket to Success," *Washington Post*, Oct. 17, 1985) Suppose that a random sample of 1000 adults in each of the U.S., East Germany, Canada, Sweden, and Japan resulted in the data in the accompanying table. (These figures are based on summary values given in the article.) Use the test for homogeneity of proportions to determine whether the proportion of the adult population with some college education is the same for the five countries compared.

Country	Some College Education	No College Education
U.S.	320	680
East Germany	173	827
Canada	172	828
Sweden	155	845
Japan	145	855

14.16 An experiment to determine the response of soybeans to long-term ozone exposure was described in the paper "Injury and Yield Response of Soybean to Chronic

Doses of Ozone and Soil Moisture Deficit'' (*Crop Science* (1987): 1016–24). Soybean seeds were exposed to one of seven different ozone treatments and the number of exposed seeds that produced normal seedlings was noted. The resulting data appears in the accompanying table. Is there evidence that the proportion of normal seedlings is not the same for all seven treatments? Test the relevant hypotheses using a level .05 test.

Treatment	Normal Seedlings	Abnormal Seedlings	Total
1	1747	410	2157
2	2099	526	2625
3	1292	386	1678
4	1676	446	2122
5	1238	413	1651
6	863	678	1541
7	1364	408	1772

14.17 A study financed by the U.S. Dept. of Justice (*Los Angeles Times*, Sept. 7, 1986) looked at the records of convicts released from prison in 1978 in three states—California, Michigan, and Texas. The number who had been rearrested within three years of release was noted, resulting in the accompanying table. Is there evidence that the proportions of those rearrested differs for the three states? Test the relevant hypotheses using a .01 level of significance.

State	Rearrested	Not Rearrested	Total
California	217	69	286
Texas	278	185	463
Michigan	145	129	274

14.18 The article ''Catholic: New Challenges for Church'' (*Los Angeles Times*, May 20, 1985) classified survey respondents according to religious preference and level of education to obtain the following two-way table. Does the data provide support for the existence of an association between level of education and religion? Test the relevant hypotheses using $\alpha = .05$

Education	Religion	
	Catholic	Protestant
Some high school	79	68
Some college	22	15
Some postgraduate study	9	5

14.19 Elbow dislocation is a common injury. The given frequency table appeared in the paper ''Elbow Dislocations'' (*J. of Amer. Med. Assoc.* (1965): 113). Samples of 61 men and 44 women suffering elbow dislocations were classified into six categories according to age. Does the data suggest that the true age distribution (as given by the proportions falling into each of the six age categories) differs for men and women who dislocate an elbow? Use a significance level of .01.

	Age					
Sex	1–10	11–20	21–30	31–40	41–50	Over 50
Male	7	21	9	13	6	5
Female	7	6	2	3	10	16

14.20   Are the educational aspirations of students related to family income? This question was investigated in the article "Aspirations and Expectations of High School Youth" (*Int. J. of Comp. Soc.* (1975): 25). The given 4 × 3 table resulted from classifying 273 high school students according to expected level of education and family income. Does the data indicate that educational aspirations and family income are not independent? Use a .10 level of significance.

	Income		
Aspired Level	Low	Middle	High
Some high school	9	11	9
High school graduate	44	52	41
Some college	13	23	12
College graduate	10	22	27

14.21   The effect of copper on earthworms was investigated in the paper "Sublethal Toxic Effects of Copper on Growh, Reproduction and Litter Breakdown Activity in the Earthworm *Lumbricus rubellus*, with Observations on the Influence of Temperature and Soil pH" (*Environ. Pollution* (1984): 207–19). Each of four concentrations of copper (14, 54, 131, and 372 mg/kg soil) was applied to a sandy soil containing a known number of worms. After six weeks, the number of surviving worms was recorded. Data compatible with that given in the paper appears in the accompanying table. Does this data strongly suggest that the mortality rate (proportion not surviving) differs for the four concentrations? Use a .01 significance level.

	Concentration Level			
	I	II	III	IV
Survived	80	74	78	66
Died	0	6	2	14

14.22   The accompanying frequency table appeared in the paper "Commitment to Work in Immigrants: Its Functions and Peculiarities" (*J. of Vocational Behavior* (1984): 329–39). The data resulted from classifying 175 workers according to two variables: *job type* (with three categories: immigrant (I), white-collar nonimmigrant (W), and executive nonimmigrant (E)) and *attitude toward authority* (with two categories: positive (P) and negative (N)). Does the data support the theory that there is an association between attitude toward authority and job type? Use a significance level of .01.

Job Type	Attitude	
	P	N
I	51	23
W	34	32
E	25	10

14.23 The *Los Angeles Times* (July 29, 1983) conducted a survey to find out why some Californians don't register to vote. Random samples of 100 Latinos, 100 Anglos, and 100 Blacks who were not registered to vote were selected. The resulting data is summarized in the accompanying table. Does the data suggest that the true proportion falling into each response category is not the same for Latinos, Anglos, and Blacks? Use a .05 significance level.

Reason for Not Registering	Latino	Anglo	Black
Not a citizen	45	8	0
Not interested	19	33	19
Can't meet residency requirements	9	35	23
Distrust of politics	5	10	8
Too difficult to register	10	10	27
Other reason	12	4	23

14.24 The relative importance attached to work and home life by high school students was examined in "Work Role Salience as a Determinant of Career Maturity in High School Students" (*J. Vocational Behavior* (1984): 30–44). Does the data summarized in the accompanying two-way frequency table suggest that sex and relative importance assigned to work and home are not independent? Test using a .05 level of significance.

Sex	Relative Importance		
	Work > Home	Work = Home	Work < Home
Female	68	26	94
Male	75	19	57

14.25 In a study of 2989 cancer deaths, the location of death (home, acute-care hospital, or chronic-care facility) and age at death were recorded, resulting in the given two-way frequency table. ("Where Cancer Patients Die" *Public Health Reports* (1983): 173). Using a .01 significance level, test the null hypothesis that age at death and location of death are independent.

Age	Location		
	Home	Acute-Care	Chronic-Care
15–54	94	418	23
55–64	116	524	34
65–74	156	581	109
Over 74	138	558	238

CHAPTER 14 SUMMARY OF KEY CONCEPTS AND FORMULAS

TERM OR FORMULA	PAGE	COMMENT
One-way frequency table	491	A compact way of summarizing data on a categorical variable, it gives the number of times each of the possible categories in the data set occurs (the frequencies.)
Goodness-of-fit statistic, $$X^2 = \sum_{\substack{\text{all} \\ \text{cells}}} \frac{\left(\begin{array}{c}\text{observed} \\ \text{cell count}\end{array} - \begin{array}{c}\text{expected} \\ \text{cell count}\end{array}\right)^2}{\text{expected cell count}}$$	494	A statistic used to provide a comparison between observed counts and those expected when a given hypothesis is true. When none of the expected counts are too small X^2 has approximately a *chi-squared distribution*.
X^2 test in a one-way frequency table: H_0: π_1 = hypothesized proportion for category 1 $\vdots$ π_k = hypothesized proportion for category k	495	The hypothesis test performed to determine whether the true category proportions are different from those specified by the given null hypothesis
Two-way frequency table	500	A rectangular table used to summarize a bivariate categorical data set; two-way tables are used to compare several populations on the basis of a categorical variable or to identify whether an association exists between two categorical variables.
X^2 test for comparing two or more populations: H_0: the true category proportions are the same for all of the populations	504	The hypothesis test performed to determine whether the true category proportions are the same for all of the populations to be compared
X^2 Test for independence: H_0: the two variables defining the table are independent	507	The hypothesis test performed to determine whether an association exists between two categorical variables

SUPPLEMENTARY EXERCISES   14.26–14.34

14.26 Each driver in a sample of size 1024 was classified according to both seat-belt usage and sex to obtain the accompanying 2 × 2 frequency table ("What Kinds of People Do Not Use Seat Belts" *Amer J. of Public Health* (1977): 1043–9). Does the data strongly suggest an association between sex and seat-belt usage? Use a .05 significance level.

	Seat-Belt Usage	
Sex	Don't Use	Use
Male	192	272
Female	284	276

14.27 One important factor that affects the quality of sorghum, a major world cereal crop, is presence of pigmentation. The paper "A Genetic and Biochemical Study on Pericarp Pigments in a Cross between Two Cultivars of Grain Sorghum, Sorghum Bicolor" (*Heredity* (1976): 413–16) reported on a genetic experiment in which three different pigmentations—red, yellow, or white—were possible. A particular genetic model predicted that these colors would appear in the ratios 9:3:4 (i.e., proportions $\frac{9}{16}$, $\frac{3}{16}$, and $\frac{4}{16}$). The experiments yielded 195 seeds with red pigmentation, 73 with yellow, and 100 with white. Does this data cast doubt on the appropriateness of this genetic theory? Carry out a goodness-of-fit test at significance level .10.

14.28 The Japanese farming community of Achihara was the focus of a study described in the article "Part-Time Farming: A Japanese Example" (*J. of Anthro. Research* (1984): 293–305). A random sample of farms for which the head of household was between 45 and 72 years old was selected, and each member of the sample was classified according to size of farm and residence of the farmer's oldest child. Data compatible with that given in the paper is summarized in the given two-way table. Using a .05 significance level, test the null hypothesis of no association between farm size and child's residence.

	Oldest Child's Residence	
Size (ha)	With Parents	Separate Residence
0–<.3	28	8
.3–<1.0	10	8
1.0–<9.0	18	20

14.29 The accompanying 2 × 2 frequency table is the result of classifying random samples of 112 librarians and 108 faculty members of the California State University system with respect to sex ("Job Satisfaction among Faculty and Librarians: A Study of Gender, Autonomy, and Decision Making Opportunities" *J. of Library Admin.* (1984): 43-56). Does the data strongly suggest that librarians and faculty members differ with respect to the proportion of males and females? Test using a .05 level of significance.

	Faculty	Librarians
Male	56	59
Female	52	53

14.30 Is there any relationship between the age of an investor and the rate of return that the investor expects from an investment? A sample of 972 common stock investors was selected and each was placed in one of four age categories and in one of four *rate believed attainable* categories ("Patterns of Investment Strategy and

Behavior among Individual Investors" *J. of Business* (1977): 296–333). The resulting data is given. Does there appear to be an association between age and rate believed attainable? Test the appropriate hypotheses using a .01 significance level.

Investor Age	Rate Believed Attainable			
	0–5%	6–10%	11–15%	Over 15%
Under 45	15	51	51	29
45–54	31	133	70	48
55–64	59	139	35	20
65 or over	84	157	32	18

14.31 The paper "Participation of Senior Citizens in the Swine Flu Inoculation Program" (*J. of Gerentology* (1979): 201–8) described a study of the factors thought to influence a person's decision to obtain a flu vaccination. Each member of a sample of 122 senior citizens was classified according to belief about the likelihood of getting the flu and vaccine status to obtain the given two-way frequency table. Using a .05 significance level, test to determine if there is an association between belief and vaccine status.

Belief	Vaccine Status	
	Received Vaccine	Didn't Receive Vaccine
Very unlikely	25	24
Unlikely	30	11
Likely	6	8
Don't know	5	13

14.32 The paper "An Instant Shot of 'Ah': Cocaine Use among Methadone Clients" (*J. of Psychoactive Drugs* (1984): 217–27) reported the accompanying data on frequency of cocaine use for individuals in three different treatment groups. Does the data suggest that the true proportion of individuals in each of the different cocaine-use categories differs for the three treatments? Carry out an appropriate test at level .05.

Cocaine Use	Treatment		
	A	B	C
None	149	75	8
1–2 times	26	27	15
3–6 times	6	20	11
At least 7 times	4	10	10

14.33 The paper "Identification of Cola Beverages" (*J. of Appl. Psych.* (1962): 356–60) reported on an experiment in which each of 79 subjects was presented with glasses of cola in pairs and asked to identify which glass contained a specific brand of cola. The accompanying data appeared in the paper. Does this data suggest that individuals' abilities to make correct identifications differ for the different brands of cola?

| | **Number of Correct Identifications** | | | |
Cola	0	1	2	3 or 4
Coca-Cola	13	23	24	19
Pepsi Cola	12	20	26	21
Royal Crown	18	28	19	14

14.34 Many shoppers have expressed unhappiness over plans by grocery stores to stop putting prices on individual grocery items. The paper "The Impact of Item Price Removal on Grocery Shopping Behavior" (*J. of Marketing* (1980): 73–93) reported on a study in which each shopper in a sample was classified by age and by whether he or she felt the need for item pricing. Based on the accompanying data, does the need for item pricing appear to be independent of age? (*Hint:* Construct the appropriate two-way frequency table.)

| | **Age** | | | | |
	<30	30–39	40–49	50–59	≥60
Number in sample	150	141	82	63	49
Number who want item pricing	127	118	77	61	41

REFERENCES

Agresti, Alan, and Barbara Agresti. *Statistical Methods for the Social Sciences* 2/e. New York: Dellen-Macmillan 1986. (This book includes a good discussion of measures of association for two-way frequency tables.)

Everitt, B. S. *The Analysis of Contingency Tables.* New York: Halstead Pres, 1977. (A compact but informative survey of methods for analyzing categorical data.)

Mosteller, Frederick, and Robert Rourke. *Sturdy Statistics.* Reading, Mass.: Addison-Wesley, 1973. (Contains several very readable chapters on the varied uses of the chi-squared statistic.)

STATISTICAL TABLES

TABLE I

Binomial Probabilities

n = 5

						π							
x	0.05	0.1	0.2	0.25	0.3	0.4	0.5	0.6	0.7	0.75	0.8	0.9	0.95
0	.774	.590	.328	.237	.168	.078	.031	.010	.002	.001	.000	.000	.000
1	.203	.329	.409	.396	.360	.259	.157	.077	.029	.015	.007	.000	.000
2	.022	.072	.205	.263	.309	.346	.312	.230	.132	.088	.051	.009	.001
3	.001	.009	.051	.088	.132	.230	.312	.346	.309	.263	.205	.072	.022
4	.000	.000	.007	.015	.029	.077	.157	.259	.360	.396	.409	.329	.203
5	.000	.000	.000	.001	.002	.010	.031	.078	.168	.237	.328	.590	.774

n = 10

						π							
x	0.05	0.1	0.2	0.25	0.3	0.4	0.5	0.6	0.7	0.75	0.8	0.9	0.95
0	.599	.349	.107	.056	.028	.006	.001	.000	.000	.000	.000	.000	.000
1	.315	.387	.268	.188	.121	.040	.010	.002	.000	.000	.000	.000	.000
2	.075	.194	.302	.282	.233	.121	.044	.011	.001	.000	.000	.000	.000
3	.010	.057	.201	.250	.267	.215	.117	.042	.009	.003	.001	.000	.000
4	.001	.011	.088	.146	.200	.251	.205	.111	.037	.016	.006	.000	.000
5	.000	.001	.026	.058	.103	.201	.246	.201	.103	.058	.026	.001	.000
6	.000	.000	.006	.016	.037	.111	.205	.251	.200	.146	.088	.011	.001
7	.000	.000	.001	.003	.009	.042	.117	.215	.267	.250	.201	.057	.010
8	.000	.000	.000	.000	.001	.011	.044	.121	.233	.282	.302	.194	.075
9	.000	.000	.000	.000	.000	.002	.010	.040	.121	.188	.268	.387	.315
10	.000	.000	.000	.000	.000	.000	.001	.006	.028	.056	.107	.349	.599

n = 15

						π							
x	0.05	0.1	0.2	0.25	0.3	0.4	0.5	0.6	0.7	0.75	0.8	0.9	0.95
0	.463	.206	.035	.013	.005	.000	.000	.000	.000	.000	.000	.000	.000
1	.366	.343	.132	.067	.030	.005	.000	.000	.000	.000	.000	.000	.000
2	.135	.267	.231	.156	.092	.022	.004	.000	.000	.000	.000	.000	.000
3	.031	.128	.250	.225	.170	.064	.014	.002	.000	.000	.000	.000	.000
4	.004	.043	.188	.225	.218	.126	.041	.007	.001	.000	.000	.000	.000
5	.001	.011	.103	.166	.207	.196	.092	.025	.003	.001	.000	.000	.000
6	.000	.002	.043	.091	.147	.207	.153	.061	.011	.003	.001	.000	.000
7	.000	.000	.014	.040	.081	.177	.196	.118	.035	.013	.003	.000	.000
8	.000	.000	.003	.013	.035	.118	.196	.177	.081	.040	.014	.000	.000
9	.000	.000	.001	.003	.011	.061	.153	.207	.147	.091	.043	.002	.000
10	.000	.000	.000	.001	.003	.025	.092	.196	.207	.166	.103	.011	.001
11	.000	.000	.000	.000	.001	.007	.041	.126	.218	.225	.188	.043	.004
12	.000	.000	.000	.000	.000	.002	.014	.064	.170	.225	.250	.128	.031
13	.000	.000	.000	.000	.000	.000	.004	.022	.092	.156	.231	.267	.135
14	.000	.000	.000	.000	.000	.000	.000	.005	.030	.067	.132	.343	.366
15	.000	.000	.000	.000	.000	.000	.000	.000	.005	.013	.035	.206	.463

Binomial Probabilities

n = 20

x							π						
	0.05	0.1	0.2	0.25	0.3	0.4	0.5	0.6	0.7	0.75	0.8	0.9	0.95
0	.358	.122	.012	.003	.001	.000	.000	.000	.000	.000	.000	.000	.000
1	.377	.270	.058	.021	.007	.000	.000	.000	.000	.000	.000	.000	.000
2	.189	.285	.137	.067	.028	.003	.000	.000	.000	.000	.000	.000	.000
3	.060	.190	.205	.134	.072	.012	.001	.000	.000	.000	.000	.000	.000
4	.013	.090	.218	.190	.130	.035	.005	.000	.000	.000	.000	.000	.000
5	.002	.032	.175	.202	.179	.075	.015	.001	.000	.000	.000	.000	.000
6	.000	.009	.109	.169	.192	.124	.037	.005	.000	.000	.000	.000	.000
7	.000	.002	.055	.112	.164	.166	.074	.015	.001	.000	.000	.000	.000
8	.000	.000	.022	.061	.114	.180	.120	.035	.004	.001	.000	.000	.000
9	.000	.000	.007	.027	.065	.160	.160	.071	.012	.003	.000	.000	.000
10	.000	.000	.002	.010	.031	.117	.176	.117	.031	.010	.002	.000	.000
11	.000	.000	.000	.003	.012	.071	.160	.160	.065	.027	.007	.000	.000
12	.000	.000	.000	.001	.004	.035	.120	.180	.114	.061	.022	.000	.000
13	.000	.000	.000	.000	.001	.015	.074	.166	.164	.112	.055	.002	.000
14	.000	.000	.000	.000	.000	.005	.037	.124	.192	.169	.109	.009	.000
15	.000	.000	.000	.000	.000	.001	.015	.075	.179	.202	.175	.032	.002
16	.000	.000	.000	.000	.000	.000	.005	.035	.130	.190	.218	.090	.013
17	.000	.000	.000	.000	.000	.000	.001	.012	.072	.134	.205	.190	.060
18	.000	.000	.000	.000	.000	.000	.000	.003	.028	.067	.137	.285	.189
19	.000	.000	.000	.000	.000	.000	.000	.000	.007	.021	.058	.270	.377
20	.000	.000	.000	.000	.000	.000	.000	.000	.001	.003	.012	.122	.358

n = 25

x							π						
	0.05	0.1	0.2	0.25	0.3	0.4	0.5	0.6	0.7	0.75	0.8	0.9	0.95
0	.277	.072	.004	.001	.000	.000	.000	.000	.000	.000	.000	.000	.000
1	.365	.199	.023	.006	.002	.000	.000	.000	.000	.000	.000	.000	.000
2	.231	.266	.071	.025	.007	.000	.000	.000	.000	.000	.000	.000	.000
3	.093	.227	.136	.064	.024	.002	.000	.000	.000	.000	.000	.000	.000
4	.027	.138	.187	.118	.057	.007	.000	.000	.000	.000	.000	.000	.000
5	.006	.065	.196	.164	.103	.020	.002	.000	.000	.000	.000	.000	.000
6	.001	.024	.163	.183	.148	.045	.005	.000	.000	.000	.000	.000	.000
7	.000	.007	.111	.166	.171	.080	.015	.001	.000	.000	.000	.000	.000
8	.000	.002	.062	.124	.165	.120	.032	.003	.000	.000	.000	.000	.000
9	.000	.000	.030	.078	.134	.151	.061	.009	.000	.000	.000	.000	.000
10	.000	.000	.011	.042	.091	.161	.097	.021	.002	.000	.000	.000	.000
11	.000	.000	.004	.019	.054	.146	.133	.044	.004	.001	.000	.000	.000
12	.000	.000	.002	.007	.027	.114	.155	.076	.011	.002	.000	.000	.000
13	.000	.000	.000	.002	.011	.076	.155	.114	.027	.007	.002	.000	.000
14	.000	.000	.000	.001	.004	.044	.133	.146	.054	.019	.004	.000	.000
15	.000	.000	.000	.000	.002	.021	.097	.161	.091	.042	.011	.000	.000
16	.000	.000	.000	.000	.000	.009	.061	.151	.134	.078	.030	.000	.000
17	.000	.000	.000	.000	.000	.003	.032	.120	.165	.124	.062	.002	.000
18	.000	.000	.000	.000	.000	.001	.015	.080	.171	.166	.111	.007	.000
19	.000	.000	.000	.000	.000	.000	.005	.045	.148	.183	.163	.024	.001
20	.000	.000	.000	.000	.000	.000	.002	.020	.103	.164	.196	.065	.006
21	.000	.000	.000	.000	.000	.000	.000	.007	.057	.118	.187	.138	.027
22	.000	.000	.000	.000	.000	.000	.000	.002	.024	.064	.136	.227	.093
22	.000	.000	.000	.000	.000	.000	.000	.000	.007	.025	.071	.266	.231
24	.000	.000	.000	.000	.000	.000	.000	.000	.002	.006	.023	.199	.365
25	.000	.000	.000	.000	.000	.000	.000	.000	.000	.001	.004	.072	.277

TABLE II

Standard Normal Probabilities

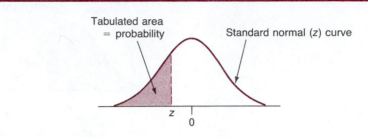

z	.00	.01	.02	.03	.04	.05	.06	.07	.08	.09
−3.4	.0003	.0003	.0003	.0003	.0003	.0003	.0003	.0003	.0003	.0002
−3.3	.0005	.0005	.0005	.0004	.0004	.0004	.0004	.0004	.0004	.0003
−3.2	.0007	.0007	.0006	.0006	.0006	.0006	.0006	.0005	.0005	.0005
−3.1	.0010	.0009	.0009	.0009	.0008	.0008	.0008	.0008	.0007	.0007
−3.0	.0013	.0013	.0013	.0012	.0012	.0011	.0011	.0011	.0010	.0010
−2.9	.0019	.0018	.0018	.0017	.0016	.0016	.0015	.0015	.0014	.0014
−2.8	.0026	.0025	.0024	.0023	.0023	.0022	.0021	.0021	.0020	.0019
−2.7	.0035	.0034	.0033	.0032	.0031	.0030	.0029	.0028	.0027	.0026
−2.6	.0047	.0045	.0044	.0043	.0041	.0040	.0039	.0038	.0037	.0036
−2.5	.0062	.0060	.0059	.0057	.0055	.0054	.0052	.0051	.0049	.0048
−2.4	.0082	.0080	.0078	.0075	.0073	.0071	.0069	.0068	.0066	.0064
−2.3	.0107	.0104	.0102	.0099	.0096	.0094	.0091	.0089	.0087	.0084
−2.2	.0139	.0136	.0132	.0129	.0125	.0122	.0119	.0116	.0113	.0110
−2.1	.0179	.0174	.0170	.0166	.0162	.0158	.0154	.0150	.0146	.0143
−2.0	.0228	.0222	.0217	.0212	.0207	.0202	.0197	.0192	.0188	.0183
−1.9	.0287	.0281	.0274	.0268	.0262	.0256	.0250	.0244	.0239	.0233
−1.8	.0359	.0351	.0344	.0336	.0329	.0322	.0314	.0307	.0301	.0294
−1.7	.0446	.0436	.0427	.0418	.0409	.0401	.0392	.0384	.0375	.0367
−1.6	.0548	.0537	.0526	.0516	.0505	.0495	.0485	.0475	.0465	.0455
−1.5	.0668	.0655	.0643	.0630	.0618	.0606	.0594	.0582	.0571	.0559
−1.4	.0808	.0793	.0778	.0764	.0749	.0735	.0721	.0708	.0694	.0681
−1.3	.0968	.0951	.0934	.0918	.0901	.0885	.0869	.0853	.0838	.0823
−1.2	.1151	.1131	.1112	.1093	.1075	.1056	.1038	.1020	.1003	.0985
−1.1	.1357	.1335	.1314	.1292	.1271	.1251	.1230	.1210	.1190	.1170
−1.0	.1587	.1562	.1539	.1515	.1492	.1469	.1446	.1423	.1401	.1379
−0.9	.1841	.1814	.1788	.1762	.1736	.1711	.1685	.1660	.1635	.1611
−0.8	.2119	.2090	.2061	.2033	.2005	.1977	.1949	.1922	.1894	.1867
−0.7	.2420	.2389	.2358	.2327	.2296	.2266	.2236	.2206	.2177	.2148
−0.6	.2743	.2709	.2676	.2643	.2611	.2578	.2546	.2514	.2483	.2451
−0.5	.3085	.3050	.3015	.2981	.2946	.2912	.2877	.2843	.2810	.2776
−0.4	.3446	.3409	.3372	.3336	.3300	.3264	.3228	.3192	.3156	.3121
−0.3	.3821	.3783	.3745	.3707	.3669	.3632	.3594	.3557	.3520	.3483
−0.2	.4207	.4168	.4129	.4090	.4052	.4013	.3974	.3936	.3897	.3859
−0.1	.4602	.4562	.4522	.4483	.4443	.4404	.4364	.4325	.4286	.4247
−0.0	.5000	.4960	.4920	.4880	.4840	.4801	.4761	.4721	.4681	.4641

Standard Normal Probabilities

z	.00	.01	.02	.03	.04	.05	.06	.07	.08	.09
0.0	.5000	.5040	.5080	.5120	.5160	.5199	.5239	.5279	.5319	.5359
0.1	.5398	.5438	.5478	.5517	.5557	.5596	.5636	.5675	.5714	.5753
0.2	.5793	.5832	.5871	.5910	.5948	.5987	.6026	.6064	.6103	.6141
0.3	.6179	.6217	.6255	.6293	.6331	.6368	.6406	.6443	.6480	.6517
0.4	.6554	.6591	.6628	.6664	.6700	.6736	.6772	.6808	.6844	.6879
0.5	.6915	.6950	.6985	.7019	.7054	.7088	.7123	.7157	.7190	.7224
0.6	.7257	.7291	.7324	.7357	.7389	.7422	.7454	.7486	.7517	.7549
0.7	.7580	.7611	.7642	.7673	.7704	.7734	.7764	.7794	.7823	.7852
0.8	.7881	.7910	.7939	.7967	.7995	.8023	.8051	.8078	.8106	.8133
0.9	.8159	.8186	.8212	.8238	.8264	.8289	.8315	.8340	.8365	.8389
1.0	.8413	.8438	.8461	.8485	.8508	.8531	.8554	.8577	.8599	.8621
1.1	.8643	.8665	.8686	.8708	.8729	.8749	.8770	.8790	.8810	.8830
1.2	.8849	.8869	.8888	.8907	.8925	.8944	.8962	.8980	.8997	.9015
1.3	.9032	.9049	.9066	.9082	.9099	.9115	.9131	.9147	.9162	.9177
1.4	.9192	.9207	.9222	.9236	.9251	.9265	.9279	.9292	.9306	.9319
1.5	.9332	.9345	.9357	.9370	.9382	.9394	.9406	.9418	.9429	.9441
1.6	.9452	.9463	.9474	.9484	.9495	.9505	.9515	.9525	.9535	.9545
1.7	.9554	.9564	.9573	.9582	.9591	.9599	.9608	.9616	.9625	.9633
1.8	.9641	.9649	.9656	.9664	.9671	.9678	.9686	.9693	.9699	.9706
1.9	.9713	.9719	.9726	.9732	.9738	.9744	.9750	.9756	.9761	.9767
2.0	.9772	.9778	.9783	.9788	.9793	.9798	.9803	.9808	.9812	.9817
2.1	.9821	.9826	.9830	.9834	.9838	.9842	.9846	.9850	.9854	.9857
2.2	.9861	.9864	.9868	.9871	.9875	.9878	.9881	.9884	.9887	.9890
2.3	.9893	.9896	.9898	.9901	.9904	.9906	.9909	.9911	.9913	.9916
2.4	.9918	.9920	.9922	.9925	.9927	.9929	.9931	.9932	.9934	.9936
2.5	.9938	.9940	.9941	.9943	.9945	.9946	.9948	.9949	.9951	.9952
2.6	.9953	.9955	.9956	.9957	.9959	.9960	.9961	.9962	.9963	.9964
2.7	.9965	.9966	.9967	.9968	.9969	.9970	.9971	.9972	.9973	.9974
2.8	.9974	.9975	.9976	.9977	.9977	.9978	.9979	.9979	.9980	.9981
2.9	.9981	.9982	.9982	.9983	.9984	.9984	.9985	.9985	.9986	.9986
3.0	.9987	.9987	.9987	.9988	.9988	.9989	.9989	.9989	.9990	.9990
3.1	.9990	.9991	.9991	.9991	.9992	.9992	.9992	.9992	.9993	.9993
3.2	.9993	.9993	.9994	.9994	.9994	.9994	.9994	.9995	.9995	.9995
3.3	.9995	.9995	.9995	.9996	.9996	.9996	.9996	.9996	.9996	.9997
3.4	.9997	.9997	.9997	.9997	.9997	.9997	.9997	.9997	.9997	.9998

TABLE III

t **Critical Values**

Central area · *t* curve
Lower-tail area · Upper-tail area
−*t* critical value · *t* critical value
0

Central Area Captured Confidence Level	.80 80%	.90 90%	.95 95%	.98 98%	.99 99%	.998 99.8%	.999 99.9%
1	3.08	6.31	12.71	31.82	63.66	318.31	636.62
2	1.89	2.92	4.30	6.97	9.93	23.33	31.60
3	1.64	2.35	3.18	4.54	5.84	10.21	12.92
4	1.53	2.13	2.78	3.75	4.60	7.17	8.61
5	1.48	2.02	2.57	3.37	4.03	5.89	6.86
6	1.44	1.94	2.45	3.14	3.71	5.21	5.96
7	1.42	1.90	2.37	3.00	3.50	4.79	5.41
8	1.40	1.86	2.31	2.90	3.36	4.50	5.04
9	1.38	1.83	2.26	2.82	3.25	4.30	4.78
10	1.37	1.81	2.23	2.76	3.17	4.14	4.59
11	1.36	1.80	2.20	2.72	3.11	4.03	4.44
12	1.36	1.78	2.18	2.68	3.06	3.93	4.32
13	1.35	1.77	2.16	2.65	3.01	3.85	4.22
14	1.35	1.76	2.15	2.62	2.98	3.79	4.14
15	1.34	1.75	2.13	2.60	2.95	3.73	4.07
16	1.34	1.75	2.12	2.58	2.92	3.69	4.02
17	1.33	1.74	2.11	2.57	2.90	3.65	3.97
18	1.33	1.73	2.10	2.55	2.88	3.61	3.92
19	1.33	1.73	2.09	2.54	2.86	3.58	3.88
20	1.33	1.73	2.09	2.53	2.85	3.55	3.85
21	1.32	1.72	2.08	2.52	2.83	3.53	3.82
22	1.32	1.72	2.07	2.51	2.82	3.51	3.79
23	1.32	1.71	2.07	2.50	2.81	3.49	3.77
24	1.32	1.71	2.06	2.49	2.80	3.47	3.75
25	1.32	1.71	2.06	2.49	2.79	3.45	3.73
26	1.32	1.71	2.06	2.48	2.78	3.44	3.71
27	1.31	1.70	2.05	2.47	2.77	3.42	3.69
28	1.31	1.70	2.05	2.47	2.76	3.41	3.67
29	1.31	1.70	2.05	2.46	2.76	3.40	3.66
30	1.31	1.70	2.04	2.46	2.75	3.39	3.65
40	1.30	1.68	2.02	2.42	2.70	3.31	3.55
60	1.30	1.67	2.00	2.39	2.66	3.23	3.46
120	1.29	1.66	1.98	2.36	2.62	3.16	3.37
z critical values	1.28	1.645	1.96	2.33	2.58	3.09	3.29
Level of significance for a *two*-tailed test	.20	.10	.05	.02	.01	.002	.001
Level of significance for a *one*-tailed test	.10	.05	.025	.01	.005	.001	.0005

Degrees of Freedom

Handwritten notes: "For d.f. Note - For df. 34 use 30, for 38, use 40. etc." and "For d.f. >120, use z crit. values"

TABLE IV(a)

F Distribution Critical Values for Tests with Significance Level .05

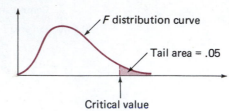

	Numerator Degrees of Freedom									
	1	2	3	4	5	6	7	8	9	10
1	161.4	199.5	215.7	224.6	230.2	234.0	236.8	238.9	240.5	241.9
2	18.51	19.00	19.16	19.25	19.30	19.33	19.35	19.37	19.38	19.40
3	10.13	9.55	9.28	9.12	9.01	8.94	8.89	8.85	8.81	8.79
4	7.71	6.94	6.59	6.39	6.26	6.16	6.09	6.04	6.00	5.96
5	6.61	5.79	5.41	5.19	5.05	4.95	4.88	4.82	4.77	4.74
6	5.99	5.14	4.76	4.53	4.39	4.28	4.21	4.15	4.10	4.06
7	5.59	4.74	4.35	4.12	3.97	3.87	3.79	3.73	3.68	3.64
8	5.32	4.46	4.07	3.84	3.69	3.58	3.50	3.44	3.39	3.35
9	5.12	4.26	3.86	3.63	3.48	3.37	3.29	3.23	3.18	3.14
10	4.96	4.10	3.71	3.48	3.33	3.22	3.14	3.07	3.02	2.98
11	4.84	3.98	3.59	3.36	3.20	3.09	3.01	2.95	2.90	2.85
12	4.75	3.89	3.49	3.26	3.11	3.00	2.91	2.85	2.80	2.75
13	4.67	3.81	3.41	3.18	3.03	2.92	2.83	2.77	2.71	2.67
14	4.60	3.74	3.34	3.11	2.96	2.85	2.76	2.70	2.65	2.60
15	4.54	3.68	3.29	3.06	2.90	2.79	2.71	2.64	2.59	2.54
16	4.49	3.63	3.24	3.01	2.85	2.74	2.66	2.59	2.54	2.49
17	4.45	3.59	3.20	2.96	2.81	2.70	2.61	2.55	2.49	2.45
18	4.41	3.55	3.16	2.93	2.77	2.66	2.58	2.51	2.46	2.41
19	4.38	3.52	3.13	2.90	2.74	2.63	2.54	2.48	2.42	2.38
20	4.35	3.49	3.10	2.87	2.71	2.60	2.51	2.45	2.39	2.35
21	4.32	3.47	3.07	2.84	2.68	2.57	2.49	2.42	2.37	2.32
22	4.30	3.44	3.05	2.82	2.66	2.55	2.46	2.40	2.34	2.30
23	4.28	3.42	3.03	2.80	2.64	2.53	2.44	2.37	2.32	2.27
24	4.26	3.40	3.01	2.78	2.62	2.51	2.42	2.36	2.30	2.25
25	4.24	3.39	2.99	2.76	2.60	2.49	2.40	2.34	2.28	2.24
26	4.23	3.37	2.98	2.74	2.59	2.47	2.39	2.32	2.27	2.22
27	4.21	3.35	2.96	2.73	2.57	2.46	2.37	2.31	2.25	2.20
28	4.20	3.34	2.95	2.71	2.56	2.45	2.36	2.29	2.24	2.19
29	4.18	3.33	2.93	2.70	2.55	2.43	2.35	2.28	2.22	2.18
30	4.17	3.32	2.92	2.69	2.53	2.42	2.33	2.27	2.21	2.16
40	4.08	3.23	2.84	2.61	2.45	2.34	2.25	2.18	2.12	2.08
60	4.00	3.15	2.76	2.53	2.37	2.25	2.17	2.10	2.04	1.99
120	3.92	3.07	2.68	2.45	2.29	2.17	2.09	2.02	1.96	1.91
∞	3.84	3.00	2.60	2.37	2.21	2.10	2.01	1.94	1.88	1.83

Denominator Degrees of Freedom

TABLE IV(b)

F **Distribution Critical Values for Tests with Significance Level .01**

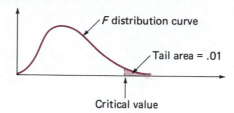

					Numerator Degrees of Freedom					
	1	2	3	4	5	6	7	8	9	10
1	4,052	4,999.5	5,403	5,625	5,764	5,859	5,928	5,982	6,022	6,056
2	98.50	99.00	99.17	99.25	99.30	99.33	99.36	99.37	99.39	99.40
3	34.12	30.82	29.46	28.71	28.24	27.91	27.67	27.49	27.35	27.23
4	21.20	18.00	16.69	15.98	15.52	15.21	14.98	14.80	14.66	14.55
5	16.26	13.27	12.06	11.39	10.97	10.67	10.46	10.29	10.16	10.05
6	13.75	10.92	9.78	9.15	8.75	8.47	8.26	8.10	7.98	7.87
7	12.25	9.55	8.45	7.85	7.46	7.19	6.99	6.84	6.72	6.62
8	11.26	8.65	7.59	7.01	6.63	6.37	6.18	6.03	5.91	5.81
9	10.56	8.02	6.99	6.42	6.06	5.80	5.61	5.47	5.35	5.26
10	10.04	7.56	6.55	5.99	5.64	5.39	5.20	5.06	4.94	4.85
11	9.65	7.21	6.22	5.67	5.32	5.07	4.89	4.74	4.63	4.54
12	9.33	6.93	5.95	5.41	5.06	4.82	4.64	4.50	4.39	4.30
13	9.07	6.70	5.74	5.21	4.86	4.62	4.44	4.30	4.19	4.10
14	8.86	6.51	5.56	5.04	4.69	4.46	4.28	4.14	4.03	3.94
15	8.68	6.36	5.42	4.89	4.56	4.32	4.14	4.00	3.89	3.80
16	8.53	6.23	5.29	4.77	4.44	4.20	4.03	3.89	3.78	3.69
17	8.40	6.11	5.18	4.67	4.34	4.10	3.93	3.79	3.68	3.59
18	8.29	6.01	5.09	4.58	4.25	4.01	3.84	3.71	3.60	3.51
19	8.18	5.93	5.01	4.50	4.17	3.94	3.77	3.63	3.52	3.43
20	8.10	5.85	4.94	4.43	4.10	3.87	3.70	3.56	3.46	3.37
21	8.02	5.78	4.87	4.37	4.04	3.81	3.64	3.51	3.40	3.31
22	7.95	5.72	4.82	4.31	3.99	3.76	3.59	3.45	3.35	3.26
23	7.88	5.66	4.76	4.26	3.94	3.71	3.54	3.41	3.30	3.21
24	7.82	5.61	4.72	4.22	3.90	3.67	3.50	3.36	3.26	3.17
25	7.77	5.57	4.68	4.18	3.85	3.63	3.46	3.32	3.22	3.13
26	7.72	5.53	4.64	4.14	3.82	3.59	3.42	3.29	3.18	3.09
27	7.68	5.49	4.60	4.11	3.78	3.56	3.39	3.26	3.15	3.06
28	7.64	5.45	4.57	4.07	3.75	3.53	3.36	3.23	3.12	3.03
29	7.60	5.42	4.54	4.04	3.73	3.50	3.33	3.20	3.09	3.00
30	7.56	5.39	4.51	4.02	3.70	3.47	3.30	3.17	3.07	2.98
40	7.31	5.18	4.31	3.83	3.51	3.29	3.12	2.99	2.89	2.80
60	7.08	4.98	4.13	3.65	3.34	3.12	2.95	2.82	2.72	2.63
120	6.85	4.79	3.95	3.48	3.17	2.96	2.79	2.66	2.56	2.47
∞	6.63	4.61	3.78	3.32	3.02	2.80	2.64	2.51	2.41	2.32

Denominator Degrees of Freedom (row labels at left)

TABLE V

Bonferroni 95% t Critical Values

Number of df	Number of Intervals						
	2	3	4	5	6	10	15
2	6.21	7.65	8.86	9.92	10.89	14.09	17.28
3	4.18	4.86	5.39	5.84	6.23	7.45	8.58
4	3.50	3.96	4.31	4.60	4.85	5.60	6.25
5	3.16	3.53	3.81	4.03	4.22	4.77	5.25
6	2.97	3.29	3.52	3.71	3.86	4.32	4.70
7	2.84	3.13	3.34	3.50	3.64	4.03	4.36
8	2.75	3.02	3.21	3.36	3.48	3.83	4.12
9	2.69	2.93	3.11	3.25	3.36	3.69	3.95
10	2.63	2.87	3.04	3.17	3.28	3.58	3.83
11	2.59	2.82	2.98	3.11	3.21	3.50	3.73
12	2.56	2.78	2.93	3.05	3.15	3.43	3.65
13	2.53	2.75	2.90	3.01	3.11	3.37	3.58
14	2.51	2.72	2.86	2.98	3.07	3.33	3.53
15	2.49	2.69	2.84	2.95	3.04	3.29	3.48
16	2.47	2.67	2.81	2.92	3.01	3.25	3.44
17	2.46	2.66	2.79	2.90	2.98	3.22	3.41
18	2.45	2.64	2.77	2.88	2.96	3.20	3.38
19	2.43	2.63	2.76	2.86	2.94	3.17	3.35
20	2.42	2.61	2.74	2.85	2.93	3.15	3.33
21	2.41	2.60	2.73	2.83	2.91	3.14	3.31
22	2.41	2.59	2.72	2.82	2.90	3.12	3.29
23	2.40	2.58	2.71	2.81	2.89	3.10	3.27
24	2.39	2.57	2.70	2.80	2.88	3.09	3.26
25	2.38	2.57	2.69	2.79	2.86	3.08	3.24
26	2.38	2.56	2.68	2.78	2.86	3.07	3.23
27	2.37	2.55	2.68	2.77	2.85	3.06	3.22
28	2.37	2.55	2.67	2.76	2.84	3.05	3.21
29	2.36	2.54	2.66	2.76	2.83	3.04	3.20
30	2.36	2.54	2.66	2.75	2.82	3.03	3.19
40	2.33	2.50	2.62	2.70	2.78	2.97	3.12
60	2.30	2.46	2.58	2.66	2.73	2.91	3.06
120	2.27	2.43	2.54	2.62	2.68	2.86	3.00
∞	2.24	2.39	2.50	2.58	2.64	2.81	2.94

TABLE VI

Chi-Squared Distribution Critical Values

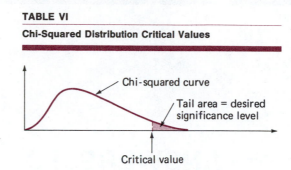

,025

		\.10	.05	.01	.001
		Significance Level			
	1	2.71	3.84	6.64	10.83
	2	4.61	5.99	9.21	13.82
	3	6.25	7.82	11.34	16.27
	4	7.78	9.49	13.28	18.47
	5	9.24	11.07	15.09	20.52
	6	10.64	12.59	16.81	22.46
	7	12.02	14.07	18.48	24.32
	8	13.36	15.51	20.09	26.12
Degrees of	9	14.68	16.92	21.67	27.88
Freedom	10	15.99	18.31	23.21	29.59
	11	17.28	19.68	24.72	31.26
	12	18.55	21.03	26.22	32.91
	13	19.81	22.36	27.69	34.53
	14	21.06	23.68	29.14	36.12
	15	22.31	25.00	30.58	37.70
	16	23.54	26.30	32.00	39.25
	17	24.77	27.59	33.41	40.79
	18	25.99	28.87	34.81	42.31
	19	27.20	30.14	36.19	43.82
	20	28.41	31.41	37.57	45.31

ANSWERS TO SELECTED ODD NUMBERED EXERCISES

CHAPTER ONE

1.1 Descriptive statistics is made up of those methods whose purpose is to organize and summarize a data set. Inferential statistics refers to those procedures or techniques whose purpose is to generalize or make an inference about the population based on the information in the sample.

1.3 The population of interest is the entire student body (the 15,000 students). The sample consists of the 200 students interviewed.

1.5 The population consists of all single family homes in Pasadena. The sample consists of the 100 homes selected for inspection.

1.7 The population consists of all 5000 bricks in the lot. The sample consists of the 100 bricks selected for inspection.

1.9 **a.** categorical **b.** categorical
    c. numerical (discrete) **d.** numerical (continuous)
    e. categorical (each zip code identifies a geographical region)
    f. numerical (continuous)

1.11 **a.** Japan, Sweden, Germany, Great Britain, United States
    b. 3.23, 2.92, 4.0, 2.8
    c. 2, 0, 1, 4, 3
    d. 49.2, 48.84, 50.3, 50.23
    e. 10, 15.5, 17, 3, 6.5
    f. Froot Loops, Cheerios, Frosted Flakes
    g. 0, 1, 0, 2, 4

CHAPTER TWO

2.1

4	00, 01, 02, 05, 07, 17, 17, 19, 24, 33, 40, 48, 50, 58, 68, 84, 92
5	01, 09, 12, 24, 53, 54, 59, 65, 69, 99
6	03, 24, 34, 44, 47
7	23, 30, 55, 77, 91, 92
8	56, 85

stems: hundreds
leaves: ones

2.3

64	64, 70, 35, 33
65	26, 83, 06, 27
66	14, 05, 94
67	70, 70, 90, 00, 98, 45, 13
68	50, 73, 70, 90
69	36, 27, 00, 04
70	05, 40, 22, 11, 51, 50
71	31, 69, 68, 05, 65, 13
72	09, 80

stems: hundreds
leaves: ones

2.5

MARINA DEL REY		LOS ANGELES-LONG BEACH
50	4	35, 50, 50, 60, 70, 75, 95
30, 60, 95	5	00, 00, 20, 40, 50
00, 27, 37, 45, 49, 60, 60, 64, 82, 82, 96	6	00, 00, 00, 50
04, 05, 05, 16	7	
	8	75

The rental rates for Marina del Rey are higher than for the Los Angeles-Long Beach Harbor. Most of the marinas in Marina del Rey charge $6.00 or more per foot, while those in the Los Angeles-Long Beach Harbor charge $6.00 or less per foot.

2.7 a.

2	6 8 8
3	0 0 0 0 1 2 2 2 2 3 3 3 3 4 4 4 4 4 4 4 4 4 4 4 4 4
3	5 5 5 5 6 6 6 7 7 7 7 7 7 7 8 8 9 9 9 9 9
4	0 0 1 1 1 1 1 2 2 2 2 2 3 3 3 3 3 3 3 3 3 3 3 3 3 3 3 4 4 4 4 4 4 4 4
4	5 5 5 5 6 6 6 7 7 8 8 8 8 8 8 9 9 9 9 9
5	0 0 0 2 2 2 2 2 3 3 3 3 3 4 4 4
5	5 5 6 6 7 7 7 8 8 8 8 8 9 9 9 9 9 9 9 9 9
6	0 2 2 3
6	5 6 6 7
7	0

b. $9/149 = .06$

2.9 b. .227   **c.** .272, .136

2.11 a. .1705   **b.** .7883

2.13 b.

CONCENTRATION	FREQUENCY
20–<30	1
30–<40	8
40–<50	8
50–<60	6
60–<70	16
70–<80	7
80–<90	2
90–<100	2
	50

c. .34, .54

2.15 a.

CLASS INTERVALS	FREQUENCY	RELATIVE FREQUENCY
0–<100	21	.21
100–<200	32	.32
200–<300	26	.26
300–<400	12	.12
400–<500	4	.04
500–<600	3	.03
600–<700	1	.01
700–<800	0	.00
800–<900	1	.01
	$n = 100$	1.00

b. .242 = 24.2%

2.17 c. .2360   **d.** .8651   **e.** .5506   **f.** 20.46 months   **g.** 29.21 months

2.25

CLASS	REL. FREQ.	FREQUENCY
1	.030	9
2	.080	24
3	.115	34.5
4	.110	33
5	.170	51
6	.115	34.5
7	.105	31.5
8	.035	10.5
9	.060	18
10	.015	4.5
11	.045	13.5
12	.035	10.5
13	.015	4.5
14	.015	4.5
15	.025	7.5
16	.010	3
17	.000	0
18	.000	0
19	.010	3
20	.010	3
		$n = 300$

2.29 b. .0908   **c.** .2382

2.35 c. Both of the histograms are skewed positively.

2.37

```
                      GAS                                    ELECTRICITY

                                   8 || 1H || 9
                0  1  2  2  3  3  3  3 || 2L || 4
     5  5  5  5  6  6  6  6  6  7  7  8  8  9  9 || 2H || 5  5  6  7
                                   1 || 3L || 0  0  1  1  2  2  3  3  3  4  4
                                     || 3H || 5  9
                                     || 4L || 0  1  3  4
                                     || 4H || 8          stems = tens
                                     || 5L || 1          leaves = ones
```

2.39 a. .5556   **b.** .3778   **c.** .4000   **d.** 53.99

2.41 d. i. .2979   **ii.** .4043   **iii.** .2978

CHAPTER THREE

3.1 a. 54.59   **b.** 55   **c.** 91.18

3.3 a. 13.875   **b.** 14   **c.** 14.07   **d.** $\bar{x} = 11.5$, median $= 12$

3.5 treated: $\bar{x} = 1.23$, median $= 1.14$
untreated: $\bar{x} = 1.76$, median $= 1.74$

3.11 a. .7   **b.** .7   **c.** 13

3.15 $\bar{x} = 58.125$, $s^2 = 83.84$, $s = 9.16$

3.17 a. $\sum x = 56.8$, $\sum x^2 = 197.804$ **b.** $\bar{x} = 3.3412$, $s^2 = .5014$, $s = .7081$

3.19 a. 48.364   **b.** $s^2 = 327.05$, $s = 18.08$

3.21 Adding the same number to each observation has no effect on the variance or standard deviation.

3.25 a. $\bar{x} = 10.554$, $s^2 = 9.061$, $s = 3.01$
b. lower quartile $= 7.95$
upper quartile $= 13.05$
iqr $= 5.1$
iqr/1.35 $= 3.778$

3.27   **a.** at least 75%   **b.** no more than 11%   **c.** roughly 95% between 25 and 45

3.33   **a.** lower quartile = 11, upper quartile = 18.5, iqr = 7.5
      b. There are no mild or extreme outliers in this data set.

3.35   For the first test the student's z-score is $(625 - 475)/100 = 1.5$ and for the second test it is $(45 - 30)/8 = 1.875$. Since the student's z-score is larger for the second test than for the first test, the student's performance was better on the second exam.

3.37   **b. i.** approximately 21   **ii.** approximately 18   **iii.** approximately 21.67   **iv.** 25.375   **v.** 17.615

3.39   no more than .16.

3.41   at most 10.6%.

3.43   **a.** -1.86 **b.** 0.36   **c.** -0.43 **d.** 2.29

3.49   **a.** $\bar{x} = 103.83$, median = 82.5, $s^2 = 3497.8091$, $s = 59.14$
      b. lower quartile = 58, upper quartile = 137.5, iqr = 79.5

3.51   **a.** $\bar{x} = 8.006$, $s^2 = .0824$, $s = .287$
      b. 10% trimmed mean = 7.976
        median = 7.91
        upper quartile = 8.01, lower quartile = 7.82, iqr = 0.19

3.53   **a.** $\bar{x} = 22.15$, $s^2 = 129.183$, $s = 11.366$ **b.** 19.4375   **c.** upper quartile = 20.5, lower quartile = 18, iqr = 2.5
      d. 25 and 28 are mild outliers and 69 is an extreme outlier

3.59 $248

CHAPTER FOUR

4.1   **a.** Sample Space = {AA, AM, MA, MM}
      c. $B = $ {MA, AM, AA}; $C = $ {AM, MA}; $D = $ {MM}
        D is a simple event.

4.3   **b.** $A = $ {(1, 2), (1, 4), (2, 1), (2, 3), (2, 4), (3, 2), (3, 4), (4, 1), (4, 2), (4, 3)}
      c. $B = $ {(1, 3), (1, 4), (2, 3), (2, 4), (3, 1), (3, 2), (4, 1), (4, 2)}

4.5   **a.** N, DN, DDN, $DDDN$, $DDDDN$
      b. There are countably infinitely many.
      c. $E = $ {DN, $DDDN$, $DDDDDN$, . . .}

4.7   **b.** $A = $ {(1, 1, 1), (2, 2, 2), (3, 3, 3)}
      c. $B = $ {(1, 2, 3), (1, 3, 2), (2, 1, 3), (2, 3, 1), (3, 1, 2), (3, 2, 1)}
      d. $C = $ {(1, 1, 1), (1, 1, 3), (1, 3, 1), (1, 3, 3), (3, 1, 1), (3, 1, 3), (3, 3, 1), (3, 3, 3)}

4.11   **a.** .0119   **b.** .00000238   **c.** .012

4.13   **a.** .40   **b.** .81   **c.** .94   **d.** .21

4.15   **a.** There are 52 simple events.   **b.** 1/52   **c.** P(heart) = 1/4, P(face card) = 3/13   **d.** 3/52   **e.** 11/26

4.17   **b.** .2500   **c.** .3333   **d.** 0   **e.** .2917

4.19   **c.** .4   **d.** .3

4.21   **a.** 1/10   **b.** 3/10   **c.** 7/10   **d.** 6/10

4.23   **a.** .55   **b.** .45   **c.** .40   **d.** .25

4.25   **a.** $p = 1/9$ **b.** P(odd) = 1/3, P(at most 3) = 4/9
      c. P(odd) = 3/7, P(at most 3) = 2/7

4.27   **a.** 3/13   **b.** $P(E|F) = 1/4$, $P(F|E) = 3/13$
      c. E and F are independent since $P(E|F) = P(E)$

4.29 A and E are independent, A and F are not independent, E and F are independent

4.31   **a.** .316   **b.** .80

4.33   **a.** .12   **b.** .42   **c.** .58

4.35   **a.** The expert assumed that the positions of the two valves were independent.
      b. If the car were driven in a straight line, the relative positions of the valves would remain unchanged. Thus, the probability of them ending up at a one o'clock, six o'clock position would be 1/12, not 1/144. The value 1/144 is smaller than the correct probability of occurrence.

4.37   **a.** E_1 and E_2 are dependent events   **b.** .992
      c. $P(E_2|E_1) = .0078$, $P(E_2|\text{not } E_1) = .0080$

4.41   **b.** .025   **c.** .069

4.43   **a.** .018   **b.** $P(E_1|L) = .444$, $P(E_2|L) = .278$, $P(E_3|L) = .278$

4.47   **a.** .2401   **b.** .2418   **c.** .084
4.51   **a.** .05   **b.** .12   **c.** P(short sleeve) = .56, P(long-sleeve) = .44   **d.** P(medium) = .49, P(print) = .25
4.53   **a.** .436   **b.** .582
4.55   **a.** .59049   **b.** .40951   **c.** .00001
4.57   **a.** .622   **b.** .124   **c.** .864
4.59   **a.** .10   **b.** E, F are not independent events
4.61   **a.** .2308   **b.** .2157   **c.** .20   **d.** .00996
4.63   **a.** .48   **b.** .256   **c.** .016   **d.** .136
4.67   They cannot be disjoint because $P(A \text{ and } B) = P(A)P(B) = .5(.4) = .2 \neq 0$.

CHAPTER FIVE

5.1   **a.** discrete   **b.** continuous   **c.** discrete   **d.** discrete   **e.** continuous
5.3   Possible y values are the positive integers.
5.5 y is a continuous variable with values from 0 to 100 ft.
5.9   **a.** .82   **b.** .18   **c.** $P(x \leq 99) = .65$, $P(x \leq 97) = .27$

5.13   **a.**

x	0	1	2	3	4
$p(x)$	.4096	.4096	.1536	.0256	.0016

    b. The most likely values for x are 0 and 1. Each has a probability of .4096.
    c. .1808

5.15

w-values	1	10	25
$p(w)$	.3	.3	.4

5.17

x-value	1	2	3	4	5	6
$p(x)$	1/12	1/12	1/12	1/4	1/4	1/4

5.19

y-value	0	1	2	3
$p(y)$	.16	.33	.32	.19

5.21   **a.** 46.5   **b.** y = cost

y	500	800	1000
$p(y)$	.10	.30	.6

$\mu = 890$

5.25 $\mu = 3.114$, $\sigma = .6364$
5.27   **a.** discrete   **b.** $P(y > 1.00) = .40$, $P(y < 1.20) = .86$ **c.** $\mu = 106.728$, $\sigma = 13.30$

5.29   **b.**

y-value	80	85	90	95
$p(y)$	.1	.3	.4	.2

    c. 88.5

5.31   10
5.33   **a.** .2637   **b.** .6328   **c.** .3672   **d.** .7363
5.37   **a.** .735   **b.** .392   **c.** .07
5.39   200, 13.42
5.41   **a.** x has a binomial distribution with $n = 100$ and $\pi = .2$.
    b. 20   **c.** $\sigma = 4$
    d. Most scores would be within three standard deviations of the mean (between 8 and 32). Thus, it would be highly unlikely for a person to score over 50.
5.43   **a.** .0434   **b.** .0023   **c.** .8453
5.45   **a.** .902   **b.** .9666   **c.** $\mu = 22.5$; $\sigma = 1.5$ **d.** .0334
5.47   **a.** $\mu = 2.64$, $\sigma = 1.54$ **b.** $P(x < \mu - 3\sigma \text{ or } x > \mu + 3\sigma) = P(x < -1.98 \text{ or } x > 7.26) = 0$

5.51

x-value	1	2	3
$p(x)$	1/2	1/3	1/6

5.55 **a.**

y	.25	1.25	2.25
$p(y)$	1/15	2/15	12/15

b.

y	0	1	2	3
$p(y)$	1/15	2/15	3/15	9/15

$\mu = 1.983$ $\qquad\qquad\qquad\qquad$ $\mu = 2.33$

5.59 **a.** .4 **b.** .10, .4333

5.61 **a.** .1804 **b.** .594 **c.** .15629 **d.** 60

CHAPTER SIX

6.1 **a.** .5 **b.** .2 **c.** $\mu = 5$

6.3 **b.** .08 **c.** .36 **d.** $P(10 \le x \le 15) = .40$, $P(12 \le x \le 17) = .40$ **e.** 13.75 **f.** 3.608

6.5 **a.** $P(x \le 10) = .5$, $P(x \ge 15) = .25$ **b.** .25 **c.** 18

6.7 **a.** 1 **b.** .84375 **c.** .6875 **d.** $\mu = 0$

6.9 **a.** .9599 **b.** .2483 **c.** .1151 **d.** .9976 **e.** .6887 **f.** .6826 **g.** 1

6.11 **a.** .9909 **b.** .9909 **c.** .1093 **d.** .1267 **e.** .0706 **f.** .8730 **g.** .0228 **h.** .9996 **i.** 1

6.13 **a.** .23 **b.** $-.23$ **c.** 2.75 **d.** 1.16

6.15 **a.** 1.96 **b.** 1.28 **c.** -1.28 **d.** 2.58 **e.** 2.58 **f.** 3.10

6.17 **a.** .5 **b.** .9772 **c.** .9772 **d.** .8185 **e.** .9938 **f.** 1

6.19 **a.** .0228 **b.** .8400

6.21 **a.** .9452 **b.** .9452 **c.** .9370 **d.** .0456

6.23 **a.** 28.602 **b.** 1.176

6.25 **a.** .9938 **b.** $P(30.5 < x < 31.5) = .9876$, $P(30 < x < 32) = 1$
c. P(one tire being underinflated) = .0013
P(at least one of the four tires is underinflated) = .0052

6.27 .0730

6.29 **a.** .9901 **b.** 31.466 **c.** 31.256 **d.** 30.534

6.31 **a.** .0016 **b.** .9946 **c.** .7925

6.33 **a.** .1114 **b.** .0409 **c.** .0968 **d.** .9429 **e.** .9001

6.35 **a.** .7960 **b.** .7016 **c.** .0143 **d.** .05

6.37 **a.** $n\pi = 50(.05) = 2.5$. The normal approximation to the binomial should not be used. **b.** .8708

6.39 It is reasonable to conclude that the normal distribution provides an adequate description.

6.41 Plot is curved, distribution not normal.

6.45 **a.** .1359 **b.** .0228
c. $P(.457 < x) = .8413$, P(all three have at least .475 ounces of cheese) = .5955

6.47 **a.** .9332 **b.** 72.8 **c.** $60

6.49 $P(x < 4.9) = .0228$, $P(5.2 < x) \approx 0$

6.51 **a.** .6915

6.53 **a.** .7745 **b.** .1587 **c.** .3085 **d.** 6.1645

6.55 **a.** $\mu = 32$; $\sigma = 5.1846$ **b.** .8760 **c.** 0

6.57 .3174

6.59 **a.** 88th percentile **b.** 98th percentile **c.** 628

6.61 15.66

6.63 **a.** .8647 **b.** .6321 **c.** .4712

CHAPTER SEVEN

7.3 **a.** population characteristic **b.** statistic
c. population characteristic **d.** statistic
e. statistic

7.9 **a.**

t value	6	11	15	21	25	30
probability	1/6	1/6	1/6	1/6	1/6	1/6

7.11 Sampling distribution of statistic #1.

value of $\bar{x}$	2.67	3	3.33	3.67
probability	.1	.4	.3	.2

Sampling distribution of statistic #2.

value	3	4
probability	.7	.3

Sampling distribution of statistic #3.

value	2.5	3	3.5
probability	.1	.5	.4

7.13 **a.**

$\bar{x}$ value	281	295.5	364.5	393.5	462.5	477
probability	1/6	1/6	1/6	1/6	1/6	1/6

$\mu_{\bar{x}} = 379$

b.

$\bar{x}$ value	281	364.5	393.5	477
probability	.25	.25	.25	.25

$\mu_{\bar{x}} = 379$

7.15 For $n = 36, 50, 100,$ and 400

7.17 **a.** $\mu_{\bar{x}} = 40, \sigma_{\bar{x}} = .625$
Since $n = 64$, which exceeds 30, the shape of the sampling distribution will be approximately normal.
b. .5762   **c.** .2628

7.19 **a.** $\mu_{\bar{x}} = 2$ and $\sigma_{\bar{x}} = .267$
b. For $n = 20$, $\mu_{\bar{x}} = 2$ and $\sigma_{\bar{x}} = .179$.
For $n = 100$, $\mu_{\bar{x}} = 2$ and $\sigma_{\bar{x}} = .08$.

7.21 **a.** .9544   **b.** Approximately 95%; Approximately .3%

7.25 **a.** Because the x distribution is normal, the $\bar{x}$ distribution will be normal (even for n less than 30).
$\mu_{\bar{x}} = 3; \sigma_{\bar{x}} = .0125$
c. .0026   **d.** .8413

7.31 .1841

7.33 $P(\bar{x} < .79) = .1587, P(\bar{x} < .77) = .0013$

7.35 .1056

7.37 **a.** .8185

7.39 .0793

CHAPTER EIGHT

8.1 Statistic II would be preferred because it is unbiased and has smaller variance than the other two.

8.3 **a.** .484   **b.** .3

8.5 .70

8.7 73.2

8.9 **a.** As the confidence level increases, the width of the large sample confidence interval also increases.
b. As the sample size increases, the width of the large sample confidence interval decreases.
c. As the population standard deviation increases, the width of the large sample confidence interval also increases.

8.11 **a.** The 90% confidence interval would have been narrower.   **b.** The statement is incorrect.   **c.** This statement is incorrect.

8.13 (349.42, 354.58)

8.15 **a.** (333.66, 364.24)   **b.** (69.24, 79.62)

8.17 (68.52, 79.48)

8.19 **a.** The interval for boys is (99.486, 103.914)
The interval for girls is (99.423, 104.177)
b. (83.05, 90.35)

8.21 385

8.23 **a.** yes   **b.** no   **c.** yes   **d.** no   **e.** yes   **f.** yes   **g.** yes   **h.** yes

8.25 (.1612, .1836)

8.27 **a.** (.492, .608)   **b.** (.411, .549)

8.29 (.3941, .7171)

8.31 (.35, .51)

8.33 **a.** (.411, .549)

8.35 97

8.39 **a.** 2.12   **b.** 1.80   **c.** 2.81   **d.** 1.71   **e.** 1.78   **f.** 2.26

8.41 **a.** (5.2, 8.0)   **b.** (31.78, 48.82)

8.43 **a.** (394.66, 461.34)   **b.** narrower
   c. For convection oven (2074.67, 2323.33)
   For food dehydrator (3708.65, 4131.35)
   For microwave oven (1393.7, 1468.3)

8.45 **a.** (232.45, 294.95)   **b.** (51.2, 65.8)
   c. The distribution of the number of operations performed in one year must be (at least approximately) normal for each of the two populations.

8.47 (5.77, 28.23)

8.49 (4.79, 6.21)

8.51 (.455, .517)

8.53 **a.** (6.88, 9.38)

8.55 **a.** (.96, 1.02)   **b.** (1.28, 1.66)   **c.** (.90, 1.08)

8.57 (.211, .259)

8.59 (4.2, 7.65)

CHAPTER NINE

9.1 $\bar{x} = 50$ is not a legitimate hypothesis, because $\bar{x}$ is a statistic, not a population characteristic.

9.5 $H_0: \pi = .6$ versus $H_a: \pi > .6$

9.9 $H_0: \mu = 40$, $H_a: \mu \neq 40$

9.11 **a.** A type I error is accepting a shipment of inferior quality. A type II error is returning to the supplier a shipment which is not of inferior quality.
   b. The calculator manufacturer would most likely consider a type I error more serious, since they would then end up producing defective calculators.
   c. From the supplier's point of view a type II error would be more serious, because the supplier would end up having lost the profits from the sale of the good printed circuits.

9.13 **a.** $H_0: \mu = 900$ versus $H_a: \mu > 900$
   b. A type I error is deciding to use the new brand when in fact its mean lifetime does not exceed that of the old brand. A type II error is deciding not to use the new brand when in fact its mean lifetime does exceed that of the old brand.

9.17 **a.** $H_0: \pi = .02$ versus $H_a: \pi < .02$
   b. A type I error is changing to robots when in fact they are not superior to humans. A type II error is not changing to robots when in fact they are superior to humans.
   c. Since a type I error means substantial loss to the company as well as to the human employees who would become unemployed a small α should be used. Therefore $\alpha = .01$ is preferred.

9.19 **a.** $z > 1.96$ or $z < -1.96$ **b.** $z > 2.58$ or $z < -2.58$
   c. $z > 1.645$ or $z < -1.645$ **d.** $z > 1.175$ or $z < -1.175$

9.21 $z = 3.316$, reject H_0

9.23 $z = -1.80$, do not reject H_0

9.27 $z = 2.96$, reject H_0

9.29 $z = -1.13$, do not reject H_0

9.31 $z = -5$, reject H_0

9.33 $z = 1.41$, reject H_0

9.35 $z = 3.55$, reject H_0

9.37 $z = 7.58$, reject H_0

9.39 $z = 2.73$, reject H_0

9.41 $z = -1.26$, do not reject H_0

9.43 **a.** $z = 4.91$, reject H_0
9.45 $z = 1.03$, do not reject H_0
9.49 $t = 1.875$, reject H_0
9.51 $t = .77$, do not reject H_0
9.53 $t = -.59$, do not reject H_0
9.59 **a.** .0358 **b.** .0892 **c.** .6170 **d.** .1616 **e.** 0
9.61 **a.** 3.0 **b.** .0013 **c.** reject H_0
9.63 $z = -6.91$, P-value $< .0002$, reject H_0
9.65 **a.** $z = 1.04$, do not reject H_0 **b.** .1492
9.67 **a.** $.10 > P$-value $> .05$ **b.** $.01 > P$-value $> .002$ **c.** $.001 > P$-value **d.** P-value $> .20$
9.69 $.05 > P$-value $> .025$
9.71 $z = .198$, do not reject H_0
9.73 **a.** $z = 4.19$, reject H_0 **b.** P-value $< .0002$
9.75 $z = -6.62$, reject H_0
9.77 $t = .16$, do not reject H_0
9.79 $z = -21.6$, reject H_0
9.81 **a.** $z = 6.43$, reject H_0 **b.** P-value $< .0002$

CHAPTER TEN

10.1 $\mu_{\bar{x}_1 - \bar{x}_2} = 5$, $\sigma_{\bar{x}_1 - \bar{x}_2} = .529$
10.3 $z = 6.74$, reject H_0
10.5 $z = -7.91$, reject H_0
10.7 $z = 15.79$, reject H_0
10.9 **a.** $(-1.32, 3.32)$ **b.** $z = -2.00$, reject H_0 **c.** $z = -2.58$, reject H_0
10.11 $(-2.176, -1.804)$
10.13 **a.** $z = -2.98$, reject H_0 **b.** $(-4.71, -0.97)$
10.15 $z = 4.63$, reject H_0 **b.** P-value $< .0002$ **c.** $(-0.9, 4.26)$
10.17 $z = -.41$, do not reject H_0
10.19 $t = -1.20$, do not reject H_0
10.21 $t = 3.48$, reject H_0
10.25 **a.** $t = .88$, do not reject H_0 **b.** $t = .90$, do not reject H_0
10.27 $t = -1.51$, do not reject H_0
10.29 **a.** $(-1.95, 5.35)$ **b. i.** $(-19.99, 16.19)$ **ii.** $(-8.72, 6.52)$
10.31 $(-.033, -.021)$
10.33 $t = 3.04$, reject H_0
10.35 $t = 2.629$, reject H_0
10.39 $t = .36$, do not reject H_0
10.41 $t = 2.63$, reject H_0
10.43 **a.** $t = 3.95$, reject H_0 **b.** $t = -1.17$, do not reject H_0
10.45 **a.** $t = 1.04$, do not reject H_0 **b.** $t = 1.71$, do not reject H_0
10.47 $t = 1.92$, do not reject H_0
10.49 **a.** the large-sample procedure is applicable **b.** $z = -2.00$, reject H_0 **c.** P-value $= .0456$
10.51 $z = -2.63$, reject H_0
10.55 $z = -.403$, do not reject H_0
10.57 $z = -1.86$, do not reject H_0
10.59 **a.** $z = 2.41$, P-value $= .0160$
10.61 $t = 4.92$, reject H_0
10.63 $z = 1.79$, do not reject H_0
10.65 $(-42.18, 69.98)$
10.67 for glucose: $z = .95$, do not reject H_0
for potassium: $z = 1.60$, do not reject H_0
for protein: $z = -.31$, do not reject H_0
for cholesterol: $z = -.55$, do not reject H_0
10.69 **a.** $z = 3.41$, reject H_0 **b.** P-value $= .0003$
10.71 **a.** $t = -1.94$, do not reject H_0 **b.** $(-5.3, .5)$
10.73 **a.** $t = -4.92$, reject H_0 **b.** $t = -4.09$, reject H_0

CHAPTER ELEVEN

11.3 **b.** not deterministic

c. There is a tendency for yield to decrease as upslope distance increases.

11.7 **a.** There appears to be a linear trend with positive slope.

b. There does not appear to be a linear relationship.

11.9 **c.** 86.2   **d.** No, the value $x = 80$ is well outside the range of x values.

11.11 **a.** (12, 12) and (12, 23) have the same x value but different y-values.

c. $\hat{y} = -19.67 + 3.2847x$ **d.** 46.02

11.13 **a.** not deterministic, (1.5, 23) and (1.5, 24.5) have the same x values but different y-values.

c. $\hat{y} = 6.4487 + 10.6026x$ **d.** 32.955

11.15 **a.** $\hat{y} = 0.38126 + 0.15173x$

b. The equation of the least squares line computed with the point (9.8, 1.9) omitted is $\hat{y} = 0.38939 + .1483x$. The data point (9.8, 1.9) does not appear to have greatly influenced the equation of the least squares line.

11.17

Observed Y:	110	123	119	86	62
Predicted $\hat{Y}$:	126.6	113.3	100	86.7	73.4
Residual $Y - \hat{Y}$:	-16.6	9.7	19	-0.7	-11.4

11.19 **a.** $\hat{y} = 94.33 - 15.389x$ **b.** SSResid = 285.66

11.21 0.9512

11.23 **a.** SSResid = 36036.93283, SSTo = 93942.85731   **b.** 0.6164

11.25 **a.** positive   **b.** negative   **c.** positive   **d.** no correlation

e. positive   **f.** weak to moderate positive   **g.** negative   **h.** no correlation

11.29 .355

11.31 **a.** .966   **b.** .9335

11.33 No, because x, artist, is not a numerical variable.

11.35 **b.** .9214

11.41 **a.** $\hat{y} = 13.9617 + 3.1703x$ **c.** 34.5687

11.43 **a.** $\hat{y} = 75.5824 - 15.5501x$ **b.** 1400.8586   **c.** SSTo = 2129.8683, $r^2 = 0.3423$

11.47 **a.** 0

b. If $y = 1$, when $x = 6$, then $r = .509$. (Comment: Any y value greater than .973 will work.)

c. If $y = -1$, when $x = 6$, then $r = -.509$. (Comment: Any y value less than $-.973$ will work.)

CHAPTER TWELVE

12.1 **a.** $y = -5.0 + .017x$ **c.** 30.7   **d.** .017   **e.** 1.7

12.3 **b.** .121

c. $P(\text{serum Mn over } 5) = .5$, $P(\text{serum Mn below } 3.8) = .1587$

12.5 **a.** $y = \alpha + \beta x$ is the equation of the population regression line and $\hat{y} = a + bx$ is the equation of the least squares line (the estimated regression line.)

b. The quantity b is a statistic. It is the slope of the estimated regression line. The quantity β is a population characteristic. It is the slope of the population regression line. The quantity b is an estimate of β.

12.7 **a.** $\hat{y} = 1.41122 - .00736x$

12.9 **a.** .883   **b.** 13.682

12.11 **b.** $\hat{y} = -.00227 + 1.2467x$ **c.** $r^2 = .436$ **d.** .0263

12.17 **a.** .766   **b.** $s_e = 6.5719$, $s_b = 8.0529$ **c.** (38.313, 66.821)

12.19 (.4, 1.2)

12.21 **a.** $t = 6.53$, reject H_0 **b.** $t = 1.56$, do not reject H_0

12.23 $t = 2.64$, reject H_0

12.27 $t = .45$, do not reject H_0

12.31 **a.** 4.038   **b.** 4.038   **c.** 3.817   **d.** s_{a+bx} is smallest when $x^* = \bar{x}$.

12.37 **a.** $\hat{y} = 2.78551 + .04462x$ **b.** $t = 10.85$, reject H_0 **c.** (3.671, 4.577)

12.39 **a.** $\hat{y} = 40.31 - .44256x$ **b.** (30.108, 34.58)   **c.** (28.32, 32.83)

12.43 $t = -2.80$, do not reject H_0

12.47 $t = 3.32$, reject H_0

12.49 $r = .574$, $t = 1.85$, do not reject H_0 **b.** $.20 > P\text{-value} > .10$

12.51 $r = .728$, $t = 3.52$, reject H_0

12.53 **a.** .8777 **b.** $t = 6.07$, reject H_0 **c.** P-value $< .0005$
12.55 $t = 2.2$, reject H_0
12.61 The residuals are positive in 1963 and 1964, then they are negative from 1965 through 1972, followed by a positive residual in 1973. The residuals exhibit a pattern in the plot and thus the plot casts doubt on the appropriateness of the simple linear regression model.
12.63 **a.** 103.11 **b.** 96.87
12.65 **a.** 9.364 **b.** increase
12.67 **a.** $R^2 = .862$, $s_e = 2.83$ **c.** $F = 51.53$, reject H_0
12.71 $F = 96.64$, reject H_0
12.73 **a.** $F = 9.06$, reject H_0 **b.** (156.60, 197.60) **c.** (188.4, 198.8)
12.75 **a.** $F = 15.58$, reject H_0 **b.** 10.031 **c.** (.0283, .0857)
12.77 **a.** (4.903, 5.632) **b.** (4.4078, 6.1272)
12.79 $t = .73$, do not reject H_0
12.81 $t = 2.87$, do not reject H_0
12.85 **a.** $\hat{y} = -8.779 + 18.874x$ **b.** $t = 21.22$, reject H_0
12.89 **a.** $\hat{y} = 57.964 + .0357x$ **b.** $t = .023$, do not reject H_0
d. The residual plot has a very distinct curvilinear pattern which indicates that a simple linear regression model is not appropriate.
12.91 **a.** $r = -.0164$, $t = -.06$, do not reject H_0

CHAPTER THIRTEEN

13.1 Each of the k population or treatment response distributions is normal and their variances are equal.
13.3 **b.** H_0 is not rejected
13.5 **a.** $F = 4.12$, reject H_0
13.7 $F = 1.85$, do not reject H_0
13.9 $F = 5.07$, do not reject H_0
13.11 $F = 2.46$, do not reject H_0
13.13 $F = .83$, do not reject H_0

13.15

Source of Variation	Degrees of Freedom	Sum of Squares	Mean Square	F
Treatments	3	75081.72	25027.24	1.70
Error	16	235419.04	14713.69	
Total	19	310500.76		

13.17

Source of Variation	Degrees of Freedom	Sum of Squares	Mean Square	F
Treatments	3	136.14	45.38	5.12
Error	60	532.26	8.871	
Total	63	668.40		

13.19 $F = 2.17$, do not reject H_0

13.21

Source of Variation	Degrees of Freedom	Sum of Squares	Mean Square	F
Treatments	2	1.6297	.8148	4.589
Error	21	3.7289	.1776	
Total	23	5.3586		

13.25 $\mu_1 - \mu_2$: $(-.9854, .1104)$
$\mu_1 - \mu_3$: $(-1.1692, -.0734)$
$\mu_2 - \mu_3$: $(-.7317, .3641)$

13.29

Source of Variation	Degrees of Freedom	Sum of Squares	Mean Square	F
Treatments	4	0.26	.065	2.24
Error	20	0.58	.029	
Total	24	0.84		

13.33 The single-factor ANOVA should not be employed in this experiment. Each type of fabric should be considered a block and then the data analyzed using the analysis for a randomized block design.

13.35 $(-7.03, 2.3187)$

13.39

Source of Variation	Degrees of Freedom	Sum of Squares	Mean Square	F
Treatments	2	11.7	5.85	.37
Blocks	4	113.5	28.375	
Error	8	125.6	15.7	
Total	14	250.8		

b. $F = .37$, do not reject H_0

CHAPTER FOURTEEN

14.1 **a.** Reject H_0 if $X^2 > 9.49$ **b.** Reject H_0 if $X^2 > 13.28$ **c.** Reject H_0 if $X^2 > 21.67$

14.5 $X^2 = 4.03$, do not reject H_0

14.7 **a.** $X^2 = 4.63$, do not reject H_0

14.9 $X^2 = 4.8$, do not reject H_0

14.11 **a.** Reject H_0 if $X^2 > 21.03$ **b.** $X^2 = 7.2$, do not reject H_0 **c.** $X^2 = 14.5$, do not reject H_0

14.13 $X^2 = 20.155$, reject H_0

14.15 $X^2 = 133.02$, reject H_0

14.17 $X^2 = 33.769$, reject H_0

14.19 $X^2 = 23.67$, reject H_0

14.21 $X^2 = 22.45$, reject H_0

14.23 $X^2 = 113.88$, reject H_0

14.25 $X^2 = 197.62$, reject H_0

14.27 $X^2 = 1.63$, do not reject H_0

14.29 $X^2 = .02$, do not reject H_0

14.31 $X^2 = 11.93$, reject H_0

14.33 $X^2 = 5.4$, do not reject H_0

INDEX

t Critical Values

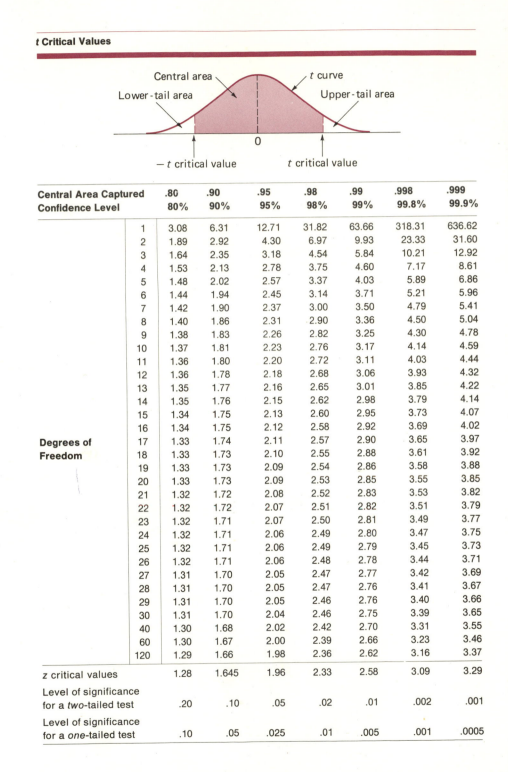

Central area · t curve
Lower-tail area · Upper-tail area
0
− t critical value · t critical value

Central Area Captured Confidence Level		.80 80%	.90 90%	.95 95%	.98 98%	.99 99%	.998 99.8%	.999 99.9%
Degrees of Freedom	1	3.08	6.31	12.71	31.82	63.66	318.31	636.62
	2	1.89	2.92	4.30	6.97	9.93	23.33	31.60
	3	1.64	2.35	3.18	4.54	5.84	10.21	12.92
	4	1.53	2.13	2.78	3.75	4.60	7.17	8.61
	5	1.48	2.02	2.57	3.37	4.03	5.89	6.86
	6	1.44	1.94	2.45	3.14	3.71	5.21	5.96
	7	1.42	1.90	2.37	3.00	3.50	4.79	5.41
	8	1.40	1.86	2.31	2.90	3.36	4.50	5.04
	9	1.38	1.83	2.26	2.82	3.25	4.30	4.78
	10	1.37	1.81	2.23	2.76	3.17	4.14	4.59
	11	1.36	1.80	2.20	2.72	3.11	4.03	4.44
	12	1.36	1.78	2.18	2.68	3.06	3.93	4.32
	13	1.35	1.77	2.16	2.65	3.01	3.85	4.22
	14	1.35	1.76	2.15	2.62	2.98	3.79	4.14
	15	1.34	1.75	2.13	2.60	2.95	3.73	4.07
	16	1.34	1.75	2.12	2.58	2.92	3.69	4.02
	17	1.33	1.74	2.11	2.57	2.90	3.65	3.97
	18	1.33	1.73	2.10	2.55	2.88	3.61	3.92
	19	1.33	1.73	2.09	2.54	2.86	3.58	3.88
	20	1.33	1.73	2.09	2.53	2.85	3.55	3.85
	21	1.32	1.72	2.08	2.52	2.83	3.53	3.82
	22	1.32	1.72	2.07	2.51	2.82	3.51	3.79
	23	1.32	1.71	2.07	2.50	2.81	3.49	3.77
	24	1.32	1.71	2.06	2.49	2.80	3.47	3.75
	25	1.32	1.71	2.06	2.49	2.79	3.45	3.73
	26	1.32	1.71	2.06	2.48	2.78	3.44	3.71
	27	1.31	1.70	2.05	2.47	2.77	3.42	3.69
	28	1.31	1.70	2.05	2.47	2.76	3.41	3.67
	29	1.31	1.70	2.05	2.46	2.76	3.40	3.66
	30	1.31	1.70	2.04	2.46	2.75	3.39	3.65
	40	1.30	1.68	2.02	2.42	2.70	3.31	3.55
	60	1.30	1.67	2.00	2.39	2.66	3.23	3.46
	120	1.29	1.66	1.98	2.36	2.62	3.16	3.37
z critical values		1.28	1.645	1.96	2.33	2.58	3.09	3.29
Level of significance for a *two*-tailed test		.20	.10	.05	.02	.01	.002	.001
Level of significance for a *one*-tailed test		.10	.05	.025	.01	.005	.001	.0005